LONDON MATHEMATICAL SOCIETY LECTURE NOTE SERIES

Managing Editor: Professor N.J. Hitchin, Mathematical Institute,
University of Oxford, 24–29 St Giles, Oxford OX1 3LB, United Kingdom

The titles below are available from booksellers, or from Cambridge University Press at www.cambridge.org

152 Oligomorphic permutation groups, P. CAMERON
153 L-functions and arithmetic, J. COATES & M.J. TAYLOR (eds)
155 Classification theories of polarized varieties, TAKAO FUJITA
158 Geometry of Banach spaces, P.F.X. MÜLLER & W. SCHACHERMAYER (eds)
159 Groups St Andrews 1989 volume 1, C.M. CAMPBELL & E.F. ROBERTSON (eds)
160 Groups St Andrews 1989 volume 2, C.M. CAMPBELL & E.F. ROBERTSON (eds)
161 Lectures on block theory, BURKHARD KÜLSHAMMER
163 Topics in varieties of group representations, S.M. VOVSI
164 Quasi-symmetric designs, M.S. SHRIKANDE & S.S. SANE
166 Surveys in combinatorics, 1991, A.D. KEEDWELL (ed)
168 Representations of algebras, H. TACHIKAWA & S. BRENNER (eds)
169 Boolean function complexity, M.S. PATERSON (ed)
170 Manifolds with singularities and the Adams-Novikov spectral sequence, B. BOTVINNIK
171 Squares, A.R. RAJWADE
172 Algebraic varieties, GEORGE R. KEMPF
173 Discrete groups and geometry, W.J. HARVEY & C. MACLACHLAN (eds)
174 Lectures on mechanics, J.E. MARSDEN
175 Adams memorial symposium on algebraic topology 1, N. RAY & G. WALKER (eds)
176 Adams memorial symposium on algebraic topology 2, N. RAY & G. WALKER (eds)
177 Applications of categories in computer science, M. FOURMAN, P. JOHNSTONE & A. PITTS (eds)
178 Lower K- and L-theory, A. RANICKI
179 Complex projective geometry, G. ELLINGSRUD *et al*
180 Lectures on ergodic theory and Pesin theory on compact manifolds, M. POLLICOTT
181 Geometric group theory I, G.A. NIBLO & M.A. ROLLER (eds)
182 Geometric group theory II, G.A. NIBLO & M.A. ROLLER (eds)
183 Shintani zeta functions, A. YUKIE
184 Arithmetical functions, W. SCHWARZ & J. SPILKER
185 Representations of solvable groups, O. MANZ & T.R. WOLF
186 Complexity: knots, colourings and counting, D.J.A. WELSH
187 Surveys in combinatorics, 1993, K. WALKER (ed)
188 Local analysis for the odd order theorem, H. BENDER & G. GLAUBERMAN
189 Locally presentable and accessible categories, J. ADAMEK & J. ROSICKY
190 Polynomial invariants of finite groups, D.J. BENSON
191 Finite geometry and combinatorics, F. DE CLERCK *et al*
192 Symplectic geometry, D. SALAMON (ed)
194 Independent random variables and rearrangement invariant spaces, M. BRAVERMAN
195 Arithmetic of blowup algebras, WOLMER VASCONCELOS
196 Microlocal analysis for differential operators, A. GRIGIS & J. SJÖSTRAND
197 Two-dimensional homotopy and combinatorial group theory, C. HOG-ANGELONI *et al*
198 The algebraic characterization of geometric 4-manifolds, J.A. HILLMAN
199 Invariant potential theory in the unit ball of C^n, MANFRED STOLL
200 The Grothendieck theory of dessins d'enfant, L. SCHNEPS (ed)
201 Singularities, JEAN-PAUL BRASSELET (ed)
202 The technique of pseudodifferential operators, H.O. CORDES
203 Hochschild cohomology of von Neumann algebras, A. SINCLAIR & R. SMITH
204 Combinatorial and geometric group theory, A.J. DUNCAN, N.D. GILBERT & J. HOWIE (eds)
205 Ergodic theory and its connections with harmonic analysis, K. PETERSEN & I. SALAMA (eds)
207 Groups of Lie type and their geometries, W.M. KANTOR & L. DI MARTINO (eds)
208 Vector bundles in algebraic geometry, N.J. HITCHIN, P. NEWSTEAD & W.M. OXBURY (eds)
209 Arithmetic of diagonal hypersurfaces over finite fields, F.Q. GOUVÉA & N. YUI
210 Hilbert C*-modules, E.C. LANCE
211 Groups 93 Galway / St Andrews I, C.M. CAMPBELL *et al* (eds)
212 Groups 93 Galway / St Andrews II, C.M. CAMPBELL *et al* (eds)
214 Generalised Euler-Jacobi inversion formula and asymptotics beyond all orders, V. KOWALENKO *et al*
215 Number theory 1992–93, S. DAVID (ed)
216 Stochastic partial differential equations, A. ETHERIDGE (ed)
217 Quadratic forms with applications to algebraic geometry and topology, A. PFISTER
218 Surveys in combinatorics, 1995, PETER ROWLINSON (ed)
220 Algebraic set theory, A. JOYAL & I. MOERDIJK
221 Harmonic approximation, S.J. GARDINER
222 Advances in linear logic, J.-Y. GIRARD, Y. LAFONT & L. REGNIER (eds)
223 Analytic semigroups and semilinear initial boundary value problems, KAZUAKI TAIRA
224 Computability, enumerability, unsolvability, S.B. COOPER, T.A. SLAMAN & S.S. WAINER (eds)
225 A mathematical introduction to string theory, S. ALBEVERIO *et al*
226 Novikov conjectures, index theorems and rigidity I, S. FERRY, A. RANICKI & J. ROSENBERG (eds)
227 Novikov conjectures, index theorems and rigidity II, S. FERRY, A. RANICKI & J. ROSENBERG (eds)
228 Ergodic theory of Z^d actions, M. POLLICOTT & K. SCHMIDT (eds)
229 Ergodicity for infinite dimensional systems, G. DA PRATO & J. ZABCZYK
230 Prolegomena to a middlebrow arithmetic of curves of genus 2, J.W.S. CASSELS & E.V. FLYNN

231 Semigroup theory and its applications, K.H. HOFMANN & M.W. MISLOVE (eds)
232 The descriptive set theory of Polish group actions, H. BECKER & A.S. KECHRIS
233 Finite fields and applications, S. COHEN & H. NIEDERREITER (eds)
234 Introduction to subfactors, V. JONES & V.S. SUNDER
235 Number theory 1993–94, S. DAVID (ed)
236 The James forest, H. FETTER & B. GAMBOA DE BUEN
237 Sieve methods, exponential sums, and their applications in number theory, G.R.H. GREAVES *et al*
238 Representation theory and algebraic geometry, A. MARTSINKOVSKY & G. TODOROV (eds)
240 Stable groups, FRANK O. WAGNER
241 Surveys in combinatorics, 1997, R.A. BAILEY (ed)
242 Geometric Galois actions I, L. SCHNEPS & P. LOCHAK (eds)
243 Geometric Galois actions II, L. SCHNEPS & P. LOCHAK (eds)
244 Model theory of groups and automorphism groups, D. EVANS (ed)
245 Geometry, combinatorial designs and related structures, J.W.P. HIRSCHFELD *et al*
246 p-Automorphisms of finite p-groups, E.I. KHUKHRO
247 Analytic number theory, Y. MOTOHASHI (ed)
248 Tame topology and o-minimal structures, LOU VAN DEN DRIES
249 The atlas of finite groups: ten years on, ROBERT CURTIS & ROBERT WILSON (eds)
250 Characters and blocks of finite groups, G. NAVARRO
251 Gröbner bases and applications, B. BUCHBERGER & F. WINKLER (eds)
252 Geometry and cohomology in group theory, P. KROPHOLLER, G. NIBLO, R. STÖHR (eds)
253 The q-Schur algebra, S. DONKIN
254 Galois representations in arithmetic algebraic geometry, A.J. SCHOLL & R.L. TAYLOR (eds)
255 Symmetries and integrability of difference equations, P.A. CLARKSON & F.W. NIJHOFF (eds)
256 Aspects of Galois theory, HELMUT VÖLKLEIN *et al*
257 An introduction to noncommutative differential geometry and its physical applications 2ed, J. MADORE
258 Sets and proofs, S.B. COOPER & J. TRUSS (eds)
259 Models and computability, S.B. COOPER & J. TRUSS (eds)
260 Groups St Andrews 1997 in Bath, I, C.M. CAMPBELL *et al*
261 Groups St Andrews 1997 in Bath, II, C.M. CAMPBELL *et al*
262 Analysis and logic, C.W. HENSON, J. IOVINO, A.S. KECHRIS & E. ODELL
263 Singularity theory, BILL BRUCE & DAVID MOND (eds)
264 New trends in algebraic geometry, K. HULEK, F. CATANESE, C. PETERS & M. REID (eds)
265 Elliptic curves in cryptography, I. BLAKE, G. SEROUSSI & N. SMART
267 Surveys in combinatorics, 1999, J.D. LAMB & D.A. PREECE (eds)
268 Spectral asymptotics in the semi-classical limit, M. DIMASSI & J. SJÖSTRAND
269 Ergodic theory and topological dynamics, M.B. BEKKA & M. MAYER
270 Analysis on Lie groups, N.T. VAROPOULOS & S. MUSTAPHA
271 Singular perturbations of differential operators, S. ALBEVERIO & P. KURASOV
272 Character theory for the odd order theorem, T. PETERFALVI
273 Spectral theory and geometry, E.B. DAVIES & Y. SAFAROV (eds)
274 The Mandlebrot set, theme and variations, TAN LEI (ed)
275 Descriptive set theory and dynamical systems, M. FOREMAN *et al*
276 Singularities of plane curves, E. CASAS-ALVERO
277 Computational and geometric aspects of modern algebra, M.D. ATKINSON *et al*
278 Global attractors in abstract parabolic problems, J.W. CHOLEWA & T. DLOTKO
279 Topics in symbolic dynamics and applications, F. BLANCHARD, A. MAASS & A. NOGUEIRA (eds)
280 Characters and automorphism groups of compact Riemann surfaces, THOMAS BREUER
281 Explicit birational geometry of 3-folds, ALESSIO CORTI & MILES REID (eds)
282 Auslander-Buchweitz approximations of equivariant modules, M. HASHIMOTO
283 Nonlinear elasticity, Y. FU & R.W. OGDEN (eds)
284 Foundations of computational mathematics, R. DEVORE, A. ISERLES & E. SÜLI (eds)
285 Rational points on curves over finite fields, H. NIEDERREITER & C. XING
286 Clifford algebras and spinors 2ed, P. LOUNESTO
287 Topics on Riemann surfaces and Fuchsian groups, E. BUJALANCE *et al*
288 Surveys in combinatorics, 2001, J. HIRSCHFELD (ed)
289 Aspects of Sobolev-type inequalities, L. SALOFF-COSTE
290 Quantum groups and Lie Theory, A. PRESSLEY (ed)
291 Tits buildings and the model theory of groups, K. TENT (ed)
292 A quantum groups primer, S. MAJID
293 Second order partial differential equations in Hilbert spaces, G. DA PRATO & J. ZABCZYK
294 Introduction to the theory of operator spaces, G. PISIER
295 Geometry and integrability, LIONEL MASON & YAVUZ NUTKU (eds)
296 Lectures on invariant theory, IGOR DOLGACHEV
297 The homotopy category of simply connected 4-manifolds, H.-J. BAUES
299 Kleinian groups and hyperbolic 3-manifolds, Y. KOMORI, V. MARKOVIC, & C. SERIES (eds)
300 Introduction to Möbius differential geometry, UDO HERTRICH-JEROMIN
301 Stable modules and the D(2)-problem, F.E.A. JOHNSON
302 Discrete and continuous nonlinear Schrödinger systems, M.J. ABLORWITZ, B. PRINARI, & A.D. TRUBATCH
303 Number theory and algebraic geometry, MILES REID & ALEXEI SKOROBOGATOV (eds)
304 Groups St Andrews 2001 in Oxford Vol. 1, COLIN CAMPBELL, EDMUND ROBERTSON, & GEOFF SMITH (eds)
305 Groups St Andrews 2001 in Oxford Vol. 2, C.M. CAMPBELL, E.F. ROBERTSON, & G.C. SMITH (eds)
307 Surveys in combinatorics 2003, C.D. WENSLEY (ed)
309 Corings and comodules, TOMASZ BRZEZINSKI & ROBERT WISBAUER
310 Topics in dynamics and ergodic theory, SERGEY BEZUGLYI & SERGIY KOLYADA (eds)
312 Foundations of computational mathematics, Minneapolis 2002, FELIPE CUCKER *et al* (eds)

London Mathematical Society Lecture Note Series. 319

Double Affine Hecke Algebras

IVAN CHEREDNIK
*University of North Carolina,
Chapel Hill*

CAMBRIDGE
UNIVERSITY PRESS

CAMBRIDGE UNIVERSITY PRESS
Cambridge, New York, Melbourne, Madrid, Cape Town, Singapore,
São Paulo, Delhi, Dubai, Tokyo, Mexico City

Cambridge University Press
32 Avenue of the Americas, New York, NY 10013-2473, USA

www.cambridge.org
Information on this title: www.cambridge.org/9780521609180

First published 2005

A catalog record for this publication is available from the British Library

Library of Congress Cataloging in Publication data

Cherednik, Ivan.
Double affine Hecke algebras / Ivan Cherednik.
p. cm. – (London Mathematical Society lecture note series ; 319)
Includes bibliographical references and index.
ISBN 0-521-60918-6 (alk. paper)
1. Hecke algebras. 2. Affine algebraic groups. 3. Harmonic analysis.
4. Knizhnik-Zamoldchikov equations. 5. Orthogonal polynomials. I. Title.
II. Series.
QA176.C49 2005 2004
512/.22 22 2004054567

ISBN 978-0-521-60918-0 Paperback

Dedicated to Ian Macdonald

PREFACE

This book is based on a series of lectures delivered by the author in Kyoto in 1996–97 and at Harvard University in 2001. The first chapter was written in collaboration with T. Akasaka, E. Date, K. Iohara, M. Jimbo, M. Kashiwara, T. Miwa, M. Noumi, Y. Saito, and K. Takemura. V. Ostrik is the coauthor of the second chapter. The author owes them a lot, as well as P. Etingof, D. Kazhdan, M. Nazarov, and E. Opdam for help and encouragement. The book was supported in part by the National Science Foundation and the Clay Mathematics Institute.

In many ways, this book began with one man, Ian Macdonald. I am deeply indebted to him.

After a comprehensive introduction, the classical and quantum Knizhnik–Zamolodchikov equations attached to root systems are studied, and their relations to the affine Hecke algebras, Kac–Moody algebras, and harmonic analysis discussed. These equations are of key importance in the analytic theory of Coxeter groups.

In Chapter 2, we switch to a systematic theory of the one-dimensional double affine Hecke algebra and its representations. It is the simplest case, but far from trivial. This algebra is closely connected with $\mathfrak{sl}_2$, $\widehat{\mathfrak{sl}}_2$, the Heisenberg and Weyl algebras, and has impressive applications.

The third chapter is about DAHA in full generality, including the Macdonald polynomials, Fourier transform, Gauss–Selberg integrals, Verlinde algebras, Gaussian sums, and diagonal coinvariants. The transition to this abstract level will be smooth for readers familiar with root systems. Only reduced root systems are considered.

This book is essentially self-contained. The chapters are relatively independent. I hope that it will be helpful for both mathematicians and physicists who want to master the new double Hecke algebra technique.

Contents

Chapter 0

Introduction

0.0 Universality of Hecke algebras

0.0.1 Real and imaginary

Before a systematic exposition, I will try to outline the connections of the representation theory of Lie groups, Lie algebras, and Kac–Moody algebras with the Hecke algebras and the Macdonald theory.

The development of mathematics may be illustrated by Figure 0.1.

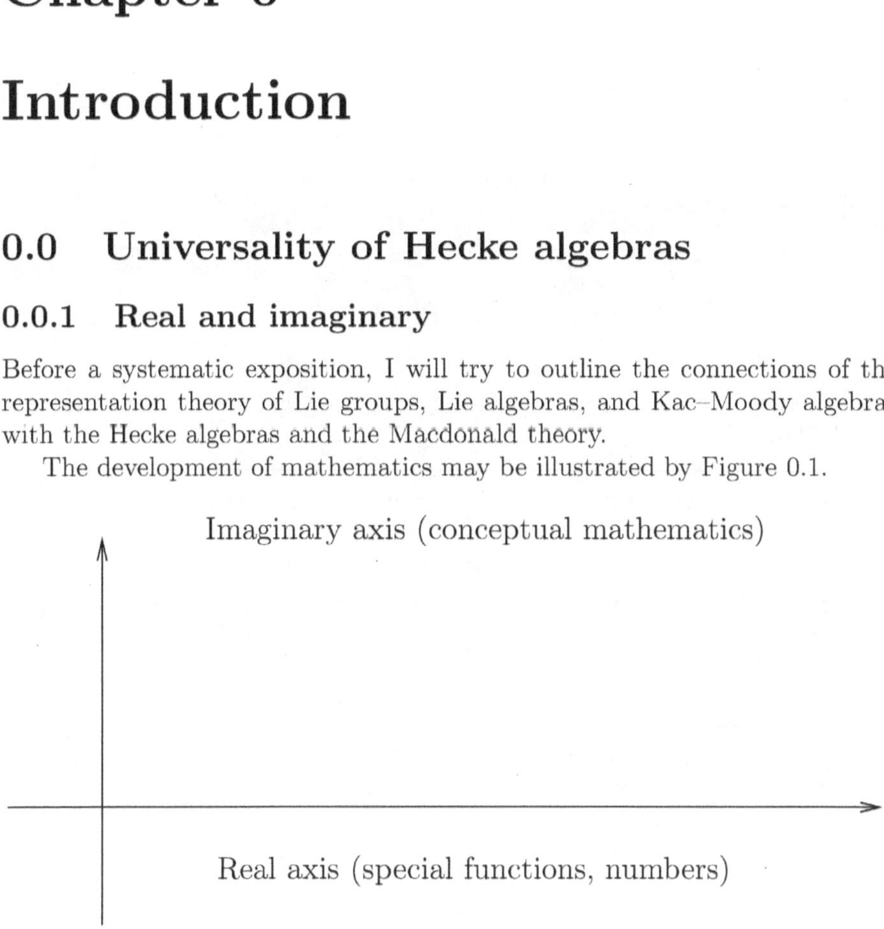

Figure 0.1: Real and Imaginary

Mathematics is fast in the imaginary (conceptual) direction but, generally, slow in the real direction (fundamental objects). Mainly I mean modern mathematics, but it may be more universal. For instance, the ancient Greeks created a highly conceptual axiomatic geometry with modest "real output." I do not think that the ratio $\Re/\Im$ is much higher now.

Let us try to project representation theory on the real axis. In Figure 0.2 we focus on Lie groups, Lie algebras, and Kac–Moody algebras, omitting the arithmetic direction (adèles and automorphic forms). The theory of special functions, arithmetic, and related combinatorics are the classical objectives of representation theory.

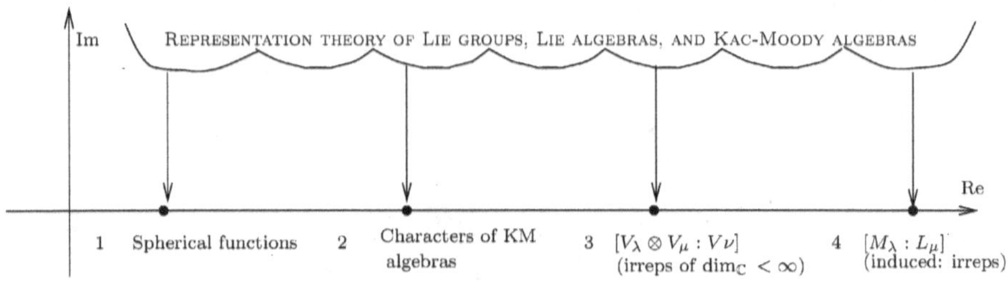

Figure 0.2: Representation Theory

Without going into detail and giving exact references, the following are brief explanations.

(1) I mean the zonal spherical functions on $K\backslash G/K$ for maximal compact K in a semisimple Lie group G. The modern theory was started by Berezin, Gelfand, and others in the early 1950s and then developed significantly by Harish-Chandra [HC]. Lie groups greatly helped to make the classical theory multidimensional, although they did not prove to be very useful for the hypergeometric function.

(2) The characters of Kac–Moody (KM) algebras are not far from the products of classical one-dimensional θ–functions and can be introduced without representation theory (Looijenga, Kac, Saito). See [Lo]. However, it is a new and important class of special functions with various applications. Representation theory explains some of their properties, but not all.

(3) This arrow gives many combinatorial formulas. Decomposing tensor products of finite dimensional representations of compact Lie groups and related problems were the focus of representation theory in the 1970s and early 1980s. They are still important, but representation theory moved toward infinite dimensional objects.

(4) Calculating the multiplicities of irreducible representations of Lie algebras in the BGG–Verma modules or other induced representations belongs to conceptual mathematics. The Verma modules were designed

as a technical tool for the Weyl character formula (which is "real"). It took time to understand that these multiplicities are "real" too, with strong analytic and modular aspects.

0.0.2 New vintage

Figure 0.3 is an update of Figure 0.2. We add the results which were obtained in the 1980s and 1990s, inspired mainly by a breakthrough in mathematical physics, although mathematicians had their own strong reasons to study generalized hypergeometric functions and modular representations.

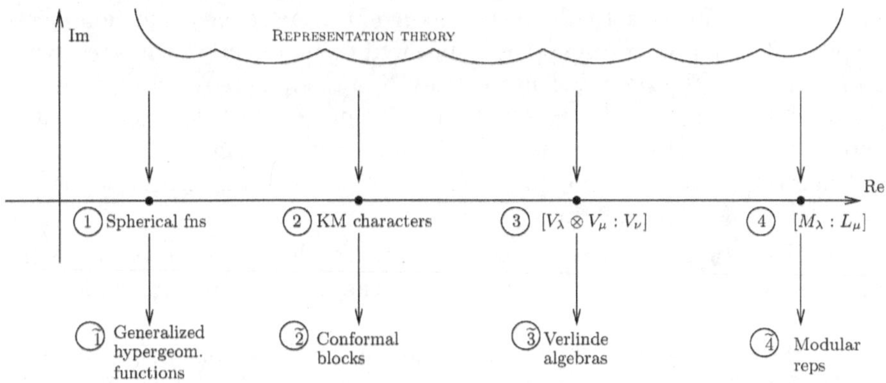

Figure 0.3: New Vintage

$(\tilde{1})$ These functions will be the main subject of the first chapter. We will study them in the differential and difference cases. The interpretation and generalization of the hypergeometric functions via representation theory was an important problem of the so-called Gelfand program and remained unsolved for almost three decades.

$(\tilde{2})$ Actually, the conformal blocks belong to the (conceptual) imaginary axis as well as their kin, the τ–function. However, they extend the hypergeometric functions, theta functions, and Selberg's integrals. They attach the hypergeometric function to representation theory, but affine Hecke algebras serve this purpose better.

$(\tilde{3})$ The Verlinde algebras were born from conformal field theory. They are formed by integrable representations of Kac–Moody algebras of a given level with "fusion" instead of tensoring. These algebras can be also defined using quantum groups at roots of unity and interpreted via operator algebras.

($\tilde{4}$) Whatever you may think about the "reality" of $[M_\lambda : L_\mu]$, these multiplicities are connected with the representations of Lie groups and Weyl groups over finite fields (modular representations). Nothing can be more real than finite fields!

0.0.3 Hecke algebras

The Hecke operators and later the Hecke algebras were introduced in the theory of modular forms, actually in the theory of GL_2 over the p–adic numbers. In spite of their p–adic origin, they appeared to be directly connected with the K–theory of the *complex* flag varieties [KL1] and, more recently, with the Harish-Chandra theory. It suggests that finite and p–adic fields are of greater fundamental importance for mathematics and physics than we think.

Concerning the great potential of p–adics, let me mention the following three well-known confirmations:

(i) The Kubota–Leopold p–adic zeta function, which is a p–adic analytic continuation of the values of the classical Riemann zeta function at negative integers.
(ii) My theorem about "switching local invariants" based on the p–adic uniformization (Tate–Mumford) of the modular curves which come from the quaternion algebras.
(iii) The theory of p–adic strings due to Witten, which is based on the similarity of the Frobenius automorphism in arithmetic to the Dirac operator.

Observation. *The real projection of representation theory goes through Hecke-type algebras.*

The arrows in Figure 0.4 are as follows.

(a) This arrow is the most recognized now. Quite a few aspects of the Harish-Chandra theory in the zonal case were covered by representation theory of the degenerate (graded) affine Hecke algebras, introduced in [Lus1] ([Dr1] for GL_n). The radial parts of the invariant differential operators on symmetric spaces, the hypergeometric functions and their generalizations arise directly from these algebras [C10].

The difference theory appeared even more promising. It was demonstrated in [C19] that the q–Fourier transform is self-dual like the classical Fourier and Hankel transforms, but not the Harish-Chandra transform. There are connections with the quantum groups and quantum

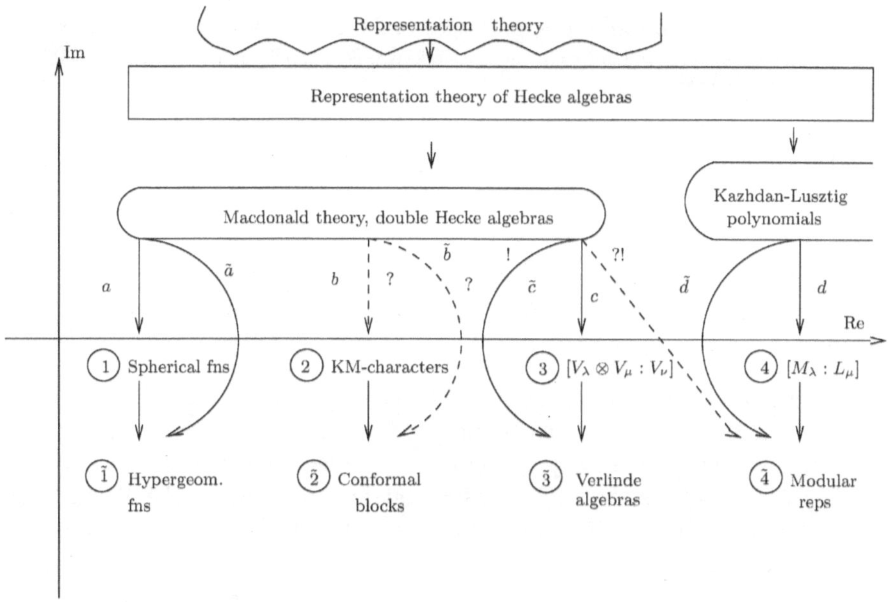

Figure 0.4: Hecke Algebras

symmetric spaces (Noumi, Olshansky, and others; see [No1]). However, the double Hecke algebra technique is simpler and more powerful.

(*b*) The conformal blocks are solutions of the KZ–Bernard equation (KZB). The double Hecke algebras lead to certain elliptic generalizations of the Macdonald polynomials [C17, C18, C23] (other approaches are in [EK1, C17, FV3], and the recent [Ra]). These algebras govern the monodromy of the KZB equation and "elliptic" Dunkl operators (Kirillov Jr., Felder–Tarasov–Varchenko, and the author).

The monodromy map is the inverse of arrow $(\tilde{b})$. The simplest examples are directly related to the Macdonald polynomials and those at roots of unity.

(*c*) Hecke algebras and their affine generalizations give a new approach to the classical combinatorics, including the characters of the compact Lie groups. The natural setting here is the theory of the Macdonald polynomials, although the analytic theory seems more challenging.

Concerning $(\tilde{c})$, the Macdonald polynomials at the roots of unity give a simple approach to the Verlinde algebras [Ki1, C19, C20]. The use of the nonsymmetric Macdonald polynomials here is an important development. Generally, these polynomials are beyond the Lie and Kac–Moody theory, although they are connected with the Heisenberg–Weyl and p–adic Hecke algebras.

(d) This arrow is the Kazhdan–Lusztig conjecture proved by Brylinski–Kashiwara and Beilinson–Bernstein and then generalized to the Kac–Moody case by Kashiwara–Tanisaki.

By $(\tilde{d})$, I mean the modular Lusztig conjecture (partially) proved by Anderson, Jantzen, and Soergel. There is recent significant progress due to Bezrukavnikov.

The arrow from the Macdonald theory to modular representations is marked by "?!." It seems to be the most challenging now (there are already first steps in this direction). It is equivalent to extending the Verlinde algebras and their nonsymmetric variants from the alcove (the restricted category of representations of Lusztig's quantum group) to the parallelogram (all representations).

If such an extension exists, it would give a k–extension of Lusztig's conjectures, formulas for the modular characters (not only those for the multiplicities), a description of modular representations for arbitrary Weyl groups, and more.

0.1 KZ and Kac–Moody algebras

In this section we comment on the role of the Kac–Moody algebras and their relations (real and imaginary) to the spherical functions and the double Hecke algebras.

0.1.1 Fusion procedure

I think that the penetration of double Hecke algebras into the fusion procedure and related problems of the theory of Kac–Moody algebras is a convincing demonstration of their potential. The fusion procedure was introduced for the first time in [C3]. On the physics side, let me also mention a contribution of Louise Dolan.

Given an integrable representation of the n–th power of a Kac–Moody algebra and two sets of points on a Riemann surface (n points and m points), I constructed an integrable representation of the m–th power of the same Kac–Moody algebra. The construction does not change the "global" central charge, the sum of the local central charges over the components. It was named later "fusion procedure."

I missed that in the special case of this correspondence, when $n = 2$ and $m = 1$, the multiplicities of irreducibles in the resulting representation are the structural constants of a certain commutative algebra, the Verlinde algebra [Ver].

Now we know that the Verlinde algebra and all its structures can be readily extracted from the simplest representation of the double affine Hecke algebra at roots of unity. Thus the Kac–Moody algebras are undoubtedly connected with the double Hecke algebras.

Double Hecke algebras dramatically simplify and generalize the algebraic theory of Verlinde algebras, including the inner product and the (projective) action of $PSL(2, \mathbb{Z})$, however, excluding the integrality and positivity of the structural constants. The latter properties require $k = 1$ and are closely connected with the Kac–Moody interpretation (although they can also be checked directly).

I actually borrowed the fusion procedure from Y. Ihara's papers "On congruence monodromy problem." A similar construction is a foundation of his theory. I changed and added some things (the central charge has no counterpart in his theory), but the procedure is basically the same. Can we go back and define Verlinde algebras in adèles' setting?

0.1.2 Symmetric spaces

The classification of Kac–Moody algebras very much resembles that of symmetric spaces. See [Ka], [He2]. It is not surprising, because the key technical point in both theories is the description of the involutions and automorphisms of finite order for the semisimple finite dimensional Lie algebras. The classification lists are similar *but do not coincide.* For instance, the BC_n–symmetric spaces have no Kac–Moody counterparts. Conversely, the KM algebra of type, say, $D_4^{(3)}$ is not associated (even formally) with any symmetric space. Nevertheless one could hope that this parallelism is not incidental.

Some kind of correspondence can be established using the isomorphism of the quantum many-body problem [Ca, Su, HO1], a direct generalization of the Harish-Chandra theory, and the affine KZ equation. The isomorphism was found by A. Matsuo and developed further in my papers. It holds when the parameter k, given in terms of the root multiplicity in the context of symmetric spaces, is an arbitrary complex number. In the Harish-Chandra theory, it equals 1/2 for $SL_2(\mathbb{R})/SO_2$, 1 in the so-called group case $SL_2(\mathbb{C})/SU_2$, and $k = 2$ for the Sp_2. The k–generalized spherical functions are mainly due to Heckman and Opdam; see Chapter 1.

Once k was made an arbitrary number, it could be expected a counterpart of the central charge c, the level, in the theory of Kac–Moody algebras. Indeed, it has some geometric meaning. However, generally, it is *not* connected with the central charge. Indeed, the number of independent k–parameters can be from 1 (A, D, E) to 5 $(C^\vee C$, the so-called Koornwinder case), but we have only one (global) central element c in the Kac–Moody theory. Also, the k–spherical functions are eigenfunctions of differential operators generalizing the radial parts of the invariant operators on symmetric spaces. These operators have no counterparts for the Kac–Moody characters. Also, the spherical functions are orthogonal polynomials; the Kac–Moody charactes are not. In addition, the latter are of elliptic type, the spherical functions are of trigonometric type.

We will discuss the elliptic quantum many-body problem (QMBP) in the first chapter. It gives a kind of theory of spherical functions in the Kac–Moody setting (at critical level). However, it supports the *unification* of c and k rather than the *correspondence* between them.

The elliptic QMBP in the GL_N–case was introduced by Olshanetsky and Perelomov [OP]. The classical root systems were considered in the paper [OOS]. The Olshanetsky–Perelomov operators for arbitrary root systems were constructed in [C17].

We see that an exact match cannot be expected. However, a map from the Kac–Moody algebras to spherical functions exists. It is for GL_N only and not exactly for the KM characters, but it does exist.

0.1.3 KZ and r–matrices

The KZ equation is the system of differential equations for the matrix elements (using physical terminology, the correlation functions) of the representations of the Kac–Moody algebras in the n–point case. The matrix elements are simpler to deal with than the characters. For instance, they satisfy differential equations with respect to the positions of the points.

The most general "integrable" case, is described by the so-called r–matrix Kac–Moody algebras from [C1] and the corresponding *r–matrix KZ equations* introduced in [C6].

It was observed in the latter paper that the classical Yang–Baxter equation can be interpreted as the compatibility of the corresponding KZ system, which dramatically enlarged the number of examples. An immediate application was a new class of KZ equations with *trigonometric* and *elliptic* dependence on the points.

It was demonstrated in [C6] that the abstract τ–function, also called the coinvariant, is a generic solution of the r–matrix KZ with respect to the action of the Sugawara (-1)–operators.

More generally, the r–matrices and the corresponding KZ equations attached to arbitrary*root systems* were defined in [C6]. For instance, the dependence on the points is via the differences (the A–case) of the points and also via the *sums* for B, C, D. The BC–case is directly related to the so-called *reflection equations* introduced in [C2].

The results due to Drinfeld–Kohno on the monodromy of the KZ equations (see [Ko]) can be extended to the r–matrix equations. In some cases, the monodromy can be calculated explicitly, for instance, for the *affine KZ* [C6, C7, C8].

0.1.4 Integral formulas for KZ

The main applications of the interpretation of KZ as a system of equations for the coinvariant were: (i) a simplification of the algebraic part of the Schechtman–Varchenko construction [SV] of integral formulas for the rational KZ, (ii) a generalization of their formulas to the trigonometric case [C9]. Paper [SV] is based on direct algebraic considerations without using the theory of Kac–Moody algebras.

There is another important "integrable" case, the so-called Knizhnik–Zamolodchikov–Bernard equation usually denoted by KZB [Be, FW1]. We will see in Chapter 1 that it can be obtain in the same abstract manner as a system of differential equations for the corresponding "elliptic" coinvariant. There must be an implication of this fact toward the integral formulas for KZB, but this has not been checked so far.

We do not discuss the integral formulas for KZB in this book, as well as the integral formulas for QKZ, the quantum Knizhnik–Zamolodchikov equation. See, e.g., [TV], [FV1], and [FTV].

Generally, the KZ equations can be associated with arbitrary algebraic curves. Then they involve the derivatives with respect to the moduli of curves and vector bundles. However, in this generality, the resulting equations are non-integrable in any reasonable sense.

Summarizing, we have the following major cases, when the Knizhnik–Zamolodchikov equation have integral formulas, reasonably simple monodromy representations, special symmetries, and other important properties:
(a) the KZ for Yang's rational r–matrix (see [SV]),
(b) the trigonometric KZ equation introduced in [C9],
(c) the elliptic KZ–Bernard equation (see [Be, FW1]).

Given a Lie algebra $\mathfrak{g}$, one may define the integrand for the KZ integral formulas is derectly connected with the coinvariants of $U(\widehat{\mathfrak{g}})$ for the Weyl modules [C9]. The contours (cycles) of integration are governed by the quantum $U_q(\mathfrak{g})$. See [FW2], [Va] and references therein. We will not discuss the contours and the q–topology of the configuration spaces in this book.

The later topic was started by Aomoto [A1, AKM] and seems an endless story. We have no satisfactory formalization of the q–topology so far. It is especially needed for QKZ. Generally, in mathematics, the contours of integration (the homology) must be dual to the differential forms (the cohomology). It gives an approach to the problem.

We note that the integral KZ formulas are directly connected with the equivalence of the $U(\widehat{\mathfrak{g}})_c$ and the quantum group $U_q(\mathfrak{g})$ due to Kazhdan, Lusztig, and Finkelberg (see [KL2]). It is for a proper relation $c \leftrightarrow q$.

0.1.5 From KZ to spherical functions

Let us discuss what the integral formulas could give for the theory of spherical functions and its generalizations. There are natural limitations.

First, only the spherical functions of type A may apper (for either choice of $\mathfrak{g}$) if we begin with the KZ integral formulas of type A.

Second, one needs an r–matrix KZ of *trigonometric* type because the Harish-Chandra theory is on the torus.

Third, only $\mathfrak{g} = \mathfrak{gl}_N$ may result in *scalar* differential operators due to the analysis by Etingof and Kirillov Jr.

Summarizing, the integral formulas for the affine KZ (AKZ) of type A are the major candidates. The AKZ is *isomorphic* to the quantum many-body problem, that is exactly the k–Harish-Chandra theory [Mat, C11].

Note that the "basic" trigonometric n–point KZ taking values in the 0–weight component of $(\mathbb{C}^n)^{\otimes n}$, which is isomorphic to the group algebra $\mathbb{C}\mathbb{S}_n$, must be considred for AKZ. The integral AKZ formula is likely to be directly connected with the Harish-Chandra formula. I did not check it, but calculations due to Mimachi, Felder, Varchenko confirm this. For instance, the dimension of the contours (cycles) of integration for such KZ is $n(n-1)/2$, which coincides with that in the Harish-Chandra integral representation for spherical functions of type A_{n-1}. His integral is over $K = SO_n \subset SL_n(\mathbb{R})$.

Establishing a direct connection with the Harsh-Chandra integral representation for the spherical functions does not seem too difficult. However it is of obvious importance, because his formula is for *all* root systems, and one can use it as an initial point for the general theory of integral formulas of the KZ equations associated with root systems.

We note that the integral KZ formulas can be justified without Kac–Moody algebras. A straightforward algebraic combinatorial analysis is complicated but possible [SV]. The proof presented in this book is based on the Kac–Moody coinvariant [C9]. However, I use the Kac–Moody algebras at the critical level only, as a technical tool, and then extend the resulting formulas to all values of the center charge.

There is another approach to the same integral formulas based on the coinvariant for the Wakimoto modules instead of that for the Weyl modules [FFR]. The calculations with the coinvariant are in fact similar to mine, but the Kac–Moody theory are used to greater potential and the combinatorial part gets simpler.

Thus one may expect the desired relation between the conformal blocks and spherical functions at the level of integral formulas for the trigonometric KZ of type A, i.e., in terms of the differences of the points, and with values in the simplest representations of $\mathfrak{g} = \mathfrak{gl}_n$.

I do not think that this correspondence is really general and can be extended to arbitrary symmetric spaces, though it certainly indicates that there must be a *unification* that combines the Kac–Moody and Harish-Chandra theories.

0.2 Double Hecke algebras

Double affine Hecke algebras (DAHA) were initially designed to clarify the classical and quantum Knizhnik–Zamolodchikov equation (for the simplest fundamental representation of $\mathfrak{g} = \mathfrak{gl}_N$) and analogous KZ equations for arbitrary Weyl groups. The first applications were to the Dunkl operators, differential and difference. The most natural way to introduce these operators is via the induced representations of DAHA.

Eventually, through applications to the theory of Macdonald polynomials, DAHAs led to a unification of the Harish-Chandra transform in the zonal case and the p–adic spherical transform in one general q–theory, which is one of the major applications.

The DAHA–Fourier transform depends on a parameter k, which generalizes the root multiplicity in the Harish-Chandra theory. This parameter becomes $1/(c+g)$ in the KZ theory, if KZ is interpreted via Kac–Moody algebras of level c and the isomorphism between AKZ and QMBP is used. The parameter q comes from the Macdonald polynomials and QKZ, the quantum KZ equation introduced by F.Smirnov and I.Frenkel–Reshetikhin [Sm, FR] and then generalized to all root systems in [C12].

The limiting cases as $q \to 1$ and $q \to \infty$ are, respectively, the Harish-Chandra and the p–adic Macdonald–Matsumoto spherical transforms.

It is not just a unification of the latter transforms. The q–transform is *self-dual* in contrast to its predecessors. The self-duality collapses under the limits above. However, there is a limiting procedure preserving the self-duality. If $q \to 1$ and we represent the torus in the form of $\{q^x\}$ instead of $\{e^x\}$, then the correspomding limit becomes self-dual. It is the multidimensional generalization from [O2, Du1, Je] of the classical Hankel transform in terms of the Bessel functions.

The new q–transform shares many properties with the generalized Hankel transform. For instance, the q–Mehta–Macdonald integrals, generalizing the classical Gauss–Selberg integrals, are in the focus of the new theory. The case when q is a root of unity is of obvious importance because of the applications to Verlinde algebras, Gaussian sums, diagonal coinvariants, and more.

0.2.1 Missing link?

There are reasons to consider DAHA as a candidate for the "missing link" between representation theory and theory of special functions. Let me explain why something seems missing.

The representation theory of finite dimensional Lie groups mainly serves multidimensional functions and gives only a little for the one-dimensional functions (with a reservation about the arithmetic direction). The fundamental objects of the modern representation theory are $\widehat{\mathfrak{sl}}_2$ and its quantum counterpart. They have important applications to the hypergeometric and theta functions, but managing these functions via $\widehat{\mathfrak{sl}}_2$ is far from simple (especially in the q–case).

Generally, the Kac–Moody theory is too algebraic to be used in developing the theory of functions. There were interesting attempts, but still we have no consistent harmonic Kac–Moody analysis. The theory of operator algebras can be considered as an analytic substitute of the Kac–Moody theory, but it does not help much with the special functions either.

On the other hand, the Heisenberg and Weyl algebras are directly related to the theory of special functions. Unfortunately they are too simple, as is the affine Hecke algebra of type A_1.

The double affine Hecke algebra $\mathcal{HH}$ ("double H") of type A_1 seems just right to incorporate major classical special functions of one variable into representation theory.

This algebra has a simple definition, but its representation theory is rich enough. I think that if we combine what is already known about $\mathcal{HH}_{A_1}$, its applications and representations, it would be a book as big as "SL_2" by Lang or "$GL(2)$" by Jacquet–Langlands. One-dimensional DAHA (already) has surprisingly many applications.

(a) There are direct relations to $\mathfrak{sl}(2)$ and $\mathfrak{osp}(2|1)$ and their quantum counterparts. The rational degeneration of $\mathcal{HH}_{A_1}$ as $q \to 1$ is a quotient of $U(\mathfrak{osp}(2|1))$. Rational DAHAs are very interesting; they are a kind of Lie algebras in the general q–theory.

(b) The Weyl and Heisenberg algebras are its limits when t, the second parameter of $\mathcal{HH}$, tends to 1. For instance, the N–dimensional representation of the Weyl algebra, as q is a primitive N–th root of unity, has a direct counterpart for $\mathcal{HH}$, namely, the nonsymmetric Verlinde algebra, with various applications, including a new theory of Gaussian sums.

(*c*) It covers the theory of Rogers' (q–ultraspherical) polynomials. Its $C^\vee C_1$–modification governs major remarkable families of one-dimensional orthogonal polynomials and has applications to the Bessel functions (the rational degeneration) and to the classical and basic hypergeometric functions.

(*d*) Generally, the duality from [VV1] connects $\mathcal{HH}$ of type A with the toroidal (double q–Kac–Moody) algebras in a sense of Ginzburg, Kapranov, and Vasserot. The $\mathcal{HH}_{A_1}$ has important applications to the representation theory of $\widehat{\mathfrak{sl}}_2$ and in the higher ranks via this duality.

(*e*) DAHAs also result from the K–theory of affine flag varieties and are related to the q–Schur algebras. The connections with the so-called double arithmetic must be mentioned too. Kapranov and then Gaitsgory–Kazhdan interpreted $\mathcal{HH}_{A_1}$ as a Hecke algebra of the "p–adic–loop group" of $SL(2)$.

(*f*) In the very first paper [C13] on DAHA, its topological interpretation was used. The algebra $\mathcal{HH}_{A_1}$ is directly connected with the fundamental group of an elliptic curve with one puncture. This establishes a link to the Grothendieck–Belyi program in the elliptic case (Beilinson–Levin) and to some other problems of modern arithmetic.

(*g*) The representation theory of $\mathcal{HH}_{A_1}$ is far from trivial, especially at roots of unity. For instance, the description of its center at roots of unity (and in the $C^\vee C_1$–case) leads to the quantization of the cubic surfaces (Oblomkov). Some relation to the Fourier–Mukai transform are expected.

(*h*) DAHA unifies the Harish-Chandra spherical transform and the p–adic Macdonald–Matsumoto transforms, The corresponding q–transform is self-dual. For A_1, it leads to a deep analytic theory of the basic hypergeometric function.

The theory is now essentially algebraic. There are some results in the analytic direction in [C21] (the construction of the general spherical functions), [C26] (analytic continuations in terms of k with applications to q–counterparts of Riemann's zeta function), and [KS1, KS2]. The latter two papers are devoted to analytic theory of the one-dimensional Fourier transform in terms of the basic hypergeometric function, a $C^\vee C_1$–extension of the q–transform considred in the book.

0.2.2 Gauss integrals and sums

The starting point of many mathematical and physical theories is the celebrated formula:

$$2\int_0^\infty e^{-x^2}x^{2k}\mathrm{d}x \;=\; \Gamma(k+1/2),\ \Re k > -1/2. \tag{0.2.1}$$

Let us give some examples.

(a) Its generalization with the product of two Bessel functions added to the integrand or, equivalently, the formula for the Hankel transform of the Gaussian e^{-x^2} multiplied by a Bessel function, is one of the main formulas in the classical theory of Bessel functions.

(b) The following "perturbation" for the same $\Re k > -1/2$,

$$2\int_0^\infty (e^{x^2}+1)^{-1}x^{2k}\ \mathrm{d}x = (1-2^{1/2-k})\Gamma(k+1/2)\zeta(k+1/2),$$

is fundamental in analytic number theory. For instance, it readily gives the functional equation for ζ.

(c) The multidimensional extension due to Mehta [Meh], when we integrate over $\mathbb{R}^n$ with the measure $\prod_{1\le i<j\le n}(x_i-x_j)^{2k}$ instead of x^{2k}, gave birth to the theory of matrix models with various applications in mathematics and physics. Its generalization to arbitrary roots (Macdonald and Opdam) is a major formula in the modern theory of Hankel transform.

(d) Switching to the roots of unity, we have an equally celebrated Gauss formula:

$$\sum_{m=0}^{2N-1} e^{\frac{\pi m^2}{2N}i} \;=\; (1+i)\sqrt{N},\quad N\in\mathbb{N}. \tag{0.2.2}$$

It is a counterpart of (0.2.1) at $k=0$, although there is no direct connection.

To fully employ modern mathematics we need to go from the Bessel to the hypergeometric functions. In contrast to the former, the latter can be studied, interpreted, and generalized by a variety of methods from representation theory and algebraic geometry to integrable models and string theory.

Technically, the measure $x^{2k}dx$ has to be replaced by $\sinh(x)^{2k}dx$ and the Hankel transform by the Harish-Chandra transform, to be more exact, by its k–extension. However, the latter is no longer self-dual, the formula (0.2.1) has no sinh–counterpart, and the Gaussian looses its Fourier–invariance. Thus a straightforward substitution creates problems. We need a more fine-tuned approach.

0.2.3 Difference setup

The main observation is that the self-duality of the Hankel transform is restored for the kernel

$$\delta(x;q,k) \stackrel{\mathbf{def}}{=\!=} \prod_{j=0}^{\infty} \frac{(1-q^{j+2x})(1-q^{j-2x})}{(1-q^{j+k+2x})(1-q^{j+k-2x})}, \quad 0<q<1, \ k\in\mathbb{C}.$$

Here δ, the Macdonald truncated theta function, is a certain unification of $\sinh(x)^{2k}$ and the measure in terms of the Gamma function serving the inverse Harish-Chandra transform (A_1). Therefore the self-duality of the resulting transform can be expected a priori.

As to (0.2.1), setting $q = \exp(-1/a), \ a>0$, we have

$$(-i)\int_{-\infty i}^{\infty i} q^{-x^2}\delta(x;q,k)\ \mathrm{d}x = 2\sqrt{a\pi}\prod_{j=0}^{\infty}\frac{1-q^{j+k}}{1-q^{j+2k}}, \quad \Re k>0. \tag{0.2.3}$$

Here both sides are well defined for all k except for the poles but coincide only when $\Re k > 0$.

One can make (0.2.3) entirely algebraic by replacing the Gaussian $\gamma^{-1} = q^{-x^2}$ by its expansion

$$\tilde{\gamma}^{-1} = \sum_{-\infty}^{+\infty} q^{n^2/4} q^{nx}$$

and using Const Term$(\sum c_n q^{nx}) = c_0$ instead of the imaginary integration:

$$\text{Const Term } (\tilde{\gamma}^{-1}\delta) = 2\prod_{j=0}^{\infty}\frac{1-q^{j+k}}{1-q^{j+2k}}.$$

Jackson integrals. A promising feature of special q–functions is the possibility of replacing the integrals by sums over $\mathbb{Z}^n$, the so-called Jackson integrals.

Technically, we switch from the imaginary integration to that for the path, which begins at $z = \epsilon i + \infty$, moves to the left till ϵi, then down through the origin to $-\epsilon i$, and then returns down the positive real axis to $-\epsilon i + \infty$ (for small ϵ). Then we apply Cauchy's theorem under the assumption that $|\Im k| < 2\epsilon, \ \Re k > 0$.

We obtain the following counterpart of (0.2.3):

$$\sum_{j=0}^{\infty} q^{\frac{(k-j)^2}{4}} \frac{1-q^{j+k}}{1-q^k} \prod_{l=1}^{j} \frac{1-q^{l+2k-1}}{1-q^l} =$$
$$q^{\frac{k^2}{4}} \prod_{j=1}^{\infty} \frac{(1-q^{j/2})(1-q^{j+k})(1+q^{j/2-1/4+k/2})(1+q^{j/2-1/4-k/2})}{(1-q^j)},$$

which is convergent for all k.

When $q = \exp(2\pi i/N)$ and k is a positive integer $\leq N/2$, we come to a Gauss–Selberg-type sum:

$$\sum_{j=0}^{N-2k} q^{\frac{(k-j)^2}{4}} \frac{1-q^{j+k}}{1-q^k} \prod_{l=1}^{j} \frac{1-q^{l+2k-1}}{1-q^l} = \prod_{j=1}^{k}(1-q^j)^{-1} \sum_{m=0}^{2N-1} q^{m^2/4}.$$

The left-hand side resembles the so-called modular Gauss–Selberg sums. However, the difference is dramatic. The modular sums are calculated in the finite fields and are embedded into roots of unity right before the final summation [Ev]. Our sums are defined entirely in cyclotomic fields. Substituting $k = [N/2]$, we arrive at (0.2.2).

0.2.4 Other directions

There are other projects involving the double Hecke algebras. We will mention only some of them.

(1) *Macdonald's q–conjectures* [M2, M3, M4]. Namely, the constant term, norm, duality, and evaluation conjectures [BZ, Kad],[C14, C16, C19]. See also [A2, It, M6, C23] about the discrete variant of the constant term conjecture, the Aomoto conjecture. My proof of the norm formula was based on the shift operators and is similar to that from [O1] in the differential case (the duality and evaluation conjectures collapse as $q \to 1$). I would add to this list the Pieri rules. As to the nonsymmetric Macdonald polynomials, the references are [O3, M5, C20]. See also [DS, Sa, M8], and the recent [St, Ra] about the case of $C^\vee C$, the Koornwinder polynomials and generalizations.

(2) *K–theoretic interpretation.* See papers [KL1, KK], then [GG, GKV], and the important recent paper [Vas]. The latter leads to the Langlands-type description of irreducible representations of double Hecke algebras. The case of generic parameters q, t is directly connected with the affine theory [KL1]. For the special parameters, the corresponding geometry becomes significantly more complicated. The Fourier transform remains unclear in this approach. I also mention here the strong Macdonald conjecture (Hanlon) and the recent [FGT].

(3) *Elementary methods.* The theory of induced, semisimple, unitary, and spherical representations can be developed successfully without K–theory. The main tool is the technique of intertwiners from [C23], which is similar to that for the affine Hecke algebras. The nonsymmetric Macdonald polynomials generate the simplest spherical representation, with the intertwiners serving as creation operators (the case of GL is due to [KnS]). For GL, this technique gives a reasonably complete classification of irreducible representations similar to the theorem by Bernstein and Zelevinsky in the affine case. Relations to [HO2] are expected.

(4) *Radial parts via Dunkl operators.* The main references are [Du1, H1] and [C11]. In the latter, it was observed that the trigonometric differential Dunkl operators can be obtained from the degenerate (graded) affine Hecke algebra from [Lus1] ([Dr1] for GL_n). The difference, elliptic, and difference-elliptic generalizations were introduced in [C12, C13, C17, C18]. The connections with the KZ equation play an important role here [Mat, C11, C22, C13]. The radial parts of the Laplace operators of symmetric spaces and their generalizations are symmetrizations of the Dunkl operators. The symmetric Macdonald polynomials are eigenfunctions of the difference radial parts. The nonsymmetric Opdam–Macdonald polynomials appear as eigenfunctions of the Dunkl operators.

(5) *Harmonic analysis.* The Dunkl operators in the simplest rational-differential setup lead to the definition of the generalized Bessel functions and the generalized Hankel transform (see [O2, Du2, Je] and also [He3]). In contrast to the Harish-Chandra and the p–adic spherical transforms, it is self-dual. The self-duality resumes in the difference setting [C28, C21]. The Mehta–Macdonald conjecture, directly related to the transform of the Gaussian, was checked in [M2, O1] in the differential case, and extended in [C21] to the difference case. It was used there to introduce the q–spherical functions. Concerning the applications to the Harish-Chandra theory, see [HO1, O3, C24] and also [HS].

(6) *Roots of unity.* The construction from [C28] generalizes and, at the same time, simplifies the Verlinde algebras, including the projective action of $PSL(2,\mathbb{Z})$ (cf. [Ka], Theorem 13.8, and [Ki1]), the inner product, and a new theory of Gaussian sums. In [C20], the nonsymmetric Verlinde algebras were considered. The symmetric elements of such algebras form the k–generalized Verlinde algebras. The simplest example is the classical N–dimensional representation of the Weyl algebra at $q = \exp(2\pi i/N)$. Let me also mention the recent [Go, C29] about the Haiman conjecture [Ha1] on the structure of the so-called diagonal coinvariants, which appeared to be directly connected with rational DAHAs and DAHAs at roots of unity.

(7) *Topology.* The group $PSL(2,\mathbb{Z})$ acts projectively on the double Hecke algebra itself. The best explanation (and proof) is based on the interpretation of this algebra as a quotient of the group algebra of the π_1 of the elliptic configuration space from [C13]. The calculation of π_1 in the GL–case is essentially due to [Bi, Sc]. For arbitrary root systems, it is similar to that from [Le], but our configuration space is different. Such π_1 governs the monodromy of the eigenvalue problem for the elliptic radial parts, the corresponding Dunkl operators, and the KZB equation. Switching to the roots of unity, the monodromy representation is the nonsymmetric Verlinde algebra; applications to the Witten–Reshetikhin–Turaev and Ohtshuki invariants are expected.

(8) *GL–Duality.* The previous discussion was about arbitrary root sys-

tems. In the case of GL, the theorem from [VV1] establishes the duality between the double Hecke algebras (actually its extension) and the q–toroidal (double Kac–Moody) algebras in the sense of Ginzburg–Kapranov–Vasserot. It generalizes the classical Schur–Weyl duality, Jimbo's q–duality for the non-affine Hecke algebra [Ji], and the affine Hecke analogs from [Dr1, C5]. When the center charge is nontrivial the duality explains the results from [KMS] and [STU], which were extended by Uglov to irreducible representations of the Kac–Moody $\hat{\mathfrak{gl}}_N$ of arbitrary positive integral levels.

(9) *Rational degeneration.*. The rational degeneration of the double affine Hecke algebra with trivial center charge ($q = 1$) is directly related to the Calogero–Moser varieties [EG]. The rational degenerations of the double affine Hecke algebra [CM, EG], in a sense, play the role of the Lie algebras of the q, t–DAHA. The trigonometric degeneration is also a sort of Lie algebra, but the rational one has the projective action of $PSL(2, \mathbb{Z})$ and other symmetries that make it closer to the general q, t–DAHA. The theory of the rational DAHA and its connections appeared to be an interesting independent direction (see [BEG, GGOR, Go, C29].

Some of the latest developments. There are interesting papers [GK1, GK2] that continued earlier results by Kapranov [Kap] towards using DAHA in the so-called double arithmetic, started by Parshin quite a few years ago. The general problem is to associate a double p–adic Lie group with DAHA.

Let me also mention the results by Ginzburg, Kapranov and Vasserot concerning the interpretation of DAHAs via "Hecke correspondences" over special algebraic surfaces. Presumably DAHAs are somehow related to the Nakajima surfaces [Nak]; at least, this is understood for the Calogero–Moser varieties [EG].

An important recent development is the establishment of the connection of DAHA with the Schur algebra in [GGOR] and then in [VV2]. It is proven in the latter paper that, under minor technical restrictions, DAHA of type A is Morita equivalent to the quantum affine Schur algebra at roots of unity.

The latest project so far employs DAHAs for the global quantization of algebraic surfaces. Oblomkov used the rank one $C^\vee C$–DAHA to quantize the cubic surfaces. Then Etingof, Oblomkov, Rains [EOR] extended this approach to quantize the Del Pezzo surfaces via certain generalizations of DAHA. Relations to [Ra] are expected.

Related directions. Let me also mention several directions that are not based on double Hecke algebras (and their variants) but have close relations to these algebras and to the theory of the symmetric Macdonald polynomials. Mainly they are about the GL–case and the classical root systems.

(a) The spherical functions on q–symmetric spaces due to Noumi and others, which are related to the Macdonald polynomials for certain values of t, central elements in quantum groups (Etingof and others).

(b) The so-called interpolation polynomials (Macdonald, Lassalle, Knop–Sahi, Okounkov–Olshansky, Rains), continuing the classical Lagrange polynomials, appeared to be connected to double Hecke algebras.

(c) The interpretation of the Macdonald polynomials as traces of the vertex operators, including applications to the Verlinde algebras (Etingof, Kirillov Jr.). It is mainly in the GL–case and for integral k. See also [EVa].

(d) Various related results on KZB and its monodromy, (Etingof, Felder, Kirillov Jr., Varchenko). The monodromy always satisfies the DAHA group-type relations, but the quadratic ones are valid in special cases only. See, e.g., [Ki2, TV, FTV, FV3].

(e) There are multiple relations of DAHA to the theory of the W–invariant differential operators, including recent developments due to Wallach, Levasseur, Stafford, and Joseph. In the theory of DAHA, a counterpart is concerning the centralizer of the nonaffine Hecke subalgebra.

(f) There are direct links to the theory of quantum groups defined by Drinfeld and Jimbo and dynamical Yang–Baxter equations through the classical and quantum r–matrices; we refer to [Dr2] and the books [Lus2, ES] without going into detail.

Strong connections with the affine Hecke algebra technique in the classical theory of GL_N and $\mathbb{S}_n$ must be noted. I mean [C5, Na1, Na2, NT, LNT, C28] and promising recent results towards Kazhdan–Lusztig polynomials by Rouquier and others.

The expectations are that Kazhdan–Lusztig polynomials and the canonical (crystal) bases in quantum groups are important for the theory of double Hecke algebras, although I don't know a good definition of the "double" Kazhdan–Lusztig polynomials.

The coefficients of the symmetric Macdonald polynomials in the stable GL–case have interesting combinatorial properties (Macdonald, Stanley, Hanlon, Garsia, Haiman, and others). The most celebrated is the so-called $n!$–conjecture recently proved by Haiman. See [GH] and [Ha2].

Let me mention here that the Macdonald polynomials for the classical root systems appeared for the first time in a Kadell work. He also proved the Macdonald norm conjecture for the BC systems [Kad]. The constant term conjecture in the GL_n–case was verified in [BZ]; see [C16] and [M8] for further references. The first proof of the norm formula for the GL is due to Macdonald. The elliptic counterparts of the Macdonald operators in the case of GL_n are defined in [Ru].

Quite a few constructions can be extended to arbitrary finite groups generated by complex reflections. For instance, the Dunkl operators and the KZ connection exist in this generality (Dunkl, Opdam, Malle). One can also try the affine and even the hyperbolic groups (Saito's root systems [Sai]).

0.3 DAHA in harmonic analysis

This section is an attempt to outline the core of the book in the special case of A_1 and discuss connections between harmonic analysis and mathematical physics. There are quite a few projects where DAHA is involved (see above). However, I think the Verlinde algebra is the most convincing demonstration of the power of new methods. It is the main object of this section.

0.3.1 Unitary theories

Generally, the problem is that we do not have the Kac–Moody harmonic analysis. For instance, we do not have a good definition of the category of " L^2 "–representations in this case. The theory of Kac–Moody algebras, playing a well-known role in modern theoretical physics, is too algebraic for this, and, presumably, these algebras must be developed to more analytic objects. Actually, the von-Neumann algebras are such objects, but they are, in a sense, too analytic. We need something in between. Hopefully the Verlinde algebras and DAHA can help.

Physical connection. Concerning the classical roots of the harmonic analysis on symmetric spaces, the corresponding representation theory was greatly stimulated by (a) physics, (b) the theory of special functions, and (c) combinatorics, historically, in the opposite order. In my opinion, the demand from physics played the major role.

(a) Harish-Chandra was Dirac's assistant for some time and always expressed unreserved admiration for Dirac, according to Helgason's interesting recent note "Harish-Chandra." The Lorentz group led him to the theory of infinite dimensional representations of semisimple Lie groups. However, later the mathematical goals like the Plancherel formula became preponderant in his research.

(b) The theory of special and spherical functions was the main motivation for Gelfand in his studies of infinite dimensional representations, although he always emphasized the role of physics (and physicists). It was reflected in his program (1950s) aimed at "adding" the spherical and hypergeometric functions to the Lie theory.

(c) Before the Lie theory, the symmetric group was the main "representation" tool in the theory of functions. It still remains of fundamental importance. However, using the symmetric group only is not sufficient to introduce and understand properly the differential equations and operators needed in the theory.

As for (a), not all representation theories are of physical importance. It is my understanding, that "unitary" representations are of major importance, although modern theoretical physics uses all kinds of representations. For instance,"massive" quantum theories must have a positive inner product. Mathematically, the unitarity is needed to decompose some natural spaces of functions, for instance, the spaces of square integrable functions, L^p–spaces, and so on and so forth.

Main unitary lines. There are four main sources of unitary theories:
(A) compact and finite groups,
(B) noncompact Lie groups,
(C) Heisenberg and Weyl algebras,
(D) operator and von-Neumann algebras.

The Clifford algebra, super Lie algebras, and "free fermions" are important too, but require a special discussion. The theory of automorphic forms and the corresponding representation theory, including that of the p–adic groups and Hecke algebras, definitely must be mentioned too. The arithmetic representation theory is an important part of modern mathematics attracting increasing attention in physics now.

The above unitary theories have merits and demerits.

The Heisenberg–Weyl algebra is heavily used in physics ("bosonization"), but it has essentially a unique unitary irreducible representation, the Fock representation. The von-Neumann factors actually have the same demerit. Only the pair "factor–subfactor" appeared good enough for combinatorially rich theory. Note that the operator algebras give another approach to the Verlinde algebras.

Spherical functions. The theory (A) is plain and square, but only finite dimensional representations can appear in this way. The representation theory of noncompact Lie groups is infinite dimensional (which is needed in modern physics), but the Harish-Chandra transform is not self-dual and its analytic theory is far from being complete. Also (B) is not very fruitful from the viewpoint of applications in the theory of special functions. For instance, the spherical functions in the so-called group case ($k = 1$) "algebraically" coincide with the characters of compact Lie groups. The other two values $k = 1/2, 2$ in the Figure 0.5 below correspond, respectively, to the orthogonal case $(SL(n, \mathbb{R})/SO(n))$ and the symplectic case.

The left column of the top block of Figure 0.5 shows the classical theory of characters and spherical functions of compact and noncompact Lie groups extended towards the orthogonal polynomials, Jack–Heckman–Opdam polynomials. The latter are generally beyond the Lie theory. We can define them as orthogonal polynomials (k must be assumed real positive or even "small" negative), or as eigenfunctions of the Sutherland–Heckman–Opdam operators, generalizing the radial parts of the Laplace operators on symmetric spaces. The latter definition works for arbitrary complex k (apart from a series of special values where the complete diagonalization is impossible).

0.3.2 From Lie groups to DAHA

Considering the left column of Figure 0.5 as a sample harmonic analysis program, not much is known in the Kac–Moody case (the right column).

At the level of "Compact Characters" in the classical harmonic analysis, we have the theory of Kac–Moody characters which is reasonably complete, in spite of combinatorial difficulties with the so-called string functions. The next level is supposed to be the theory of KM spherical functions. There are several approaches (let me mention Dale Peterson, Lian and Zuckerman, and the book of Etingof, I.Frenkel, Kirillov Jr.), but we have no satisfactory general theory so far with a reservation about the group case, where the technique of vertex operators and conformal blocks can be used. Extending the theory from the group case ($k = 1$) to arbitrary root multiplicities is a problem.

However, in my opinion, the key problem is that the Kac–Moody characters are not pairwise orthogonal functions in a way that one can expect taking the classical theory (the left column) as a sample. Analytically, they are given in terms of theta functions and cannot be integrated over noncompact regions in any direction unless special algebraic tools are used. It is exactly how Verlinde algebras enter the game.

Verlinde algebras. We can identify the characters of integrable representations of the Kac–Moody algebra of a given level (central charge) c with the corresponding classical characters treated as functions at a certain alcove of P/NP for the weight lattice P and $N = c + h^\vee$; $h^\vee$ is the dual Coxeter number. The P/NP is naturally a set of vectors with the components in the N–the roots of unity. The images of the characters form a linear basis in the algebra of all functions on this alcove, called the Verlinde algebra. The fusion product corresponds to the pointwise multiplication, and the images of the characters become pairwise orthogonal with respect to the Verlinde inner product.

The identification of the KM characters with the classical characters at roots of unity was used by Kac for the first time, when he calculated the action of the $SL(2,\mathbb{Z})$ on the KM-characters. The automorphism $\tau \mapsto -1/\tau$ transforms the images of the characters into the delta functions of the points of the alcove.

We mention that the interpretation of the Verlinde algebras in terms of quantum groups due to Kazhdan–Lusztig–Finkelberg leads to the classical characters at roots of unity as well.

The general drawback of this approach is that the c, the levels, must be positive integers. The corresponding Verlinde algebras are totally disconnected for different levels, unless a special p–adic limiting procedure is used, similar to the one due to Ohtshuki in the theory of invariants of knots and links. It is why a uniform theory for all (unimodular) q is needed.

LIE GROUPS | **KAC-MOODY**
Compact Characters | KM-Characters
⇓ | ⇓
Spherical Functions | Symmetric Spaces?
$(k = 1, 2, 1/2)$ | *ellip radial parts* !
⇓ | ⇓
Orthog Polynomials | Measure&Integrtn?
(k *is arbitrary*) | *Verlinde Algebras* !

\- -

DEGENERATE DAHA & **DUNKL OPERATORS**
Nonsym Harish-Chandra | Elliptic Nonsym Theory
H-Ch transform : ↓ | ↓ **: not finished yet**
| Rational-Difference Nonsymmetric Theory |

\- -

GENERAL q,t–DAHA & **ACTION OF PSL$_2$(Z)**
q-Fourier Transform becomes self-dual !

Figure 0.5: Harmonic Analysis and DAHA

Another drawback is of a technical nature. The Verlinde algebras are too combinatorial apart form the $\mathfrak{sl}_2$–case. For instance, there is a very difficult problem of finding the $PSL(2,\mathbb{Z})$–invariants in the tensor product of the Verlinde algebra and its dual. There is a solution of this problem, important in physical applications, only in the cases of $\mathfrak{sl}_2$ and $\mathfrak{sl}_3$. The list of the "invariants" is sophisticated in the latter case.

It is expected that the theory at generic q would be simpler to deal with. Many physicists now are working in the setting of unimodular q ($|q| = 1$) that are *not* roots of unity.

Difference theory. The bottom block of Figure 0.5 shows the difference theory and the general double affine Hecke algebra (DAHA). In contrast to the previous "Harish-Chandra level," the q–Fourier transform is self-dual, as holds for the classical Fourier transform and in the group case. Thus we are back to "normal" Fourier theory.

This, however, does not mean that the analytic difficulties dissappear. *Six* different analytic settings are known in the q, t–case. Namely, there are one compact and two noncompact theories, and each exists in two variants: for real q and unimodular q, not counting the choice of the analytic spaces that can be used for the direct and inverse transforms.

There are two other theories of an algebraic nature. The *seventh* setup is the theory at the roots of unity. All irreducible representations are finite dimensional as q is a root of unity. The major example is the nonsymmetric Verlinde algebra, which is Fourier-invariant and, moreover, invariant with respect to the projective action of $PSL(2,\mathbb{Z})$. The *eighth* is the theory of p–adic integration, where the connection with DAHA has not yet been established.

The projective action of $PSL(2,\mathbb{Z})$ is one of the most important parts of the theory of DAHA. The Fourier transform corresponds to the matrix $\begin{pmatrix} 0 & 1 \\ -1 & 0 \end{pmatrix}$ and is related to the transposition of the periods of an elliptic curve.

Special functions. There is a fundamental reason to expect much from DAHA in analysis: it is a *direct* generalization of the Heisenberg–Weyl algebra. DAHA was designed to fulfill a gap between the representation theory and classical theory of the special functions. It naturally incorporates the Bessel function, the hypergeometric and basic (difference) hypergeometric functions, and their multidimensional generalizations into the representation theory. This is especially true for the basic hypergeometric function.

DAHA is heavily involved in the *multidimensional* theory of hypergeometric-type functions. This direction is relatively recent. The Dunkl operators and the rational DAHA are the main tools in the theory of the multidimensional Bessel functions. This theory is relatively recent; the first reference seems [O2]. Respectively, the degenerate DAHA and the trigonometric Dunkl operators serve the multidimensional hypergeometric functions.

The one-dimensional DAHA is closely related to the super Lie algebra $\mathfrak{osp}(2|1)$ and to its "even part," $\mathfrak{sl}(2)$ via the "exponential map" from DAHA to its rational degeneration. The one-dimensional Dunkl operator is the square root of the radial part of the Laplace operator in the rank one case, similar to the Dirac operator, although it is a special feature of the one-dimensional case.

For instance, the finite dimensional representations of $\mathfrak{osp}(2|1)$ have a natural action of DAHA of type A_1. Their even parts, the classical finite dimensional representations of $\mathfrak{sl}(2)$, are modules over the subalgebra of symmetric elements of DAHA:

DAHA can be viewed as a natural successor of $\mathfrak{sl}(2)$.

0.3.3 Elliptic theory

A solid part of the Kac–Moody harmonic analysis is the construction of the "elliptic radial parts" due to Olshanetsky–Perelomov (the GL_N–case), Ochiai–Oshima–Sekiguchi (the BC–type), and from [C17] for all reduced root systems. They exist only at the *critical level*, which does not make them too promising. We certainly need the theory for an arbitrary central charge. The degenerate ("trigonometric," to be more exact) double affine Hecke algebra presumably provides such a theory.

Nonsymmetric theory. The most recognized applications of the degenerate DAHA so far are in classical harmonic analysis, namely, this algebra is very helpful in the Harish-Chandra theory of spherical functions and, more importantly, adds the nonsymmetric spherical functions and polynomials to the Lie–Harish-Chandra theory.

The *same* degenerate DAHA also serves the elliptic nonsymmetric Harish-Chandra theory. The "nonsymmetric" is the key here, because the "elliptic" Dunkl operators (infinite trigonometric, to be more exact), commute at *arbitrary levels*, in contrast to the elliptic radial parts. The latter can be always defined as symmetrizations of the Dunkl operators, but commute only at the critical level [C17].

We also mention the recent [Ra] devoted to the interpolation and orthogonal "elliptic polynomials" in the $C^\vee C$–case. The connection with the DAHA of type $C^\vee C$ (defined by Noumi–Sahi) has not been established yet.

In the theory of DAHA, the "transforms" become homomorphisms of representations. For instance, the nonsymmetric Harish-Chandra transform appears as an analytic homomorphism from the trigonometric-differential polynomial representation of the degenerate DAHA ("Nonsym Harish-Chandra") to its rational-difference polynomial representation.

There exists a third elliptic-differential polynomial representation of DAHA ("Elliptic nonsym theory") which is analytically isomorphic to the rational-difference representation. The corresponding isomorphism can be called a nonsymmetric elliptic Harish-Chandra transform. See the middle block of Figure 0.5.

The analytic part of the elliptic theory is not finished yet. The general q, t–DAHA governs the monodromy of the eigenfunctions of the elliptic Dunkl eigenvalue problem, corresponding to the action of the Weyl group and the translations by the periods of an elliptic curve. Therefore, besides the "elliptic transform," shown in the figure as the arrow to the "difference-rational theory," there is another arrow from the elliptic theory to the q, t–block.

Similar to the Verlinde algebras, that describe the action of $PSL(2, \mathbb{Z})$ on the Kac–Moody characters, the monodromy representation can be used to describe the projective action of the $PSL(2, \mathbb{Z})$ on the eigenfunctions of the elliptic Dunkl eigenvalue problem.

Deformation. A natural objective is to define the generalized Verlinde algebras for *arbitrary* unimodular q. Since DAHA describes the monodromy of the elliptic nonsymmetric theory (the previous block), such an extension of the Verlinde theory is granted. However, what can be achieved via DAHA appears more surprising.

Under minor technical restrictions, DAHA gives a *flat* deformation of the

Verlinde algebra to arbitrary unimodular q. To be exact, *little* Verlinde algebras must be considered here, associated with the root lattices instead of the weight lattices.

The dimension of the little Verlinde algebra remains the same under this deformation. All properties are preserved but the integrality and positivity of the structural constants, which are lost for generic q.

These deformations were constructed in [C27, C28] using *singular* k, rational numbers with the Coxeter number as the denominator. See Section 0.5.1 below and the end of Chapter 3.

It triggers the question about the classification of all representations of DAHA for such special fractional k. Indeed, if a "unitary" Kac–Moody representation theory exists, it could be associated with such representations, by analogy with the interpretation of the Verlinde algebras via the integrable representations of the Kac–Moody algebras. We also note a relation of the finite dimensional representations of DAHAs (for singular k) to the generalized Dedekind–Macdonald η–identities from [M1]; see [C28] and Chapter 3.

There is an important application. We establish a connection between little Verlinde algebras of type A_1 and finite dimensional representations of $\mathfrak{sl}(2)$, since the deformation above followed by the limit $q \to 1$ to the rational DAHA equip the little Verlinde algebras with the action of $\mathfrak{sl}(2)$.

Toward KM harmonic analysis. We conclude, that generalizations of the Verlinde algebra are reasonable candidates for categories of representations of unitary type, especially for those associated with the Kac–Moody algebras. If we expect the reduced category of integrable Kac–Moody representations to be a part of such a category, then it must be an extension of the Verlinde algebra, or, more likely, must have the reduced category as a quotient. I think that it is unlikely that the L^2–KM theory, if it exists, can be assumed to contain " 1 ", the basic representation, and other integrable representations from [Ka].

Using DAHA for this project, its unitary representations at $|q| = 1$ are natural candidates for such categories, especially for singular k, although generic q, k are interesting too. They are expected to have some extra structures. The most important are a) "fusion multiplication", b) hopefully, the action of $PSL(2, \mathbb{Z})$, and c) "good restrictions" to the affine Hecke subalgebras similar to the classical decomposition with respect to the maximal compact subgroup.

The existence of multiplication indicates that the quotients (if we include 1) and constituents (generally) of the polynomial representation of DAHA or those for the functional variants of the polynomial representation must be examined first. All of them are commutative algebras. The most interesting "AHA–restriction theory" occurs at the singular k, when the polynomial representation becomes reducible.

0.4 DAHA and Verlinde algebras

The Lie groups formalize the concept of symmetry in the theory of special functions, combinatorics, geometry, and, last but not least, physics. In a similar way, *abstract* Verlinde algebras "describe" Fourier transforms, especially the theories with the Gaussian, satisfying the following fundamental property. The Fourier image of the Gaussian must be to be proportional to its inverse or the Gaussian itself, depending on the setting, similar to the properties of the Laplace and Fourier transforms in the classical theory. Thus, the Verlinde algebras (are expected to) formalize an important portion of the classical Fourier analysis.

Note that the Fourier transform has an interpretation in the Lie theory as a reflection from the Weyl group. There many such reflections apart from the $\mathfrak{sl}(2)$–case. This does not match well the natural expectation that there must be a unique Fourier transform, or, at least, the major one if there are several. It is really unique in the DAHA theory.

0.4.1 Abstract Verlinde algebras

In the finite dimensional semisimple variant, the *abstract Verlinde algebra* is the algebra of $\mathbb{C}$–valued functions V =Funct($\bowtie$) on a finite set $\bowtie$ equipped with a linear automorphism σ, *the Fourier transform.* The algebra V has a unit, which is 1 considered as a constant function. Note that σ is not supposed to preserve the (pointwise) multiplication. As a matter of fact, it never does.

The space V has a natural basis of the *characteristic functions* $\chi_i(j) = \delta_{ij}$, where $i, j \in\bowtie$, δ_{ij} is the Kronecker delta.

The first two assumptions are that

(a) $\sigma^{-1}(1) = \chi_o$ for the *zero-point* $o \in\bowtie$, and

(b) the numbers $\mu_i \stackrel{\mathbf{def}}{=\!=} \sigma(\chi_i)(o)$ are nonzero.

Since the latter constants are interpreted as masses of particles in conformal field theory, let us impose further conditions $\mu_i > 0$.

Then the *spherical functions* are $p_i \stackrel{\mathbf{def}}{=\!=} \sigma(\chi_i)\mu_i^{-1}$ for $i \in\bowtie$. In other words, p_i is proportional to $\sigma(\chi_i)$ and satisfies the normalizing condition $p_i(o) = 1$.

Introducing the *inner product* as $\langle f, g\rangle \stackrel{\mathbf{def}}{=\!=} \sum_i \mu_i f(i)\overline{g(i)}$, the corresponding delta functions are $\delta_i = \mu_i^{-1}\chi_i = \sigma^{-1}(p_i)$. Indeed, they are obviously dual to the characteristic functions with respect to the inner product.

Concerning taking the weights $\mu_i = \sigma(\chi_i)(o)$ in the inner product, such choice is equivalent to the following classical property, which holds in all variants of the Fourier theory:

the Fourier-images of the delta functions are spherical functions.

The next assumption is that

(c) σ is unitary up to proportionality with respect to $\langle\, , \rangle$.

It readily gives the *norm formula:*

$$\langle p_i, p_j\rangle = \delta_{ij}\mu_i^{-1}\langle 1, 1\rangle.$$

In this approach, the latter formula is a result of a simple sequence of formal definitions. However, it is really fruitful. It leads to the best-known justification of the norm formula for the Macdonald polynomials, including the celebrated constant term conjecture, which is the formula for $\langle 1, 1\rangle$.

To be more exact, it provides a deduction of the norm formula from the so-called evaluation formula for the values of the Macdonald polynomials at the "zero-point." In its turn, the latter formula results from the self-duality of the double affine Hecke algebra (the existence of the self-dual Fourier transform).

The self-duality and the evaluation formula are directly connected with the following important symmetry:

(d) $p_i(j) = p_j(i)$ for $i \in\bowtie\ni j$.

Note that the theory of Macdonald polynomials requires an infinite dimensional variant of the definitions considered above. Namely, σ^{-1} becomes an isomorphism from the algebra of Laurent polynomials to its dual, the space of the corresponding delta functions. Hence it becomes a map from one algebra to another algebra.

Nevertheless it is possible to deduce the major formulas for the Macdonald polynomials within the finite dimensional "self-dual" setting above, by using the consideration at the roots of unity. Without going into detail, it goes as follows.

The number of Macdonald polynomials, which are well defined when q is a *root of unity*, grows together with the order of q. One checks, say, the Pieri rules for such Macdonald polynomials using the duality argument, then tends the order of q to ∞, and, finally, obtain the desired formula for arbitrary q and *all* Macdonald polynomials. The Pieri rules for multiplication of the Macdonald polynomials by the monomial symmetric functions were, indeed, justified in [C19, C20] by using the roots if unity.

Concluding the discussion of abstract Verlinde algebras, the *Gaussian* appears as a function $\gamma \in V$ such that

(e) $\sigma(\gamma) = \text{const}\cdot\gamma^{-1}$, $\gamma(i) \neq 0$ for $i \in\bowtie$.

The constant here is the *abstract Gauss–Selberg sum.* It is necessary to fix the normalization of σ and γ to make this sum well defined. The normalization of σ has already been fixed by the condition $\sigma(\chi_o) = 1$. The natural normalization of the Gaussian is $\gamma(o) = 1$.

The assumptions (a-e) are more than sufficient to make the Verlinde algebras rigid enough. However, in my opinion, the key axiom is PBW, the Poincare–Birkhoff–Witt property, which requires the operator approach, to be discussed next.

0.4.2 Operator Verlinde algebras

The above discussion is actually about a "good self-dual" invertible linear operator acting in a commutative algebra. It is obviously too general, and more structures have to be added. We need to go to the operator level, switching from the characteristic functions and the spherical functions to the corresponding *commutative* algebras $\mathcal{X}, \mathcal{Y}$ of the operators, which are diagonal at the corresponding sets of functions.

The *operator Verlinde algebra* $\mathcal{A}$ (the main example will be the double affine Hecke algebra) is generated by commutative algebras $\mathcal{X}$ and $\mathcal{Y}$, and the algebra $\mathbf{H}$ controlling the symmetries of the X–operators and the Y–operators. In the main examples, the Weyl groups (or somewhat more general groups) are the groups of symmetries upon degenerations; $\mathbf{H}$ are the corresponding Hecke algebras serving the non-degenerate q, t–case.

The key and the most restrictive assumption is the *PBW property*, which states that

(A) the natural map from the tensor product $\mathcal{X} \otimes \mathcal{Y} \otimes \mathbf{H}$ to $\mathcal{A}$ is an isomorphism of the linear spaces, as well as the other five maps corresponding to the other orderings of $\mathcal{X}, \mathcal{Y}, \mathbf{H}$.

The Fourier transform and the Gaussian are formalized as follows. The *projective* $PSL_2(\mathbb{Z})$ must act in $\mathcal{A}$ by outer automorphisms, i.e.,

(B) τ_+, τ_- act in $\mathcal{A}$ as algebra automorphisms and satisfy the Steinberg relation $\tau_+\tau_-^{-1}\tau_+ = \tau_-^{-1}\tau_+\tau_-^{-1}$;

(C) they preserve the elements from $\mathbf{H}$, the element $\sigma \stackrel{\mathbf{def}}{=\!=} \tau_+\tau_-^{-1}\tau_+$ maps $\mathcal{X}$ onto $\mathcal{Y}$, and τ_+ is the identity on $\mathcal{X}$.

Note that (B)–(C) provide that $\tau_- = \sigma\tau_+^{-1}\sigma^{-1}$ is identical on $\mathcal{Y}$. The automorphism τ_+ becomes the multiplication by the Gaussian in Verlinde algebras, which connects (C) with the assumption (e) above.

The remaining feature of abstract Verlinde algebras to be interpreted using the operator approach is the existence of the inner product. We will postulate the existence of the corresponding involution. We assume that:

(D) $\mathcal{A}$ is a quotient of the group algebra of the group $\mathcal{B}$ such that the anti-involution $\mathcal{B} \ni g \mapsto g^{-1}$ of $\mathcal{B}$ becomes an anti-involution of $\mathcal{A}$;

(E) $\tau_\pm$ and σ come from automorphisms of the group $\mathcal{B}$ and therefore commute with the anti-involution from (D).

We note that the inner products can be associated with somewhat different involutions, for instance, in the case of *real* harmonic analysis ($q > 0$). The assumption that there is a system of *unitary* generators is a special feature of the unimodular theory ($|q| = 1$). The anti-involution in the Verlinde case and those in the main generalizations do satisfy (D)–(E).

The Verlinde algebras can be now *re-defined* as σ–invariant unitary irreducible representations V of $\mathcal{A}$ that are $\mathcal{X}$–spherical, i.e., are some quotients of the (commutative) algebra $\mathcal{X}$. A representation is called unitary if it has a hermitian inner product inducing the anti-involution of $\mathcal{A}$ from (D). Note that (D)–(E) *automatically* guarantee that the Fourier transform σ is "projectively unitary" in σ–invariant unitary irreducible representations.

Thanks to irreducibility, both σ and τ_+ are fixed uniquely in the group $\mathrm{Aut}_{\mathbb{C}}(V)/\mathbb{C}^*$ (if they act in V) and induce the corresponding automorphisms of $\mathcal{A}$.

Strictly speaking, assumptions (D)–(E) can be replaced by a more general property that $\mathcal{A}$ has an anti-involution which commutes with the $\tau_\pm$ and σ. However the double Hecke algebras (the main examples so far) really appear as quotients of the group algebras and $\tau_\pm$ and σ act in this group, the elliptic braid group.

An important advantage of the operator approach is that we can relax the constraints for the abstract Verlinde algebras defined above by considering *non-unitary* and even *non-semisimple* projective $PSL(2,\mathbb{Z})$–invariant X–spherical irreducible representations of $\mathcal{A}$. They do appear in applications.

Another advantage is that it is not necessary to impose the Fourier-invariance. Generally, the main problem of the Fourier analysis is in calculating the Fourier images $\sigma(V)$ of *arbitrary* $\mathcal{A}$–modules V and the corresponding transforms $V \ni v \mapsto \sigma(v) \in \sigma(V)$, which induce σ in $\mathcal{A}$.

Similarly, τ_+ becomes multiplication by the Gaussian given by a variant of the classical formula e^{x^2} in the main examples. The operator approach makes it possible to define the Gaussian for any irreducible $\mathcal{A}$-module V; it as an operator with values in $\tau_+(V)$.

Let us discuss now the DAHA of type A_1 in detail, where the theory is already quite interesting. The transition to arbitrary root system is sufficiently smooth. See Chapter 2 and, especially, Chapter 3.

0.4.3 Double Hecke Algebra

The most natural definition goes through the *elliptic braid group* $\mathcal{B}_q \stackrel{\mathbf{def}}{=\!=} \langle T, X, Y, q^{1/4}\rangle/$ with the relations

$$TXT = X^{-1},\ TY^{-1}T = Y,$$
$$Y^{-1}X^{-1}YXT^2 \;=\; q^{-1/2}.$$

Here

$$\mathcal{B}_1 = \pi_1^{\mathrm{orb}}(\{E \setminus 0\}/\mathbb{S}_2)$$

$\cong \pi_1(\{E \times E \setminus \mathrm{diag}\}/\mathbb{S}_2)$, where E is an elliptic curve. Using the orbifold fundamental group here makes it possible to "divide" by the symmetric group $\mathbb{S}_2$ without removing the ramification completely, i.e., removing all four points of second order. Only one puncture is needed to obtain the above relations.

Actually, in the case of A_n, it is sufficient to consider the product of $n+1$ copies of E and remove the "diagonal" before dividing by $\mathbb{S}_{n+1}$ instead of using the orbifold group. The corresponding braid group is *isogenous* to the one with the relations above; it was calculated by Birman and Scott.

To complete the definition of DAHA, we impose the quadratic T–relation:

$$\mathcal{H} \stackrel{\mathbf{def}}{=\!=} \mathbb{C}[\mathcal{B}_q]/((T-t^{1/2})(T+t^{-1/2})).$$

If $t = 1$, $\mathcal{H}$ becomes the Weyl algebra extended by the reflection; T becomes s satisfying $sXs = X^{-1}$, $sYs = Y^{-1}$, $s^2 = 1$.

The *Fourier transform*, which plays a major role in the theory, is the following outer automorphism of DAHA:

$$\sigma : X \mapsto Y^{-1}, Y \mapsto XT^2,\ T \mapsto T.$$

Thus the DAHA Fourier transforms finds a conceptual interpretation as the transposition of the periods of the elliptic curve, This is not surprising from the viewpoint of CFT, KZB, and the Verlinde algebras, but such a connection with topology still remains challenging when we deal with the applications of DAHA in harmonic analysis.

The representations where σ acts (i.e., becomes inner) are called *Fourier-invariant* or self-dual. The nonsymmetric Verlinde algebra and the Schwartz space are examples.

More generally, the topological interpretation above readily gives that the group $PSL(2,\mathbb{Z})$ acts projectively in $\mathcal{H}$:

$$\begin{aligned}
&\tau_+ : Y \mapsto q^{-1/4}XY,\ X \mapsto X,\ T \mapsto T,\\
&\tau_- : X \mapsto q^{1/4}YX,\ \ Y \mapsto Y,\ T \mapsto T,\\
&\sigma = \tau_+\tau_-^{-1}\tau_+ = \tau_-^{-1}\tau_+\tau_-^{-1},
\end{aligned}$$

$$\text{where } \begin{pmatrix}11\\01\end{pmatrix} \mapsto \tau_+,\ \ \begin{pmatrix}10\\11\end{pmatrix} \mapsto \tau_-.$$

We will use k such that $t = q^k$ and (sometimes) set $\mathcal{H}(k)$ as $t = q^k$. The algebra $\mathcal{H}(k)$ acts in $\mathcal{P}$ = *Laurent polynomials* in terms of $X = q^x$, namely,

$$\begin{aligned}
&T \mapsto t^{1/2}s + \frac{t^{1/2}-t^{-1/2}}{q^{2x}-1}(s-1),\\
&Y \mapsto spT,\ \ sf(x) \stackrel{\mathbf{def}}{=\!=} f(-x),\\
&pf(x) \stackrel{\mathbf{def}}{=\!=} f(x+1/2),\ \ t = q^k.
\end{aligned}$$

The operator Y is called the *difference Dunkl operator.*

It is important to note that τ_- preserves $\mathcal{P}$. On the other hand, τ_+ does not act there for a very simple reason. If it acts then it must be multiplication by the Gaussian q^{x^2}, which does not belong to $\mathcal{P}$.

The "radial part" appears as follows. One checks that $Y+Y^{-1}$ preserves $\mathcal{P}_{sym} \stackrel{\text{def}}{=}$ symmetric (even) Laurent polynomials (the Bernstein lemma in the theory of affine Hecke algebras). The restriction $H = Y+Y^{-1}\mid_{sym}$ is the *q–radial part* and can be readily calculated:

$$H = \frac{t^{1/2}X - t^{-1/2}X^{-1}}{X - X^{-1}}p + \frac{t^{1/2}X^{-1} - t^{-1/2}X}{X^{-1} - X}p^{-1}.$$

0.4.4 Nonsymmetric Verlinde algebras

The key new development in the theory of orthogonal polynomials, algebraic combinatorics, and related harmonic analysis is the definition of the nonsymmetric Opdam–Macdonald polynomials. The main references are [O3, M5, C20]. Opdam mentions in [O3] that this definition (in the differential setup) was given in Heckman's unpublished lectures.

These polynomials are expected to be, generally speaking, beyond quantum groups and Kac–Moody algebras because of the following metamathematical reason. Major special functions in the Lie and Kac–Moody theory, including the classical and Kac–Moody characters, spherical functions, and conformal blocks, are W–invariant.

However, this definition is not quite new in representation theory. The limits of the nonsymmetric polynomials as $q \to \infty$ are well known. They are the spherical functions due to Matsumoto. The nonsymmetric Verlinde algebras are *directly* connected to the nonsymmetric polynomials evaluated at the roots of unity,

Let $q = \exp(2\pi i/N),\ 0 < k < N/2,\ k \in \mathbb{Z}$. The *nonsymmetric Verlinde algebra* V is defined as the algebra of functions of the set

$$\bowtie = \{-\frac{N-k+1}{2}, \ldots, -\frac{k+1}{2}, -\frac{k}{2}, \frac{k+1}{2}, \ldots, \frac{N-k}{2}\}. \tag{0.4.1}$$

It has the unique structure of a $\mathcal{H\!H}$–module that makes the map $q^x(z) = q^z$ from $\mathcal{P}$ to V a $\mathcal{H\!H}$–homomorphism. The above set is not s–invariant; it is sp–invariant. Nevertheless, the formula for T can be used in V, because the contributions of the "forbidden" points $k/2 = s(-k/2)$ and $(N-k+1)/2 = s(-(N-k+1)/2)$ come with zero coefficients.

The operators X, Y, T are unitary in V with respect to the positive hermitian form which will not be discussed here (it generalizes the inner product for conformal blocks). See Chapter 2. The positivity requires choosing the "minimal" primitive N–root of unity q above.

The whole $PSL(2,\mathbb{Z})$ acts in V projectively as well as in the image V_{sym} of $\mathcal{P}_{sym}$. The latter image can of course be defined without any reference to the polynomial representation. A general definition is as follows: $V_{sym} = \{f \mid Tf = t^{1/2}f\}$. Here it simply means that the function $f(z)$ must be s–invariant (even) for the points z that do not leave the above set under the action of s (recall that it is not s–invariant). Therefore:

$$\dim_{\mathbb{C}} V = 2N - 4k, \ \dim_{\mathbb{C}} V_{sym} = N - 2k + 1.$$

We call V a *perfect representation.* By perfect, we mean that it has all major features of the irreducible representations of the Weyl algebras at roots of unity, i.e., cannot be better. Formally, it means that it has a perfect duality pairing, directly realted to the Fourier transform, and a projective action of $PSL(2, \mathbb{Z})$.

The nonsymmetric *characters* in V and the symmetric ones in V_{sym} are, respectively, the eigenfunctions of Y and $Y + Y^{-1}$.

When $k = 0$, we come to the well-known definitions in the theory of Weyl algebra. As $k = 1$, $V_{sym}^{k=1}$ is the usual *Verlinde algebra* [Ver]:

τ_+ becomes the Verlinde T–operator and σ becomes the S–operator.

0.4.5 Topological interpretation

Let E be an elliptic curve over $\mathbb{C}$, i.e., $E = \mathbb{C}/\Lambda$, where $\Lambda = \mathbb{Z} + \mathbb{Z}\imath$. Topologically, the lattice can be arbitrary. Let $0 \in E$ be the zero-point, and -1 the automorphism $x \mapsto -x$ of E. We are going to define the orbifold fundamental group of the space $(E \setminus 0)/\pm 1 = \mathbb{P}^1_{\mathbb{C}} \setminus 0$. Since this space is contractible, its usual fundamental group is trivial. One could take the quotient after removing all (four) ramification points of -1. However, it would enlarge the fundamental group dramatically. Therefore we need to understand this space in a more refined way.

We take the *base (starting) point* $\star = -\varepsilon - \varepsilon\imath \in \mathbb{C}$ for small $\varepsilon > 0$. The *orbifold fundamental group* $\pi_1^{orb}((E \setminus 0)/\pm 1)$ is defined following [C13]. We switch from E to its universal cover $\mathbb{C}$ and define the *paths* as curves $\gamma \in \mathbb{C} \setminus \Lambda$ from $\star$ to $\widehat{w}(\star)$, where $\widehat{w} \in \widehat{W} = \{\pm 1\} \ltimes \Lambda$.

The *composition of the paths* is via $\widehat{W}$: we add the image of the second path under $\widehat{w}$ to the first path if the latter ends at $\widehat{w}(\star)$.

The fundamental group $\pi_1^{orb}(\cdot)$ of the above paths modulo homotopy is isomorphic to

$$\mathcal{B}_1 = \langle T, X, Y \rangle / \langle TXT = X^{-1}, TY^{-1}T = Y, Y^{-1}X^{-1}YXT^2 = 1 \rangle,$$

where T is the half-turn, i.e., the clockwise half-circle from $\star$ to $s(\star)$, X, Y are, respectively, 1 and $\imath$ considered as vectors from $\star$.

The definition we used is close to the calculation of the fundamental group of $\{E \times E \setminus \text{diagonal}\}$ divided by the transposition of the components, however, there is no exact coincidence. Let me also mention the relation to the elliptic braid group due to v.d. Lek, although he removes all points of second order.

The topological interpretation is the best way to understand why the group $PSL_2(\mathbb{Z})$ acts in $\mathcal{B}_1$ projectively.

Its elements act in $\mathbb{C}$ naturally, by the corresponding real linear transformations. On E, they commute with the reflection -1, preserve 0, and permute the other three points of second order. Given $g \in SL_2(\mathbb{Z})$, we set $g = \exp(h), g_t = \exp(th)$ for the proper $h \in \mathfrak{sl}_2(\mathbb{R})$, $0 \le t \le 1$.

The position of the base point $\star$ will become $g(\star)$, so we need to go back, i.e., connect the image with the base point by a path. To be more exact, the g–image of $\gamma \in \pi_1^{orb}$ will be the union of the paths

$$\{g_t(\star)\} \cup g(\gamma) \cup \{\widehat{w}(g_{-t}(\star)\},$$

where the path for γ goes from $\star$ to the point $\widehat{w}(\star)$.

We can always choose the base point sufficiently close to 0 and connect it with its g–image in a small neighborhood of zero. This makes the corresponding automorphism of $\mathcal{B}_1$ unique up to powers of T^2. All such automorphisms fix T, because they preserve zero and the orientation.

Thus we have constructed a homomorphism

$$\alpha : SL_2(\mathbb{Z}) \to \mathrm{Aut}_T(\mathcal{B}_1)/T^{2\mathbb{Z}},$$

where $\mathrm{Aut}_T(\mathcal{B}_1)$ is the group of automorphisms of $\mathcal{B}_1$ fixing T. The elements from $T^{2\mathbb{Z}} = \{T^{2n}\}$ are considered here as inner automorphisms.

Let τ_+, τ_- be the α–images of the matrices $\begin{pmatrix} 1 & 1 \\ 0 & 1 \end{pmatrix}$ and $\begin{pmatrix} 1 & 0 \\ 1 & 1 \end{pmatrix}$. Then $\sigma = \tau_+\tau_-^{-1}\tau_+$ corresponds to $\begin{pmatrix} 0 & 1 \\ -1 & 0 \end{pmatrix}$.

Taking the "simplest" pullbacks for $\tau_\pm$, we arrive at the Steinberg relation:

$$\tau_+\tau_-^{-1}\tau_+ = \tau_-^{-1}\tau_+\tau_-^{-1}.$$

Abstract construction. Generalizing, let $\mathcal{E}$ be an algebraic, or complex analytic, or symplectic, or real analytic manifold, or similar. It may be noncompact and singular. We assume that there is a continuous family of *topological isomorphisms* $\mathcal{E} \to \mathcal{E}_t$ for manifolds $\mathcal{E}_t$ as $0 \le t \le 1$, and that $\mathcal{E}_1$ is isomorphic to $\mathcal{E}_0 = \mathcal{E}$. The path $\{\mathcal{E}_t\}$ in the moduli space $\mathcal{M}$ of $\mathcal{E}$ induces an outer automorphism ε of the fundamental group $\pi_1(\mathcal{E}, \star)$ defined as above. Namely, we take the image of $\gamma \in \pi_1(\mathcal{E}, \star)$ in $\pi_1(\mathcal{E}_1, \star_1)$ for the image $\star_1$ of the base point $\star \in \mathcal{E}$ and conjugate it by the path from $\star_1$ to $\star$. We obtain that *the fundamental group* $\pi_1(\mathcal{M})$ (whatever it is) *acts in* $\pi_1(\mathcal{E})$ *by outer automorphisms modulo inner automorphisms.*

The above considerations correspond to the case when a group G acts in $\mathcal{E}$ preserving a submanifold D. Then $\pi_1(\mathcal{M})$ acts in $\pi_1^{orb}((\mathcal{E}\setminus D)/G)$ by outer automorphisms.

Another variant is with the Galois group taken instead of $\pi_1(\mathcal{M})$, assuming that $\mathcal{E}$ is an algebraic variety over a field that is not algebraically closed.

The action of $\pi_1(\mathcal{M})$ on an *individual* $\pi_1(\mathcal{E})$ extends, in a way, the celebrated *Kodaira–Spencer map* and is of obvious importance. However, calculating the fundamental groups of algebraic (or similar) varieties, generally, is difficult. The main examples are the products of algebraic curves and related configuration spaces. Not much can be extracted from the action above without an explicit description of the fundamental group.

0.5 Applications

The Verlinde algebras and, more generally, the finite dimensional representations of DAHA, have quite a few applications. We will mainly discuss the non-cyclotomic Gaussian sums and the diagonal coinvariants. Both constructions are based on the DAHA deformation–degeneration technique. We also discuss a classification of the Verlinde algebras.

0.5.1 Flat deformation

Recall that we set $\mathcal{H}(k)$ when $t = q^k$. We will need the *little nonsymmetric Verlinde algebra* $\widetilde{V}^k$ which is the unique nonzero irreducible quotient of V^k upon the restriction to the *little DAHA* $\widetilde{\mathcal{H}} \stackrel{\mathbf{def}}{=\!=} \langle T, X^2, Y^2\rangle$, corresponding to the subspace of Laurent polynomials in terms of X^2. Its dimension is $\dim\widetilde{V}^k = \dim V^k/2 = N-2k$.

Generally, $\widetilde{V}_{sym}^{k=1}$ is a subalgebra of the Verlinde algebra defined for the radical weights (from the root lattice) instead of all weights.

Let $N = 2n+1$, $q^{1/2} = -\exp(\frac{\pi i}{N})$. Such a choice of q is necessary for the positivity of the corresponding inner product.

We set $m = n-k$, $\dim_{\mathbb{C}}\widetilde{V}^k = N-2k = 2m+1$. The aim is to deform $\widetilde{V}$, $\widetilde{V}_{sym}$, the projective action of $PSL(2,\mathbb{Z})$, and all other structures making q arbitrary unimodular. The construction is as follows.

For any q, $\bar{k} = -\frac{1}{2} - m$, $m \in \mathbb{Z}_+$, $\mathcal{P}$ considered as an $\mathcal{H}(\bar{k})$–module has a unique irreducible quotient $\overline{V} \neq 0$:

$$\overline{V} = Funct(\frac{\bar{k}+1}{2}, \ldots, -\frac{\bar{k}+1}{2}, -\frac{\bar{k}}{2}),$$

$$dim_{\mathbb{C}}\,\overline{V} = 2m+1, \;\; dim_{\mathbb{C}}\,\overline{V}_{sym} = m+1.$$

It is unitary as $|q| = 1$, $0 < arg(q) < \pi/m$.

The representation $\overline{V}$, the desired deformation, becomes $\widetilde{V}^k$ as

$$q = e^{\frac{2\pi i}{N}},\ arg(q) = \frac{2\pi}{2n+1},\ \bar{k} = k - \frac{1}{2} - n,\ i.e.\ ,\ m = n - k.$$

Concerning the positivity of the inner product, note that $\frac{2\pi}{2n+1} < \pi/m$, since $k \geq 0$.

The deformation construction makes it possible to connect the Verlinde algebras and their nonsymmetric k–generalizations with the classical representation theory. Since the parameter q is now generic, one may expect relations to $\mathfrak{sl}(2)$.

Indeed, the representation $\overline{V}$ is a q–deformation of an irreducible representation of $\mathfrak{osp}(2|1)$ (see Chapter 2). Respectively, $\overline{V}_{sym}$ deforms the irreducible representation of $\mathfrak{sl}(2)$ of spin$=m/2$. We need the rational degeneration of DAHA to clarify it.

0.5.2 Rational degeneration

The *rational DAHA* is the following quasi-classical limit:

$$Y = e^{-\sqrt{\kappa}y/2},\ X = e^{\sqrt{\kappa}x},\ q = e^{\kappa},\ \kappa \to 0$$

of $\mathcal{H}$. Explicitly,

$$\mathcal{H}'' \stackrel{\text{def}}{=\!=} \langle x, y, s\rangle/\text{relations:}$$

$$[y, x] = 1 + 2ks,\ s^2 = 1,\ sxs = -x,\ sys = -y.$$

Note that the notation $\mathcal{H}'$ is reserved for trigonometric degeneration (see Chapter 2).

The algebra $\mathcal{H}''$ acts in the polynomial representation, which is

$$\mathbb{C}[x],\ s(x) = -x,\ \ y \mapsto D = \frac{d}{dx} - \frac{k}{x}(s-1).$$

The square of the *Dunkl operator* D is the radial part of the Laplace operator:

$$D^2\ |_{sym} = \frac{d^2}{dx^2} + \frac{2k}{x}\frac{d}{dx}.$$

The DAHA Fourier automorphism σ becomes the outer automorphism corresponding to the *Hankel transform* in the theory of Bessel functions.

Let $\bar{k} = -\frac{1}{2} - m$. Then

$$\lim_{\kappa\to 0} \overline{V} = \overline{V}'' \stackrel{\text{def}}{=\!=} \mathbb{C}[x]/(x^{2m+1}).$$

We call this module a *perfect rational* representation. The automorphism σ acts there. Its symmetric part $\overline{V}''_{sym}$ is nothing else but the irreducible representation of $\mathfrak{sl}(2)$ of dim$=m+1$. The formulas for the generators of $\mathfrak{sl}(2)$ in terms of x, y and the action of σ are

$$h = (xy + yx)/2,\ e = x^2, f = -y^2,\ \ \sigma \text{ becomes } w_0.$$

0.5.3 Gaussian sums

The new approach to the Gaussian sums based on the deformation construction above is as follows.

The τ_- acts in $\overline{V}$ and $\overline{V}_{sym}$, since it acts in the polynomial representation $\mathcal{P}$. In contrast to the polynomial representation, τ_+ also acts in $\overline{V}$ and its symmetric part. It is the multiplication by

$$q^{x^2} \stackrel{\mathbf{def}}{=\!=} q^{\frac{(\pm\bar{k}+j)^2}{4}},$$

where $\pm$ =plus for $j > 0$ and minus otherwise. Similarly, without going into detail, σ is essentially the matrix $\left(\mathrm{char}_i(\frac{\pm\bar{k}+j}{2})\right)$ for the nonsymmetric characters char_i (Y–eigenfunctions), where i and j belongs to the same set (0.4.1).

The interpretation of the Gausssian as τ_+ is of key importance for the calculation of the *non-cyclotomic Gaussian sum.* Generally speaking, a Gauss–Selberg sum is a summation of the Gaussian with respect to a certain "measure." Here it becomes

$$\sum_{j=0}^{m} q^{j^2-\bar{k}j}\frac{1-q^{2j+\bar{k}}}{1-q^{\bar{k}}}\prod_{l=1}^{2j}\frac{1-q^{l+2\bar{k}-1}}{1-q^{l}}$$

for *any* q, $\bar{k} = -1/2 - m$. It equals

$$\prod_{j=1}^{m}\frac{1-q^{2\bar{k}+2j}}{1+q^{\bar{k}+2j}}.$$

Under the reduction considered above,

$$N = 2n+1,\ m = n-k,\ 0 \le k \le n,$$
$$q^{1/2} = -\exp(\pi i/N),\ \overline{V} = \widetilde{V}^k,$$

and the product can be somewhat simplified:

$$\prod_{j=1}^{m}\frac{1-q^{2\bar{k}+2j}}{1+q^{\bar{k}+2j}} = q^{-\frac{m(m+1)}{4}}\prod_{j=0}^{m-1}(1-q^{n-j}).$$

Now let $m{=}n$, i.e., let m be the maximal possible. Then $k = n - m = 0$, $\bar{k} = -1/2 - n = 0 \bmod N$, and the "measure" in the Gaussian sum becomes trivial. Setting $l = \frac{n}{2} \bmod N$, we arrive at the identity:

$$\sum_{j=0}^{N-1} q^{j^2} = q^{l^2}\prod_{j=1}^{n}(1-q^{j}).$$

The product can be readily calculated using Galois theory. It equals $\sqrt{N}$ for $n = 2l$ and $i\sqrt{N}$ otherwise. It gives a new proof of the classical Gauss formulas.

0.5.4 Classification

We are going to describe *all* nonsymmetric Verlinde algebras of type A_1, i.e., irreducible quotients of the polynomial representation $\mathcal{P}$ that are $PSL(2,\mathbb{Z})$–invariant. The assumption is that $q^{1/2}$ is a primitive $2N$–th root of unity, $t = q^k$. All σ–invariant quotients of $\mathcal{P}$ must be through the $\mathcal{H}$–module $V_{4N} = \mathcal{P}/(X^{2N} + X^{-2N} - (t^N + t^{-N}))$.

Indeed, the invariance gives that the element $X^{2N} + X^{-2N}$, central in $\mathcal{H}$, must act as its σ–image $Y^{2N} + Y^{-2N}$, which is $t^N + t^{-N}$ in $\mathcal{P}$.

The module V_{4N} is irreducible unless k is integral or half-integral. Note that for generic k, it is *not* σ–invariant. Instead, the involution

$$X \leftrightarrow Y,\ T \leftrightarrow -T^{-1},\ t^{1/2} \mapsto t^{1/2},\ q^{1/2} \mapsto q^{-1/2}$$

is an inner automorphism of V_{4N}. Let us discuss special ("singular") k in detail.

The module V_{4N} becomes, respectively,

$$V^{-2} = \mathcal{P}/(X^{2N} + X^{-2N} - 2) \text{ or } V^{2} = \mathcal{P}/(X^{2N} + X^{-2N} + 2),$$

as $k \in \mathbb{Z}$ or $k \in 1/2 + \mathbb{Z}$. Respectively, $t^N = q^{kN} = \pm 1$. The module V^2 is an extension of $V_{2N} \stackrel{\mathbf{def}}{=\!=} \mathcal{P}/(X^N + X^{-N})$ by its ς_y–image $\varsigma_y(V_{2N})$.

The representations $V = V_{2N-4k}$ of dimension $2N - 4k$ defined above for the integral $0 < k < N/2$ is a quotient of V^{-2}. The representation $\overline{V} = \overline{V}_{2|k|}$ of dimension $2|k|$ for the half-integral $-N/2 < k = -1/2 - m < 0$ (the notation was $\bar{k}$ was such k) is a quotient of V^2.

The construction of V_{2N-4k} holds without changes for the positive *half-integral* $k < N/2$, but then V_{2N-4k} becomes a quotient of V^2. The same notation V_{2N-4k} will be used.

Let $k \in \mathbb{Z}/2, |k| < N/2$. The substitution $T \mapsto -T,\ t^{1/2} \mapsto -t^{1/2}$ identifies the polynomial representations for $t^{1/2}$ and $-t^{1/2}$. Thus it is sufficient to decompose $\mathcal{P}$ upon the transformation $k \mapsto k + N$, and we can assume that $N/2 \le k < N/2$. We will also need the outer involutions of $\mathcal{H}$:

$$\begin{aligned}
\iota &: T \mapsto -T,\ X \mapsto X,\ Y \mapsto Y,\ q^{1/2} \mapsto q^{1/2},\ t^{1/2} \mapsto t^{-1/2},\\
\varsigma_x &: T \mapsto T,\ X \mapsto -X,\ Y \mapsto Y,\ q^{1/2} \mapsto q^{1/2},\ t^{1/2} \mapsto t^{1/2},\\
\varsigma_y &: T \mapsto T,\ X \mapsto X,\ Y \mapsto -Y,\ q^{1/2} \mapsto q^{1/2},\ t^{1/2} \mapsto t^{1/2}.
\end{aligned}$$

Let $V^{+}_{2|k|} \stackrel{\mathbf{def}}{=\!=} \overline{V}_{2|k|},\ V^{-}_{2|k|} \stackrel{\mathbf{def}}{=\!=} \varsigma_x(\overline{V}_{2|k|})$.

Finally, up to ι, ς, there are *three different series of nonsymmetric Verlinde algebras*, namely, V_{2N-4k} (integral $N/2 > k > 0$), $V_{2|k|}$ (half-integral $-N/2 < k < 0$), and $V_{2N+4|k|}$ for *integral* $-N/2 < k < 0$.

The latter module is defined as a unique nonzero irreducible quotient of $\mathcal{P}$ for such k and is also isomorphic to the $\iota\varsigma_y$–image of the kernel of the map

$V^{-2} \to V_{2N-4|k|}$. It is *non-semisimple.* The previous two (series of modules) are semisimple. All three are projective $PSL(2,\mathbb{Z})$–invariant.

The modules $V^{\pm 2}$ are decomposed as follows. There are four exact sequences

$$0 \to \iota\varsigma_y(V_{2N+4k}) \to V^{-2} \to V_{2N-4k} \to 0 \quad \text{for} \quad k \in \mathbb{Z}_+,$$
$$0 \to \iota(V_{2k}^+ \oplus V_{2k}^-) \to V_{2N} \to V_{2N-4k} \to 0 \quad \text{for} \quad k \in 1/2 + \mathbb{Z}_+.$$

The arrows must be reversed for $k < 0$:

$$0 \to \iota\varsigma_y(V_{2N-4|k|}) \to V^{-2} \to V_{2N+4|k|} \to 0 \quad \text{for} \quad k \in -1 - \mathbb{Z}_+,$$
$$0 \to \iota(V_{2N-4|k|}) \to V_{2N} \to V_{2|k|}^+ \oplus V_{2|k|}^- \to 0 \quad \text{for} \quad k \in -1/2 - \mathbb{Z}_+.$$

Otherwise V^{-2} and V_{2N} are irreducible.

Recently, a certain non-semisimple variant of the Verlinde algebra appeared in [FHST] in connection with the fusion procedure for the $(1,p)$ Virasoro algebra, although this connection is still not justified in full. Generally speaking, the fusion procedure for the Virasoro-type algebras and the so-called W–algebras can lead to non-semisimple Verlinde algebras. There are no reasons to expect the existence of a positive hermitian inner product there like the Verlinde pairing for the conformal blocks, because the corresponding physics theories are *massless.* Surprisingly, the algebra from [FHST] is defined using the usual (*massive*) Verlinde algebra under certain degeneration. Presumably it coincides with the algebra of even elements of $V_{2N+4|k|}$ for $k = -1$ or, at least, is very close to this algebra.

Mathematically, the module $V_{2N+4|k|}$ and its multidimensional generalizations could be expected to be connected with the important problem of describing the complete tensor category of the representations of Lusztig's quantum group at roots of unity [Lus3]. The Verlinde algebra, the symmetric part of $V_{2N-4|k|}$ for $k = 1$, describes the so-called *reduced category* and, in a sense, corresponds to the Weyl chamber. The non-semisimple modules of type $V_{2N+4|k|}$ are expected to appear in the so-called *case of the parallelogram.*

0.5.5 Weyl algebra

Before turning to the diagonal invariants, let us first discuss the specialization $t = 1$. One has:

$$\mathcal{H}\!\!\mathcal{H}^{(t=1)} = \mathcal{W} \rtimes \mathbb{S}_2,\ T \mapsto s \in \mathbb{S}_2,\ s^2 = 1,$$

where the Weyl algebra, denoted here by $\mathcal{W}$, is a quotient of the algebra of noncommutative polynomials $\mathbb{C}[X^{\pm 1}, Y^{\pm 1}]$ with the relations

$$sXs = X^{-1},\ sYs = Y^{-1},\ XY = YXq^{1/2}.$$

Given $N \in \mathbb{N}$, we set $q^{1/2} = \exp(\frac{2\pi i}{N})$,

$$\mathcal{W}^\bullet = \mathcal{W}/(X^N = 1 = Y^N),\ \mathcal{H}^\bullet \stackrel{\mathbf{def}}{=} \mathcal{H}^{(t=1)}/(X^N = 1 = Y^N) = \mathcal{W}^\bullet \rtimes \mathbb{S}_2.$$

The algebra $\mathcal{W}^\bullet$ has a unique irreducible representation

$$V^\bullet \stackrel{\mathbf{def}}{=} \mathbb{C}[X, X^{-1}]/(X^N = 1),\ Y(X^m) = q^{-m/2}X^m, m \in \mathbb{Z},$$

which is also a unique irreducible $\mathcal{H}^\bullet$–module.

Moreover, τ and σ act in $V^\bullet$. Recall that the action of σ in $\mathcal{H}^{(t=1)}$ is as follows:

$$X \mapsto Y^{-1},\ Y \mapsto X,\ s \mapsto s.$$

The problem is that the above description of $V^\bullet$ doesn't make the σ–invariance clear and is inconvenient when studying the σ–action, playing a major role in the theory of theta functions and automorphic forms.

To make the X and the Y on equal footing, we can take a σ–eigenvector $v \in V^\bullet$ and set

$$V^\bullet = (\mathcal{W}/\mathcal{J}_v)(v) \text{ for the } \sigma - \text{invariant ideal } \mathcal{J}_v = \{H \mid H(v) = 0, H \in \mathcal{W}\}.$$

This presentation makes the σ–invariance of $V^\bullet$ obvious, but finding "natural" σ–eigenvectors in $V^\bullet$ with reasonably explicit $\mathcal{J}_v$ is not an easy problem. The simplest case is $N = 3$, $\dim V^\bullet = 3$, where we can proceed as follows.

The space $\{v \in V^\bullet \mid s(v) = -v\}$ is one-dimensional for $N = 3$. It is nothing but the space of odd characteristics in the classical theory of one-dimensional theta functions. It equals $\mathbb{C}d$ for $d = X - X^{-1}$. Since $\sigma(s) = s$, this space must be σ–invariant. Then, calculating $\mathcal{J}_d$ is simple:

$$\mathcal{J}_d = \mathcal{W}J_o, \quad \text{where } J_o = \{H \in \mathcal{W}^\bullet \mid sH = Hs,\ H(d) = 0 \in V^\bullet.$$

One has:

$$\begin{aligned}\{Y + Y^{-1}\}(d) &= \{(Y + Y^{-1})(X - X^{-1})\}(1) = \{q^{-1/2}(XY - X^{-1}Y^{-1}) \\ &\quad + q^{1/2}(XY^{-1} - X^{-1}Y)\}(1) = (q^{1/2} + q^{-1/2})(d) = -d, \\ \{X + X^{-1}\}(d) &= X^2 - X^{-2} = X^{-1} - X = -d.\end{aligned}$$

Thus, $Y + Y^{-1} + 1,\ X + X^{-1} + 1 \in J_o$, and we can continue:

$$\{YX + Y^{-1}X^{-1}\}(d) = q^{-1}X^2 - 1 + 1 - q^{-1}X^{-2} = -q^{-1}d.$$

Actually, this calculation is not needed, since $Y + Y^{-1} + 1$ and $X + X^{-1} + 1$ algebraically generate J_o, (it is not true in the commutative polynomials!) and we can use the Weyl relations.

Finally, we arrive at the equality $\mathcal{J}_d = \mathcal{W}J_o$. The next section contains a generalization of this construction for $\mathcal{H}$.

Using the non-commutativity of X and T in the latter argument is directly connected with the recent results on W–invariant polynomial differential operators due to Wallach, Levasseur, Stafford, and Joseph. They prove that the invariant operators are generated by the W–invariant polynomials and the W–invariant differential operators with constant coefficients, using that they *do not* commute. Of course it is not true for the polynomials in terms of two sets of variables.

0.5.6 Diagonal coinvariants

The previous construction can be naturally generalized to the case of an arbitrary Weyl group W acting in $\mathbb{R}^n$. We set

$$w^{-1}f(x,y) = f(wx, wy) \text{ for a function } f \text{ in terms of } x, y \in \mathbb{R}^n,$$

$$\mathbf{I}_o \stackrel{\mathbf{def}}{=} \mathrm{Ker}(\mathbb{C}[x,y]_{sym} \ni f \mapsto f(0,0)), \quad \text{and} \quad \nabla \stackrel{\mathbf{def}}{=} \mathbb{C}[x,y]/(\mathbb{C}[x,y]\mathbf{I}_o).$$

Haiman conjectured that ∇ has a natural quotient $\mathbf{V}$ of dim$=(h+1)^n$ for the Coxeter number h. It must be isomorphic to the space Funct$(Q/(h+1)Q)$ as a W–module for the root lattice Q and have a proper character with respect to the degree in terms of x and y. Note that $Q/(h+1)Q$ is isomorphic to $P/(h+1)P$ for the weight lattice P, since the order $[P:Q]$ is relatively prime with $(h+1)$.

He proved that ∇=$\mathbf{V}$ and dim$\nabla = (n+2)^n$ in the case of A_n for W=$\mathbb{S}_{n+1}$. The coincidence is a very special feature of A_n. As far as I know, no general uniqueness claims about $\mathbf{V}$ as a graded vector space and as a W–module were made. One needs the double Hecke algebra to make this quotient "natural."

Haiman's conjecture was recently justified by Gordon. He proves that

$$\mathbf{V} = gr\overline{V}'' \text{ for } \bar{k} = -1/h - 1,$$

where $\overline{V}''$ is a W–generalization of the *perfect rational* representation $\overline{V}''$ considered above as $m=1$. By gr, we mean taking the graded vector space of $\overline{V}''$ with respect to the degree in terms of x and y; see Section 2.12.5 from Chapter 2 and Section 3.11 from Chapter 3.

Recall that $h=2$ and, generally, $\dim_{\mathbb{C}} \overline{V}'' = 2m+1$ in the case of A_1. Here we make $m=1$, so the representation considered by Gordon is of dimension 3 for A_1. Even in this case, the coincidence $gr\overline{V}'' = \nabla$ is an instructional exercise. It is not immediate because ∇ is given in terms of *double* polynomials and $\overline{V}''$ was defined as a quotient of the space of *single* polynomials; cf. the previous section.

Adding q to the construction, we obtain the following theorem, describing a W–generalization of the perfect module $\overline{V}$ considered above for the simplest nontrivial $\bar{k} = -1 - 1/h$.

(a) There exist Lusztig-type (exp-log) isomorphisms $\overline{V}'' \rightleftharpoons \overline{V} \rightleftharpoons V^\bullet$ for $\bar{k}$=$-1/h-1$. Here $V^\bullet$ is a unique $\mathcal{H}^\bullet$–quotient of $\mathcal{P} = \mathbb{C}[X, X^{-1}]$ for the reduction:

$$\bullet = \{q^{1/[P:Q]} = \exp(\frac{2\pi i}{N}),\ N \stackrel{\mathbf{def}}{=\!=} h+1,\ t = q^{k} = 1\}.$$

(b) Concerning the coinvariants,

$$\overline{V} = \mathcal{H}/\mathcal{H}I_o(\Delta),\ I_o = Ker(\mathcal{H}_{sym} \ni H \mapsto H(\Delta) \in \overline{V}),$$

where $\mathcal{H}_{sym}$ is a subalgebra of the elements of $\mathcal{H}$ commuting with T. Here Δ is the discriminant $\Delta = t^{1/2}X - t^{-1/2}X^{-1}$ for $n = 1$; generally, it is a product of such binomials over positive roots.

The second part demonstrates how double polynomials appear (the general definition of $\overline{V}$ is given in terms of single polynomials). Using that Δ generates the one-dimensional "sign-representation" of the nonaffine Hecke algebra $\langle T\rangle$, we can naturally identify $\overline{V}$ with quotients of the space of double Laurent polynomials.

Since $t = 1$ for the $\bullet$–reduction, one has: $\mathcal{H}^\bullet$=(Weyl algebra)$\rtimes W$. Thus $V^\bullet$ is its unique irreducible representation under the relations X^N=1=Y^N imposed, namely, $V^\bullet \simeq \mathbb{C}[X, X^{-1}]/(X^N = 1)$. Generally, its dimension is dim=N^{rank} (rank = the number of X–generators), exactly what Haiman conjectured. Note that given n, part (b) holds only thanks to a very special choice of N.

In Gordon's proof, the $(h+1)^n$–formula requires a construction of the resolution of $\overline{V}''$ and more. The theorem above gives an explanation of why the dimension is so simple. The dimension looks like that from the theory of Weyl algebra, although the definition of the space of coinvariants has nothing to do with the roots of unity, and Gordon's theorem really belongs to this theory!

Chapter 1

KZ and QMBP

1.0 Soliton connection

The main object of this chapter is the *r–matrix KZ equation* for the usual classical r–matrices and their generalizations attached to the root systems from [C6]. The key observation in this paper was that the classical Yang–Baxter equation (YBE) for r is *equivalent* (under minor technical constraints) to the cross-derivative integrability condition for the generalized Knizhnik–Zamolodchikov (KZ) equation in terms of r. An immediate application was a new broad class of KZ-type equations with the trigonometric and elliptic dependence on the arguments.

Another application is the *W–invariant KZ,* a W–equivariant system of the differential equations with values in the group algebra $\mathbb{C}W$ of the Weyl group. It was defined in [C6] as an example of r–matrix KZ, but, as a matter of fact, it is directly based on the so-called "reflection equations" and their generalizations from [C16], the YBE extended to arbitrary Weyl groups. The paper [C16] readily gives a justification of the cross-derivative conditions for this system, although a straightforward proof via the reduction to the rank two consideration is simple as well.

Generally, the r–matrix must have some symmetries and satisfy the rank two commutator relations, that can be immediately verified (if they hold). The problem is with finding examples and the representation theory interpretation. In the case of the W–invariant KZ from [C6], the cross-derivative integrability conditions and the W–equivariance follow from the properties of the intertwining operators of the affine Hecke algebras; see [C10, C11] and the references therein.

The W–invariant KZ is now the main tool in the analytic theory of the configuration spaces associated with the root systems. It was extended to the affine KZ equation (AKZ) for the *affine* Weyl group in [C7, C8], and then to the groups generated by complex reflections by Dunkl, Opdam and Malle. A significant part of this chapter is devoted to the calculation of the monodromy representation of AKZ.

Note that one can formally associate Hecke algebras with abstract Coxeter groups and similar groups. The most universal proof of the fact, if it holds, that the resulting Hecke algebra satisfies the Poincare–Birkhoff–Witt property, i.e., that this algebra is a flat deformation of the group algebra, goes via the consideration of a KZ-type connection and obtaining the Hecke algebra as its monodromy.

For instance, the monodromy of the W–invariant KZ identifies the corresponding Hecke algebra with $\mathbb{C}W$ for generic values of the parameter k in the KZ equation directly related to the parameter t of the Hecke algebra.

1.0.1 Classical r–matrices

The classical r–matrices were derived from the quantum inverse method due to Faddeev, Sklyanin, and Takhtajan (see [FT]). They allowed us to standardize the study of soliton equations, to come close to their classification, and to understand the genuine role of soliton theory in mathematics. Without going into detail, let me give the following general references [KuS, BD, Se, C3] and mention the paper [Sk], which influenced my research in soliton theory a great deal.

In the definition of the r–matrix KZ, I combined the *r–matrices* with the theory of the *τ–function* from [DJKM1, DJKM2] and some previous papers of the same authors. As for the τ–function, let me also mention [Sat] and the exposition [Ve]. The third important ingredient was the approach to *affine flag varieties* from the papers [KP, PK].

Let $\mathfrak{g}$ be a simple Lie algebra and $r(\lambda)$ a function of $\lambda \in \mathbb{C}$ taking values in $\mathfrak{g} \otimes \mathfrak{g}$. We assume that $r(\lambda) = t/\lambda + \tilde{r}(\lambda)$ for some analytic function $\tilde{r}$ in a neighborhood of $\lambda = 0$, where $t = \sum_\alpha I_\alpha \otimes I_\alpha, \{I_\alpha\} \subset \mathfrak{g}$ is an orthonormal basis with respect to the Killing form $(\, ,\,)_K$ on $\mathfrak{g}$. The notation, standard in the theory of r–matrices, will be used:

$$ {}^1a = a\otimes 1\otimes 1\otimes \ldots, {}^2a = 1\otimes a\otimes 1\otimes \ldots, \ldots, {}^{ij}(a\otimes b) = {}^i a\, {}^j b, ((a\otimes b)c)_K = a(b,c)_K $$

for a, b, c being from the universal enveloping algebra $U(\mathfrak{g})$ of $\mathfrak{g}$ (or from any of its quotient algebra). We also set $\{A \overset{\otimes}{,} B\}$ if the products of the entries ab in $A \otimes B$ are replaced everywhere by $\{a, b\}$.

A classical r–matrix is a solution of the classical Yang–Baxter equation

$$ [\,{}^{12}r(\lambda), {}^{13}r(\lambda+\mu) + {}^{23}r(\mu)] + [\,{}^{13}r(\lambda+\mu), {}^{23}r(\mu)] = 0. \qquad (1.0.1) $$

Theorem 1.0.1. *Assuming that ${}^{12}r(\lambda) + {}^{21}r(-\lambda) = 0$, the following three assertions are equivalent to (1.0.1) and to each other:*

(a) The relation

$$ \{M(\lambda) \overset{\otimes}{,} M(\mu)\} \;=\; [r(\mu-\lambda), {}^1M(\lambda) + {}^2M(\mu)] $$

defines a Poisson bracket on the indeterminate coefficients of $M(\lambda)$ that is the generic element of the Lie algebra $\widetilde{\mathfrak{g}}$ of $\mathfrak{g}$–valued meromorphic functions.

(b) The functions in the form

$$M_r(\lambda) = \sum Res\ ({}^{12}r(\lambda-\mu)^2 M(\mu))_K d\mu), \tag{1.0.2}$$

where the sum is over the set of poles of $M \in \widetilde{\mathfrak{g}}$, constitute a Lie subalgebra $\widetilde{\mathfrak{g}}_r \subset \widetilde{\mathfrak{g}}$.

(c) The equations

$$\partial G/\partial \lambda_i = \kappa \sum_{j \neq i} {}^{ij}r(\lambda_i - \lambda_j)G \tag{1.0.3}$$

for a function G of $\lambda_1, \lambda_2, \ldots$ taking values in $U(\mathfrak{g}) \otimes \ldots \otimes U(\mathfrak{g})$ satisfy the cross-derivative integrability conditions.

The references are [C3, C6]. It readily follows from (1.0.2) that the formal differentiations

$$D_g F = (FgF^{-1})_0 F, \quad M_0 = M - M_r, \quad g \in \widetilde{\mathfrak{g}} \tag{1.0.4}$$

satisfy the relations

$$[D_g, D_f] = D_{[g,f]}, \quad g, f \in \widetilde{\mathfrak{g}}. \tag{1.0.5}$$

Here $F = \exp(M_0)$ for the generic element M_0 of the Lie subalgebra $\widetilde{\mathfrak{g}}_0 \subset \widetilde{\mathfrak{g}}$ of holomorphic functions; differentiations act in the algebra of functionals of the coefficients of M_0. These $\{D_g, g \in \widetilde{\mathfrak{g}}\}$ are the generalized Bäcklund–Darboux infinitesimal transformations, while formula (1.0.5) and its group analog embrace many concrete results (from the 1850 to the present) on composing Bäcklund transforms for nonlinear differential equations; see [C1, C3] and the references therein.

Now we can introduce the τ*–function* as a solution of the system ($g \in \widetilde{\mathfrak{g}}, \kappa \in \mathbb{C}$)

$$(D_g\tau)\tau^{-1} = \kappa(\sum \text{Res}\ (F^{-1}dF, g)_K). \tag{1.0.6}$$

Formula (1.0.2) induces the decomposition

$$\widehat{\mathfrak{g}} = \mathfrak{g} \oplus \mathbb{C}c, \quad [x + \zeta c, y + \xi c] = [x, y] + \sum \text{Res}\ (dx, y)_K c.$$

In this decomposition, c acts as a multiplication by κ and $\widetilde{\mathfrak{g}}_0 \subset \widehat{\mathfrak{g}}$ annihilates τ. If $\kappa \in k\mathbb{N}$ for an appropriate integer $k > 0$ (depending on the r–matrix under consideration; see [C3]), then the τ–function is "integrable," for instance, satisfies infinitely many Hirota-type relations.

All these are very close to the theory of modular forms and representations of *adèle groups* in arithmetic, with $\widetilde{\mathfrak{g}}_r \subset \widetilde{\mathfrak{g}} \supset \widetilde{\mathfrak{g}}_0$ playing the role of principal and, respectively, integer adèles, and τ being something like the Tamagawa measure.

1.0.2 Tau function and coinvariant

System (1.0.3) for $1 \leq i, j \leq n$ is a natural generalization of the Aomoto–Kohno system with $r(\lambda) = t/\lambda$, which, in its turn, is a direct generalization of the Knizhnik–Zamolodchikov equation in the two-dimensional conformal field theory (with $\mathfrak{g} = \mathfrak{gl}_N$). I came to (1.0.3) while performing the following calculation in [C6].

Let us generalize the definition of τ to allow arbitrary representations of $\widehat{\mathfrak{g}}$ (not only the representations with the trivial action of $\widetilde{\mathfrak{g}}_0$ on τ). We also allow the elements from $\widetilde{\mathfrak{g}}$ to have poles at (pairwise distinct) points $\lambda_1, \ldots, \lambda_n$.

Given a finite dimensional representation V of $\mathfrak{g}$, $\widetilde{\mathfrak{g}}_0$ naturally acts on $V^{\otimes n}$ by the projection onto the Lie algebra $\mathfrak{g} \times \cdots \times \mathfrak{g}$ (n times). Let us introduce the *BGG–Verma module* $\mathcal{M}$ as the universal $\widehat{\mathfrak{g}}$–module generated by $V^{\otimes n}$, where c acts by multiplication by κ and $\widetilde{\mathfrak{g}}_0$ acts on $V^{\otimes n}$ as above. For every $x \in \mathcal{M}$, there exists a unique element $\tau(x) \subset V^{\otimes n} \subset \mathcal{M}$ such that $\tau(x) - x \in \widetilde{\mathfrak{g}}_r \mathcal{M}$. Here we use the decomposition $\widehat{\mathfrak{g}} = \mathbb{C}c \oplus \widetilde{\mathfrak{g}}_0 \oplus \widetilde{\mathfrak{g}}_r$.

This construction (from [C1]) gives an important interpretation of the τ–function as the Kac–Moody *coinvariant.* In soliton theory, we examine its transformations with respect to Kac–Moody commutative subalgebras (the D_g–flows for pairwise commutative g). In conformal field theory (CFT), the tau function appears as the *n-point correlation function* in a way similar to the above calculation. The dependence on the positions of the points, the celebrated Knizhnik–Zamolodchikov equation, is of key importance for CFT. As a matter of fact, the following reasoning is similar to the original deduction of KZ in [KZ].

We set formally

$$(l_i\tau)(x) = \tau(l_i x), \quad l_i = \sum_{\alpha, j \geq 0} (I_\alpha \tilde{\lambda}_i^{-1-j})(I_\alpha \tilde{\lambda}_i^j),$$

where $\tilde{\lambda}_i = \lambda - \lambda_i$ is a local parameter in a neighborhood of λ_i and l_i is the Sugawara element of degree -1, namely, the generator L_{-1} of the Virasoro algebra embedded into the i–th component of $\widehat{\mathfrak{g}}$. Then one can verify that

$$l_i\tau + \kappa' \partial\tau/\partial u_i = (\rho_i + \sum_{j \neq i} {}^{ij}r_V(u_i - u_j))\tau$$

for a proper κ' and some ρ (trivial in the most interesting cases), where ${}^{ij}r_V$ is the image of r in $\mathrm{End}_{\mathbb{C}}(V^{\otimes n})$. This leads to the compatibility condition for equations (1.0.3) and eventually provides the integral formulas for the τ–function.

This derivation of the r–matrix KZ equation was used in [C9] to simplify the algebraic part of the Schechtman–Varchenko construction [SV] of the integral formulas for the rational KZ ([DJMM] for SL_2) and generalize their formulas to the trigonometric case.

1.0.3 Structure of the chapter

We will reproduce the above calculation in detail and consider the integral formulas in the last two sections. The r–matrices there are "of type A," i.e., depending on the differences of the arguments $(u_i - u_j)$ as above.

In the first section, we introduce the affine KZ equations using a different approach, based on the properties of the Weyl groups and root systems; the differences $(u_i - u_j)$ are replaced by the roots. The r–matrices attached to the root systems and the corresponding KZ are studied in Section 1.5, including the elliptic case.

The affine KZ is a special example of the general r–matrix trigonometric KZ. Its "integrability" reveals itself in explicit formulas in terms of the hypergeometric function for the monodromy representation. The later plays the key role when establishing an isomorphism with another well-known "integrable" system of differential equations, the QMBP, also called the Heckman–Opdam system (Section 1.2).

In the next Sections 1.3–1.4, we switch to the quantum theory including QAKZ and Macdonald polynomials (in the GL–case). The exposition is based on the affine and double affine Hecke algebras.

Section 1.6 contains applications of DAHA to the Harish-Chandra theory and QMBP, mainly, a new theory of the inverse Harish-Chandra transform. It has its own mini-introduction.

Acknowledgments. *This chapter is based on two series of lectures that I gave at IIAS and RIMS Kyoto University, and the paper [C22]. I thank the participants of the lectures for useful questions, comments, and discussion, which made for a highly stimulating atmosphere. I am very grateful to those who took notes and helped to prepare [C22] for publication. This chapter is a truly joint venture. I acknowledge my special indebtedness to M. Kashiwara and T. Miwa.*

I also used my notes of the lectures on the r–matrix KZ and the affine and double affine Hecke algebras delivered in Paris. I am thankful to P. Gerardin, R. Rentschler, and especially M. Duflo. I am grateful to M. Kashiwara, who helped me with [C24], which is the basis of the section on the Harish-Chandra inversion. This paper was inspired by [O3]. I thank E. Opdam for his comments.

1.1 Affine KZ equation

We discuss in this section the degenerate affine Hecke algebra and the corresponding affine Knizhnik–Zamolodchikov equation. It is demonstrated that the W–equivariance of the latter (we say W–invariance, abbusing a little the standard terminology) and the cross-derivative integrability conditions ("compatibility") are equivalent to the definition of the degenerate affine Hecke algebra. This observation (see [C7, C11]) is an important link between the affine Hecke theory and the theory of special functions, especially, the theory of the hypergeometric function.

We begin with the rank one and rank two examples.

1.1.1 Hypergeometric equation

We introduce the affine Knizhnik–Zamolodchikov equation (AKZ) associated with the root system of type A_1.

First, we consider the differential equation:

$$\frac{\partial \Phi}{\partial u} = \left(k \frac{s}{e^u - 1} + x \right) \Phi. \tag{1.1.1}$$

Here Φ is a function of a complex variable u, $k \in \mathbb{C}$ is a parameter, and s and x are operators acting in the vector space or algebra, where Φ takes its value.

Second, we impose the following two relations:

$$s^2 = 1, \tag{1.1.2}$$

$$sx + xs = k. \tag{1.1.3}$$

These relations are necessary and sufficient to make (1.1.1) invariant. By invariance, we mean that if Φ satisfies (1.1.1), then so does

$$\tilde{\Phi}(u) = s\Phi(-u). \tag{1.1.4}$$

Let us check it.

We plug in

$$\begin{aligned} -\frac{\partial \tilde{\Phi}(u)}{\partial u} &= s \left(k \frac{s}{e^{-u} - 1} + x \right) \Phi(-u) \\ &= \left(k \frac{s}{e^{-u} - 1} + sxs \right) \tilde{\Phi}(u) \end{aligned}$$

and use

$$\frac{1}{1 - e^{-u}} = \frac{1}{e^u - 1} + 1.$$

Finally,

$$\frac{\partial \tilde{\Phi}(u)}{\partial u} = \left(k \frac{s}{e^u - 1} + ks - sxs \right) \tilde{\Phi}(u),$$

where $ks - sxs = x$.

Third, the **affine Knizhnik–Zamolodchikov equation** is an equation in the form (1.1.1) with values in the **degenerate affine Hecke algebra** $\mathcal{H}'_{A_1}$ of type A_1 generated by the elements s and x subject to (1.1.2) and (1.1.3):

$$\mathcal{H}'_{A_1} \stackrel{\text{def}}{=\!=} \langle s, x\rangle/\{(1.1.2),(1.1.3)\}. \tag{1.1.5}$$

Let $\Phi(u)$ be a function of u with values in $\mathcal{H}'_{A_1}$. Note that one can multiply a solution $\Phi(u)$ of (1.1.1) by an arbitrary constant element on the right, i.e., $\Phi(u)a$ for $a \in \mathcal{H}'_{A_1}$ is a solution as well.

We claim that (1.1.1) is integrable in terms of the classical hypergeometric function or elementary functions in arbitrary irreducible representations of $\mathcal{H}'_{A_1}$.

Let us start with a more general equation

$$\frac{\partial\Phi}{\partial z} = \left(\frac{A}{1-z} + \frac{B}{z}\right)\Phi. \tag{1.1.6}$$

It becomes (1.1.1) for $z = e^{-u}$, $A = -ks$, and $B = -x$.

Generally speaking, equation (1.1.6) is non-integrable. There is an interesting "new" theory of this equation started by Drinfeld with application in arithmetic, but we are not going to consider this direction.

If A, B are 2×2 matrices acting on the 2–component vector-function Φ, then (1.1.6) is nothing but a variant of the hypergeometric differential equation. This includes the case of irreducible representations of $\mathcal{H}'_{A_1}$, because the latter exist only in dimension 1 or 2.

Explicitly, the generic irreducible representation ρ of $\mathcal{H}'_{A_1}$ is given by

$$\rho(s) = \begin{pmatrix}1 & 0\\ 0 & -1\end{pmatrix}, \qquad \rho(x) = k\left(\frac{s}{2} + \begin{pmatrix}0 & \zeta\\ \xi & 0\end{pmatrix}\right). \tag{1.1.7}$$

Because of the gauge transformation $\zeta \to c\zeta$, $\xi \to c^{-1}\xi$, it is characterized by $\zeta\xi$ up to isomorphisms. It will be more convenient to switch to

$$\mu = \left(\zeta\xi + \frac{1}{4}\right)^{1/2}. \tag{1.1.8}$$

Equation (1.1.6) for $A = -k\rho(s)$, $B = -\rho(x)$ has a solution $\Phi = \binom{\Phi_1}{\Phi_2}$ in terms of the Gauss hypergeometric function with the first component

$$\Phi_1(u) = z^{-k\mu}(1-z)^k F(k(1-2\mu), k, 1-2k\mu; z), \tag{1.1.9}$$

where $z = e^{-u}$ and

$$F(\alpha,\beta,\gamma;z) \stackrel{\text{def}}{=\!=} \sum_{n=0}^{\infty} \frac{(\alpha)_n(\beta)_n}{(\gamma)_n n!} z^n \quad \text{for} \quad (x)_n = x(x+1)\cdots(x+n-1).$$

If the first component is known, we automatically know the second.

We note that the parameters α, β, γ in (1.1.9) obey the constraint $\alpha + 1 = \beta + \gamma$. If $\zeta\xi = 0$, then the representation ρ from (1.1.7) is reducible. In this case, the solutions are given in terms of elementary functions.

1.1.2 AKZ equation of type *GL*

This equation is a natural abstract variant of the original Knizhnik–Zamolodchikov (KZ) equation from conformal field theory. Its consistency and invariance lead to the defining relations of the degenerate affine Hecke algebra $\mathcal{H}'_{GL_n}$ introduced by Drinfeld [Dr1].

The KZ equation reads as

$$\frac{\partial \Phi}{\partial z_i} = k \left(\sum_{0 \le j \le n,\, j \ne i} \frac{\Omega_{ij}}{z_i - z_j} \right) \Phi \quad (0 \le i \le n). \tag{1.1.10}$$

In the least sophisticated case from [KZ], the Ω_{ij} are the permutation matrices in the tensor square of some $\mathbb{C}^N$, where N has nothing to do with n, the number of variables. Let us generalize it assuming that the Ω_{ij} are symbols, constant matrices, or operators and $\Omega_{ij} = \Omega_{ji}$. We will treat them formally and consider the function $\Phi(z)$ in terms of $z = (z_0, \ldots, z_n)$, taking values in the abstract algebra generated by the elements Ω_{ij}. The self-consistency of the system of equations (1.1.10) means that

$$\frac{\partial A_j}{\partial z_i} - \frac{\partial A_i}{\partial z_j} = [A_i, A_j], \tag{1.1.11}$$

where

$$A_i = k \sum_{j \ne i} \frac{\Omega_{ij}}{z_i - z_j}. \tag{1.1.12}$$

It holds for any complex parameter k if and only if

$$[\Omega_{ij}, \Omega_{kl}] = 0, \tag{1.1.13}$$

$$[\Omega_{ij}, \Omega_{ik} + \Omega_{jk}] = 0, \tag{1.1.14}$$

where the indices i, j, k, l are pairwise distinct. The KZ in this abstract form is, essentialy, due to Aomoto. It was also studied by Kohno [Ko].

Comment. Generally, we need a*trigonometric* KZ equation to obtain the degenerate (graded) affine Hecke algebra. Namely, we must start with the r–matrix KZ for the simplest trigonometric classical r–matrices. The GL_n–case is special, since there exists a substitution transforming the trigonometric KZ that is needed (there are many), to a rational Aomoto-type KZ, although the output is not exactly in the Aomoto–Kohno–Drinfeld form.

The trigonometric KZs (and their elliptic generalizations) appeared for the first time in [C6]. The so-called KZB [Be] and more general KZ associated with the curves of arbitrary genus a not *closed* system in terms of the positions of the points. The reference [C6] is definitely the first concerning the elliptic and trigonometric *r–matrix* KZ, although it is possible that the trigonometric KZ-type formulas appeared before.

In the "standard" conformal field theory, the dependence on the positions of the points z_i is rational. I remember that my physicists friends considered the trigonometric KZ a pure mathematical speculation (indeed, they were), when I demonstrated them. Now they are quite common for both mathematicians and physicists. ❑

Consider the group algebra $\mathbb{C}\mathbb{S}_n$ of the **permutation group** $\mathbb{S}_n$ of the set $\{1,\ldots,n\}$. We denote the transposition of i and j by s_{ij}. Then, if we make $\Omega_{ij} = s_{ij}$ $(0 \le i,j \le n)$, the relations (1.1.13)–(1.1.14) are satisfied.

Let us treat the zero-index separately. Setting

$$z_0 = 0, \tag{1.1.15}$$
$$\Omega_{ij} = s_{ij} \quad (i,j \neq 0), \tag{1.1.16}$$
$$\Omega_{0i} = k^{-1}\Omega_i, \tag{1.1.17}$$

equation (1.1.10) turns into

$$\frac{\partial\Phi}{\partial z_i} = \left[k\left(\sum_{1\le j\le n,\, j\neq i}\frac{s_{ij}}{z_i - z_j}\right) + \frac{\Omega_i}{z_i}\right]\Phi \quad (1 \le i \le n), \tag{1.1.18}$$

and relations (1.1.13)–(1.1.14) read as follows:

$$[s_{ij}, \Omega_i + \Omega_j] = 0, \tag{1.1.19}$$
$$[ks_{ij} + \Omega_i, \Omega_j] = 0, \tag{1.1.20}$$
$$[s_{ij}, \Omega_l] = 0, \tag{1.1.21}$$

where the indices i, j, l are pairwise distinct.

Substituting

$$z_i = e^{v_i}, \tag{1.1.22}$$

we come to

$$\frac{\partial\Phi}{\partial v_i} = \left(k\sum_{j\neq i}\frac{s_{ij}}{1 - e^{v_j - v_i}} + \Omega_i\right)\Phi. \tag{1.1.23}$$

Using the elements

$$y_i = \Omega_i + k\sum_{j>i} s_{ij}, \tag{1.1.24}$$

$$\frac{\partial\Phi}{\partial v_i} = \left(k\sum_{j>i}\frac{s_{ij}}{e^{v_i - v_j} - 1} - k\sum_{j<i}\frac{s_{ij}}{e^{v_j - v_i} - 1} + y_i\right)\Phi. \tag{1.1.25}$$

The elements $\{y\}$ are convenient here, since in the limit

$$v_1 \gg v_2 \gg \cdots \gg v_n,$$

we obtain the system

$$\frac{\partial \Phi}{\partial v_i} = y_i \Phi. \tag{1.1.26}$$

The consistency of these equations is equivalent to the commutativity

$$[y_i, y_j] = 0, \tag{1.1.27}$$

which of course can be deduced algebraically from (1.1.19), and (1.1.20), and (1.1.21).

We claim that (1.1.27), together with the relations

$$[s_i, y_j] = 0 \quad \text{if } j \neq i, i+1, \tag{1.1.28}$$

$$s_i y_i - y_{i+1} s_i = k, \tag{1.1.29}$$

where $s_i = s_{i\,i+1}$ $(1 \le i \le n-1)$, ensure (1.1.19)-(1.1.21); it is a simple algebraic calculation.

Let us introduce the **degenerate affine Hecke algebra of type** GL_n as an algebraic span of $\mathbb{C}\mathbb{S}_n$ and y_i $(1 \le i \le n)$ subject to (1.1.27), (1.1.28), and (1.1.29). It will be denoted by $\mathcal{H}'_{GL_n}$, or simply by $\mathcal{H}'_n$. We call the system (1.1.25) with values in $\mathcal{H}'_n$ the **AKZ of type** GL_n. It is well defined, i.e., self-consistent, and $\mathbb{S}_n$–**invariant** in the following sense.

The group $\mathbb{S}_n$ acts on $\mathbb{C}^n$ naturally by

$$v = (v_1, \ldots, v_n) \in \mathbb{C}^n \mapsto w(v) = (v_{i_1}, \ldots, v_{i_n}) \in \mathbb{C}^n$$

for $w^{-1} = (i_1, i_2, \ldots, i_n) \in \mathbb{S}_n$. Given a function $\Phi(v)$ of $v \in \mathbb{C}^n$ with the values in $\mathcal{H}'_n$, we define the action of $w \in \mathbb{C}\mathbb{S}_n$ on $\Phi(v)$ by

$$\left(w(\Phi)\right)(v) = w \cdot \Phi\Big(w^{-1}(v)\Big). \tag{1.1.30}$$

Here the dot means the product in $\mathcal{H}'_n$. It follows from (1.1.28) and (1.1.29) that if Φ satisfies (1.1.25), then so does $w(\Phi)$.

The invariance alone is equivalent to relations (1.1.28) and (1.1.29). We conclude that, by imposing the invariance, the "limiting self-consistency" (1.1.27) results in the general self-consistency of our system for all k.

Example. The basic variant of the Knizhnik–Zamolodchikov equation is as follows. Let $\mathfrak{gl}_N$ be the matrix algebra with the standard set of generators $\{e_{lm}, 1 \le l, m \le n\}$ with the entries $e^{ab}_{lm} = \delta_{la}\delta_{mb}$. Given any representations $V_0, V_1 \ldots, V_n$ of $\mathfrak{gl}_N$, the tensor product $V = V_0 \otimes V_1 \otimes \cdots \otimes V_n$ has the natural structure of a $\mathfrak{gl}_N$–module. We set $\Omega_{ij} = \sum_{lm} e^{(i)}_{lm} e^{(j)}_{ml}$, where $e^{(i)}_{lm}$ act in V_i, $i \neq j$. Then (1.1.14) and (1.1.13) are fulfilled and we get a self-consistent KZ equation. This example was generalized by Kohno to arbitrary reductive Lie algebras and extensively used by Drinfeld in his theory of quasi-Hopf algebras.

If V_i coincide with the fundamental N–dimensional representation for all $i > 0$, then Ω_{ij} can be identified with the transposition s_{ij} for $i, j > 0$. The representation V_0 still can be arbitrary. This results in an $\mathbb{S}_n$–invariant self-consistent equation in the form of (1.1.18). According to the above analysis we must obtain a representation of $\mathcal{H}'_n$ in V. The formulas coincide with (1.1.25):

$$s_{ij} \mapsto \Omega_{ij},\ y_i \mapsto \Omega_{0i} + k\sum_{j>i}\Omega_{ij},\ 1 \le i, j \le n.$$

Note that, given a weight of $\mathfrak{gl}_N$, the KZ equation under consideration can be restricted to the subspace of the corresponding highest vectors in V. Therefore the latter is a $\mathcal{H}'_n$–submodule.

1.1.3 Degenerate affine Hecke algebra

In this subsection we define the degenerate affine Hecke algebra and the corresponding affine KZ for an arbitrary root system.

Let Σ be a root system in $\mathbb{R}^n$ with the inner product $(\ ,\ )$. We choose a system of simple roots $\alpha_1, \dots, \alpha_n$ of Σ and denote the set of positive roots by Σ_+. For a root $\alpha \in \Sigma$, the corresponding coroot is

$$\alpha^\vee = \frac{2\alpha}{(\alpha, \alpha)},$$

and the corresponding reflection is

$$s_\alpha(u) = u - (\alpha^\vee, u)\alpha \quad (u \in \mathbb{R}^n).$$

We will denote s_{α_i} simply by s_i. The fundamental coweights b_i are as follows:

$$(b_i, \alpha_j) = \delta_{ij}.$$

Sometimes the notation $a_i = \alpha_i^\vee$ will be used. For $u \in \mathbb{R}^n$, the coordinates will be $u_i = (u, \alpha_i)$. We also set $u_\alpha = (u, \alpha)$ for $\alpha \in \Sigma$. Check that

$$\frac{\partial u_\alpha}{\partial u_i} = \nu_\alpha^i = (b_i, \alpha) = \text{the multiplicity of } \alpha_i \text{ in } \alpha.$$

Let W be the **Weyl group** of Σ: $W = \langle s_\alpha, \alpha \in \Sigma\rangle = \langle s_1, \dots, s_n\rangle$. It acts on functions on $\mathbb{R}^n$ by

$$^w f(u) = f(w^{-1}(u)) \quad (u \in \mathbb{R}^n). \tag{1.1.31}$$

One then has:

$$^w u_\alpha = (w^{-1}(u), \alpha) = u_{w(\alpha)}.$$

Definition 1.1.1. *The* **degenerate affine Hecke algebra** $\mathcal{H}'_\Sigma$ *is the associative algebra generated by* $\mathbb{C}W$ *and* $x_1,\ldots,x_n$ *with the following relations:*

$$[x_i, x_j] = 0, \quad \forall i, j, \tag{1.1.32}$$

$$[s_i, x_j] = 0, \quad \text{if} \quad i \neq j, \tag{1.1.33}$$

$$s_i x_i - \hat{x}_i s_i = k. \tag{1.1.34}$$

Here k *is a complex number and*

$$\hat{x}_i = x_i - \sum_{j=1}^{n} (\alpha_i^\vee, \alpha_j) x_j. \tag{1.1.35}$$

❑

This definition is due to Lusztig [Lus1]. He calls it the graded affine Hecke algebra because k is a formal parameter in his paper. In this chapter, we mainly treat k as a complex number. Drinfeld introduced this algebra in the GL_n–case in [Dr1]. These algebras are natural degenerations of the affine Hecke algebras.

Introducing

$$x_b = \sum_{i=1}^{n} (b, \alpha_i) x_i = \sum_{i=1}^{n} k_i x_i \quad \text{for} \quad b = \sum_{i=1}^{n} k_i b_i, \tag{1.1.36}$$

we can express the right-hand side of (1.1.35) as $x_{s_i(b_i)} = x_{b_i} - x_{a_i}$. Therefore, in the degenerate Hecke algebra:

$$s_i x_b - x_{s_i(b)} s_i = x_b s_i - s_i x_{s_i(b)} = k(b, \alpha_i). \tag{1.1.37}$$

The AKZ. We will use the following partial derivatives:

$$\partial_b(u_\alpha) = (\alpha, b), \ \frac{\partial}{\partial u_i} = \partial_i = \partial_{b_i}. \tag{1.1.38}$$

Let us consider the following system of partial differential equations:

$$\frac{\partial \Phi}{\partial u_i} = \left(k \sum_{\alpha \in \Sigma_+} \nu_\alpha^i \frac{s_\alpha}{e^{u_\alpha} - 1} + x_i \right) \Phi \qquad (1 \leq i \leq n). \tag{1.1.39}$$

Here k is a complex number. We denote the right-hand side of (1.1.39) by $A_i \Phi$. Here we assume that Φ takes values in an associative algebra generated by $\mathbb{C}W$ and $x_1,\ldots, x_n$. We call (1.1.39) **self-consistent** if

$$[\frac{\partial}{\partial u_i} - A_i, \frac{\partial}{\partial u_j} - A_j] = 0. \tag{1.1.40}$$

It is called **invariant** if, for any solution Φ of (1.1.39) and any element w of W, the function $w(\Phi)$ defined in (1.1.30) is a solution of (1.1.39) as well.

Theorem 1.1.2. *The system* (1.1.39) *is self-consistent and invariant if and only if* $s_1, \ldots, s_n, x_1, \ldots, x_n$ *satisfy* (1.1.32), (1.1.33), *and* (1.1.34) *from the definition of* $\mathcal{H}'_\Sigma$. □

The **AKZ equation associated with** Σ is the system (1.1.39) for the function Φ with values in $\mathcal{H}'_\Sigma$.

Using x_b and ∂_b from (1.1.36) and (1.1.38), the system (1.1.39) can be expressed as

$$\partial_b \Phi = \left(k \sum_{\alpha \in \Sigma_+} (b, \alpha) \frac{s_\alpha}{e^{u_\alpha} - 1} + x_b \right) \Phi. \tag{1.1.41}$$

Comment. The parameter k may depend on the length of roots. Generally speaking, the AKZ equation is as follows:

$$\frac{\partial \Phi}{\partial u_i} = \left(\sum_{\alpha \in \Sigma_+} k_{|\alpha|} \nu^i_\alpha \frac{s_\alpha}{e^{u_\alpha} - 1} + x_i \right) \Phi. \tag{1.1.42}$$

1.1.4 Examples

Let us write down the explicit forms of the AKZ equation in the simplest cases.

When $\Sigma = A_1$, the AKZ equation is exactly (1.1.1).

For A_2, the AKZ equation is

$$\begin{aligned} \frac{\partial \Phi}{\partial u_1} &= \left\{ k \left(\frac{s_{12}}{e^{u_1} - 1} + \frac{s_{13}}{e^{u_1 + u_2} - 1} \right) + x_1 \right\} \Phi, \\ \frac{\partial \Phi}{\partial u_2} &= \left\{ k \left(\frac{s_{23}}{e^{u_2} - 1} + \frac{s_{13}}{e^{u_1 + u_2} - 1} \right) + x_2 \right\} \Phi, \end{aligned}$$

where s_{ij} denotes the transposition of i and j. In this case, $\hat{x}_1 = x_2 - x_1$, $\hat{x}_2 = x_1 - x_2$.

The root system B_2 appears in the following way. Let ϵ_1 and ϵ_2 form an orthonormal basis of $\mathbb{R}^2$. Then the set of positive roots consists of the following vectors:

$$\begin{aligned} \alpha_1 &= \epsilon_1 - \epsilon_2, \\ \alpha_2 &= \epsilon_2, \\ \alpha_1 + \alpha_2 &= \epsilon_1, \\ \alpha_1 + 2\alpha_2 &= \epsilon_1 + \epsilon_2. \end{aligned}$$

Let $s = s_1$ and $t = s_2$. Then s and t satisfy $tsts = stst$ (the Coxeter relation for $W_{B_2} = W_{C_2}$) and $s^2 = 1$, $t^2 = 1$. One has $\hat{x}_1 = x_2 - x_1$, and $\hat{x}_2 = 2x_1 - x_2$.

The AKZ equation reads as

$$\frac{\partial\Phi}{\partial u_1} = \left\{k\left(\frac{s}{e^{u_1}-1}+\frac{sts}{e^{u_1+u_2}-1}+\frac{tst}{e^{u_1+2u_2}-1}\right)+x_1\right\}\Phi$$

$$\frac{\partial\Phi}{\partial u_2} = \left\{k\left(\frac{t}{e^{u_2}-1}+\frac{sts}{e^{u_1+u_2}-1}+2\frac{tst}{e^{u_1+2u_2}-1}\right)+x_2\right\}\Phi.$$

Note the appearance of the multiplier 2 in the last equation. Generally, the multipliers can be from 1 to 6. The coefficient 6 occurs for E_8 only.

The A_{n-1}–case. We will show how the AKZ equation of type GL_n discussed in Section 1.1.2 reduces to the AKZ equation for the root system $\Sigma \subset \mathbb{R}^{n-1}$ of type A_{n-1}.

First,

$$x = y_1 + \ldots + y_n \tag{1.1.43}$$

is central in the algebra $\mathcal{H}'_n$. Setting

$$x_i = y_1 + \ldots + y_i - \frac{i}{n}x, \tag{1.1.44}$$

we have an embedding of $\mathcal{H}'_\Sigma$, where Σ is the root system of type A_{n-1}, into $\mathcal{H}'_n$. Concerning the AKZ equations, let

$$u_i = v_i - v_{i+1} \quad (1 \le i \le n-1). \tag{1.1.45}$$

The space $\mathbb{R}^{n-1}$ will be identified with the quotient space of $\mathbb{R}^n = \oplus_{i=1}^n \mathbb{R}\epsilon_i$:

$$\mathbb{R}^{n-1} \simeq \oplus_{i=1}^n \mathbb{R}\epsilon_i/\mathbb{R}\epsilon,$$

where $\{\epsilon_i\}_{1\le i\le n}$ is the orthonormal basis and $\epsilon = \epsilon_1+\ldots+\epsilon_n$. Due to (1.1.25),

$$\sum_{i=1}^n \frac{\partial\Phi}{\partial v_i} = x\Phi.$$

Therefore, the function

$$\Phi'(v) = e^{-x\cdot\frac{1}{n}(v_1+\cdots+v_n)}\Phi(v)$$

satisfies the AKZ equation of type A_{n-1}:

$$\frac{\partial\Phi'}{\partial u_i} = \left(k\sum_{j\le i<l}\frac{s_{jl}}{e^{u_j+\cdots+u_{l-1}}-1}+x_i\right)\Phi' \qquad (1 \le i \le n-1).$$

1.2 Isomorphism theorems for AKZ

We now introduce the general affine Hecke algebra $\mathcal{H}^t_\Sigma$ and connect it with the degenerate affine Hecke algebra $\mathcal{H}'_\Sigma$ using the monodromy of the AKZ equation. An isomorphism also will be established between the solution space of the AKZ equation and that of the quantum many-body problem.

1.2.1 Induced representations

For $\lambda = (\lambda_1, \ldots, \lambda_n) \in \mathbb{C}^n$, the corresponding character of $\mathbb{C}[x_1, \ldots, x_n]$ is the ring homomorphism $\mathbb{C}[x_1, \ldots, x_n] \to \mathbb{C}$ given by $x_i \mapsto \lambda_i$. We will denote it by λ too.

Definition 1.2.1. *Given λ, the* **induced** *$\mathcal{H}'_\Sigma$–module is*

$$I_\lambda \stackrel{\text{def}}{=} \operatorname{Ind}_{\mathbb{C}[x_1,\ldots,x_n]}^{\mathcal{H}'_\Sigma}(\lambda) = \mathcal{H}'_\Sigma \otimes_{\mathbb{C}[x_1,\ldots,x_n]} \mathbb{C}_\lambda. \tag{1.2.1}$$

Here $\mathbb{C}_\lambda$ is endowed with the $\mathbb{C}[x_1, \ldots, x_n]$–module structure by the character λ.

We have the Poincaré–Birkhoff–Witt-type theorem for $\mathcal{H}'_\Sigma$. That is, any $h \in \mathcal{H}'_\Sigma$ is expressed uniquely in either of the following two ways:

$$\sum_{w \in W} p_w(x) w \;=\; h \;=\; \sum_{w \in W} w q_w(x) \tag{1.2.2}$$

with $p_w, q_w \in \mathbb{C}[x_1, \ldots, x_n]$. The existence readily results from relations (1.1.32)–(1.1.34) in $\mathcal{H}'_\Sigma$. The uniqueness is equivalent to

$$I_\lambda = \mathbb{C}W = \oplus_{w \in W} \mathbb{C}w. \tag{1.2.3}$$

Thus I_λ is isomorphic to $\mathbb{C}W$ as a W–module, where the action of x_i is uniquely determined by the formulas $x_i(e) = \lambda_i e$ for the identity $e \in W$. The action of x_i can be extended to other elements of $\mathbb{C}W$ using the defining relation. It is similar to the calculations in the Fock representation.

There is another important construction. Let J be the $\mathcal{H}'_\Sigma$–module induced from the trivial character $+ : W \longrightarrow \mathbb{C},\ w \to 1$, i.e.,

$$J = \operatorname{Ind}_{\mathbb{C}W}^{\mathcal{H}'_\Sigma}(+). \tag{1.2.4}$$

It is isomorphic to $\mathbb{C}[x_1, \ldots, x_n]$ as a vector space, and, moreover, they are isomorphic as $\mathbb{C}[x_1, \ldots, x_n]$–modules, thanks to the Poincare–Birkhoff–Witt property (PBW). One gets finite dimensional representations from J, using the coincidence of the center of $\mathcal{H}'_\Sigma$ with the algebra of W–invariant polynomials in terms of x_i. This theorem is due to Bernstein. The procedure is as follows.

Let us fix an element $\lambda = (\lambda_1, \ldots, \lambda_n) \in \mathbb{C}^n$ and introduce the ideal L_λ in $\mathbb{C}[x_1, \ldots, x_n]$ generated by $p(x) - p(\lambda)$ for all W–invariant polynomials p. Then $J_\lambda = J/L_\lambda$ has a structure of $\mathcal{H}'_\Sigma$–module due to Bernstein. Its dimension is $|W|$, the cardinality of W.

In the following theorem, we will also need the **trivial anti-involution** $^\circ$ on $\mathcal{H}'_\Sigma$:

$$x_i^\circ = x_i, \quad s_i^\circ = s_i, \ (ab)^\circ = b^\circ a^\circ, \ k^\circ = k. \tag{1.2.5}$$

The formulas can be extended to the whole $\mathcal{H}'_\Sigma$ because the relations of $\mathcal{H}'_\Sigma$ are self-dual. For an $\mathcal{H}'_\Sigma$–module V, we can consider its dual $\mathrm{Hom}_{\mathbb{C}}(V,\mathbb{C})$, which has a natural "right" anti-action of $\mathcal{H}'_\Sigma$. Composing it with the anti-automorphism $^\circ$, we make it a standard "left" $\mathcal{H}'_\Sigma$–module, which will be denoted V°.

Later on, $\lambda_b = \sum l_i\lambda_i$ for $b = \sum l_i b_i$.

Theorem 1.2.2. *(a) I_λ is irreducible if and only if $\lambda_{\alpha^\vee} \neq \pm k$ for any $\alpha \in \Sigma_+$.*
(b) There exists $\lambda' = w(\lambda)$ for $w \in W$ such that $\lambda'_{\alpha^\vee} \neq -k$ for any $\alpha \in \Sigma_+$. Then

$$J_\lambda \simeq I_{\lambda'}. \tag{1.2.6}$$

(c) For the longest element w_0 in W,

$$I^\circ_\lambda = I_{w_0(\lambda)}. \tag{1.2.7}$$

❑

The key fact in proving Theorem 1.2.2 is

Lemma 1.2.3. *$I_{(0,\dots,0)}$ is irreducible.*

The proof of this lemma is from [C19] and is based on the intertwining operators of degenerate affine Hecke algebras (to be defined below). See also [KL1, Kat, Ro] and the references therein (the affine case). In [C19], the affine case is deduced from the case of the degenerate Hecke algebra using the exponential-type map, which is closely connected with the Lusztig isomorphisms (to be discussed later).

Definition 1.2.4. *For $1 \le i \le n$, we set*

$$f_i = f_{s_i} = s_i - \frac{k}{x_{a_i}}. \tag{1.2.8}$$

Given a reduced decomposition $W \ni w = s_{i_n}\cdots s_{i_1}$, the elements $f_w \overset{\mathbf{def}}{=\!=} f_{i_n}\cdots f_{i_1}$ are well defined. We call them the **intertwiners**.

The elements f_w belong to the localization of the degenerate affine Hecke algebra $\mathcal{H}'_\Sigma$ by the W–invariant polynomials in terms of x (they are central). They give a certain "baxterization" of w, i.e., a deformation preserving the Coxeter relations, and are closely related to the well-known Yang R–matrix.

Let us check that f_w does not depend on the choice of the reduced decomposition of the element $w \in W$. We have

$$f_{s_i}x_b = x_{s_i(b)}f_{s_i}, \tag{1.2.9}$$
$$f_w x_b = x_{w(b)}f_w. \tag{1.2.10}$$

Indeed,

$$\left(s_i - \frac{k}{x_{a_i}}\right) x_b = x_{s_i(b)} \left(s_i - \frac{k}{x_{a_i}}\right), \tag{1.2.11}$$

which can be rewritten as follows:

$$s_i x_b - x_{s_i(b)} s_i = -k \frac{x_{s_i(b)} - x_b}{x_{a_i}}. \tag{1.2.12}$$

Using the notation x_b from (1.1.36), the right-hand side of (1.2.12) is $k(b, \alpha_i)$. So we come to (1.1.37). Relations (1.2.10) fix f_w uniquely up to the multiplication on the right by functions in x. The longest element of f_w is w. It has no x–multiplier. Therefore f_w does not depend on a particular choice of the reduced decomposition.

To demonstrate the role of intertwiners in the theory, let us check the irreducibility of I_λ for generic λ. First, the vectors $\{f_w(e) \in I_\lambda\}$ are common eigenvectors of x_b, because

$$x_b f_w(e) = f_w x_{w^{-1}(b)}(e) = \lambda_{w^{-1}(b)} f_w(e).$$

For a generic λ, their eigenvalues are simple and these vectors are linearly independent. Any nonzero $\mathcal{H}'_\Sigma$–submodule A of I_λ contains at least one eigenvector of x_b which has to be proportional to some $f_w(e)$, since all eigenvalues are simple. On the other hand, f_w are invertible elements:

$$f_i^{-1} = (1 - \frac{k^2}{{x_{a_i}}^2})^{-1} f_i.$$

Therefore $e \in A$. Since I_λ is generated by e, we conclude that $A = I_\lambda$.

A variant of this reasoning for arbitrary λ leads to the proof of the theorem. However, if λ is arbitrary, one must operate with the intertwiners much more carefully. It is necessary to multiply them by the denominators and remember that the invertibility does not hold for special λ.

Comment. The $\mathcal{H}'$–quotients A of J°_λ will be interpreted below as certain quotients of the D–module representing the corresponding quantum many-body eigenvalue problem. A solution of the AKZ with values in J°_λ induces solutions in any of its $\mathcal{H}'$–quotients (unless I is reducible). It gives a one-to-one correspondence between the $\mathcal{H}'$–submodules, quotients, or constituents of J and those of the D–module representing the quantum many-body eigenvalue problem. The description of the latter is an analytical problem. It is very interesting to combine the two approaches.

The classification of the irreducible constituents of the induced representations is a well-known problem in the theory of (degenerate) affine Hecke algebras. Their multiplicities are described in terms of the Kazhdan–Lusztig polynomials.

1.2.2 Monodromy of AKZ

In this subsection we discuss the monodromy of the AKZ equation, which is a key ingredient in establishing the isomorphism between the AKZ equation in the representation J_λ° and the **quantum many-body problem** (QMBP) for the eigenvalue λ.

Let U' be the open subset of $\mathbb{C}^n$ given by

$$U' = \{u \in \mathbb{C}^n | \prod_{\alpha\in\Sigma_+} (e^{u_\alpha} - 1) \neq 0\}. \tag{1.2.13}$$

The lattice generated by fundamental coweights $b_1, \ldots, b_n$ is denoted by $P^\vee$. It is isomorphic to $\mathbb{Z}^n$ and acts on $\mathbb{C}^n$ by the translations, namely, $b(u) = u + 2\pi\sqrt{-1}b$, where $b \in P^\vee$. The semidirect product $\widehat{W} = W \ltimes P^\vee$ is the so-called **extended affine Weyl group**, acting on $\mathbb{C}^n$ and leaving U' invariant. Picking $u^0 \in U'$, we set

$$\pi_1 = \pi_1(U'/\widehat{W}, u^0).$$

The group structure of π_1 is described as follows. Given an element $w \in \widehat{W}$, let γ_w be a path from u^0 to $w^{-1}(u^0)$ in U'. For elements $w_1, w_2 \in \widehat{W}$, we define the **composition of paths** $\gamma_{w_2} \circ \gamma_{w_1}$ of γ_{w_1} and γ_{w_2} as the path formed by γ_{w_1} and the path γ_{w_2} mapped by w_1^{-1}. See Figure 1.1. The class of γ in the fundamental group will be denoted by $\bar{\gamma}$. The map $\bar{\gamma}_w \to w$ is a homomorphism onto $\widehat{W}$.

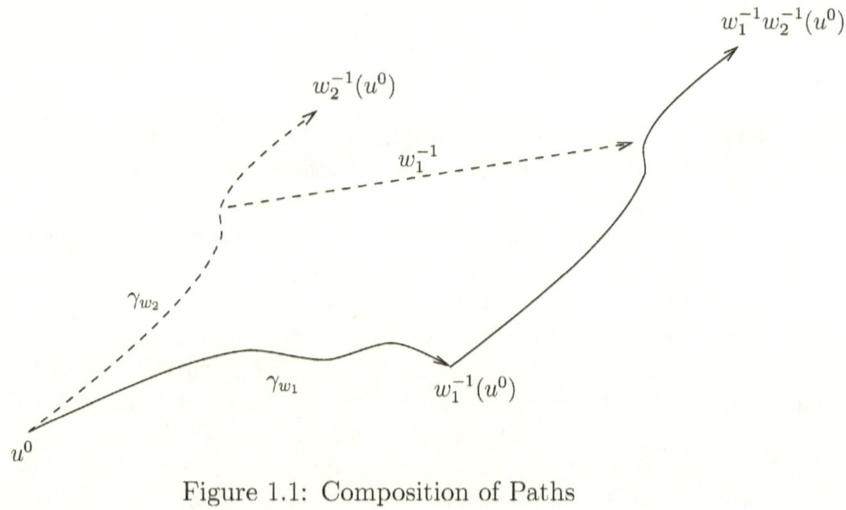

Figure 1.1: Composition of Paths

It is convenient to choose u^0 and the generators of π_1 as follows. For $\Re = \mathrm{Re}, \Im = \mathrm{Im}$,

$$C = (\sqrt{-1}\mathbb{R})^n \setminus \{u \in (\sqrt{-1}\mathbb{R})^n \,|\, 0 < \Im u_\alpha < 2\pi \text{ for every } \alpha \in \Sigma_+\}.$$

Then $\mathbb{C}^n \setminus C$ is a simply connected open subset of U'. Let us take $u^0 \in \mathbb{C}^n$ such that $\Re u^0_\alpha \gg 0$ for $\alpha \in \Sigma_+$. For any element $w \in \widehat{W}$, we denote a path from u^0 to $w^{-1}(u^0)$ in $\mathbb{C}^n \setminus C$ by γ_w. This condition simply means that whenever $u_\alpha \in i\mathbb{R}$ intersects the imaginary axis, it must go through the "window" $0 < \Im u_\alpha < 2\pi$.

For any element $w \in \widehat{W}$, we define an element $\bar{\gamma}_w$ of π_1 to be the image of γ_w. Since $\mathbb{C}^n \setminus C$ is simply connected, $\bar{\gamma}_w$ depends only on w. We set $\tau_i = \gamma_{s_i}$ and choose χ_i to be a path from u^0 to the point u' with the same coordinates $u'_j = u^0_j$ for $j \neq i$ and $u'_i = u^0_i + 2\pi\sqrt{-1}$. The structure of π_1 is described in the following theorem from [Le, C8].

Theorem 1.2.5. *In the notation above,*

$$\pi_1 = \langle \bar{\tau}_1, \ldots, \bar{\tau}_n, \bar{\chi}_1, \ldots, \bar{\chi}_n \rangle, \tag{1.2.14}$$

$$\bar{\tau}_i \text{ satisfy the Coxeter relations,} \tag{1.2.15}$$

$$[\bar{\chi}_i, \bar{\chi}_j] = [\bar{\tau}_i, \bar{\chi}_j] = 0 \quad (i \neq j), \tag{1.2.16}$$

$$\bar{\tau}_i^{-1} \bar{\chi}_i \bar{\tau}_i^{-1} = \bar{\chi}_{s_i(b_i)}. \tag{1.2.17}$$

Here, for $b = \sum_{i=1}^n l_i b_i$, *we put*

$$\bar{\chi}_b = \prod_{i=1}^n \bar{\chi}_i^{l_i}. \tag{1.2.18}$$

Figure 1.2 proves the relation (1.2.17). It shows the u_i–coordinate only, which is sufficient for this relation. ❑

Let us introduce the **affine Hecke algebra** $\mathcal{H}^t_\Sigma$ associated with a root system Σ as a quotient of the group algebra of π_1 by the quadratic relations.

Definition 1.2.6. *The affine Hecke algebra associated with an irreducible reduced root system* Σ *is an associative* $\mathbb{C}$*–algebra generated by* $1, T_1, \ldots, T_n, X_1, \ldots, X_n$ *such that:*

$$T_i \text{ satisfy the Coxeter relations,} \tag{1.2.19}$$

$$[X_i, X_j] = [T_i, X_j] = 0 \quad i \neq j, \tag{1.2.20}$$

$$T_i^{-1} X_i T_i^{-1} = X_{s_i(b_i)}, \tag{1.2.21}$$

$$(T_i - t^{1/2})(T_i + t^{-1/2}) = 0. \tag{1.2.22}$$

The monomials X_b *are defined as in (1.2.18),* $t^{1/2} \in \mathbb{C}^*$. *Here and above, by Coxeter relations, we mean the* **homogeneous Coxeter relations**: $T_iT_jT_i\ldots = T_jT_iT_j\ldots$, *with* m_{ij} *factors on each side, where* $m_{ij} = 2, 3, 4, 6$ *whenever the corresponding vertices in the Dynkin diagram are connected by* $0, 1, 2, 3$ *laces.*

Let Φ be an invertible solution of the AKZ equation associated with Σ, defined in a neighborhood of u^0. Then, for $w \in \widehat{W}$, the transform $w^{-1}(\Phi)$ is defined near $w^{-1}(u^0)$ (see (1.1.30)). Let γ be a path in U' from u^0 to $w^{-1}(u^0)$. Denote by $(w^{-1}(\Phi))_{\bar{\gamma}}$ the analytic continuation of $w^{-1}(\Phi)$ back to u^0 along the path γ, where $\bar{\gamma}$ is the class of γ in the fundamental group π_1. We will also use the projection homomorphism $\widehat{W} \to W$ sending $w = \bar{w}b$ to $\bar{w}$ for $b \in P^\vee, \bar{w} \in W$. Using this homomorphism, we can extend the action of W from (1.1.30) to $\widehat{W}$, multiplying Φ on the left by $\bar{w}$.

Let us define the **monodromy** $T_{\bar{\gamma}}$ to be the ratio

$$T_{\bar{\gamma}} = \left(w^{-1}(\Phi)\right)_{\bar{\gamma}}^{-1} \Phi = \left(\Phi(w(u))\right)_{\bar{\gamma}}^{-1} \cdot \bar{w} \cdot \Phi, \tag{1.2.23}$$

where the dot means the product in $\mathcal{H}'_\Sigma$.

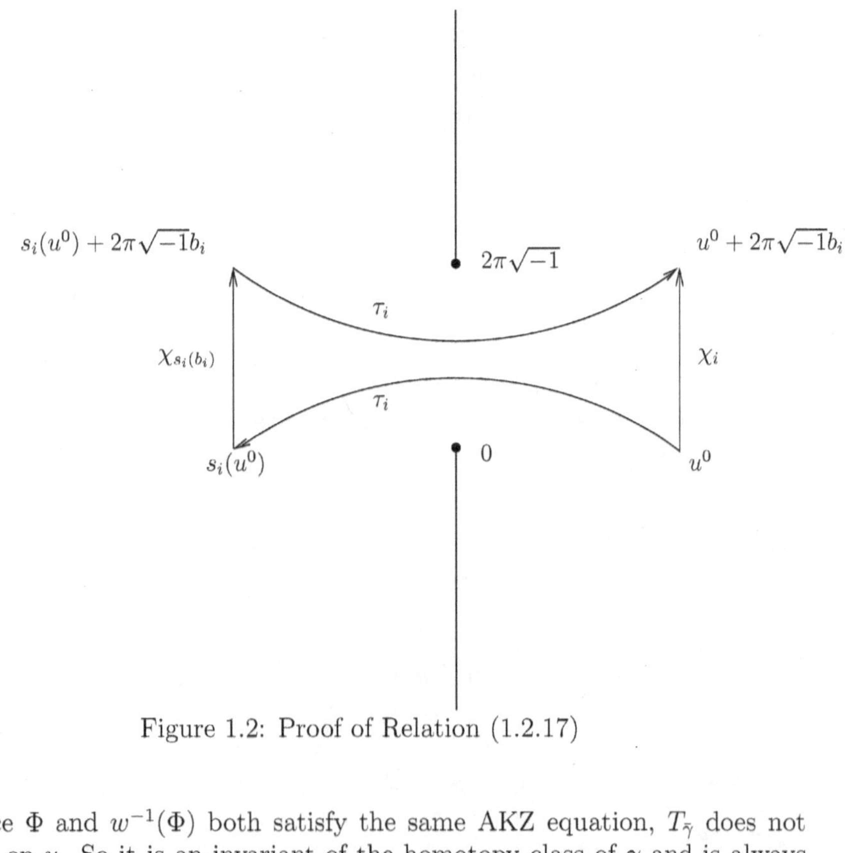

Figure 1.2: Proof of Relation (1.2.17)

Since Φ and $w^{-1}(\Phi)$ both satisfy the same AKZ equation, $T_{\bar{\gamma}}$ does not depend on u. So it is an invariant of the homotopy class of γ and is always invertible. If we choose u^0 and the paths γ_w in $\mathbb{C}^n \setminus C$ as above, then T_w for $w \in \widehat{W}$ are well defined. The monodromy is a homomorphism from π_1 (but not from $\widehat{W}$), which readily results from the definition.

As a preparation for an explicit computation of $\{T_w\}$, we shall introduce a special class of solutions Φ.

Proposition 1.2.7. *For generic λ, there exists a unique solution $\Phi_{as}(u)$ of the AKZ equation such that*

$$\Phi_{as}(u) = \widehat{\Phi}(u)e^{\sum_{i=1}^{n} u_i x_i} \quad \text{for} \tag{1.2.24}$$

$$\widehat{\Phi}(u) = 1 + \sum_{m=(m_1,\ldots,m_n)}^{m_i\geq 0,\, m\neq 0} \Phi_m e^{-\sum_{i=1}^{n} m_i u_i}, \tag{1.2.25}$$

where $\Re u \to \infty$, and Φ_m are independent of u. ❑

We call the solution in the proposition the **asymptotically free solution**. To be more exact, we need either to complete $\mathcal{H}'_\Sigma$, or restrict ourselves to finite dimensional representations of this algebra to introduce such solutions. Establishing the (local) convergence is easy in either case. In this chapter we will follow the second way.

We will give general formulas, which are rigorous in finite dimensional representations (mainly in the induced representations).

Let us examine the condition necessary for the existence of the asymptotically free solutions in the case of A_1. A general consideration follows the same lines. In this case,

$$\Phi_{as}(u) = \left(1 + \sum_{m>0} \Phi_m e^{-mu}\right) e^{ux} = \hat{\Phi}(u)e^{ux}. \tag{1.2.26}$$

Equation (1.1.39) leads to

$$\frac{\partial \hat{\Phi}(u)}{\partial u} = k\frac{s}{e^u - 1}\hat{\Phi}(u) + [x, \hat{\Phi}(u)]. \tag{1.2.27}$$

Comparing the coefficients of e^{-mu}:

$$-m\Phi_m = [x, \Phi_m] + (\textit{terms with } \Phi_j, j < m). \tag{1.2.28}$$

Given a representation of $\mathcal{H}'_\Sigma$ we find Φ_m, assuming that $m + \mathrm{ad}(x)$ is invertible for any $m > 0$ in this representation. Therefore, setting $\mathrm{Spec}(x) = \{\mu_j\}$, the conditions $m + \mu_i - \mu_j \neq 0,\ m = 1, 2, \ldots,$ ensure the existence of the asymptotically free solutions. These inequalities are fulfilled in generic induced representations. The convergence estimates are straightforward.

1.2.3 Lusztig's isomorphisms

In this subsection we establish an isomorphism between $\mathcal{H}^t_\Sigma$ and $\mathcal{H}'_\Sigma$ using the monodromy of the AKZ equation.

Let us fix an invertible solution $\Phi(u)$ of the AKZ system in a neighborhood of $u^0 \in U^* = \mathbb{C}^n \setminus C \subset U'$. The functions $\Phi(w(u))$ can be continued analytically from $w(u^0)$ back to u^0. Since $\mathcal{H}_\Sigma$ is infinite dimensional, we have to consider all formulas in its finite dimensional representations. The final expressions can be treated as elements in a proper completion of the degenerate Hecke algebra as well, but we are not going to discuss the corresponding formal theory in detail.

Theorem 1.2.8 ([C9]). *There exists a homomorphism from $\mathcal{H}^t_\Sigma$ to $\mathcal{H}'_\Sigma$ given by*

$$T_j \longmapsto T'_j, \qquad X_j \longmapsto X'_j, \qquad t^{1/2} \stackrel{\text{def}}{=\!=} \exp(\pi\sqrt{-1}k),$$

where

$$T'_j = \Phi(s_j(u))^{-1} s_j \Phi(u), \qquad X'_j = \Phi(u - 2\pi\sqrt{-1}b_j)^{-1}\Phi(u).$$

If t is sufficiently general (for instance, not a root of unity), then this map is an isomorphism at the level of finite dimensional representations or in proper completions.

Under the notation in (1.2.23), $T'_j = T_{\mathcal{T}_j}$ and $X'_j = T_{\mathcal{X}_j}$. Hence the relations (1.2.19)–(1.2.21) result from Theorem 1.2.5, and only the quadratic relations (1.2.22) need to be proved. We omit a simple direct proof, since these relations follow from the exact formulas below. ❑

Let us find the formulas for T'_j and X'_j for the asymptotically free solution $\Phi_{as}(u)$. Given $b \in P^\vee$, we set

$$X_b = \prod_{j=1}^{n} X_j^{k_j} \qquad \text{for } b = \sum_{j=1}^{n} k_j b_j,$$

and define X'_b analogously.

Theorem 1.2.9 ([C8, C7]). *Let us choose the asymptotically free solution $\Phi_{as}(u)$ as $\Phi(u)$. Then*

(a) $X'_j = \exp(2\pi\sqrt{-1}x_j)$,

(b) $s_i - \dfrac{k}{x_{a_i}} = g(x_{a_i})\left(T'_i + \dfrac{t^{1/2} - t^{-1/2}}{{X'_{a_i}}^{-1} - 1}\right)$,

where the function $g(v)$ is defined by

$$g(v) = \frac{\Gamma(1+v)^2}{\Gamma(1+k+v)\Gamma(1-k+v)},$$

and (b) is in fact a formula for T'_i in terms of $\{s, x\}$.

We will sketch the proof of Theorem 1.2.9. Statement (a) is immediate, since

$$\Phi_{as}(u-2\pi\sqrt{-1}b_j)=\widehat{\Phi}(u)e^{\sum u_i x_i-2\pi\sqrt{-1}x_j}=\Phi_{as}(u)e^{-2\pi\sqrt{-1}x_j}.$$

To prove statement (b), we reduce the problem to the A_1–case. Let us fix the index i $(1\le i\le n)$. Set $E(u)=e^{\sum_{i=1}^n u_i x_i}$, so that $\Phi_{\mathrm{as}}(u)=\hat{\Phi}(u)E(u)$. Let us define $\Phi^{(i)}(u)$ as follows:

$$\Phi^{(i)}(u)=\Phi^{\infty(i)}(u_i)E(u),$$

where $\Phi^{\infty(i)}(u_i)=\lim_{\Re u_j\to+\infty(j\neq i)}\hat{\Phi}(u)$. The AKZ system for $\Phi^{(i)}(u)$ is

$$\frac{\partial\Phi^{(i)}}{\partial u_i}=\left(k\frac{s_i}{e^{u_i}-1}+x_i\right)\Phi^{(i)},\tag{1.2.29}$$

$$\frac{\partial\Phi^{(i)}}{\partial u_j}=x_j\Phi^{(i)}\qquad(j\neq i).\tag{1.2.30}$$

Reduction procedure. Since the monodromy T_i' does not depend on u, the choice of the point u^0, and the path connecting u^0 and $s_i(u^0)$, the path may be replaced by any deformation in U'. We can also degenerate this system by sending the parameters of such a deformation to the limits if the resulting system is well defined. Then the resulting monodromy will remain unchanged.

Using this flexibility, we conclude that T_i' equals

$$T_i^{(i)}=(\Phi^{(i)}(s_i(u)))^{-1}s_i\Phi^{(i)}(u).$$

Indeed, the latter is the "limiting monodromy" for a path with $\Re u_j(j\neq i)$ approaching infinity. We note that $\Re u_j(s_i(u^0))\to+\infty$ if $\Re u_j(u^0)$ does.

In the reduced equations (1.2.29) and (1.2.30), we can naturally diminish the space of values by considering the subalgebra of $\mathcal{H}'_\Sigma$ generated by x_j $(1\le j\le n)$ and s_i. In this algebra, the following elements are central:

$$x_j(j\neq i),\quad x_i-\frac{1}{2}x_{a_i}.$$

Hence, if we define $E^{(i)}(u)$ by

$$E^{(i)}(u)=e^{\sum_{j=1}^n u_j x_j-u_i x_{a_i}/2},$$

it enjoys the following properties:

(i) $E^{(i)}(u)$ commutes with s_i,

(ii) $E^{(i)}(s_i(u))=E^{(i)}(u)$.

The second property can be verified directly:

$$\sum_j (s_i(u))_j x_j - \frac{1}{2}(s_i(u))_i x_{a_i} = \sum_j (u_j - (a_i,\alpha_j)u_i)\, x_j$$
$$+\frac{1}{2}u_i x_{a_i} = \sum_j u_j x_j - \frac{1}{2}u_i x_{a_i}.$$

Here we have used that $(s_i(u))_j = (\alpha_j, s_i(u))$ and $\sum_j (a_i,\alpha_j)x_j = x_{a_i}$. Setting

$$\tilde\Phi^{(i)}(u) = \Phi^{(i)}(u)E^{(i)}(u)^{-1},$$

the system of equations (1.2.29,1.2.30) becomes precisely the AKZ equation for $\tilde\Phi^{(i)}(u)$ in the A_1–case:

$$\frac{\partial\tilde\Phi^{(i)}(u)}{\partial u_i} = \left(k\frac{s_i}{e^{u_i}-1} + \frac{1}{2}x_{a_i}\right)\tilde\Phi^{(i)}(u), \tag{1.2.31}$$

$$\frac{\partial\tilde\Phi^{(i)}(u)}{\partial u_j} = 0 \qquad (j \neq i). \tag{1.2.32}$$

Because of the above properties of $E^{(i)}(u)$, the monodromy of $\tilde\Phi^{(i)}(u)$ coincides with T_i'. The $\tilde\Phi^{(i)}(u)$ can be expressed in terms of the hypergeometric function (a straightforward calculation), which concludes the proof.

To explain a simple structure of the formula for T', let us involve the intertwiners of $\mathcal{H}_\Sigma^t$. They are defined similarly to those in the degenerate case:

$$f_i = s_i - \frac{k}{x_{a_i}} \text{ for } \mathcal{H}'_\Sigma, \quad F_i = T_i + \frac{t^{1/2}-t^{-1/2}}{X_{a_i}^{-1}-1} \text{ for } \mathcal{H}^t_\Sigma.$$

Lemma 1.2.10. $F_iX_b = X_{s_i(b)}F_i$.

The lemma readily results from the definition of $\mathcal{H}^t_\Sigma$. Cf. formula (1.2.10). ❑

The image F_i' of F_i in $\mathcal{H}'_\Sigma$ with respect to the homomorphism constructed in Theorem 1.2.8 can be represented as $F_i' = g_i(x)f_i$ for a function g_i of x. Indeed, $f_iX_b' = X'_{s_i(b)}f_i$, which gives the proportionality. Recall that $X_b' = \exp(2\pi\sqrt{-1}x_b)$. Here $g_i(x)$ must be of the form $g(x_{a_i})$ for a function g in one variable and now can be calculated using the hypergeometric equation (1.2.31). See [C7] for the detail.

We note that the above reduction procedure makes the quadratic relations for T_i' quite obvious; the hypergeometric function and exact formulas are not necessary. Let us demonstrate this.

We set $i=1$ to simplify the indices and switch from (1.2.31) to (1.1.18) with the two variables z_1, z_2 and z_0 treated as a parameter:

$$\frac{\partial\Phi'}{\partial z_j} = \left[k\left(\frac{s_1}{z_j-z_k}\right) + \frac{\Omega_j}{z_j-z_0}\right]\Phi' \quad (j=1,2,\, k=3-j). \tag{1.2.33}$$

When $z_0 = 0$ the substitutions are as follows:

$$2x_1 = \Omega_1 - \Omega_2 + ks_1,\ u_1 = \log(z_1/z_2),\ \Phi' = \Phi^{(1)}(u_1)(z_1z_2)^{-1/2(\Omega_1+\Omega_2+ks_1)}.$$

The monodromy corresponding to the transposition of z_1 and z_2 for $z_0 = 0$ coincides with T_1'.

Now, using that it does not depend on z_0 up to conjugation (the same reduction argument as has been applied above), we may send z_0 to infinity and eliminate the Ω–terms. The resulting monodromy can be calculated immediately and readily gives the desired quadratic relations.

Comment. (i) We note that Heckman in [H1] used a similar reduction approach when checking the quadratic equation for the monodromy of the quantum many-body problem. It is a well known method of finding the characteristic polynomial of the monodromy matrix in terms of the "matrix-residues" of a given local system. Generally, the holomorphic terms can contribute too, so the reasoning must include the exact limiting procedure.

(ii) Our next aim is to establish an isomorphism of AKZ and the latter. For generic λ, Heckman's formulas for the monodromy and the above ones for AKZ give the same irreducible representation of $\mathcal{H}^t$. Hence we may conclude that these equations are isomorphic. However, the most interesting case is when λ is special. The isomorphism for *arbitrary* λ will be considered in the next subsection. ❑

(iii) Let us apply Theorem 1.2.9 to the standard rational KZ equation in the GL_n–case. We need to calculate the monodromy of

$$\frac{\partial\Phi}{\partial v_i} = \left(k\sum_{j>i}\frac{s_{ij}}{e^{v_i-v_j}-1} - k\sum_{j<i}\frac{s_{ij}}{e^{v_j-v_i}-1} + y_i\right)\Phi \quad (1 \le i \le n).$$

Taking special $y_i = k\sum_{j=i+1}^{n} s_{ij}$ and substituting $z_i = e^{v_i}$, we come to

$$\frac{\partial\Phi}{\partial z_i} = k\sum_{j\neq i}\frac{s_{ij}}{z_i - z_j}\Phi \quad (1 \le i \le n),$$

which corresponds to the simplest $\Omega_{ij} = 0$ in (1.1.18). Algebraically, these $\{y_i\}$, the so-called **Jucys–Murphy generators**, induce a homomorphism from $\mathcal{H}_n'$ to $\mathbb{C}\mathbb{S}_{n+1}$ due to Drinfeld [Dr1]. Diagonalizing them [C4], we, for instance, recover the monodromy formulas computed by Tsuchiya-Kanie [TK]. It gives an *explicit* special case for the the general results on the monodromy "up to conjugation" of the rational KZ over the Lie algebras and their representations due to Drinfeld and Kohno (see [Ko]). ❑

In Theorems 1.2.8 and 1.2.9, we established the isomorphism

$$\mathcal{H}_\Sigma^t \simeq \mathcal{H}_\Sigma', \qquad X_j \longmapsto e^{2\pi\sqrt{-1}x_j},$$

where $t = e^{2\pi\sqrt{-1}k}$. It was represented as a proportionality of the intertwiners of the degenerate and non-degenerate affine Hecke algebras:

$$F_j = T_j + \frac{t^{1/2} - t^{-1/2}}{X_{a_j}^{-1} - 1} \longmapsto g(x_{a_j}) \left(s_j - \frac{k}{x_{a_j}} \right).$$

This construction can be naturally generalized. There are many homomorphisms of this kind for different g. In fact, we need only a very mild restriction on $g(x)$.

For instance, normalizing the intertwiners to make them "unitary" ($f^2 = 1 = F^2$), we come to the simplest possible map:

$$X_j \longmapsto e^{2\pi\sqrt{-1}x_j}, \quad \frac{F_j}{t^{1/2} + \frac{t^{1/2}-t^{-1/2}}{X_{a_j}^{-1}-1}} \longmapsto \frac{s_j - \frac{k}{x_{a_j}}}{1 - \frac{k}{x_{a_j}}}.$$

Actually, here we have four formulas in one, since we can put the denominators on the right and on the left. One of them was found and used by Lusztig in [Lus1].

1.2.4 AKZ is isomorphic to QMBP

Here we present the isomorphism between the AKZ equation and the quantum many-body problem (QMBP). The latter will appear as a "trace" of the first. We will need a variant of the general concept of monodromy by A. Grothendieck, *without choosing the initial point.*

Let us fix the notations:

$${}^w\Phi(u) = \Phi(w^{-1}(u)), \quad w = \bar{w}b \in \widehat{W} = W \ltimes P^\vee, u \in \mathbb{C}^n.$$

Given a finite union C of affine real closed half-hyperplanes, we set $U = \mathbb{C}^n \backslash C$ assuming that

(i) U does not contain "bad hyperplanes" $\prod_{\alpha \in \Sigma_+}(e^{u_\alpha} - 1) = 0$,

(ii) U is simply connected,

(iii) $\left(\mathbb{C}^n \backslash \cup_{w \in \widehat{W}} w(C)\right) / \widehat{W}$ is connected.

We shall refer to such C as a **system of cutoffs**. In Section 1.2.2, a special system of cutoffs (U^*) has already been used to compute the monodromy.

Let us fix a system of cutoffs C and U. Then for each $w \in \widehat{W}$ there is a path γ_w (unique up to homotopy) joining u^0 and $w^{-1}(u^0)$. So the choice of C implies a choice of a system of representatives $\bar{\gamma}_w$ in the fundamental group $\pi_1(U'/\widehat{W}, u^0)$. Here U' is complementary to the union of "bad hyperplanes" (1.2.13).

We pick a solution Φ of the AKZ equation in U and define the **monodromy function** $\mathcal{T}_w$ $(w \in \widehat{W})$:

$$\bar{w}\Phi = {}^{w^{-1}}\Phi \cdot \mathcal{T}_w \quad w = \bar{w}b \in \widehat{W}. \tag{1.2.34}$$

Here Φ is invertible at least at one point in U and is extended analytically to the whole U. The values are in the endomorphisms of a finite dimensional representation of $\mathcal{H}'_\Sigma$ (we will later apply the construction to the induced representations).

The monodromy $\{\mathcal{T}_w\}_{w \in \widehat{W}}$ functions satisfy the following:

(a) (1–cycle condition) ${}^{v^{-1}}(\mathcal{T}_w)\mathcal{T}_v = \mathcal{T}_{wv} \quad \forall w, v \in \widehat{W}$,

(b) $\dfrac{\partial}{\partial u_i}\mathcal{T}_w = 0$, and hence $\mathcal{T}_w$ is locally constant.

Property (b) holds, since both Φ and $w(\Phi) = \bar{w}^w\Phi$ satisfy the same differential equation of the first order (the AKZ equation). It readily results in the invertibility of $\mathcal{T}_w$ on $\mathbb{C}^n - \cup_{w \in \widehat{W}} w(C)$. The latter set is not connected, so $\mathcal{T}$ is not just a constant.

Next, let us introduce the operators σ_w, σ'_w $(w \in \widehat{W})$, acting on functions F on U:

$$(\sigma_w F)(u) = ({}^{w^{-1}}F)(u) = F(w(u)), \quad \sigma_i = \sigma_{s_i},$$
$$(\sigma'_w F)(u) = ({}^{w^{-1}}F)(u)\mathcal{T}_w, \quad \sigma'_i = \sigma'_{s_i}.$$

The relations for the operators σ'_w are the same as for the permutations σ_w:

$$\begin{aligned} &(1) \qquad \sigma'_w\sigma'_v = \sigma'_{vw}, \qquad \sigma'_w u_b = u_{w^{-1}(b)}\sigma'_w, \\ &(2) \qquad \sigma'_w\partial_b = \partial_{w^{-1}(b)}\sigma'_w, \quad \partial_b(u_a) = (b, a). \end{aligned} \tag{1.2.35}$$

Note that property (1) follows from the 1–cocycle condition for $\{\mathcal{T}_w\}_{w \in \widehat{W}}$. Indeed,

$$\begin{aligned} (\sigma'_w\sigma'_v)(F) &= \sigma'_w(\sigma'_v(F)) \\ &= \sigma'_w({}^{v^{-1}}F\mathcal{T}_v) \\ &= {}^{w^{-1}}({}^{v^{-1}}F\mathcal{T}_v)\mathcal{T}_w \\ &= {}^{w^{-1}v^{-1}}F({}^{w^{-1}}\mathcal{T}_v)\mathcal{T}_w \\ &= {}^{(vw)^{-1}}F\mathcal{T}_{vw} \\ &= \sigma'_{vw}(F). \end{aligned}$$

Let $\mathcal{S}ol_{AKZ}$ be the space of solutions of the AKZ equation with values in $\mathcal{H}'_\Sigma$. When we consider the AKZ equation with values in a finite dimensional $\mathcal{H}'_\Sigma$–module V, we will denote the space of its solutions by $\mathcal{S}ol_{AKZ}(V)$.

Starting with AKZ, let us go to QMBP. In what follows, $\Phi \in Sol_{AKZ}$ or $\Phi \in Sol_{AKZ}(\mathrm{End}(V))$. In the latter case, all operators act on $\mathrm{End}(V)$–valued functions.

There will be three main steps.

(1) Using $s_\alpha \Phi = \sigma'_{s_\alpha}\Phi$, we rewrite the AKZ equation:

$$\begin{aligned} x_i\Phi &= \left(\frac{\partial}{\partial u_i} - k\sum_{\alpha\in\Sigma_+}\nu_\alpha^i\frac{s_\alpha}{e^{u_\alpha}-1}\right)\Phi \\ &= \left(\frac{\partial}{\partial u_i} - k\sum_{\alpha\in\Sigma_+}\nu_\alpha^i(e^{u_\alpha}-1)^{-1}\sigma'_{s_\alpha}\right)\Phi \qquad (1\le i\le n). \end{aligned}$$

Let us denote:

$$\mathcal{D}'_i = \frac{\partial}{\partial u_i} - k\sum_{\alpha\in\Sigma_+}\nu_\alpha^i(e^{u_\alpha}-1)^{-1}\sigma'_{s_\alpha}.$$

The local invertibility of Φ and the relations $\mathcal{D}'_i\Phi = x_i\Phi$ result in the commutativity

$$[\mathcal{D}'_i,\mathcal{D}'_j] = 0 \quad \forall i,j.$$

Here one can use that the commutators do not contain the derivatives, which readily results from the relations for σ'. The commutativity of this operators is not difficult to check algebraically. It was done in [C13]. See [C11] for a more conceptual proof based on the induced representations. It also follows from the corresponding difference theory, where this and similar statements are actually much simpler (and completely conceptual).

(2) Since the multiplication by x_i commutes with $\mathcal{D}'_j$, we obtain

$$p(x_1,\dots,x_n)\Phi = p(\mathcal{D}'_1,\dots,\mathcal{D}'_n)\Phi$$

for any polynomial $p\in\mathbb{C}[x_1,\dots,x_n]$.

(3) For $\lambda = (\lambda_1,\dots,\lambda_n)\in\mathbb{C}^n$, let us take an $\mathcal{H}'_\Sigma$–module V_λ with the following properties:

(i) $p(x_1,\dots x_n) = p(\lambda_1,\dots,\lambda_n)$ (1.2.36)
on V_λ for any $p\in\mathbb{C}[x_1,\dots,x_n]^W$,

(ii) there exists a linear map $\mathrm{tr}: V_\lambda \longrightarrow \mathbb{C}$ satisfying

$$\mathrm{tr}(wa) = \mathrm{tr}(a) \quad \forall w\in W, a\in V_\lambda. \qquad (1.2.37)$$

Let $p(x_1, \ldots, x_n)$ be a polynomial. Using the commutation relations from (1.2.35), we can write

$$p(\mathcal{D}'_1, \ldots, \mathcal{D}'_n) = \sum_{w \in W} \mathcal{D}'^{(p)}_w \sigma'_w,$$

where the $\mathcal{D}'^{(p)}_w$ are differential operators (they do not contain σ'). They are scalar and commute with $\mathcal{H}'_\Sigma$. Thus

$$p(x_1, \ldots, x_n)\Phi = \sum_{w \in W} \mathcal{D}'^{(p)}_w \sigma'_w \Phi = \sum_{w \in W} \mathcal{D}'^{(p)}_w w\Phi.$$

Now we assume that p is W–invariant. Applying tr (see (1.2.36) and (1.2.37)), we come to

$$p(\lambda_1, \ldots, \lambda_n)\psi = L'_p\psi \text{ for } L'_p = \sum_{w \in W} \mathcal{D}'^{(p)}_w,$$

where

$$\psi(u) = \operatorname{tr}(\Phi(u))$$

is a $\mathbb{C}$–valued function. The differential operators L'_p are W–invariant, which follows from the same construction (we will reprove this fact algebraically below).

Let us introduce the **trigonometric Dunkl operators** $\mathcal{D}_i$ $(1 \le i \le n)$ replacing σ' by σ:

$$\mathcal{D}_i = \frac{\partial}{\partial u_i} - k \sum_{\alpha \in \Sigma_+} \nu^i_\alpha (e^{u_\alpha} - 1)^{-1} \sigma_{s_\alpha}.$$

Repeating the above construction, define $\mathcal{D}^{(p)}$ for a W–invariant polynomial p by

$$p(\mathcal{D}_1, \ldots \mathcal{D}_n) = \sum_{w \in W} \mathcal{D}^{(p)}_w \sigma_w.$$

Since in the construction of L'_p and L_p we use only the commutation relations (1.2.35) for σ'_w and σ_i, these operators coincide.

Comment. The trigonometric Dunkl operators are from [C10]. The Dunkl operators from [Du1] are rational (see also [Du2] and [C11] and the references therein). When defining these operators, I used [H2]. Heckman's trigonometric "global Dunkl operators" also lead to the QMBP, but they do not commute. Note that the operators above do not involve the *divided differences*, and therefore do not preserve the space of Laurent polynomials. Generally, the Dunkl operators are considered in the normalization in terms of the divided differences. Algebraically, the choice of the normalization is insignificant, but it influences the diagonalization process and the corresponding representation theory. ❑

We are now in a position to introduce the QMBP with the eigenvalue $\lambda = (\lambda_1, \dots, \lambda_n) \in \mathbb{C}^n$. It is the following system of the differential equations for a $\mathbb{C}$–valued function ψ:

$$L_p\psi = p(\lambda_1, \dots, \lambda_n)\psi \qquad (p \in \mathbb{C}[x_1, \dots, x_n]^W).$$

It is known [HO1] (and easy to see by looking at the leading terms of L_p) that the dimension of the space of solutions ψ is $|W|$.

Summarizing what has been done above, we come to the following theorem.

Theorem 1.2.11. *Applying* tr *we obtain a homomorphism*

$$\mathrm{tr} : \mathcal{S}ol_{AKZ}(V_\lambda) \longrightarrow \mathcal{S}ol_{QMBP}(\lambda).$$

Here $\mathcal{S}ol_{QMBP}(\lambda)$ denotes the space of solutions to the QMBP with the eigenvalue λ.

We can say more for concrete representations, especially for the induced representations J_λ° (see (1.2.4)). We define the **trace**

$$\mathrm{tr} : J_\lambda^\circ \longrightarrow \mathbb{C}$$

as the map dual to the embedding

$$\mathbb{C} \longrightarrow J_\lambda = \mathbb{C}[x_1, \dots, x_n]/L_\lambda$$

sending 1 to $1 \in \mathbb{C}[x_1, \dots, x_n]$. Here L_λ denotes the ideal generated by $p(x) - p(\lambda)$, $p \in \mathbb{C}[x_1, \dots, x_n]^W$. One easily checks that such tr satisfies the conditions (1.2.37).

Theorem 1.2.12 ([C9]). *For any $\lambda \in \mathbb{C}^n$,* tr *gives an isomorphism*

$$\mathrm{tr} : \mathcal{S}ol_{AKZ}(J_\lambda^\circ) \xrightarrow{\sim} \mathcal{S}ol_{QMBP}(\lambda).$$

Proof. The key observation is that

$$\text{for any } \mathcal{H}'_\Sigma\text{-submodule } M \neq \{0\} \text{ in } J_\lambda^\circ, \text{ we have } \mathrm{tr}\,|_M \neq 0. \tag{1.2.38}$$

Indeed, if $0 \neq f \in M$, then there exists a polynomial $p(x) \in \mathbb{C}[x_1, \dots, x_n]$ such that $f(p) \neq 0$. However, $f(p) = tr(p(f)) \in \mathrm{tr}(M)$.

To prove Theorem 1.2.12, it is enough to show the injectivity of tr, since the surjectivity will then follow by comparing the dimensions of the solution spaces (both of them are $|W|$). So let us suppose that, for $\varphi(u) \in \mathcal{S}ol_{AKZ}(J_\lambda^\circ)$, identically

$$\mathrm{tr}(\varphi) = 0. \tag{1.2.39}$$

We will show that

$$\operatorname{tr}(\mathcal{H}'_\Sigma\varphi) = 0. \tag{1.2.40}$$

Differentiating (1.2.39),

$$0 = \operatorname{tr}(\frac{\partial\varphi}{\partial u_i}) = k\sum_{\alpha\in\Sigma_+}\nu_\alpha^i \operatorname{tr}\left(\frac{s_\alpha}{e^{u_\alpha}-1}\varphi\right) + \operatorname{tr}(x_i\varphi).$$

By the W–invariance of tr, $\operatorname{tr}(s_\alpha\varphi) = \operatorname{tr}(\varphi) = 0$. Hence

$$\operatorname{tr}(x_i\varphi) = 0. \tag{1.2.41}$$

Differentiating this equation by u_j we have

$$0 = k\sum_{\alpha\in\Sigma_+}\nu_\alpha^j \operatorname{tr}\left(x_i\frac{s_\alpha}{e^{u_\alpha}-1}\varphi\right) + \operatorname{tr}(x_ix_j\varphi).$$

Using the commutation relations of x_j and s_α, we deduce from (1.2.39) and (1.2.41) that

$$\operatorname{tr}(x_ix_j\varphi) = 0.$$

Proceeding in the same way, we establish that

$$\operatorname{tr}(x_{i_1}\cdots x_{i_l}\varphi) = 0$$

for any $i_1, \ldots, i_l$. Combining this with the W–invariance of tr, we obtain (1.2.40).

For each u^0, consider the submodule $M = \mathcal{H}'_\Sigma\varphi(u^0) \subset J^\circ_\lambda$. Then $\operatorname{tr}|_M = 0$, and we deduce that $M = 0$ from the key observation above. This completes the proof of Theorem 1.2.12. ❑

Comment. The map from Theorem 1.2.12 was found by Matsuo [Mat] for induced representations I_λ. He proved his theorem algebraically (without the passage through the trigonometric Dunkl operators discussed above) using an explicit presentation for AKZ in I_λ.

The isomorphism for J°_λ (or for I_λ with properly ordered λ, "dominate"; see (1.2.6)) was established independently and simultaneously by Matsuo and the author. He proved that a certain determinant is nonzero for dominant λ. I used the modules J in [C11]. My theorem included the rational. His approach was entirely "trigonometric."

Matsuo was the first to conjecture that the QMBP (the Heckman–Opdam system) and the corresponding variant of the trigonometric KZ from [C6] are isomorphic. The affine KZ were defined in full generality a bit later (in [C7]). ❑

Let us give the formula for the simplest L_p. If $p_2(x_1, \ldots, x_n) = \sum_{i=1}^n x_{\alpha_i}x_i$, then

$$L_2 = L_{p_2} = \sum_{i=1}^n \partial_{\alpha_i}\partial_i + \sum_{\alpha\in\Sigma_+}(\alpha,\alpha)\frac{k(1-k)}{(e^{u_\alpha/2}-e^{-u_\alpha/2})^2}.$$

This operator is called Sutherland operator or Heckman–Opdam operator. It was studied in many papers. The Calogero operator is its rational degeneration. In harmonic analysis, these operators are, respectively, due to Harish-Chandra and Helgason (for special k), See also below.

More generally, let A be a $\mathbb{C}W$–module and

$$V_{A,\lambda} = \left(\mathrm{Ind}_{\mathbb{C}W}^{\mathcal{H}'_\Sigma}(A)/L_\lambda\right)^\circ.$$

As before, L_λ is the ideal generated by $p(x) - p(\lambda)$, $p \in \mathbb{C}[x_1, \ldots, x_n]^W$. Then the following holds:

$$\mathcal{S}ol_{AKZ}(V_{A,\lambda}) \xrightarrow{\sim} \mathcal{S}ol_{QMBP_A}(\lambda),$$

where now the right-hand side means a matrix version of QMBP, called the **spin–QMBP**. It was introduced in [C11] for the first time. It is a unification of the Haldane–Shastry model and that by Calogero–Sutherland.

For example, the L–operator corresponding to p_2 above reads as

$$L_2 = \sum_{i=1}^{n} \partial_{\alpha_i}\partial_i + \sum_{\alpha\in\Sigma_+} (\alpha,\alpha)\frac{k(s_\alpha^* - k)}{(e^{u_\alpha/2} - e^{-u_\alpha/2})^2}.$$

where by s_α^* we mean the image of s_α in $\mathrm{Aut}(A)$. See [C11] for more detail.

1.2.5 The *GL*–case

Let us describe AKZ and QMBP in the GL_n–case. In Section 1.1.3, we introduced the degenerate affine Hecke algebra of type GL_n. It is the algebra

$$\mathcal{H}'_n = \langle \mathbb{C}\mathbb{S}_n, y_1, \ldots, y_n \rangle$$

subject to the following relations:

$$s_i y_i - y_{i+1} s_i = k, \qquad s_i y_j = y_j s_i \quad (i \neq j, j+1),$$
$$y_i y_j = y_j y_i \quad (1 \leq i, j \leq n).$$

Following Section 1.1.3, we will use the coordinates v_i.

Let us conjugate the AKZ for GL_n by the function Δ^k for the discriminant $\Delta \overset{\mathbf{def}}{=\!=} \prod_{i<j}(e^{v_i} - e^{v_j})$. Later it will be necessary to prepare the passage to the difference theory. The system then becomes

$$\frac{\partial\Phi}{\partial v_i} = \Big\{k\Big(\sum_{j(>i)} \frac{s_{ij} - 1}{e^{v_i - v_j} - 1} - \sum_{j(<i)} \frac{s_{ij} - 1}{e^{v_j - v_i} - 1}\Big) + y_i + k\Big(i - \frac{n+1}{2}\Big)\Big\}\Phi. \tag{1.2.42}$$

In this form, it can be quantized (see Section 1.3.2). The system is self-consistent and $\mathbb{S}_n$–invariant.

The corresponding **trigonometric Dunkl operators** are given by the formula

$$\mathcal{D}_i = \frac{\partial}{\partial v_i} - k\Big(\sum_{j(>i)} (e^{v_i - v_j} - 1)^{-1}(\sigma_{ij} - 1) \\ - \sum_{j(<i)} (e^{v_j - v_i} - 1)^{-1}(\sigma_{ij} - 1) + i - \frac{n+1}{2}\Big).$$

Here the σ_{ij} stand for the transpositions of the coordinates:

$$\sigma_{ij} v_i = v_j \sigma_{ij}.$$

Similarly, by σ_w we mean the permutation of the coordinates corresponding to w^{-1}.

The main point of the theory is that $\{\mathcal{D}\}$ satisfy the relations for the degenerate Hecke algebra:

$$[\mathcal{D}_i, \mathcal{D}_j] = 0 = [\mathcal{D}_i, y_j],\ i \neq j, \qquad \sigma_{i\,i+1}\mathcal{D}_i - \mathcal{D}_{i+1}\sigma_{i\,i+1} = k.$$

This statement (for any reduced root systems) is from [C11]. The appearance of the degenerate Hecke algebras in the context of the Dunkl operators was a major development.

Later we will deduce these relations from the difference theory (where they are almost obvious). They readily give that $p(\mathcal{D}_1, \ldots, \mathcal{D}_n)$ and the corresponding L_p are W–invariant for the W–invariant polynomials p. Use the description of the center of $\mathcal{H}'$ to see this.

In the case of GL_n, given symmetric $p \in \mathbb{C}[x_1, \ldots, x_n]^{\mathbb{S}_n}$,

$$p(\mathcal{D}_1, \ldots, \mathcal{D}_n) = \sum_{w \in \mathbb{S}_n} D_w^{(p)} \sigma_w,$$

where the $D_w^{(p)}$ are scalar differential operators defined from the expansion

$$L_p = p(\mathcal{D}_1, \ldots, \mathcal{D}_n)\big|_{symm.poly.} = \sum_{w \in \mathbb{S}_n} D_w^{(p)}.$$

Let us take the elementary symmetric polynomials as p:

$$e_m(x) = \sum_{i_1 < \cdots < i_m} x_{i_1} \cdots x_{i_m},\ L_m \stackrel{\mathbf{def}}{=\!=} L_{e_m}.$$

The first two L–operators are

$$L_1 = \sum_{i=1}^{n} \frac{\partial}{\partial v_i},$$

$$L_2 = \sum_{i<j} \frac{\partial^2}{\partial v_i \partial v_j} - \frac{k}{2} \sum_{i<j} \coth\left(\frac{v_i - v_j}{2}\right)\left(\frac{\partial}{\partial v_i} - \frac{\partial}{\partial v_j}\right) - \frac{k^2}{4}\binom{n+1}{3}.$$

When we replace e_2 by $p = \sum_i x_i^2$, the corresponding L–operator is conjugated (by Δ^k) to the Sutherland operator [Su] up to a constant term. A particular case was considered by Koornwinder; see [HO1] and book [HS] for general algebraic and systematic analytic theories. For special values of the parameter k, these operators were well known. They are the radial parts of the Laplace operators on the symmetric spaces. The value $k = 1$ is the so-called group case $(SL(n,\mathbb{C})/SU_n)$. The values $k = 1/2$ and $k = 1$ correspond, respectively, to the orthogonal case $(SL(n,\mathbb{R})/SO_n)$ and the symplectic case.

The problem was to make k arbitrary, ensuring the invariance and the pairwise commutativity. A more difficult problem was to generalize the Harish-Chandra theory to the case of arbitrary real $k > 0$ or even small negative k. It was essentially done by Heckman and Opdam (see below). However, there are still open analytic problems to solve.

The rational counterparts of the L–operators are due to Calogero [Ca]. The corresponding eigenvalue problem is equivalent to the rational variant of AKZ. It is an extension of the rational W–valued KZ from [C6] by the operators x which are "permuted" by the elements from W. Here the J_λ–modules cannot be represented as I_λ. The isomorphism theorem fails for I_λ; it holds only for the J–modules (in the rational setting) [C11].

1.3 Isomorphisms for QAKZ

Let us now turn to the q–deformations. We introduce the quantum affine Knizhnik–Zamolodchikov (QAKZ) equation, and show that there is an isomorphism between the (spaces of) solutions of the QAKZ equation and the solutions of the generalized Macdonald eigenvalue problem.

1.3.1 Affine Hecke algebras

First, we recall the definition of the affine Hecke algebra $\mathcal{H}_n^t$ in the case of GL_n. Let $t^{1/2} \in \mathbb{C}^*$ be a parameter. Then $\mathcal{H}_n^t$ is the algebra defined over $\mathbb{C}$ by the following set of generators and relations:

$$\text{generators} \ : \ T_1, \ldots, T_{n-1}, Y_1, \ldots, Y_n,$$

$$\text{relations} \ : \ (T_i - t^{1/2})(T_i + t^{-1/2}) = 0, \quad (1 \le i \le n-1); \tag{1.3.1}$$

$$T_i T_{i+1} T_i = T_{i+1} T_i T_{i+1}, \quad (1 \le i \le n-2); \tag{1.3.2}$$

$$T_i T_j = T_j T_i, \quad (|i-j| > 1); \tag{1.3.3}$$

$$Y_i Y_j = Y_j Y_i, \quad (1 \le i, j \le n); \tag{1.3.4}$$

$$Y_i T_j = T_j Y_i, \quad (j \ne i, i-1); \tag{1.3.5}$$

$$T_i^{-1} Y_i T_i^{-1} = Y_{i+1}. \quad (1 \le i \le n-1). \tag{1.3.6}$$

The relations (1.3.1) are called the quadratic relations, (1.3.2)–(1.3.3) the Coxeter relations, (1.3.4) the commutativity relations, and (1.3.5)–(1.3.6) the cross-relations.

It follows from the defining relations (1.3.1)–(1.3.6) that

$$P = T_1 \cdots T_{i-1} Y_i T_i^{-1} \cdots T_{n-1}^{-1}$$

is independent of i $(1 \le i \le n)$. For instance,

$$P = T_1 \cdots T_{n-1} Y_n = Y_1 T_1^{-1} \cdots T_{n-1}^{-1}. \tag{1.3.7}$$

Lemma 1.3.1. *The algebra $\mathcal{H}_n^t$ can be presented as*

$$\mathcal{H}_n^t = \langle T_1, \ldots, T_{n-1}, P \rangle / \sim, \tag{1.3.8}$$

where the quotient is by the quadratic relations (1.3.1), *the Coxeter relations* (1.3.2)–(1.3.3), *and the following:*

(a) $PT_{i-1} = T_i P$ $(1 < i < n)$,

(b) P^n *is central.*

Proof. In terms of Y_i, we have $P^n = Y_1 \cdots Y_n$. Relations (a) and (b) readily follow from (1.3.7) and the defining relations (1.3.1)–(1.3.6). For example,

$$PT_1P^{-1} = Y_1T_1^{-1}(T_2^{-1}T_1T_2)T_1Y_1^{-1} = Y_1T_1^{-1}(T_1T_2^{-1}T_1^{-1})T_1Y_1^{-1} = T_2.$$

To establish (1.3.8), we start with $T_1, \ldots, T_{n-1}, P$ and introduce the elements $Y_1, \ldots, Y_n$ by

$$Y_1 \stackrel{\text{def}}{=} PT_{n-1} \cdots T_1, \quad Y_2 \stackrel{\text{def}}{=} T_1^{-1} Y_1 T_1^{-1}, \ldots .$$

We must check the relations $Y_1Y_2 = Y_2Y_1$, $T_jY_1 = Y_1T_j$ $(j > 1)$, etc., using (a) and (b). The pairwise commutativity of all Y_i and other required relations formally follow from them.

The first reads

$$Y_1T_1^{-1}Y_1T_1^{-1} = T_1^{-1}Y_1T_1^{-1}Y_1.$$

We plug in the formula for Y_1 and move P to the left. The commutativity with "distant" T is obvious. ❑

1.3.2 Definition of QAKZ

In this subsection, we introduce the QAKZ equation.

Definition 1.3.2. *For $u \in \mathbb{C}$, we define the* **intertwiners** *by*

$$F_i(u) = \frac{T_i + \dfrac{t^{1/2} - t^{-1/2}}{e^u - 1}}{t^{1/2} + \dfrac{t^{1/2} - t^{-1/2}}{e^u - 1}}. \tag{1.3.9}$$

❑

They satisfy

$$F_i(u)F_i(-u) = 1, \tag{1.3.10}$$
$$F_i(u)F_{i+1}(u+v)F_i(v) = F_{i+1}(v)F_i(u+v)F_{i+1}(u). \tag{1.3.11}$$

The second relation can be deduced from Lemma 1.2.10 in the same way as for the degenerate affine Hecke algebra.

The **affine quantum Knizhnik–Zamolodchikov** (QAKZ) equation is the following system of *difference* equations for a function $\Phi(v)$ that takes values in $\mathcal{H}_n^t$ (or any $\mathcal{H}_n^t$–module):

$$\begin{aligned}
&\Phi(v_1, \ldots, v_i + h, \ldots, v_n) \\
&\quad = F_{i-1}(v_i - v_{i+1} + h) \ldots F_1(v_i - v_1 + h)T_1 \cdots T_{i-1}Y_i \\
&\quad \times T_i^{-1} \cdots T_{n-1}^{-1} F_{n-1}(v_i - v_n) \cdots F_i(v_i - v_{i+1}) \\
&\qquad \times \Phi(v_1, \ldots, v_i, \ldots, v_n) \quad (i = 1, \ldots, n).
\end{aligned} \tag{1.3.12}$$

Here h is a new parameter.

Theorem 1.3.3. *The QAKZ system* (1.3.12) *is self-consistent. This system is W–***invariant** *in the following sense: if $\Phi(v)$ is a solution, then so is*

$$F_i(v_{i+1} - v_i)\, {}^{s_i}\Phi(v) = {}^{s_i}\left(F_i(v_i - v_{i+1})\Phi(v)\right).$$

This follows from (1.3.10), (1.3.11). Later we will make the invariance obvious "conceptually." ❑

Let us discuss the quasi-classical limit of the QAKZ system. Setting

$$t \stackrel{\mathbf{def}}{=\!=} e^{kh} = q^k, \quad q = e^h,$$

let $h \to 0$. The generators T_i, Y_i are supposed to be in the form

$$\begin{aligned}
T_i &= s_i + \frac{kh}{2} + \cdots \qquad (s_i^2 = 1), \\
Y_i &= 1 + hy_i + \cdots,
\end{aligned}$$

where by "$\cdots$," we mean here and in the following formulas the terms of order h^2.

The relations of the degenerate affine Hecke algebra for s_i, y_i can be readily verified. Using the formula

$$t^{1/2}T_i^{-1}F_i(u) = 1 + \frac{kh}{e^u - 1}(s_i - 1) + \cdots,$$

we find that

$$\begin{aligned}
&h^{-1}\left(\Phi(\ldots, v_i + h, \ldots) - \Phi(\ldots, v_i, \ldots)\right) \\
&= \{y_i + k\big(\sum_{j(>i)} \frac{s_{ij} - 1}{e^{v_i - v_j} - 1} \\
&- \sum_{j(<i)} \frac{s_{ij} - 1}{e^{v_j - v_i} - 1} + i - \frac{n+1}{2}\big)\}\Phi(\ldots, v_i, \ldots) + \cdots.
\end{aligned}$$

Hence the AKZ equation (1.2.42) is the limit of the QAKZ equation as $h \to 0$, which is called the **semi-classical limit**.

To make the QAKZ equations more transparent, we need the **affine Weyl group of type** GL_n. It is the semidirect product

$$\widehat{\mathbb{S}}_n = \mathbb{S}_n \ltimes \mathbb{Z}^n,$$

where

$$\mathbb{Z}^n = \bigoplus_{i=1}^{n} \mathbb{Z}\gamma_i$$

is a free abelian group of rank n. Define the action of $\widehat{\mathbb{S}}_n$ on a vector $v = (v_1, \ldots, v_n) \in \mathbb{R}^n$ by

$$\begin{aligned}
s_{ij}v &= (v_1, \ldots, v_j, \ldots, v_i, \ldots, v_n) = s_{ji}v, \qquad i < j, \\
\gamma_i v &= (v_1, \ldots, v_i + h, \ldots, v_n), \ {}^{\gamma_i}(v_j) = v_j - h\delta_{ij}.
\end{aligned}$$

Introducing

$$\pi = \gamma_1 s_1 \cdots s_{n-1} = s_1 \cdots s_{n-1}\gamma_n,$$

its action on $\mathbb{R}^n$ and the coordinates read as

$$\pi v = (v_n + h, v_1, \ldots, v_{n-1}), \ {}^{\pi}v_n = v_1 - h, \ {}^{\pi}v_1 = v_2, \ldots.$$

It is convenient to represent the elements γ_i, π graphically.

Figure 1.3 shows the following reduced decomposition of γ_i:

$$\gamma_i = s_{i-1} \cdots s_1 \pi s_{n-1} \cdots s_i.$$

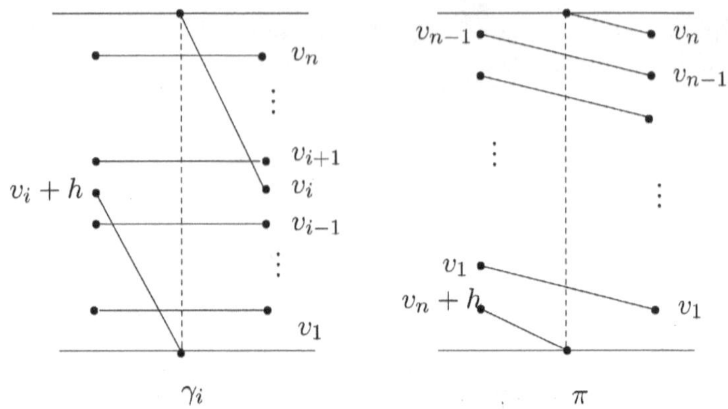

Figure 1.3: Graphs for γ and π

Lemma 1.3.4. *The affine Weyl group of the GL_n-type can be presented as*

$$\widehat{\mathbb{S}}_n = \langle s_1, \ldots, s_{n-1}, \pi \rangle / \sim,$$

where the relations are

$$s_i^2 = 1, \quad s_i s_j = s_j s_i \quad (|i-j| > 1), \quad s_i s_{i+1} s_i = s_{i+1} s_i s_{i+1},$$

$$(a)\ \pi s_{i-1} \pi^{-1} = s_i \ (1 < i < n), \quad \text{and} \ (b)\ \pi^n \text{ is central.}$$

❑

For a function $\Psi(v)$ with values in $\mathcal{H}_n^t$, let

$$\tilde{s}_i(\Psi) = F_i(v_{i+1} - v_i)\, {}^{s_i}\Psi, \tag{1.3.13}$$

$$\tilde{\pi}(\Psi) = P^{\pi}\Psi. \tag{1.3.14}$$

Theorem 1.3.5 ([C12]). *Formulas* (1.3.13) *and* (1.3.14) *can be extended to an action of* $\widehat{\mathbb{S}}_n$.

We denote this action by $\widehat{\mathbb{S}}_n \ni w : \Psi \mapsto \tilde{w}(\Psi)$. For instance,

$$\tilde{\gamma}_i(\Psi)(v_1, \ldots, v_n) = F_{i-1}(v_{i-1} - v_i)^{-1} \cdots F_1(v_1 - v_i)^{-1} P$$
$$\times F_{n-1}(v_i - v_n - h) \cdots F_i(v_i - v_{i+1} - h)\Psi(v_1, \ldots, v_i - h, \ldots, v_n).$$

Hence the QAKZ equation simply means the invariance of $\Phi(v)$ with respect to the pairwise commuting elements γ_i:

$$\text{QAKZ} \Longleftrightarrow \tilde{\gamma}_i(\Phi) = \Phi \quad (i = 1, \ldots, n). \tag{1.3.15}$$

Let us connect QAKZ with the QKZ introduced by Smirnov and Frenkel-Reshetikhin [Sm, FR]. We fix an N–dimensional complex vector space V and define the matrix $T \in \mathrm{End}(V \otimes V)$ by

$$T = (t^{1/2} - t^{-1/2}) \sum_{i<j} E_{ii} \otimes E_{jj} + \sum_{i \neq j} E_{ij} \otimes E_{ji} + t^{1/2} \sum_{i=1}^{N} E_{ii} \otimes E_{ii},$$

following Baxter and Jimbo. The algebra $\mathcal{H}_n^t$ acts on $V^{\otimes n}$ by

$$T_i(a_1 \otimes \cdots \otimes a_n) = a_1 \otimes \cdots \otimes T(a_i \otimes a_{i+1}) \otimes \cdots \otimes a_n, \quad (1.3.16)$$
$$P(a_1 \otimes \cdots \otimes a_n) = Ca_n \otimes a_1 \otimes \cdots \otimes a_{n-1}, \quad (1.3.17)$$

where $a_i \in V$ and $C = \mathrm{diag}(\lambda_1, \ldots, \lambda_n)$. One can check that this action is well defined by a direct calculation.

Assuming that N, the dimension of V, coincides with n, the number of components, let

$$\left(V^{\otimes n}\right)_0 = \mathrm{span}\{e_{w(1)} \otimes \cdots \otimes e_{w(n)} \mid w \in \mathbb{S}_n\}$$

be the 0–weight subspace. Here $e_1, \ldots, e_n$ are the standard basic vectors of V. It is easy to see that this subspace is closed under the action of $\mathcal{H}_n^t$. We state the next proposition without proof.

Proposition 1.3.6. *If $N = n$ and $\lambda = (\lambda_1, \ldots, \lambda_n)$ is generic, then the 0–weight space $(V^{\otimes n})_0$ is isomorphic to $I_\lambda = \mathrm{Ind}_{\mathbb{C}[Y_1,\ldots,Y_n]}^{\mathcal{H}_n^t}(\lambda)$.*

Writing AQKZ in $(V^{\otimes n})_0$ we obtain the QKZ for the GL_n in its fundamental representation. Combining this observation with the isomorphism of AQKZ with the Macdonald eigenvalue problem (our next aim), we can readily explain why the Macdonald polynomials appear in many calculations involving the vertex operators. The QKZ was introduced as an equation for the n-point function in conformal field theory based on quantum groups. The most universal definition of the n-point function is in terms of the vertex operators.

1.3.3 The monodromy cocycle

Let Φ be a solution of the QAKZ equation. Thanks to (1.3.15), $\tilde{w}(\Phi)$ is also a solution of the QAKZ equation for any $w \in \tilde{\mathbb{S}}_n$. We define $\mathcal{T}_w \in \mathcal{H}_n^t$ by

$${}^{w}\mathcal{T}_w = \Phi^{-1}\tilde{w}(\Phi) \quad \text{for } w \in \tilde{\mathbb{S}}_n$$

and call it the **monodromy cocycle**. It follows from (1.3.13) and (1.3.14) that

$$F_i(v_i - v_{i+1})\Phi = {}^{s_i}\Phi\mathcal{T}_i \quad (1.3.18)$$

and

$$P\Phi = {}^{\pi^{-1}}\Phi \mathcal{T}_\pi. \tag{1.3.19}$$

Here $\mathcal{T}_i$ stands for $\mathcal{T}_{s_i}$.

Lemma 1.3.7.

$${}^{w_2^{-1}}(\mathcal{T}_{w_1})\mathcal{T}_{w_2} = \mathcal{T}_{w_1 w_2} \quad \textit{for } w_1, w_2 \in \tilde{\mathbb{S}}_n.$$

Indeed,

$$\Phi^{w_1 w_2}\mathcal{T}_{w_1 w_2} = \widetilde{w_1 w_2}(\Phi) = \tilde{w}_1(\tilde{w}_2\Phi) = \tilde{w}_1(\Phi^{w_2}\mathcal{T}_{w_2}) = \Phi^{w_1}\mathcal{T}_{w_1}{}^{w_1 w_2}\mathcal{T}_{w_2}.$$

The QAKZ equation implies that $\mathcal{T}_{\gamma_i} = 1$. Hence $\mathcal{T}_w$ depends only on the image $\bar{w}$ of w in $\mathbb{S}_n$.

Let $\mathcal{F}(\mathbb{C}^n, \mathcal{H}_n^t)$ be the set of $\mathcal{H}_n^t$–valued functions on $\mathbb{C}^n$. We obtain the following two *anti-actions* of $\tilde{\mathbb{S}}_n$:

$$\sigma_w(\Psi) = {}^{w^{-1}}\Psi, \tag{1.3.20}$$

$$\sigma'_w(\Psi) = {}^{w^{-1}}\Psi\mathcal{T}_w, \tag{1.3.21}$$

where $w \in \tilde{\mathbb{S}}_n$ and $\Psi \in \mathcal{F}(\mathbb{C}^n, \mathcal{H}_n^t)$. For instance, $\sigma_{\gamma_i}(v_i) = v_i + h = \sigma'_{\gamma_i}(v_i)$. Lemma 1.3.7 means exactly that σ' is an anti-action, i.e., $\sigma'_{w_1 w_2} = \sigma'_{w_2}\sigma'_{w_1}$.

We note that, in the difference theory, the monodromy of KZ can be always made trivial. Indeed, the 1-cocycle $\{\mathcal{T}_w, w \in W\}$ is always a co-boundary because of the Hilbert Theorem 90. Hence, by conjugating solutions of AQKZ we can always get rid of the monodromy. So the above actions σ, σ' are not very much different, in contrast to the differential theory.

One can try to apply the Hilbert Theorem 90 to the AQKZ itself, though the group $\mathbb{Z}^n$ is infinite. It gives the following formal integration of QAKZ. Let $\Psi \in \mathcal{F}(\mathbb{C}^n, \mathcal{H}_n^t)$. Then the infinite sum

$$\sum_{b \in P^\vee} \tilde{b}(\Psi), \tag{1.3.22}$$

where $P^\vee = \oplus_{i=1}^n \mathbb{Z}\gamma_i \subset \tilde{\mathbb{S}}_n$, satisfies the AQKZ.

If Ψ is rapidly decreasing, then one can check that $\sum_{b \in P^\vee} \tilde{b}(\Psi)$ is convergent. Therefore, constructing *some* solutions Φ of QAKZ poses no problem. What is difficult is to ensure good analytic properties and a proper asymptotical behavior.

1.3.4 Macdonald's eigenvalue problem

In this subsection, we introduce the Macdonald eigenvalue problem and prove its equivalence to the QAKZ equation. This is a q–analog of the relation between AKZ and QMBP discussed in Section 1.2.4.

Let Φ be a solution of the QAKZ equation with values in End(V) for an $\mathcal{H}_n^t$–module V. We assume that it is invertible for sufficiently general v. Setting $\sigma_i' = \sigma_{s_i}'$, we obtain from (1.3.18) and (1.3.9):

$$F_i(v_i - v_{i+1})\Phi = \sigma_i'(\Phi),$$

$$T_i\Phi = \left(t^{1/2}\sigma_i' + \frac{t^{1/2} - t^{-1/2}}{e^{v_i - v_{i+1}} - 1}(\sigma_i' - 1)\right)\Phi.$$

Let us introduce the operator $\hat{T}_i{}'$ $(1 \le i \le n)$ by

$$\hat{T}_i{}' = t^{1/2}\sigma_i' + \frac{t^{1/2} - t^{-1/2}}{e^{v_i - v_{i+1}} - 1}(\sigma_i' - 1). \tag{1.3.23}$$

Then $\hat{T}_i{}'\Phi = T_i\Phi$ and $\sigma_\pi'\Phi = P\Phi$ (see (1.3.19)). The operators $\hat{T}_i{}'$ and σ_π' commute with the left multiplication by T_j, P, and any elements from $\mathcal{H}_n^t$. Using all these:

$$\begin{aligned}
Y_i\Phi &= T_{i-1}^{-1}\cdots T_1^{-1}PT_{n-1}\cdots T_{i+1}T_i\Phi \\
&= T_{i-1}^{-1}\cdots T_1^{-1}PT_{n-1}\cdots T_{i+1}\hat{T}_i'\Phi \\
&= \hat{T}_i'T_{i-1}^{-1}\cdots T_1^{-1}PT_{n-1}\cdots T_{i+1}\Phi \\
&\cdots \\
&= \hat{T}_i'\cdots \hat{T}_{n-1}'\sigma_\pi'(\hat{T}_1')^{-1}\cdots(\hat{T}_{i-1}')^{-1}\Phi.
\end{aligned}$$

We come to the following definition:

$$\Delta_i' \stackrel{\mathbf{def}}{=\!=} \hat{T}_i'\cdots \hat{T}_{n-1}'\sigma_\pi'(\hat{T}_1')^{-1}\cdots(\hat{T}_{i-1}')^{-1},\ 1 \le i \le n. \tag{1.3.24}$$

Since $Y_i\Phi = \Delta_i'\Phi$ and the elements Y_i commute with each other, we conclude that

$$[\Delta_i', \Delta_j'] = 0.$$

Recall that the operators Δ_i' act in End(V)–valued functions. If we understand them formally, then their commutativity follows algebraically from the relations

$$\sigma_i'v_i = v_{i+1}\sigma_i', \tag{1.3.25}$$
$$\sigma_i'\sigma_{\gamma_i} = \sigma_{\gamma_{i+1}}\sigma_i', \tag{1.3.26}$$
$$\sigma_{\gamma_i}' = \sigma_{\gamma_i}. \tag{1.3.27}$$

The last equality means that $\mathcal{T}_{\gamma_i} = 1$.

Let Q be a polynomial in n variables. Then

$$Q(Y_1, \ldots, Y_n)\Phi = Q(\Delta_1', \ldots, \Delta_n')\Phi$$

and we can represent

$$Q(\Delta'_1, \ldots, \Delta'_n) = \sum_{w \in \mathbb{S}_n} D_w^{(Q)} \sigma'_w, \tag{1.3.28}$$

where the $D_w^{(Q)}$ are pure difference operators, which do not contain σ'_w ($w \in \mathbb{S}_n$).

For symmetric Q, we introduce a difference operator M_Q of Macdonald type by

$$M_Q = \sum_{w \in \mathbb{S}_n} D_w^{(Q)}.$$

Let φ be a $\mathbb{C}$–valued function on $\mathbb{C}^n$. The system

$$M_Q \varphi = Q(\lambda_1, \ldots, \lambda_n)\varphi. \tag{1.3.29}$$

will be called the **Macdonald eigenvalue problem**. The **Macdonald operators** M_Q can be calculated for σ instead of σ'. Similar to the differential case, the result is the same.

Given $\lambda = (\lambda_1, \ldots, \lambda_n) \in \mathbb{C}^n$, we need a (left) $\mathcal{H}_n^t$–module V_λ with the following properties:

1. For any symmetric polynomial Q in n variables and all $a \in V_\lambda$,

$$Q(Y_1, \ldots, Y_n)a = Q(\lambda_1, \ldots, \lambda_n)a.$$

2. There exists a $\mathbb{C}$–linear map $\mathrm{tr} : V_\lambda \to \mathbb{C}$ such that

$$\mathrm{tr}((T_i - t^{1/2})a) = 0 \text{ for all } i \text{ and } a \in V_\lambda.$$

As in the previous sections, we fix a local invertible solution $\Phi(v)$ of the QAKZ equation with values in $\mathrm{End}(V_\lambda)$. Note that all V_λ–valued solutions $\varphi(v)$ of the QAKZ equation can be written in the form $\varphi(v) = \Phi(v)a(v)$ for a $P^\vee$–periodic V_λ–valued function $a(v)$. The periodicity means that

$$a(\ldots, v_i + h, \ldots) = a(v) \text{ for } i = 1, \ldots, n.$$

Theorem 1.3.8. *Let V_λ be an $\mathcal{H}_n^t$–module with the above properties. We denote the space of solutions of the QAKZ equation with values in V_λ by $Sol_{QAKZ}(V_\lambda)$, and the space of solutions of the Macdonald eigenvalue problem (1.3.29) by $Sol_{Mac}(\lambda)$. Then*

$$\mathrm{tr} : Sol_{QAKZ}(V_\lambda) \to Sol_{Mac}(\lambda).$$

Proof. Let $\varphi(v) = \Phi(v)a \in Sol_{QAKZ}(V_\lambda)$. Then

$$(\sigma_i' - 1)\Phi = \left(t^{1/2} + \frac{t^{1/2} - t^{-1/2}}{e^{v_i - v_{i+1}} - 1}\right)^{-1} (T_i - t^{1/2})\Phi.$$

For a reduced decomposition $w = s_{i_1} \cdots s_{i_l}$ of $w \in \mathbb{S}_n$,

$$\begin{aligned}\sigma_w' - 1 &= \sigma_{i_l}' \cdots \sigma_{i_1}' - 1 \\ &= \sigma_{s_{i_l}}' \cdots \sigma_{i_2}'(\sigma_{i_1}' - 1) + \sigma_{i_l}' \cdots \sigma_{i_2}' - 1 \\ &\cdots \\ &= \sum_{k=1}^{l} \sigma_{i_l}' \cdots \sigma_{i_{k+1}}'(\sigma_{i_k}' - 1).\end{aligned}$$

Since σ_i' commutes with the left action of $\{T\}$, we have

$$\begin{aligned}(\sigma_w' - 1)\Phi &= \sum_{k=1}^{l} \sigma_{i_l}' \cdots \sigma_{i_{k+1}}'(\sigma_{i_k}' - 1)\Phi \\ &= \sum_{k=1}^{l} \sigma_{i_l}' \cdots \sigma_{i_{k+1}}'(\text{a scalar function})(T_{i_k} - t^{1/2})\Phi \\ &= \sum_{k=1}^{l} (\text{a scalar function})(T_{i_k} - t^{1/2})\sigma_{i_l}' \cdots \sigma_{i_{k+1}}'\Phi.\end{aligned}$$

Using the commutativity of $D_w^{(Q)}$ with $T_i - t^{1/2}$, we represent $D_w^{(Q)}(\sigma_w' - 1)\Phi$ in the form of a sum $\sum (T_i - t^{1/2})\Psi_i$ for some $\mathcal{H}_n^t$–valued functions Ψ_i. Finally

$$\begin{aligned}Q(\lambda_1, \ldots, \lambda_n)\Phi &= Q(\Delta_1', \ldots, \Delta_n')\Phi \\ &= \sum_{w \in \mathbb{S}_n} D_w^{(Q)} \sigma_w' \Phi \\ &= \sum_{w \in \mathbb{S}_n} D_w^{(Q)}\Phi + \sum_{w \in \mathbb{S}_n} D_w^{(Q)}(\sigma_w' - 1)\Phi \\ &= M_Q\Phi + \sum (T_i - t^{1/2})\Psi_i.\end{aligned}$$

Applying this relation to $a \in V_\lambda$ and taking tr, we conclude:

$$Q(\lambda_1, \ldots, \lambda_n)\,\mathrm{tr}(\varphi) = M_Q\,\mathrm{tr}(\varphi).$$

□

Let us now consider $V_\lambda = J_\lambda^\circ$. The definition is quite similar to the differential case. We start with

$$J_\lambda = \mathrm{Ind}_{H_n^t}^{\mathcal{H}_n^t}(+)/L_\lambda.$$

Here $H_n^t = \langle T_1, \ldots, T_{n-1}\rangle \subset \mathcal{H}_n^t$, $+ : H_n^t \longrightarrow \mathbb{C}$ is the one-dimensional representation sending T_i to $t^{1/2}$, and L_λ is the ideal generated by $p(Y_1, \ldots, Y_n) - p(\lambda)$ $(p \in \mathbb{C}[x_1, \ldots, x_n]^{\mathbb{S}_n})$. As in Section 1.2.1, J_λ° stands for the dual module defined via the anti-involution $\circ$ of $\mathcal{H}_n^t$:

$$Y_i^\circ = Y_i, \qquad T_i^\circ = T_i.$$

The main result of this subsection is the following theorem from [C13, C15].

Theorem 1.3.9. *If $V_\lambda = J_\lambda^\circ$, then the map from $Sol_{QAKZ}(V_\lambda)$ to $Sol_{Mac}(\lambda)$ is injective.*

The theorem results from the following two lemmas.

Lemma 1.3.10. *Let K be an $\mathcal{H}_n^t$–submodule of J_λ°. Then $\mathrm{tr}(K) = 0$ implies $K = 0$.*

The proof repeats that for the differential case. See (1.2.38). ❑

Lemma 1.3.11. *Let φ be a V_λ–valued solution of the QAKZ equation. Assume $\mathrm{tr}(\varphi) = 0$. Then $\mathrm{tr}(\mathcal{H}_n^t \varphi) = 0$.*

Proof. First

$$\mathrm{tr}(T_i\varphi) = t^{1/2}\,\mathrm{tr}(\varphi) = 0$$

for all i. Then $\pi = s_1 s_2 \cdots s_{n-1}\gamma_n$,

$$\begin{aligned}
\sigma'_\pi - 1 &= \sigma_{\gamma_n}\sigma'_{n-1}\cdots\sigma'_1 - 1 \\
&= \sigma_{\gamma_n}\left(\sum_{k=1}^{n-1}\sigma'_{n-1}\cdots\sigma'_{k+1}(\sigma'_k - 1)\right) + \sigma_{\gamma_n} - 1,
\end{aligned}$$

and $\mathrm{tr}(\sigma_{\gamma_n}\varphi) = \mathrm{tr}(\varphi) = 0$. Therefore, representing $\varphi = \Phi a$ $(a \in V_\lambda)$, we have

$$\begin{aligned}
\mathrm{tr}(P\varphi) - \mathrm{tr}(\varphi) &= \mathrm{tr}\left((\sigma'_\pi - 1)\Phi a\right) \\
&= \mathrm{tr}\left(\sigma_{\gamma_n}(\sum_{k=1}^{n-1}\sigma'_{n-1}\cdots\sigma'_{k+1}(\sigma'_k - 1)\Phi)a\right) \\
&= \sum_{k=1}^{n-1}\mathrm{tr}\left(\sigma_{\gamma_n}\sigma'_{n-1}\cdots\sigma'_{k+1}f_i(v)(T_k - t^{1/2})\Phi a\right) \\
&= \sum_{k=1}^{n-1}\mathrm{tr}\left((T_k - t^{1/2})\sigma_{\gamma_n}\sigma'_{n-1}\cdots\sigma'_{k+1}f_i(v)\Phi a\right) \\
&= 0,
\end{aligned}$$

where the $f_i(v)$ are $\mathbb{C}$–valued functions. Hence $\mathrm{tr}(P\varphi) = 0$ and

$$\begin{aligned}\mathrm{tr}(Y_n\varphi) &= \mathrm{tr}(T_{n-1}^{-1}\cdots T_1^{-1}P\varphi)\\ &= t^{(1-n)/2}\,\mathrm{tr}(P\varphi)\\ &= 0.\end{aligned}$$

Now we shall prove that $\mathrm{tr}(Y_i\varphi) = 0$ for all i by induction. Assume that $\mathrm{tr}(Y_i\varphi) = 0$ for $k+1 \le i \le n$. Since $Y_k = T_{k-1}^{-1}\cdots T_1^{-1}PT_{n-1}\cdots T_k$, it is enough to see that $\mathrm{tr}(PT_{n-1}\cdots T_k\varphi) = 0$. Using that φ is a solution of QAKZ,

$$\begin{aligned}\mathrm{tr}(F_{k-1}^{-1}\cdots F_1^{-1}PF_{n-1}\cdots F_k\varphi) &= \mathrm{tr}(\gamma_k^{-1}\varphi)\\ &= \gamma_k^{-1}\,\mathrm{tr}(\varphi)\\ &= 0.\end{aligned}$$

On the other hand,

$$F_i(v) = c_i(v)(T_i + f_i(v)),$$

where $c_i(v)$ and $f_i(v)$ are some scalar functions. Therefore

$$\begin{aligned}0 = \mathrm{tr}(PF_{n-1}\cdots F_k\varphi) &= \mathrm{tr}(c_{n-1}\cdots c_kP(T_{n-1}+f_{n-1}(v))\cdots(T_k+f_k(v))\varphi)\\ &= \sum_{I=(i_1,\ldots,i_l)} \mathrm{tr}(c_I(v)PT_{i_l}\cdots T_{i_1}\varphi),\end{aligned}$$

where $I = (i_1,\ldots,i_l)$ is a sequence of integers such that $k \le i_1 < i_2 < \cdots < i_l \le n-1$, and $c_I(v)$ are some scalar function.

If $I \ne I_0 = (k, k+1,\ldots,n-1)$, then we have the following possibilities:

1. $i_l \ne n-1$,
2. $i_l = n-1$ and there exists an m $(1 \le m \le l)$ such that $i_j - i_{j-1} = 1$ for any $j = m+1, m+2,\ldots,l$ and $i_m - i_{m-1} > 1$,
3. Otherwise.

Case (1): As $i_l < n-1$, we have

$$\begin{aligned}\mathrm{tr}(PT_{i_l}\cdots T_{i_1}\varphi) &= \mathrm{tr}(T_{i_l+1}\cdots T_{i_1+1}P\varphi)\\ &= t^{l/2}\,\mathrm{tr}(P\varphi)\\ &= 0.\end{aligned}$$

Case (2): Since$[T_i,T_j] = 0$ for $|i-j| > 1$,

$$\begin{aligned}\mathrm{tr}(P(T_{i_l}\cdots T_{i_m})(T_{i_{m-1}}\cdots T_{i_1})\varphi) &= \mathrm{tr}(P(T_{i_{m-1}}\cdots T_{i_1})(T_{i_l}\cdots T_{i_m})\varphi)\\ &= \mathrm{tr}(T_{i_{m-1}+1}\cdots T_{i_1+1}PT_{i_l}\cdots T_{i_m}\varphi)\\ &= t^{(m-1)/2}\,\mathrm{tr}(PT_{i_l}\cdots T_{i_m}\varphi).\end{aligned}$$

By the induction hypothesis, $\mathrm{tr}(PT_{i_l}\cdots T_{i_m}\varphi)=0$. Hence

$$\mathrm{tr}(PT_{i_l}\cdots T_{i_1}\varphi)=0.$$

Case (3): In this case $I=(i_1,\dots,i_l)$ must be of the form $i_l=n-1,\ i_{l-1}=n-2,\dots,i_1=n-l>k$. By induction, $\mathrm{tr}(PT_{i_l}\cdots T_{i_1}\varphi)=0$. So $\mathrm{tr}(Y_i\varphi)=0$ for all i.

Because of the relations between T and Y, it remains to check that $\mathrm{tr}(Y_{i_1}\cdots Y_{i_l}\varphi)=0$ for any l. One can show this by induction on l. ❑

1.3.5 Macdonald's operators

We set

$$\begin{aligned}
\hat{T}_i &= t^{1/2}\sigma_i+\frac{t^{1/2}-t^{-1/2}}{e^{v_i-v_{i+1}}-1}(\sigma_i-1), && (1\le i\le n-1), && (1.3.30)\\
G_{ij} &= t^{1/2}+\frac{t^{1/2}-t^{-1/2}}{e^{v_i-v_j}-1}(1-\sigma_{ij}), && (1\le i,j\le n), && (1.3.31)\\
\Delta_i &= \hat{T}_i\cdots\hat{T}_{n-1}\sigma_\pi\hat{T}_1^{-1}\cdots\hat{T}_{i-1}^{-1}. &&&& (1.3.32)
\end{aligned}$$

Here the σ_w are from (1.3.20), $\sigma_{ij}=\sigma_{s_{ij}}$.

Switching from $\{T\}$ to $\{G\}$:

$$\begin{aligned}
\hat{T}_i\sigma_i &= G_{ii+1},\\
G_{ij}^{-1} &= t^{-1/2}-\frac{t^{1/2}-t^{-1/2}}{e^{v_i-v_j}-1}(1-\sigma_{ij}),\\
\Delta_i &= G_{ii+1}\cdots G_{in}\sigma_{\gamma_i}G_{1i}^{-1}\cdots G_{i-1i}^{-1}.
\end{aligned}$$

Let e_m be the m–th elementary symmetric polynomial in n variables. We represent

$$e_m(\Delta_1,\dots,\Delta_n)=\sum_{w\in\mathbb{S}_n}D_w^{(m)}\sigma_w, \tag{1.3.33}$$

for difference operators $D_w^{(m)}$ and define

$$M_m=M_{e_m}=\sum_{w\in\mathbb{S}_n}D_w^{(m)}.$$

All these operators are W–invariant, which results from the following lemmas.

Lemma 1.3.12. *Consider the algebra $\hat{\mathcal{H}}$ generated by $\hat{T}_i$ $(1\le i\le n-1)$, Δ_j $(1\le j\le n)$. Then $T_i\mapsto\hat{T}_i$, $Y_j\mapsto\Delta_j$ extends to an algebra isomorphism $\mathcal{H}_n^t\xrightarrow{\sim}\hat{\mathcal{H}}$. Moreover, if Q is a symmetric polynomial in n variables, then $Q(\Delta_1,\dots,\Delta_n)$ is a central element in $\hat{\mathcal{H}}$.*

Actually, this observation is the key point (it can be checked directly or with some representation theory). We note that the formulas for T generalize the so-called Demazure operations and the Bernstein–Gelfand–Gelfand operations. They were also used by Lusztig and in a paper by Kostant–Kumar.

From now on, we identify $\mathcal{H}_n^t$ and its image $\hat{\mathcal{H}}$.

Lemma 1.3.13. *Let $f(v_1, \dots, v_n)$ be a function on $\mathbb{C}^n$. Then f is symmetric if and only if $(\hat{T}_i - t)f = 0$ for all i.* ❑

Lemma 1.3.14. *Let Q be a symmetric polynomial in n variables. Then $Q(\Delta_1, \dots, \Delta_n)$ acts on the space of the symmetric polynomials in e^{v_i} $(1 \leq i \leq n)$.*

Proof. This follows immediately from Lemma 1.3.12 and Lemma 1.3.13. ❑

Let us calculate M_1. Since M_1 is symmetric, it suffices to find the coefficient of σ_{γ_1}. Using the G–representation, it is easy to see that σ_{γ_1} does not appear in $\Delta_2, \dots, \Delta_n$. The σ_{γ_1}–factor of Δ_1 equals $\prod_{i=2}^n \frac{t^{1/2}e^{v_1-v_i}-t^{-1/2}}{e^{v_1-v_i}-1}\sigma_{\gamma_1}$. Upon the symmetrization:

$$M_1 = \sum_{i=1}^n \prod_{j\neq i} \frac{t^{1/2}e^{v_i} - t^{-1/2}e^{v_j}}{e^{v_i} - e^{v_j}} \sigma_{\gamma_i}.$$

Similarly,

$$M_m = \sum_{I=(i_1,\dots,i_m)} \prod_{i\in I\ j\notin I} \frac{t^{1/2}e^{v_i} - t^{-1/2}e^{v_j}}{e^{v_i} - e^{v_j}} \sigma_{\gamma_{i_1}} \cdots \sigma_{\gamma_{i_m}},$$

where $I = (i_1, \dots, i_m)$ is a sequence of integers such that $1 \leq i_1 < \cdots < i_m < n$.

To recapitulate, let us consider the classical limit of the Macdonald operators. Setting $q = e^h$ and $t = q^k$, $h \to 0$, we have

$$\begin{aligned} \Delta_i &= 1 + h\mathcal{D}_i + O(h^2), \\ M_1 - n &= h \sum \frac{\partial}{\partial v_i} + O(h^2), \end{aligned}$$

$$M_2 - (n-1)M_1 + \frac{n(n-1)}{2} = h^2 L_2 + O(h^3).$$

Comment. Take a solution $\Phi = \Phi(v)$ of the QAKZ equation in an $\mathcal{H}^t$–module V, assuming that Φ has the *trivial monodromy*. Then, for any polynomial $p \in \mathbb{C}[x_1, \dots, x_n]$, we have

$$p(Y_1, \dots, Y_n)\Phi = p(\Delta_1, \dots, \Delta_n)\Phi, \qquad (1.3.34)$$

where Δ_i are the difference Dunkl operators defined previously. Note that Δ'_i can be replaced by Δ_i because the monodromy of Φ is trivial. We also need a linear functional $\mathrm{pr}: V_\lambda \to \mathbb{C}$ for a vector $\lambda = (\lambda_1, \dots, \lambda_n) \in \mathbb{C}^n$ such that

$$\mathrm{pr}(Y_i b) = \lambda_i \, \mathrm{pr}(b) \quad (i = 1, \dots, n) \tag{1.3.35}$$

for any $b \in V$. Given any element $a \in V$, let us define a scalar-valued function $\varphi = \varphi(v)$ by setting

$$\varphi(v) = \mathrm{pr}(\Phi(v)a) \in \mathbb{C}. \tag{1.3.36}$$

Then formula (1.3.34) implies

$$p(\lambda_1, \dots, \lambda_n)\varphi = p(\Delta_1, \dots, \Delta_n)\varphi. \tag{1.3.37}$$

Thus, the scalar-valued function $\varphi = \varphi(v)$ solves the **difference Dunkl eigenvalue problem.**

1.3.6 Arbitrary root systems

Let $\Sigma = \{\alpha\} \in \mathbb{R}^n$ be a reduced root system of rank n (of type A, B, C, D, E, F, or G) and

$$\mathcal{H}^t = \langle T_1, \dots, T_n, X_1, \dots, X_n \rangle \tag{1.3.38}$$

be the corresponding affine Hecke algebra. The **baxterization** (a parametric deformation satisfying the Yang–Baxter relations) of T_i will be given by

$$F_i = T_i + \frac{t^{1/2} - t^{-1/2}}{e^{u_i} - 1} \quad \text{with} \quad u_i = (u, \alpha_i) \tag{1.3.39}$$

for each $i = 1, \dots, n$. We also have to use the element

$$T_0 = X_{\theta^\vee} T_\theta^{-1} \tag{1.3.40}$$

corresponding to the simple affine root $\alpha_0 = \delta - \theta$ for θ being the highest root. Its baxterization is quite similar:

$$F_0 = T_0 + \frac{t^{1/2} - t^{-1/2}}{e^{h - u_\theta} - 1}, \tag{1.3.41}$$

where $u_\theta = (u, \theta)$. The functions $F_0, F_1, \dots, F_n$ satisfy the Yang–Baxter equations associated with the extended Dynkin diagram. For example, in the case of $\circ^1 \Rightarrow \circ^2$, we have

$$F_1(v)F_2(u+v)F_1(2u+v)F_2(u) = F_2(u)F_1(2u+v)F_2(u+v)F_1(v). \tag{1.3.42}$$

The arguments of F_i have a nice geometric interpretation as the angles in the pictures of two lines with the reflection in the x-axis. Such a plane geometric interpretation holds for arbitrary classical root systems, including the $C^\vee C$. See [C12].

Using T_0, the affine Hecke algebra $\mathcal{H}^t$ has an alternative presentation:

$$\mathcal{H}^t = \langle T_0, T_1, \ldots, T_n; \Pi \rangle, \tag{1.3.43}$$

where Π is a certain finite abelian group. The group Π is isomorphic to $P^\vee/Q^\vee$. It is the set of all elements of the extended affine Weyl group

$$\widehat{W} = W \ltimes P^\vee, \quad P^\vee = \bigoplus_{i=1}^{n} \mathbb{Z}\, b_i, \tag{1.3.44}$$

preserving the set $\{\alpha_0, \alpha_1, \ldots, \alpha_n\}$ of the simple affine roots. It gives the embedding of Π into the automorphism group of the extended Dynkin diagram. The action of $\widehat{W}$ on $\mathbb{R}^n \bigoplus \mathbb{R}\delta$ is by the affine reflections and the corresponding shifts in the δ–direction for $P^\vee$:

$$b(z + \zeta\delta) = z + (\zeta - (b, z))\delta.$$

Lemma 1.3.15. *One has:* $\widehat{W} = \langle s_0, s_1, \ldots, s_n; \Pi \rangle$ *for* $s_0 = (\theta^\vee) \cdot s_\theta$. ❑

The group Π can be embedded into the affine Hecke algebra. The images P_π of the elements $\pi \in \Pi$ permute $\{T_i\}$ in the same way as π permute $\{s_i\}$. The baxterization of the elements in Π is trivial: $F_\pi = P_\pi$ for each $\pi \in \Pi$.

Keeping the notations of the previous sections, we have the following theorem.

Theorem 1.3.16. *Given any* $\mathcal{H}^t$*–valued function* $\Psi = \Psi(u)$*, the formulas*

$$\widetilde{s}_i(\Psi) = {}^{s_i}(F_i \Psi) \tag{1.3.45}$$

for all $i = 0, 1, \ldots, n$ *and*

$$\widetilde{\pi}(\Psi) = P_\pi\, {}^{\pi}\Psi \tag{1.3.46}$$

for all $\pi \in \Pi$ *induce a representation of* $\widehat{W}$*.* ❑

The QAKZ equation for Σ is the invariance condition $\widetilde{b}(\Phi) = \Phi$ for all $b \in P^\vee$. It can be shown that this equation is equivalent to the difference QMBP associated with the root system Σ defined via similar Dunkl operators. A conceptual proof of this isomorphism theorem is given by means of the intertwiners of double affine Hecke algebras (see [C13, C15]).

1.4 DAHA and Macdonald polynomials

1.4.1 Rogers' polynomials

We will apply the Hecke algebra technique to the Macdonald polynomials. We will concentrate on the duality and the recurrence relations. The key notion will be the double affine Hecke algebra. Let us start with A_1.

The corresponding L–operator in the differential case reads as follows:

$$L^{(k)} = \frac{\partial^2}{\partial u^2} + 2k\frac{e^u + e^{-u}}{e^u - e^{-u}}\frac{\partial}{\partial u} + k^2, \tag{1.4.1}$$

where k is a complex parameter. There are two special values of k when the operator $L^{(k)}$ is very simple. For $k = 0$ we have $L^{(0)} = \partial^2/\partial u^2$. When $k = 1$,

$$L^{(1)} = d^{-1}\frac{\partial^2}{\partial u^2}d, \quad \text{with} \quad d = e^u - e^{-u}. \tag{1.4.2}$$

Similarly, we can conjugate by d^k for any k:

$$d^k L^{(k)} d^{-k} = \frac{\partial^2}{\partial u^2} - \frac{4k(k-1)}{(e^u - e^{-u})^2}. \tag{1.4.3}$$

Sometimes the latter operator is more convenient to deal with than the original L.

Let us now consider the eigenvalue problem for the operator $L^{(k)}$:

$$L^{(k)}\varphi = \lambda^2\varphi. \tag{1.4.4}$$

If $k = 1$, there is an immediate solution:

$$\varphi(u;\lambda) = \frac{\sinh(u\lambda)}{\sinh(u)\sinh(\lambda)}. \tag{1.4.5}$$

In this normalization, this solution is symmetric with respect to u and λ. Without the $\sinh(\lambda)$ in the denominator, it extends the characters of finite dimensional representations of $SL_2(\mathbb{C})$.

If $k = 1/2$, this operator is the **radial part** of the Casimir operator for the symmetric space $SL_2(\mathbb{R})/SO_2(\mathbb{R})$. Namely, it is the restriction of the Casimir operator C to the double coset space $SO_2\backslash SL_2/SO_2$, which is identified with a domain in $\mathbb{R}^*/S_2$. If $k = 1, 2$, then $L^{(k)}$ corresponds to $SL_2(\mathbb{C})/SU(2)$ and $SL_2(\mathcal{K})/SU_2(\mathcal{K})$ for the quaternions $\mathcal{K}$.

For any k, there is a family of even ($u \to -u$) solutions of (1.4.4) in the form

$$p_n = e^{nu} + e^{-nu} + \text{lower integral exponents}. \tag{1.4.6}$$

The eigenvalues are

$$L^{(k)}p_n = (n+k)^2 p_n \text{ for } n = 0, 1, 2, \ldots\ . \tag{1.4.7}$$

This family of Laurent polynomials satisfies the orthogonality relations

$$\text{Constant Term } (p_n p_m \, d^{2k}) = c_n \delta_{nm}. \tag{1.4.8}$$

They are called the **ultraspherical polynomials**.

We can also consider the rational limit

$$\ell^{(k)} = \frac{\partial^2}{\partial u^2} + \frac{2k}{u}\frac{\partial}{\partial u} \tag{1.4.9}$$

of the operator $L^{(k)}$, switching from the Sutherland model to the Calogero model. The solutions of the rational eigenvalue problem are expressed in terms of the Bessel function. In this case, the solutions can be normalized to ensure the symmetry between the variable and the eigenvalue. In the trigonometric case it is possible only for two special values, $k = 0, 1$. Apart from these special cases, there is no such symmetry, which is a serious demerit of the harmonic analysis on symmetric spaces.

In the difference theory, this symmetry holds for any k and for all root systems, which is expected to help to renew the Harish-Chandra theory. The so-called group case ($k = 1$) is an intersection point of the differential (classical) and the difference (new) theories.

Difference case. We set $x = e^u$ and introduce the "multiplicative difference" Γ acting as $\Gamma(f(x)) = f(q^{1/2}x)$ and satisfying the obvious commutation relation $\Gamma x = q^{1/2}x\Gamma$ with the coordinate.

The Macdonald operator L is as follows:

$$L = \frac{t^{1/2}x - t^{-1/2}x^{-1}}{x - x^{-1}}\Gamma + \frac{t^{1/2}x^{-1} - t^{-1/2}x}{x^{-1} - x}\Gamma^{-1}. \tag{1.4.10}$$

The parameter k in the difference setup is determined from the relation $t = q^k$. When $t = q$ (or $k = 1$), the operator L is simple:

$$L = \frac{1}{x - x^{-1}}(\Gamma + \Gamma^{-1})(x - x^{-1}). \tag{1.4.11}$$

Compare this formula with (1.4.2) in the differential case and notice that (1.4.11) is easier to check than (1.4.2).

The eigenvalue problem

$$L\varphi = (\Lambda + \Lambda^{-1})\varphi \tag{1.4.12}$$

always has a self-dual family of solutions. When $n = 0, 1, 2, \ldots$, there exists a unique family of Laurent polynomials, the so-called q, t–**ultraspherical** or **Rogers** polynomials,

$$p_n = x^n + x^{-n} + \text{lower terms}, \tag{1.4.13}$$

which are symmetric with respect to the transformation $x \to x^{-1}$ and satisfy the equation

$$Lp_n = (t^{1/2}q^{n/2} + t^{-1/2}q^{-n/2})p_n. \tag{1.4.14}$$

The following **duality formula** is proved in the next subsection by using the double affine Hecke algebra.

Theorem 1.4.1 (Duality). *$p_n(t^{1/2}q^{m/2})p_m(t^{1/2}) = p_m(t^{1/2}q^{n/2})p_n(t^{1/2})$ for any $m, n = 0, 1, 2, \ldots$.* ❑

If we set

$$\pi_n(x) = \frac{p_n(x)}{p_n(t^{1/2})}, \tag{1.4.15}$$

the duality can be rewritten as follows:

$$\pi_n(t^{1/2}q^{m/2}) = \pi_m(t^{1/2}q^{n/2}) \quad (m, n = 0, 1, 2, \ldots). \tag{1.4.16}$$

The Rogers polynomials are the Macdonald polynomials of type A_1.

Concerning the general Macdonald polynomials associated with arbitrary root systems, there are three main Macdonald conjectures (see [M4, M5] and the book [M7]):

(1) the scalar product conjecture,

(2) the evaluation conjecture,

(3) the duality conjecture.

One may also add the Pieri rules to the list. These conjectures were justified recently using the double affine Hecke algebras in [C14, C19]. The scalar product conjecture (the norm formula) was established by Opdam in the differential setup in complete generality in [O1]. He introduced the shift operators and deduced the norm formula for any $k \in \mathbb{N}$ from the trivial particular case $k = 0$. The passage to an arbitrary complex k is based on analytic continuation. I extended his approach to the q, t–case. The evaluation and the duality conjectures collapse in the differential case.

1.4.2 A Hecke algebra approach

The duality from Theorem 1.4.1 can be rephrased as the symmetry of a certain scalar product. This product is closely connected with the difference counterpart of the spherical Fourier transform. For any symmetric Laurent polynomials $f, g \in \mathbb{C}[x + x^{-1}]$, we set

$$\{f, g\} = (a(L)g)(t^{1/2}), \tag{1.4.17}$$

where a is a polynomial such that $f(x) = a(x+x^{-1})$. So we apply the operator $a(L)$ to g and then evaluate the result at $x = t^{1/2}$.

Theorem 1.4.2. *$\{f,g\} = \{g,f\}$ for any $f, g \in \mathbb{C}[x + x^{-1}]$.* ❑

Theorem 1.4.1 follows from Theorem 1.4.2. Indeed, if $f = \pi_m$ and $g = \pi_n$, we can compute the scalar product as follows:

$$\begin{aligned} \{\pi_m, \pi_n\} &= (a(L)\pi_n)(t^{1/2}) \\ = a(t^{1/2}q^{n/2} + t^{-1/2}q^{-n/2})\pi_n(t^{1/2}) &= \pi_m(t^{1/2}q^{n/2}). \end{aligned} \tag{1.4.18}$$

Use $L\pi_n = (t^{1/2}q^{n/2}+t^{-1/2}q^{-n/2})\pi_n$ and the normalization $\pi_n(t^{1/2}) = 1$. Hence Theorem 1.4.2 implies $\pi_m(t^{1/2}q^{n/2}) = \pi_n(t^{1/2}q^{m/2})$. In fact, the theorems are equivalent, since the p_n form a basis in the space of all symmetric Laurent polynomials.

Definition 1.4.3. *The* **double affine Hecke algebra** *$\mathcal{H}^{q,t}$ of type A_1 is the quotient*

$$\mathcal{H}^{q,t} = \langle X, Y, T\rangle / \sim, \tag{1.4.19}$$

by the relations for the generators X, Y, T

$$TXT = X^{-1}, \qquad T^{-1}YT^{-1} = Y^{-1}, \tag{1.4.20}$$

$$Y^{-1}X^{-1}YXT^2 = q^{-1/2}, \quad (T - t^{1/2})(T + t^{-1/2}) = 0. \tag{1.4.21}$$

❑

Now we consider q, t as numbers or parameters.

The *first* point of the theory is the following statement of the PBW-type. Any element of $H \in \mathcal{H}$ can be uniquely expressed in the form

$$H = \sum_{i,j\in\mathbb{Z}} c_{i\epsilon j} X^i T^\epsilon Y^j \quad (c_{i\epsilon j} \in \mathbb{C}, \ \epsilon = 0, 1). \tag{1.4.22}$$

The *second* important fact is the symmetry of $\mathcal{H}^{q,t}$ with respect to X and Y.

Theorem 1.4.4. *There exists an* anti-involution *$\phi : \mathcal{H} \to \mathcal{H}$ fixing $q^{1/2}, t^{1/2}$ such that $\phi(X) = Y^{-1}$, $\phi(Y) = X^{-1}$, and $\phi(T) = T$.*

Proof. Indeed, ϕ transposes the first two relations and leaves the third one invariant. ❑

Third, we introduce the **expectation value** $\{H\}_0 \in \mathbb{C}$ of an element $H \in \mathcal{H}^{q,t}$ by

$$\{H\}_0 = \sum_{i,j\in\mathbb{Z}} c_{i\epsilon j} t^{-i/2} t^{\epsilon/2} t^{j/2}, \ \epsilon = 0, 1, \tag{1.4.23}$$

using the expansion (1.4.22). The definitions of ϕ and $\{\ \}_0$ give that

$$\{\phi(H)\}_0 = \{H\}_0 \quad \text{for any } H \in \mathcal{H}^{q,t}. \tag{1.4.24}$$

Fourth, the operator counterpart of the pairing $\{f,g\}$ on $\mathcal{H}^{q,t}\times\mathcal{H}^{q,t}$ is

$$\{A,B\}_0=\{\phi(A)B\}_0 \tag{1.4.25}$$

for any $A,B\in\mathcal{H}^{q,t}$.

The ϕ-invariance of the expectation value (1.4.24) ensures that it is symmetric:

$$\{A,B\}_0=\{B,A\}_0.$$

We also remark that this pairing is non-degenerate for generic q,t.

Theorem 1.4.2 now readily results from the following lemma.

Lemma 1.4.5. *For any symmetric Laurent polynomials* $f(x)$,$g(x)\in\mathbb{C}[x+x^{-1}]$,

$$\{f(X),g(X)\}_0=\{f,g\}.$$

Proof. We need the **polynomial representation** of the double affine Hecke algebra $\mathcal{H}^{q,t}$. We start with the one-dimensional representation of the Hecke algebra $\mathcal{H}_Y=\langle T,Y\rangle$ sending $T\mapsto t^{1/2}$ and $Y\mapsto t^{1/2}$, denoted simply by $+$. The corresponding induced representation of $\mathcal{H}^{q,t}$ is

$$V=\mathrm{Ind}_{\mathcal{H}_Y}^{\mathcal{H}}(+)=\mathcal{H}/\{\mathcal{H}(T-t^{1/2})+\mathcal{H}(Y-t^{1/2})\}\simeq\mathbb{C}[x,x^{-1}]. \tag{1.4.26}$$

The last isomorphism is $x^n\leftrightarrow X^n \mod \{\mathcal{H}(T-t^{1/2})+\mathcal{H}(Y-t^{1/2})\}$. Under the identification of V with the ring $\mathbb{C}[x,x^{-1}]$ of Laurent polynomials, the element X acts on $\mathbb{C}[x,x^{-1}]$ as multiplication by x; T and Y act by the operators

$$\widehat{T}=t^{1/2}s+\frac{t^{1/2}-t^{-1/2}}{x^2-1}(s-1) \quad\text{and}\quad \widehat{Y}=s\Gamma\widehat{T}, \tag{1.4.27}$$

respectively. Here $s(f)(x)=f(x^{-1})$. We use the fact that the equiality $Hf(x)=g(x)$ in V means that $Hf(X)-g(X)$ belongs to $\mathcal{H}(T-t^{1/2})+\mathcal{H}(Y-t^{1/2})$.

The expectation value is the composition

$$\mathcal{H}\xrightarrow{\alpha}V\cong\mathbb{C}[x,x^{-1}]\xrightarrow{\beta}\mathbb{C}, \tag{1.4.28}$$

where α is the residue modulo $\{\mathcal{H}(T-t^{1/2})+\mathcal{H}(Y-t^{1/2})\}$ and $\beta(f)=f(t^{-1/2})$ is the **evaluation map** at $t^{-1/2}$.

Taking any $f,g\in\mathbb{C}[X,X^{-1}]$,

$$\begin{aligned}\{f(X),g(X)\}_0 &= \{\phi(f(X))g(X)\}_0 \\ =\{f(Y^{-1})g(X)\}_0 &= f(\widehat{Y}^{-1})(g)(t^{-1/2}).\end{aligned} \tag{1.4.29}$$

The last equality follows from (1.4.28). If f and g are symmetric and $f(X) = a(X+X^{-1})$, then

$$\{f(X), g(X)\}_0 = a(L)(g)(t^{1/2}) = \{f, g\}, \tag{1.4.30}$$

since the operator $\widehat{Y}+\widehat{Y}^{-1}$ acts on the symmetric Laurent polynomials via L. It is straightforward. The lemma and the duality are established. ❑

This method of proving the duality theorem can be generalized to any root systems.

We now discuss the application of the duality to the **Pieri rules**, which are the recurrence formulas for π_n with respect to the index n. First, we **discretize** functions and operators.

Recall that the π–polynomials $\pi_n(x)$, the renormalized Rogers polynomials, are characterized by the conditions

$$L\pi_n = (t^{1/2}q^{n/2} + t^{-1/2}q^{-n/2})\pi_n, \quad \pi_n(t^{1/2}) = 1, \tag{1.4.31}$$

where

$$L = \frac{t^{1/2}x - t^{-1/2}x^{-1}}{x - x^{-1}}\Gamma + \frac{t^{1/2}x^{-1} - t^{-1/2}x}{x^{-1} - x}\Gamma^{-1}. \tag{1.4.32}$$

As above, $\Gamma x = q^{1/2}x\Gamma$. Denote the set of $\mathbb{C}$ valued functions on $\mathbb{Z}$ by $\mathrm{Funct}(\mathbb{Z},\mathbb{C})$. For any Laurent polynomial $f \in \mathbb{C}[x, x^{-1}]$ or a more general rational function, we define $\widehat{f} \in \mathrm{Funct}(\mathbb{Z},\mathbb{C})$ by setting

$$\widehat{f}(m) = f(t^{1/2}q^{m/2}) \quad \text{for all} \quad m \in \mathbb{Z}. \tag{1.4.33}$$

Considering $\mathcal{A} = \langle \mathbb{C}(x), \Gamma\rangle$ as an abstract algebra with the fundamental relation $\Gamma x = q^{1/2}x\Gamma$, the action of $\mathcal{A}$ on $\varphi \in \mathrm{Funct}(\mathbb{Z},\mathbb{C})$ is as follows:

$$\widehat{x}\varphi(m) = t^{1/2}q^{m/2}\varphi(m), \quad \widehat{\Gamma}\varphi(m) = \varphi(m+1). \tag{1.4.34}$$

The correspondence $f \mapsto \widehat{f}$, whenever it is well defined (the functions f may have denominators), is an $\mathcal{A}$–homomorphism $\mathbb{C}(x) \to \mathrm{Funct}(\mathbb{Z},\mathbb{C})$.

Due to (1.4.31):

$$\widehat{L}\widehat{\pi}_n(m) = (t^{1/2}q^{n/2} + t^{-1/2}q^{-n/2})\widehat{\pi}_n(m). \tag{1.4.35}$$

The Pieri rules result from this equality as follows.

The duality $\widehat{\pi}_n(m) = \widehat{\pi}_m(n)$ implies:

$$\mathcal{L}\widehat{\pi}_m(n) = (t^{1/2}q^{n/2} + t^{-1/2}q^{-n/2})\widehat{\pi}_m(n) = (\widehat{x}+\widehat{x}^{-1})\widehat{\pi}_m(n). \tag{1.4.36}$$

Here $\mathcal{L}$ is $\widehat{L}$ acting on the indices m instead of n. Explicitly,

$$\begin{aligned}&(\widehat{x}+\widehat{x}^{-1})\widehat{\pi}_m(n) \qquad (1.4.37)\\ &= \frac{tq^{m/2} - t^{-1}q^{-m/2}}{t^{1/2}q^{m/2} - t^{-1/2}q^{-m/2}}\widehat{\pi}_{m+1}(n) + \frac{q^{-m/2} - q^{m/2}}{t^{-1/2}q^{-m/2} - t^{1/2}q^{m/2}}\widehat{\pi}_{m-1}(n).\end{aligned}$$

For generic q, t, the mapping $f \to \widehat{f}$ is injective. Hence one can pull (1.4.37) back, removing the hats:

$$\begin{aligned}(x+x^{-1})\pi_m =& \frac{tq^{m/2}-t^{-1}q^{-m/2}}{t^{1/2}q^{m/2}-t^{-1/2}q^{-m/2}}\pi_{m+1}\\ &+\frac{q^{m/2}-q^{-m/2}}{t^{1/2}q^{m/2}-t^{-1/2}q^{-m/2}}\pi_{m-1}.\end{aligned} \tag{1.4.38}$$

This is the **Pieri formula** in the case of A_1. See [AI].

We remark that this formula makes sense when $m = 0$, since the coefficient of π_{m-1} "automatically" vanishes at $m = 0$. Such vanishing conditions are automatically fulfilled for any generalizations of the Pieri rules in the difference theory (but not in the differential theory). See the example of GL_n below.

The Pieri rules obtained above can be used to prove the so-called **evaluation conjecture** describing the values of p_n at $x = t$. Applying (1.4.38) repeatedly, we obtain the formula

$$(x+x^{-1})^\ell\pi_m = c_{\ell,m}\pi_{m+\ell} + \text{lower terms} \tag{1.4.39}$$

for each $\ell = 0, 1, 2, \ldots$. The leading coefficient $c_{\ell,m}$ can be readily calculated:

$$c_{\ell,m} = \prod_{i=0}^{\ell-1}\frac{tq^{(m+i)/2}-t^{-1}q^{-(m+i)/2}}{t^{1/2}q^{(m+i)/2}-t^{-1/2}q^{-(m+i)/2}}. \tag{1.4.40}$$

Let us apply (1.4.39) for $m = 0$:

$$(x+x^{-1})^\ell = c_{\ell,0}\pi_\ell + \text{lower terms}. \tag{1.4.41}$$

Comparing the coefficients of $x^\ell + x^{-\ell}$, we have $1 = c_{\ell,0}/p_\ell(t^{1/2})$, since $p_\ell = x^\ell + x^{-\ell} + \cdots$. Hence

$$p_\ell(t^{1/2}) = c_{\ell,0} = \prod_{i=0}^{\ell-1}\frac{tq^{i/2}-t^{-1}q^{-i/2}}{t^{1/2}q^{i/2}-t^{-1/2}q^{-i/2}}. \tag{1.4.42}$$

This value is easy to calculate directly (the formulas for p_n are known). However, the method descibed in this subsection is applicable to arbitrary root systems. We need only the duality, which is the major advantage of the difference theory versus the differential case.

1.4.3 The *GL*–case

In this subsection, we will generalize the above considerations to the double affine Hecke algebra of GL_n–type. We will focus on the main points and skip the justification of the duality and the Pieri rules because they are very close to the case of A_1.

In the GL_n–case, the Macdonald operators $M_0 = 1, M_1, \ldots, M_n$ are as follows:

$$M_m = \sum_{I=\{i_1<\ldots<i_m\}} \prod_{i\in I, j\notin I} \frac{t^{1/2}x_i - t^{-1/2}x_j}{x_i - x_j} \Gamma_{i_1} \cdots \Gamma_{i_m}. \tag{1.4.43}$$

We set $t = q^k$ (cf. the differential case). For instance, the so-called group case $k = 1$ corresponds to $t = q$.

The **Macdonald polynomials** p_λ for GL_n satisfy the Macdonald eigenvalue problem:

$$M_m p_\lambda = e_m(t^{(n-1)/2}q^{\lambda_1}, \ldots, t^{(-n+1)/2}q^{\lambda_n})p_\lambda \quad (m = 0, 1, \ldots, n), \tag{1.4.44}$$

where $\lambda = (\lambda_1, \ldots, \lambda_n)$ are partitions, i.e., sequences of integers $\lambda_i \in \mathbb{Z}$ such that $\lambda_1 \geq \lambda_2 \geq \ldots \geq \lambda_n \geq 0$. Here e_m is the elementary symmetric function of degree m. Given λ, $p_\lambda = p_\lambda(x)$ is a symmetric polynomial in $x = (x_1, \ldots, x_n)$ of degree $|\lambda| = \sum_{i=1}^n \lambda_i$ in the form

$$p_\lambda(x) = x_1^{\lambda_1} \cdots x_n^{\lambda_n} + \text{lower order terms}. \tag{1.4.45}$$

The lower order terms are understood in the sense of the dominance ordering. Namely, a partition obtained from λ by subtracting simple roots $(0, \ldots, 0, 1, -1, 0, \ldots, 0)$ is lower than λ. For instance,

$$(\lambda_1, \lambda_2, \ldots) > (\lambda_1 - 1, \lambda_2 + 1, \ldots) > (\lambda_1 - 1, \lambda_2, \lambda_3 + 1 \ldots) > \cdots . \tag{1.4.46}$$

We will use the abbreviation

$$t^\rho q^\lambda = (t^{(n-1)/2}q^{\lambda_1}, \ldots, t^{(-n+1)/2}q^{\lambda_n}), \tag{1.4.47}$$

where $2\rho = (n-1, n-3, \ldots, -n+1)$. So $M_m p_\lambda = e_m(t^\rho q^\lambda)p_\lambda$. Using k: $t = q^k$ and $t^\rho q^\lambda = q^{k\rho+\lambda}$.

Given a partition λ, we set

$$\pi_\lambda(x) = \frac{p_\lambda(x)}{p_\lambda(t^\rho)} = \frac{p_\lambda(x_1, \ldots, x_n)}{p_\lambda(t^{(n-1)/2}, t^{(n-3)/2}, \ldots, t^{(-n+1)/2})}. \tag{1.4.48}$$

Theorem 1.4.6 (Duality). *For any partitions λ and μ, we have*

$$\pi_\lambda(t^\rho q^\mu) = \pi_\mu(t^\rho q^\lambda). \tag{1.4.49}$$

This **duality formula** implies the following **Pieri formula**.

Theorem 1.4.7. *Setting* $\epsilon_I = \sum_{i\in I}\epsilon_i$, *where* ϵ_i *are unit vectors,*

$$e_m(x)\pi_\lambda(x) = \sum_{|I|=m}\prod_{i\in I, j\notin I}\frac{t^{j-i+1/2}q^{\lambda_i-\lambda_j} - t^{-1/2}}{t^{j-i}q^{\lambda_i-\lambda_j} - 1}\pi_{\lambda+\epsilon_I}(x). \tag{1.4.50}$$

Here the summation is taken only over subsets $I \subset \{1, 2, \ldots, n\}$ such that $|I| = m$ and $\lambda + \epsilon_I$ *remain partitions* (in the case of arbitrary root systems, *remain dominant*). As a matter of fact, it happens automatically, since the coefficient of $\pi_{\lambda+\epsilon_I}(x)$ on the right vanishes unless $\lambda+\epsilon_I$ is a partition (dominant).

We can determine the value of the Macdonald polynomial $p_\lambda(x)$ at $x = t^\rho$ exactly by the method used in the A_1–case. The corresponding **evaluation formula** was conjectured by Macdonald and proved by Koornwinder. The above theorems (for GL_n) are also due to Macdonald and Koornwinder; [EK3, EK2] contain the approach via the vertex operators, which also leads to the Pieri rules.

For arbitrary roots, the above theorems were established in [C19]. Cf. Chapter 3.

Double Hecke algebra. The double Hecke algebra is the best way to introduce the difference Dunkl operators. The operators M_m are the restrictions of the polynomials e_m in terms of the operators Δ_i from (1.3.32) to the symmetric functions. The operators Δ_i describe the action of the generators Y_i of the $\mathcal{HH}$, which will be defined next, in the induced representation $\mathrm{Ind}_{\mathcal{H}_Y}^{\mathcal{HH}}(+)$, isomorphic to the algebra of Laurent polynomials $\mathbb{C}[X_1^{\pm1}, \ldots, X_n^{\pm1}]$ and called the polynomial representation. So the analogy with (1.4.26) is complete.

The **double affine Hecke algebra** (DAHA) $\mathcal{HH} = \mathcal{HH}^{q,t}$ for GL_n is the algebra generated by the following two commutative algebras of Laurent polynomials in n variables:

$$\mathbb{C}[X_1^{\pm1}, \ldots, X_n^{\pm1}] \quad \text{and} \quad \mathbb{C}[Y_1^{\pm1}, \ldots, Y_n^{\pm1}], \tag{1.4.51}$$

and the **nonaffine Hecke algebra** of type A_{n-1}:

$$\mathcal{H} = \langle T_1, \ldots, T_{n-1}\rangle \tag{1.4.52}$$

with the standard braid and quadratic relations. The other relations are as follows:

$$T_iX_iT_i = X_{i+1}\ (i = 1, \ldots, n-1), \quad T_iX_j = X_jT_i\ (j \neq i, i+1), \tag{1.4.53}$$

$$T_i^{-1}Y_iT_i^{-1} = Y_{i+1}\ (i = 1, \ldots, n-1), \quad T_iY_j = Y_jT_i\ (j \neq i, i+1), \tag{1.4.54}$$

$$Y_2^{-1}X_1Y_2X_1^{-1} = T_1^2, \tag{1.4.55}$$

$$\widetilde{Y}X_j = qX_j\widetilde{Y} \quad \text{and} \quad \widetilde{X}Y_j = q^{-1}Y_j\widetilde{X}. \tag{1.4.56}$$

Here $\widetilde{X} = \prod_{i=1}^{n} X_i$ and $\widetilde{Y} = \prod_{i=1}^{n} Y_i$. They commute with $\{T_1, \ldots, T_{n-1}\}$, thanks to (1.4.53) and (1.4.55). Actually, only one of the relations in (1.4.56) is sufficient. It follows from the $\{X, T, \pi\}$-presentation of $\mathcal{H}$ considered below.

When $q = 1, t = 1$, we come to the **elliptic Weyl group** or 2-extended Weyl group of type GL_n due to Saito [Sai]. If $q = 1$ but there are no quadratic relations, the corresponding group is the **elliptic braid group** (π_1 of the product of n elliptic curves without the diagonals divided by $\mathbb{S}_n$). It was calculated by Birman [Bi] and Scott [Sc].

Establishing the connection with (1.3.32), the formulas $X_i = e^{v_i}$, $T_i = \hat{T}_i$, and $Y_i = \Delta_i$ give the **polynomial representation** of $\mathcal{H}$.

There is another version of this definition, using the element π, introduced by the formula

$$Y_1 = T_1 \cdots T_{n-1}\pi^{-1} \tag{1.4.57}$$

and having the following commutation relations with X_i and T_i:

$$\pi X_i = X_{i+1}\pi \quad (i = 1, \ldots, n-1), \quad \pi X_n = q^{-1}X_1\pi, \tag{1.4.58}$$

and

$$\pi T_i = T_{i+1}\pi \quad (i = 1, \ldots, n-2). \tag{1.4.59}$$

In the polynomial representation, this element coincides with π from Lemma 1.3.4. It acts on the functions $X_i = e^{v_i}$ through the action of π^{-1} on vectors v. Considered formally, π is the image of the element P from Lemma 1.3.1 with respect to the Kazhdan–Lusztig involution, sending $T \to T^{-1}$, $Y \to Y^{-1}$, and $t^{1/2} \to t^{-1/2}$. We note that $\widetilde{Y} = \pi^{-n}$.

Since

$$T_{i-1} \cdots T_1 X_1 T_1 \cdots T_{i-1} = X_i, \quad T_1 \cdots T_{i-1} Y_i T_{i-1} \cdots T_1 = Y_1, \tag{1.4.60}$$

we can reduce the list of generators. Namely,

$$\mathcal{H} = \langle X_1, Y_1, T_1, \ldots, T_{n-1} \rangle \tag{1.4.61}$$

or

$$\mathcal{H} = \langle X_1, \pi, T_1, \ldots, T_{n-1} \rangle. \tag{1.4.62}$$

In terms of $\{T, \pi, X\}$, the list of defining relations of $\mathcal{H}$ is as follows:
(a) $X_iX_j = X_jX_i$ $(1 \le i, j \le n)$, $T_iX_iT_i = X_{i+1}(1 \le i < n)$;
(b) the braid relations and quadratic relations for $T_1, \ldots, T_{n-1}$;
(c) $\pi X_i = X_{i+1}\pi$ $(i = 1, \ldots n-1)$ and $\pi^n X_i = q^{-1}X_i\pi^n$ $(i = 1, \ldots, n)$;
(d) $\pi T_i = T_{i+1}\pi(i = 1, \ldots n-2)$ and $\pi^n T_i = T_i\pi^n$ $(i = 1, \ldots, n-1)$.

For instance, let us deduce (1.4.55) from these formulas. Substituting, the left-hand side equals:

$$\begin{aligned}&(T_1\pi T_{n-1}^{-1}\cdots T_2^{-1})X_1(T_2\cdots T_{n-1}\pi^{-1}T_1^{-1})X_1^{-1}\\ &= T_1\pi T_{n-1}^{-1}\cdots T_2^{-1}(T_2\cdots T_{n-1})X_1\pi^{-1}(T_1^{-1}X_1^{-1}T_1^{-1})T_1\\ &= T_1(\pi X_1\pi^{-1}X_2^{-1})T_1 = T_1^2.\end{aligned}$$

The presentation via π, however, is not convenient from the viewpoint of the symmetry between X_i and Y_i, which will be discussed now. It is better to use $\{Y\}$ instead of π.

Comment. (i) The definition of $\mathcal{H\!H}$ is adjusted to the formulas for Δ_i from (1.3.32). In the next chapters, we will replace the Y_i by their images with respect to the following *anti*-involution of $\mathcal{H\!H}$:

$$T_i \mapsto T_i,\ X_i \mapsto X_i,\ \pi \mapsto \pi^{-1},\ q \mapsto q, t^{1/2} \mapsto t^{1/2}.$$

Under this anti-involution:

$$Y_i = T_i \dots T_{n-1}\pi^{-1}T_1^{-1}\dots T_{i-1}^{-1} \ \mapsto\ T_{i-1}^{-1}\dots T_1^{-1}\pi T_{n-1}\dots T_i.$$

(ii) Concerning the case of SL_n, we set $\pi^n = 1 = \widetilde{Y}$. The corresponding double Hecke algebra $\mathcal{H\!H}_{A_n}$ is a variant of $\mathcal{H\!H}$, where the $\mathbb{C}[X]$ is replaced by the subalgebra of

$$\mathbb{C}[X_1,\dots,X_n,\widetilde{X}^{1/n}]$$

formed by the polynomials of *degree zero*. Also $q^{1/n}$ must be added. The X–generators corresponding to the fundamental weights and are as follows:

$$\widetilde{X}_i \ =\ X_1\cdots X_i\widetilde{X}^{-i/n},\ 1 \le i \le n-1.$$

The anti-dual Y–generators are $\widetilde{Y}_i = Y_1\cdots Y_i$, where the definition of Y_i remains unchanged. They satisfy Theorem 1.4.8 below. The particular case of SL_2, i.e., the case of A_1, was considered above. ❑

The algebra $\mathcal{H\!H}$ contains the following two affine Hecke algebras:

$$\mathcal{H}_X^t = \langle X_1,\dots,X_n,T_1,\dots,T_{n-1}\rangle,$$

$$\mathcal{H}_Y^t = \langle Y_1,\dots,Y_n,T_1,\dots,T_{n-1}\rangle. \tag{1.4.63}$$

They are isomorphic to each other by the correspondence $X_i \leftrightarrow Y_i^{-1}$. This map can be extended to an *anti-involution* of $\mathcal{H\!H}$. It is a general statement that holds for any root system.

Theorem 1.4.8. *There exists an anti-involution* $\phi : \mathcal{H\!H}^{q,t} \to \mathcal{H\!H}^{q,t}$ *such that* $\phi(X_i) = Y_i^{-1}$, $\phi(Y_i) = X_i^{-1}$ *for* $i = 1,\dots,n$, *and* $\phi(T_i) = T_i$ $(i = 1,\dots,n-1)$. *It preserves* q, t.

Proof. We need to check that relation (1.4.55) is self-dual with respect to ϕ. The other relations are obviously ϕ–invariant. One has:

$$\begin{aligned}1 &= T_1^{-2}Y_2^{-1}X_1Y_2X_1^{-1} = T_1^{-1}\{Y_1^{-1}(T_1X_1T_1^{-1})Y_1T_1^{-1}X_1^{-1}\}\\ &= Y_1^{-1}X_2T_1^{-2}Y_1(T_1^{-1}X_1^{-1}T_1^{-1}) = Y_1^{-1}X_2T_1^{-2}Y_1X_2^{-1}.\end{aligned}$$

The latter can be rewritten as $Y_1X_2^{-1}Y_1^{-1}X_2 = T_1^2$, which is the ϕ–image of (1.4.55). ❑

Using this involution we can establish the duality theorem for the Macdonald polynomials in the GL_n–case in the same way as we did in the A_1–case. Generalizing the theory to the case of arbitrary roots, we can prove the Macdonald conjectures and much more. This gives a very convincing example of the power of the modern difference-operator methods.

Some relations. Formula (1.4.55), the affine Hecke relations, and the duality result in a variety of identities of the commutator type. First, for $n > j \geq i \geq 1$,

$$Y_{i+1}^{-1}X_iY_{i+1}X_i^{-1} = T_i^2,\ Y_{j+1}^{-1}X_iY_{j+1}X_i^{-1} = T_j\cdots T_{i+1}T_i^2T_{i+1}^{-1}\cdots T_j^{-1}. \quad (1.4.64)$$

By the duality,

$$X_{j+1}^{-1}Y_iX_{j+1}Y_i^{-1} = T_j\cdots T_{i+1}T_i^{-2}T_{i+1}^{-1}\cdots T_j^{-1}.$$

More generally, we can calculate the group commutators for products of different Y_j and products of different X_i assuming that the Y–indices j and the X–indices i are disjoint. The commutators must be of type (1.4.64), i.e., either in the form $(Y)^{-1}(X)(Y)(X)^{-1}$ or in the form $(X)^{-1}(Y)(X)(Y)^{-1}$. For instance,

$$\begin{aligned}&(Y_{j+m}\cdots Y_{j+1})^{-1}X_i(Y_{j+m}\cdots Y_{j+1})X_i^{-1} \qquad (1.4.65)\\ &=(T_j\cdots T_{j+m-1})\cdots(T_{i+1}\cdots T_{i+m})(T_i\cdots T_{i+m-1})\\ &\quad\times(T_{i+m-1}\cdots T_i)(T_{i+m}^{-1}\cdots T_{i+1}^{-1})(T_{j+m-1}^{-1}\cdots T_j^{-1})\\ &=\{T_i^{-1}\cdots T_{j-1}^{-1}\}\{T_j\cdots T_{j+m-1}T_{j+m-1}\cdots T_j\}\{T_{j-1}\cdots T_i\},\end{aligned}$$

where $j \geq i \geq 1$, $n-j \geq m \geq 1$, and the inverse powers of T appear only as $j > i$.

The following identity contains only positive powers of T:

$$\begin{aligned}&(Y_{j+m}\cdots Y_{j+1})^{-1}(X_j\cdots X_i)(Y_{j+m}\cdots Y_{j+1})(X_j\cdots X_i)^{-1} \qquad (1.4.66)\\ &=(T_j\cdots T_{j+m-1})(T_{j-1}\cdots T_{j+m-2})\cdots(T_i\cdots T_{i+m-1})\\ &\quad\times(T_{i+m-1}\cdots T_i)\cdots(T_{j+m-2}\cdots T_{j-1})(T_{j+m-1}\cdots T_j)\end{aligned}$$

for the same range of i, j, m.

If there is a "gap" in the sequence of indices between the product of the consecutive Y and the consecutive X, we can shift the indices of Y using conjugations with the proper T:

$$
\begin{aligned}
&(Y_{j+m+l}\cdots Y_{j+l+1})^{-1}(X_j\cdots X_i)(Y_{j+m+l}\cdots Y_{j+l+1})(X_j\cdots X_i)^{-1} \qquad (1.4.67)\\
&=\{(T_{j+l}\cdots T_{j+m+l-1})\cdots(T_{j+1}\cdots T_{j+m})\}\\
&\quad\times(T_j\cdots T_{j+m-1})(T_{j-1}\cdots T_{j+m-2})\cdots(T_i\cdots T_{i+m-1})\\
&\quad\times(T_{i+m-1}\cdots T_i)\cdots(T_{j+m-2}\cdots T_{j-1})(T_{j+m-1}\cdots T_j)\\
&\quad\times\{(T_{j+m}^{-1}\cdots T_{j+1}^{-1})\cdots(T_{j+m+1-1}^{-1}\cdots T_{j+l}^{-1})\}.
\end{aligned}
$$

Note that all group commutators of considered type are elements of the **pure braid group** and obviously generate it. This group is formed by the products of $T_i^{\pm 1}$ with trivial images in the $\mathbb{S}_n$.

In the case $j=1, i=1, j+m=n$, we obtain

$$(Y_n\cdots Y_2)^{-1}X_1(Y_n\cdots Y_2)X_1^{-1} \;=\; T_1T_2\cdots T_{n-1}^2\cdots T_2T_1.$$

Therefore we can rewrite the first of the defining relations from (1.4.56) (recall that it results in the second) as follows:

$$X_1Y_1X_1^{-1}Y_1^{-1}=q^{-1}T_1^{-1}\cdots T_{n-1}^{-2}\cdots T_1^{-1}.$$

More generally,

$$X_iY_i(T_{i-1}\cdots T_1^2\cdots T_{i-1})X_i^{-1}Y_i^{-1}=q^{-1}(T_i^{-1}\cdots T_{n-1}^{-2}\cdots T_i^{-1}). \qquad (1.4.68)$$

Denoting the T–products in the left-hand side by L_i and those in the right-hand side by R_i, $L_1=1=R_n$, $L_iL_j=L_jL_i$, and $R_iR_j=R_jR_i$ for all indices. The commutativity is due to [C5]. The mapping $T_i\mapsto T_i(i<n)$, $X_i\mapsto L_i(i\le n)$ determines a surjective homomorphism of $\mathcal{H}_X^t$ onto the nonaffine Hecke subalgebra. So do the R_i.

Combined with (1.4.56), formula (1.4.67) readily gives the formulas for the group commutators of the fundamental X–weights and Y–weights. Namely, let $\widehat{X}_i=X_1\cdots X_i$, $\widehat{Y}_j=Y_1\cdots Y_j$ for $1\le i\le n$. Thanks to the duality one may assume that $j\ge i$. Then

$$
\begin{aligned}
&\widehat{Y}_j\widehat{X}_i\widehat{Y}_j^{-1}\widehat{X}_i^{-1} \qquad (1.4.69)\\
&=q^i(Y_n\cdots Y_{j+1})^{-1}(X_1\cdots X_i)(Y_n\cdots Y_{j+1}))(X_1\cdots X_i)^{-1}\\
&=q^i\{(T_j\cdots T_{n-1})\cdots(T_1\cdots T_{n-j})\}\{(T_{n-j}\cdots T_1)\cdots\\
&\quad\times(T_{n+i-j-1}\cdots T_i)\}\{(T_{n+i-j}^{-1}\cdots T_{i+1}^{-1})\cdots(T_{n-1}^{-1}\cdots T_j^{-1})\}.
\end{aligned}
$$

1.5 Abstract KZ and elliptic QMBP

We will introduce a general r–matrix KZ. It can be reduced to the affine Knizhnik–Zamolodchikov equation, which gives a simple proof of the self-consistency of AKZ. Another application is the double KZ and the elliptic quantum many-body problem for arbitrary root systems, generalizing that due to Olshanetsky–Perelomov in the GL_n–case.

1.5.1 Abstract r–matrices

We keep the notation from the previous sections. Let $\Sigma = \{\alpha\} \subset \mathbb{R}^n$ be a finite root system of rank n of type $A_n, B_n, \ldots, G_2$. We normalize the inner product of $\mathbb{R}^n$ setting $(\theta, \theta) = 2$, where $\theta \in \Sigma$ is the maximal root. Let $\{\alpha_1, \ldots \alpha_n\}$ be the simple roots, $\Sigma_+ = \{\alpha = \sum_{i=1}^n c_i\alpha_i \in \Sigma | \ \forall i,\ c_i \geq 0\}$ the set of positive roots, $\{b_1, \ldots, b_n\} \subset \mathbb{R}^n$ the dual fundamental weights (i.e., $(b_i, \alpha_j) = \delta_{ij}$), $P^\vee$ the lattice spanned by $\{b_1, \ldots, b_n\}$, and W the Weyl group generated by the reflections

$$s_\alpha(u) = u - \frac{2(\alpha, u)}{(\alpha, \alpha)}\alpha, \ u \in \mathbb{R}^n, \alpha \in \Sigma, \ s_i = s_{\alpha_i}.$$

For $u \in \mathbb{R}^n$ and $\alpha \in \Sigma$, $u_\alpha = (\alpha, u) \in \mathbb{R}$ and $u_i = (\alpha_i, u)$. Thus, if $\alpha = \sum_{i=1}^n c_i\alpha_i$, then $u_\alpha = \sum_{i=1}^n c_i u_i$. We will use the derivations:

$$\partial_v(u_\alpha) = (v, \alpha), \ \partial_i(u_\alpha) = (b_i, \alpha), \ i = 1, \ldots n, \ v \in \mathbb{R}^n. \tag{1.5.1}$$

Let us begin with the formal theory of the following partial differential equations:

$$\partial_i \Phi(u) = \left(\sum_{\alpha \in \Sigma_+} \nu_\alpha^i r_\alpha \right) \Phi(u). \tag{1.5.2}$$

Here the values of $\Phi(u)$ are in a vector space V over $\mathbb{C}$, and $r_\alpha = r_\alpha(u_\alpha)$ is a function of one variable u_α with values in $\mathrm{End}(V)$. We set:

$$\nu_\alpha^i \ = \ \mathrm{mult}_{\alpha_i}(\alpha) \ = \ (b_i, \alpha).$$

Equivalently, for $v \in \mathbb{R}^n$,

$$\partial_v \Phi(u) = \left(\sum_{\alpha \in \Sigma_+} (v, \alpha) r_\alpha \right) \Phi(u), \tag{1.5.3}$$

where $v = b_i$ in (1.5.2). Let

$$X_i = \sum_{\alpha \in R_+} \nu_\alpha^i r_\alpha, \ X_v = \sum_{\alpha \in R_+} (v, \alpha) r_\alpha. \tag{1.5.4}$$

The compatibility of system (1.5.2) (the cross-derivative integrability conditions) is equivalent to the purely algebraic commutativity of $\{X_i\}$:
$\Leftrightarrow [X_v, X_{v'}] = 0\ \forall v, v' \in \mathbb{R}^n$.
Indeed, $\partial_i(\nu_\alpha^j r_\alpha) = \partial_j(\nu_\alpha^i r_\alpha)$ for all α. We are going to establish the necessary conditions for the compatibility.

For $\alpha, \beta \in \Sigma$ ($\beta \neq \pm\alpha$), let $\langle\alpha, \beta\rangle$ be the subspace of $\mathbb{R}^n$ spanned by α and β. Then $\Sigma \cap \langle\alpha, \beta\rangle$ is a root system of rank 2.

Proposition 1.5.1. *If the operators*

$$X_v^{(\alpha,\beta)} = \sum(v, \gamma)r_\gamma,\ \gamma \in \Sigma_+, \gamma \in \Sigma \cap \langle\alpha, \beta\rangle\ \ (v \in \langle\alpha, \beta\rangle) \tag{1.5.5}$$

commute for all $v \in \langle\alpha, \beta\rangle \ni v'$, for instance, as $v = \alpha$ and $v' = \beta$, and for arbitrary pairs $\{\alpha,\ \beta\} \subset \Sigma_+$, then $[X_i, X_j] = 0$ ($\forall i, j = 1, \ldots, n$). That is, the commutativity of X for subsystems of rank 2 is sufficient for the total commutativity. ❑

By Proposition 1.5.1, it is sufficient to consider the rank 2 cases (Figure 1.4). One may assume that α, β are the **standard simple roots** from the figure, switching to proper linear combinations if necessary. The commutativity conditions are directly connected with the quantum Yang–Baxter equations from [C2]. The definition of quantum $\{R_\alpha\}$ and their relation to $\{r_\alpha\}$ are as follows.

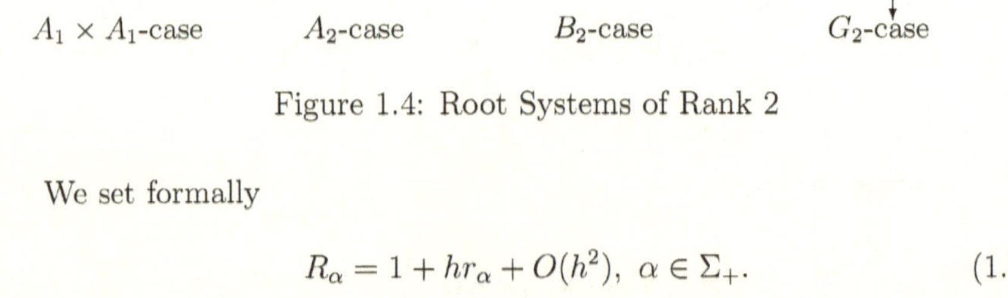

Figure 1.4: Root Systems of Rank 2

We set formally

$$R_\alpha = 1 + hr_\alpha + O(h^2),\ \alpha \in \Sigma_+. \tag{1.5.6}$$

The classical Yang–Baxter equations are the coefficient of h^2 in the quantum Yang–Baxter equations for R:
(0) $A_1 \times A_1$-case: $R_\alpha R_\beta = R_\beta R_\alpha \Rightarrow$

$$[r_\alpha, r_\beta] = 0. \tag{1.5.7}$$

(1) A_2-case: $R_\alpha R_{\alpha+\beta} R_\beta = R_\beta R_{\alpha+\beta} R_\alpha \Rightarrow$

$$[r_\alpha, r_{\alpha+\beta} + r_\beta] + [r_{\alpha+\beta}, r_\beta] = 0. \tag{1.5.8}$$

(2) B_2-case: $R_\alpha R_{\alpha+\beta} R_{\alpha+2\beta} R_\beta = R_\beta R_{\alpha+2\beta} R_{\alpha+\beta} R_\alpha \Rightarrow$

$$[r_\alpha, r_{\alpha+\beta} + r_{\alpha+2\beta} + r_\beta] + [r_{\alpha+\beta}, r_{\alpha+2\beta} + r_\beta] + [r_{\alpha+2\beta}, r_\beta] = 0. \tag{1.5.9}$$

(3) G_2-case: $R_\alpha R_{\alpha+\beta} R_{2\alpha+3\beta} R_{\alpha+2\beta} R_{\alpha+3\beta} R_\beta = R_\beta R_{\alpha+3\beta} R_{\alpha+2\beta} R_{2\alpha+3\beta} R_{\alpha+\beta} R_\alpha$

$$\begin{aligned} \Rightarrow &[r_\alpha, r_{\alpha+\beta} + r_{2\alpha+3\beta} + r_{\alpha+2\beta} + r_{\alpha+3\beta} + r_\beta] + \\ &+ [r_{\alpha+\beta}, r_{2\alpha+3\beta} + r_{\alpha+2\beta} + r_{\alpha+3\beta} + r_\beta] + \\ &+ [r_{2\alpha+3\beta}, r_{\alpha+2\beta} + r_{\alpha+3\beta} + r_\beta] + [r_{\alpha+2\beta}, r_{\alpha+3\beta} + r_\beta] + [r_{\alpha+3\beta}, r_\beta] = 0. \end{aligned} \tag{1.5.10}$$

The structure of the quantum Yang–Baxter equations is very simple: the product of all $R_{\alpha'}$ for positive α' clockwise coincides with the product taken counterclockwise. Respectively, the left-hand sides of the classical equations are the sums of the commutators $[r_{\alpha'}, r_{\beta'}]$, where α', β' are ordered clockwise.

Recall that all the roots above must be *positive* in Σ. Otherwise, the symbols $R_{\alpha'}, r_{\alpha'}$ are not defined.

The classical Yang–Baxter equations are very close to the desired commutativity relations for X_v, but not exactly what we need. Indeed, the coefficients of all $[r_\alpha, r_\beta]$ in these equations are 1, but they are not always 1 when we calculate the commutators of $X_v^{(a,b)}$. The simplest example is B_2, when $[X_\alpha, X_\beta] = 0$ *is not* the classical B_2–YBE from (1.5.9).

This discrepancy disappears in the theory of the quantum KZ. See [C12]. The limiting procedure from QKZ to KZ clarifies in full why we need to add some additional relations to the classical YBE to ensure the commutativity of X. However, it is not difficult to find these extra conditions without any reference to QKZ, as I did in my first paper devoted to the Knizhnik–Zamolodchikov equations [C6].

Proposition 1.5.2. *For any pair $\{\alpha, \beta\} \in \Sigma_+$ such that the entire intersection of Σ and $\langle \alpha, \beta \rangle$ is represented by Figure 1.4, we impose the conditions:*
(0) In the $A_1 \times A_1$–case: (1.5.7).
(1) In the A_2–case: (1.5.8).
(2) In the B_2–case: (1.5.9) and $\{[r_\alpha, r_{\alpha+2\beta}] = 0\}$.
(3) In the G_2–case: (1.5.10), $\{\ \alpha' \perp \beta' \Rightarrow [r_{\alpha'}, r_{\beta'}] = 0,\ \alpha', \beta' \in \langle \alpha, \beta \rangle \cap \Sigma_+\ \}$, and the A_2–relation (1.5.8) for the long roots.
Then the operators X are pairwise commutative. ❑

The proposition gives the conditions *necessary and sufficient* for the X–commutativity, if we consider a bit more special situation. For the same root system Σ, we set

$$x_i = \sum_{\alpha \in \Sigma_+} k_\alpha \nu_\alpha^i r_\alpha, \tag{1.5.11}$$

where $k_\alpha = k_\beta$ if there exists $w \in W$ such that $w(\alpha) = \beta$ (in other words, $k_\alpha = k_\beta$ if $(\alpha, \alpha) = (\beta, \beta)$). We call such k **invariant.**

Note that here we use the lower case x with k instead of the upper case X in the "abstract" theory of r–matrices.

We call a two-dimensional subsystem of Σ with the generators α, β **standard** if it is described by one of the pictures in Figure 1.4. Respectively, α, β are called standard. They can generate a subsystem that is smaller than the complete intersection $\Sigma \cap \langle \alpha, \beta \rangle$.

Proposition 1.5.3. *Let Σ be a root system of rank n. If a pair of positive roots α, β is standard, then we impose the conditions*
(0) $A_1 \times A_1$–case: (1.5.7); (1) A_2–case: (1.5.8);
(2) B_2–case: (1.5.9); (3) G_2–case: (1.5.10).
They are necessary and sufficient for the commutativity of the operators x_i ($i = 1, \ldots, n$) for arbitrary invariant k.

1.5.2 Degenerate DAHA

Sometimes the $\{r_\alpha\}$ exist for *all* roots in Σ, positive and negative, and the relations (0–3) from Proposition 1.5.3 hold for all standard $\alpha, \beta \in \Sigma$. Then we will call such a set, if it exists, an **extension** of $\{r_\alpha, \alpha \in \Sigma_+\}$.

Let us assume that r exists and satisfies the conditions of Proposition 1.5.3 for positive roots only. Then there are two natural extensions of r to all roots. The first is the extension by 0: $r_{-\alpha} = 0$ ($\alpha \in \Sigma_+$). The second is given by $r_{-\alpha} = r_\alpha$ ($\alpha \in \Sigma_+$). The most useful extension is the third one, which will be called the **invariant extension** of a given r–matrix. It requires a W–invariance of $\{r_\alpha, \alpha \in \Sigma_+\}$, which is defined as follows.

We need to suppose that the Weyl group W acts on the algebra $\mathcal{R}$ containing $\{r_\alpha\}$, provided that we have the relations $w(r_\alpha) = r_{w(\alpha)}$ whenever $\{\alpha,\ w(\alpha) \in \Sigma_+,\ w \in W\}$. Such r are called W**–invariant**. Given a negative root $\beta \in \Sigma$, we set $r_\beta = w(r_\alpha)$ for any $w \in W$ and $\alpha \in \Sigma_+$ satisfying $w(\alpha) = \beta$. The definition does not depend on the particular choice of w and α. The group W acts on all $\{r_\alpha |\, \alpha \in \Sigma\}$ by the same formulas as for positive roots. We will call a W–invariant r satisfying the assumption of Proposition 1.5.3 for all roots, not only positive, an **invariant** r**–matrix**.

Proposition 1.5.4. *For an invariant r–matrix, we set*

$$x_b = \sum_{\alpha \in \Sigma_+} k_\alpha (b, \alpha) r_{\alpha(u)}, \tag{1.5.12}$$

where $u \in \mathbb{R}^n$ and $k_\alpha = k_\beta$ if $\exists w \in W$ s.t. $w(\alpha) = \beta$. Then

$$[x_b, x_{b'}] = 0 \qquad (\forall b, b' \in \mathbb{R}^n), \tag{1.5.13}$$

$$s_i(x_b) - x_{s_i(b)} = k_{\alpha_i}(b, \alpha_i)(s_i(r_{\alpha_i}) + r_{\alpha_i}). \tag{1.5.14}$$

Proof. Relation (1.5.13) readily results from Proposition 1.5.3. As to (1.5.14), it follows from the relation $s_i(\Sigma_+) = (\Sigma_+ \setminus \{\alpha_i\}) \cup \{-\alpha_i\}$. ❑

If W is a subgroup of the group of invertible elements of the ring $\mathcal{R}$, and $w(a) = waw^{-1}$, for $w \in W$, $a \in \mathcal{R}$, and, moreover, $s(r_{\alpha_i}) + r_{\alpha_i} = s_i$, then we obtain

$$s_i x_b s_i - x_{s_i(b)} = k_{\alpha_i}(b, \alpha_i) s_i. \tag{1.5.15}$$

So we come back to the definition of the degenerate Hecke algebra $\mathcal{H}'_\Sigma$ from the previous sections. Let us recall it.

Definition 1.5.5. *The* **degenerate Hecke algebra** $\mathcal{H}'_\Sigma$ *for the root system* Σ *is generated by* $\{s_\alpha, x_b\ ,\ \alpha \in R,\ b \in P^\vee\}$ *with the defining relations*

$$[x_b, x_{b'}] = 0 \qquad (\forall b, b' \in P^\vee), \tag{1.5.16}$$

$$s_i x_b - x_{s_i(b)} s_i = k_{\alpha_i}(b, \alpha_i), \tag{1.5.17}$$

and those for the simple generators s_i *of the Weyl group.* ❑

Proposition 1.5.4 looks like a formal manipulation with the definitions. However, it is very useful. Thanks to it, we are able to construct quite a few representations of $\mathcal{H}'_\Sigma$ using various r–matrices. For instance, let us apply this machinery to establish the compatibility and W–invariance of AKZ.

Let us double the set of variables by adding the pairwise commutative elements $\{v_a,\ a \in P^\vee\}$, provided we have that $v_{a+b} = v_a + v_b,\ a, b \in P^\vee$. However, in contrast to the "old" set $\{u_a\}$, they will *not* commute with W. We impose:

$$wv_\alpha = v_{w(\alpha)} w \ (\alpha \in P^\vee,\ w \in W).$$

Setting

$$r^o_\alpha = -(e^{v_\alpha} - 1)^{-1} s_\alpha, \tag{1.5.18}$$

such a r^o satisfies Proposition 1.5.3.

The verification is simple. Here one can also use that this r–matrix is derived from the R–matrix

$$R_\alpha = t^{1/2} + (t^{1/2} - t^{-1/2})(e^{v_\alpha} - 1)^{-1}(1 - s_\alpha), \tag{1.5.19}$$

when $h \to 0$ $(t = 1 + h)$. Namely, r^o is the coefficient of h up to a scalar term, that does not influence the commutators. It is easy to check that this R satisfies the quantum Yang–Baxter equations (0-3) from (1.5.7)–(1.5.10).

The latter claim is due to Lusztig. There is a "conceptual" check of these relations ("without calculations") based on the explicit formulas for the generators $\{T\}$ in the polynomial representation of the affine Hecke algebra induced from the character $\{T_i \mapsto t^{1/2},\ 0 \le i \le n\}$.

Proposition 1.5.6. *The elements*

$$x_b^o = \sum_{\alpha>0} k_\alpha (b,\alpha) r_\alpha^o \quad (b \in P^\vee) \tag{1.5.20}$$

satisfy the relations of the degenerate Hecke algebra $\mathcal{H}'_\Sigma$. The representation $\mathcal{H}'_\Sigma \to End(\mathbb{C}[e^{v_a}, a \in P^\vee])$ sending $w \mapsto w, x_b \mapsto x_b^o$ is faithful. ❑

Now we will add $\{u_\alpha\}$:

$$\hat{r}_\alpha = \hat{r}_\alpha(u_\alpha) = \frac{s_\alpha}{e^{u_\alpha} - 1} + r_\alpha^o. \tag{1.5.21}$$

It is an invariant r–matrix and, moreover, unitary:

$$\hat{r}_\alpha(u_\alpha) + s_\alpha \hat{r}_\alpha(-u_\alpha) s_\alpha = 0. \tag{1.5.22}$$

This r is nothing but the W–extension of the intertwining operators of the degenerate Hecke algebra, where the generators $\{T\}$ are taken in the polynomial representation $\mathbb{C}[e^{v_a}]$.

Proposition 1.5.7. *(i) Consider the following system of partial differential equations:*

$$\partial_b \Phi(u) = \sum_{\alpha>0} k_\alpha (b,\alpha) \hat{r}_\alpha \Phi(u) \tag{1.5.23}$$

for $\hat{r}_\alpha$ from (1.5.21), $\Phi(u) \in \mathbb{C}[e^{v_a}]$, and for invariant $\{k_\alpha\}$. It is self-consistent and W–invariant.

(ii) The AKZ with values in $\mathcal{H}'_\Sigma$:

$$\frac{\partial \Phi}{\partial u_i} = \left(\sum_{\alpha \in \Sigma_+} k_\alpha \nu_\alpha^i \frac{s_\alpha}{e^{u_\alpha} - 1} + x_i \right) \Phi, \qquad 1 \le i \le n, \tag{1.5.24}$$

is self-consistent and W–invariant.

Proof. Here the W–invariance of (1.5.23) follows from (1.5.22). Recall that the system $\partial_b \Phi(u) = A_b \Phi(u)$ $(b \in P^\vee)$ is said to be self-consistent if

$$[\partial_b - A_b, \partial_{b'} - A_{b'}] = 0 \qquad (b, b' \in P^\vee). \tag{1.5.25}$$

The invariance means that if $\Phi(u)$ is a solution, so are the $s_i \Phi(s_i(u))$ for all i. The second claim readily results from the first. Collecting all r^o together and replacing them by $\{x\}$, we obtain a realization of AKZ in the $\mathbb{C}[e^{v_a}]$, which is a faithful module.

1.5.3 Elliptic QMBP

Another application of the general theory of the classical r–matrices is a generalization of the elliptic quantum many-body problem from [OP] to arbitrary root systems. Olshanetsky and Perelomov introduced it for GL_n. Ochiai, Oshima, and Sekiguchi in [OOS] generalized their construction to arbitrary classical root systems. More exactly, they considered the quantum Hamiltonian in the form

$$H = \sum_{i=1}^{n} \partial_{\alpha_i}\partial_i + \sum_{\alpha>0} V(u_\alpha) \tag{1.5.26}$$

and deduced from the existence of the higher conservation laws (differential operators commuting with H) that the potential V has to be the Weierstrass $\wp$–function or its degenerations. Arbitrary root systems were covered in [C17]. We will reproduce here the simplest variant of the construction from this paper.

Let us first recall the construction of the Calogero–Sutherland operators. We keep the notation from the previous section. The root system $\Sigma(\subset \mathbb{R}^n)$ is arbitrary. We will use the same $u_\alpha = (u, \alpha)$ imposing the relation $s_\alpha u_b s_\alpha^{-1} = u_{s_\alpha(b)}$, where $s_\alpha \in W$. So *in this subsection* $\{u_\alpha\}$ *they do not commute with* W.

Proposition 1.5.8. *The* **trigonometric Dunkl operators**

$$\mathcal{D}_b = \partial_b - \sum_{\alpha\in\Sigma_+} k_\alpha(b,\alpha)(e^{u_\alpha}-1)^{-1}s_\alpha \tag{1.5.27}$$

are pairwise commutative for $b \in P^\vee$ and satisfy the relations of the degenerate affine Hecke algebra:

$$s_i\mathcal{D}_b - \mathcal{D}_{s_i(b)}s_i = k_i(b,\alpha_i). \tag{1.5.28}$$

Proof. Here k is invariant: $k_\alpha = k_\beta$ if $(\exists w \in W$ s.t. $w(\alpha) = \beta)$. The statement is obvious, since $r^o_\alpha = -(e^{u_\alpha}-1)^{-1}s_\alpha$ obey Proposition 1.5.3 and the relation

$$r^o_\alpha + s_\alpha r^o_\alpha s_\alpha = s_\alpha. \tag{1.5.29}$$

❑

Recall that we proceeded in Section 1.2.4 as follows. Let $\mathbb{C}[x_1,\ldots,x_n]^W$ be the algebra of W–invariant polynomials. We set

$$\mathcal{L}_p = p(\mathcal{D}_{b_1},\ldots,\mathcal{D}_{b_n}), \tag{1.5.30}$$

where $p \in \mathbb{C}[x_1,\ldots,x_n]^W$. Since $w\mathcal{L}_p w^{-1} = \mathcal{L}_p$, the operators $L_p = \mathcal{L}_p|_{Sym}$ (the *restriction* $|_{Sym}$ is to the symmetric functions) are pairwise commutative and W–invariant. If $p_2 = \sum_{i=1}^n x_{\alpha_i} x_{b_i}$, then

$$L_{p_2} = \sum_{i=1}^{n} \partial_{\alpha_i}\partial_{b_i} + \sum_{\alpha\in\Sigma_+} \frac{k_\alpha(1-k_\alpha)(\alpha,\alpha)}{(e^{\frac{u_\alpha}{2}} - e^{-\frac{u_\alpha}{2}})^2}. \tag{1.5.31}$$

This operator is the Hamiltonian of the Sutherland model.

We are going to "double" this construction, switching to the *affine* root system. Proposition 1.5.3 can be readily extended to affine root systems provided we have the convergence of the products of D_i.

The affine root system is the set

$$\Sigma^a = \{[\alpha,\ j] \in \mathbb{R}^{n+1} |\ \alpha \in \Sigma, j \in \mathbb{Z}\}, \tag{1.5.32}$$
$$\Sigma^a_+ = \{[\alpha,\ j] |\ \alpha \in \Sigma, j \in \mathbb{Z}_{>0}\} \cup \{\alpha = [\alpha,\ 0] |\ \alpha \in \Sigma_+\}. \tag{1.5.33}$$

The **affine Weyl group** $\widetilde{W}$ is generated by the reflections $s_{\tilde{\alpha}}$ ($\tilde{\alpha} = [\alpha, j] \in \Sigma^a$) acting in $\mathbb{R}^{n+1}$:

$$s_{\tilde{\alpha}}([v,\xi]) = [v,\xi] + 2\frac{(\alpha, v)}{(\alpha,\alpha)}\tilde{\alpha}, \tag{1.5.34}$$

where $v \in \mathbb{R}^n$ and $\xi \in \mathbb{R}$.

The **extended affine Weyl group** $\widehat{W}$ is generated by W and the "translations" corresponding to $b \in P^\vee = \oplus_{i=1}^n \mathbb{Z} b_i$:

$$b([v,\xi]) = [v, -(b,v)+\xi], \tag{1.5.35}$$

where $v \times \xi \in \mathbb{R}^{n+1}$. On the space $\mathbb{R}^{n+1}$, W acts preserving ξ. The group $\widehat{W}$ contains $\widetilde{W}$ generated by W and $Q^\vee = \oplus_{i=1}^n \mathbb{Z} a_i$ for $a_i = \alpha_i^\vee = 2\alpha_i/(\alpha,\alpha)$. One then has:

$$\widehat{W} = W \ltimes P^\vee, \quad \widehat{W} = \Pi \ltimes \widetilde{W}, \tag{1.5.36}$$

for the group Π isomorphic to $P^\vee/Q^\vee$.

Concerning the abstract theory of affine r–matrices, one starts with an W–invariant *nonaffine* r–matrix r_α and assumes the following. The group $\widehat{W}$ must act on the algebra $\mathcal{R}$, containing $\{r_\alpha\}$, provided we have the relations $b(r_\alpha) = r_\alpha$ whenever $(b,\alpha) = 0$. This is sufficient to extend r to a $\widehat{W}$–invariant **affine r–matrix**:

$$r_{[\alpha,\xi]} = b(r_\alpha) \text{ for any } b \in P^\vee \text{ such that } (b,\alpha) = -\xi.$$

The relations from Proposition 1.5.3 and the W–invariance are necessary and sufficient to make this construction well defined.

Note that it clarifies in full the *additional relations,* that were necessary in Proposition 1.5.2 to ensure the commutativity of $\{X\}$. Recall that the commutativity of r for orthogonal long roots in the case of B_2 was added there, etc. They are *exactly* the relations which are necessary to ensure the existence of an affine extension. This connection is not by chance and results directly from the difference (quantum) theory.

Infinite Dunkl operators. Once an affine r–matrix is given, we can introduce the operators X and x, that are the special case of the X–operators "with k" (in this section).

They are now given by infinite sums. Formally they are pairwise commutative. However, the *convergence of their products must be provided.* We will manage the commutativity of the infinite Dunkl operators in this subsection and then will consider the W–symmetrizations.

The analysis from [C17] shows that, generally, the symmetrizations are not pairwise commutative because infinite sums are involved. We will consider here only the setting when this problem does not occur. Technically, we use s_α instead of the usual $(s_\alpha - 1)$ in the definition of the Dunkl operators. The general claim is that the symmetrizations of the latter ("elliptic radial parts") commute with each other only at the *critical level*; see [C17].

The **infinite affine Dunkl operators** will be differential operators with the coefficients from $\text{Funct}(\mathbb{C}^n, \mathbb{C}_\infty \widehat{W})$, where $\mathbb{C}_\infty \widehat{W} = \{\sum_{\tilde{w}\in\widehat{W}} c_{\tilde{w}}\tilde{w},\ c_{\tilde{w}} \in \mathbb{C}\}$ consists of infinite sums in contrast to the standard group algebra $\mathbb{C}\widehat{W}$.

We fix a parameter $\eta \in \mathbb{C}$, an arbitrary W–invariant function k_α on Σ, and set

$$\tilde{\alpha} = [\alpha, m] \text{ and } e^{u_{\tilde{\alpha}}} = e^{u_\alpha + m\eta}.$$

Then for each $b \in P^\vee$,

$$\mathcal{D}_b = \partial_b - \sum_{\tilde{\alpha}\in\Sigma_+^a} k_\alpha(b, \alpha)(e^{u_{\tilde{\alpha}}} - 1)^{-1} s_{\tilde{\alpha}}. \tag{1.5.37}$$

Lemma 1.5.9. *(i) If* $\text{Re}\eta > 0$, *then the products of the operators* $\mathcal{D}_c$ *are well defined and can be represented as infinite sums:*

$$\mathcal{D}_{c_1} \ldots \mathcal{D}_{c_k} \overset{\text{def}}{=\!=} \sum_{\tilde{w}\in\widehat{W}} \Psi_{\tilde{w}}(u, \eta)\tilde{w}, \tag{1.5.38}$$

where $\Psi_{\tilde{w}}(u, \eta)$ *are differential operators with the coefficients meromorphic in* $\mathbb{C}^n \ni v$.

(ii) Moreover, the absolute values of the coefficients of $\Psi_{\tilde{w}}(u, \eta)$ *are bounded pointwise (apart from the singularities) by a function* $C(u, \eta)\ \epsilon(u, \eta)^{l(\tilde{w})}$, *where* $0 \le \epsilon(u, \eta) < 1$, $C > 0$, *and* $l(\tilde{w})$ *is the length of* $\tilde{w}$ *in* $\widehat{W}$ *with respect to the generators* $\{s_i,\ 0 \le i \le n\}$.

Proof. See [C17] for the proof and the exact definition of the length. The condition $\text{Re}\eta > 0$ is replaced by a somewhat weaker inequality in [C17], but the claim holds for all positive $\text{Re}\eta$. ❑

Theorem 1.5.10. *If* $\mathrm{Re}\eta > 0$, *then we have the following relations:*

$$[\mathcal{D}_b, \mathcal{D}_{b'}] = 0, \tag{1.5.39}$$
$$s_i \mathcal{D}_b - \mathcal{D}_{s_i(b)} s_i = k_i(b, \alpha_i), \tag{1.5.40}$$
$$s_0 \mathcal{D}_b - \mathcal{D}_{s_\theta(b)} s_0 = -k_\theta(b, \theta), \tag{1.5.41}$$

where θ *is the maximal root of* Σ *and* $s_0 = s_{[-\theta,1]} \in W^a \subset \widehat{W}$. ❑

For $p \in \mathbb{C}[x_1, \ldots, x_n]^W$, we set

$$\mathcal{L}_p = p(\mathcal{D}_{b_1}, \ldots, \mathcal{D}_{b_n}). \tag{1.5.42}$$

Then $\widehat{w}\mathcal{L}_p\widehat{w}^{-1} = \mathcal{L}_p$ $(\widehat{w} \in \widehat{W})$ and the operators $L_p \stackrel{\mathbf{def}}{=\!=} \mathcal{L}_p\|_{Sym}$ are pairwise commutative, i.e., $[L_p, L_{p'}] = 0$ for all W–invariant p, p'.

Here $\|_{Sym}$ is the *affine restriction* to the $\widehat{W}$–invariant functions in contrast to the nonaffine theory. Explicitly, if we have

$$p(\mathcal{D}_{b_1}, \ldots, \mathcal{D}_{b_n}) = \sum_{\widehat{w}} \Psi_{\widehat{w}}(u, \eta)\widehat{w},$$

where $\Psi_{\widehat{w}}(u, \eta)$ is a differential operator (and does not contain the elements from $\widehat{W}$), then we have

$$L_p = \sum_{\widehat{w}} \Psi_{\widehat{w}}(u, \eta). \tag{1.5.43}$$

The convergence readily follows from the lemma. The coefficients of L_p are $\widehat{W}$–equivariant, in particular, $P^\vee$–invariant.

For instance, if we put $p_2 = \sum_{i=1}^n x_{\alpha_i} x_{b_i}$, then

$$L_{p_2} = \sum_{i=1}^n \partial_{\alpha_i}\partial_{b_i} + \sum_{\alpha\in\Sigma_+} (\alpha, \alpha)k_\alpha(k_\alpha - 1)\tilde{\zeta}'(u_\alpha), \tag{1.5.44}$$

where

$$\tilde{\zeta}(z) = \sum_{m=0}^\infty \frac{1}{e^{m\eta+z} - 1} - \sum_{m=1}^\infty \frac{1}{e^{m\eta-z} - 1}, \; \tilde{\zeta}'(z) = \frac{d\tilde{\zeta}(z)}{dt}. \tag{1.5.45}$$

Note that

$$\tilde{\zeta}(z + \eta) = \tilde{\zeta}(z) + 1, \quad \tilde{\zeta}(z) + \tilde{\zeta}(-z) = -1, \tag{1.5.46}$$
$$-\tilde{\zeta}'(z) + c = \wp(z; \Omega) = \frac{1}{z^2} + \sum_{\omega\in\Omega\setminus\{0\}} \left\{\frac{1}{(z-\omega)^2} - \frac{1}{\omega^2}\right\} \tag{1.5.47}$$

for some constant c. Here $\Omega = \{2\pi\sqrt{-1}\mathbb{Z} + \eta\mathbb{Z}\}$ is a lattice in $\mathbb{C}$.

Comment. (i) The above construction gets more interesting when we consider the Dunkl operators with $(s_{\tilde{\alpha}} - 1)$ instead of $s_{\tilde{\alpha}}$. Still they are pairwise commutative. However, relation (1.5.41) is more complicated and the reduction to the L–operators is governed by the double affine Hecke algebra with *zero central charge.* The resulting operators preserve the Looijenga space [Lo] of (formal) theta functions of "critical" level $c = -kg$ for the dual Coxeter number g (assuming that we have one k). See [C17].

(ii) When $k = 1$ (i.e., in the group case), we come to the relation $c + g = 0$, which coincides with the the critical level condition for the Kac–Moody algebras (c is the central charge). I did not check the detail, but it is likely that there may be an explanation based on the duality from [VV1] in the case of GL_n. The appearance of the Kac–Moody critical level condition as the commutativity condition for the elliptic L–operators has no explanation for other root systems at the moment. However, it obviously indicates that double Hecke algebras are on the right track.

1.5.4 Double affine KZ

Continuing the main theme of this chapter, let us establish the relations between the elliptic QMBP and the corresponding KZ equations, to be introduce. One can expect this KZ to be of elliptic type, but the most natural construction leads to a trigonometric KZ with infinite sums.

The commutativity of the parameters u_α with the extended affine Weyl group $\widehat{W}$ and $\mathcal{H}'_0$, the degenerate double affine Hecke algebra of zero level, will be assumed (contrary to the previous subsection). The $\{u_i\}$ will be the arguments of the double affine KZ. We follow [C17].

Definition 1.5.11. *The* 0**–level degenerate double affine Hecke algebra** $\mathcal{H}'_0$ *is generated by pairwise commutative elements* $\{x_b\}$ *and the group* $\widetilde{W}$ *satisfying the "cross-relations"*

$$s_i x_b - x_{s_i(b)} s_i = k_i(b, \alpha_i), \tag{1.5.48}$$
$$s_0 x_b - x_{s_\theta(b)} s_0 = -k_\theta(b, \theta), \tag{1.5.49}$$

where $s_0 = s_{[-\theta,1]} \in W^a$.

Theorem 1.5.12. *The double affine KZ (DAKZ)*

$$\partial_b(\Phi) = \Big(\sum_{\tilde\alpha\in\Sigma^a_+} k_\alpha(e^{u_{\tilde\alpha}} - 1)^{-1} s_{\tilde\alpha} + x_b\Big)\Phi \tag{1.5.50}$$

is $\widehat{W}$*–invariant and self-consistent. Here* $\tilde\alpha = [\alpha, j]$*,* k_α *is an arbitrary* W*–invariant function on* Σ*, and* $e^{u_{\tilde\alpha}} = e^{u_\alpha + j\eta}$ *for a fixed* η*.* ❑

Let us consider (1.5.50) in the representation of $\mathcal{H}'_0$ induced from the character $\{s_i \mapsto 1\}$ of $\widetilde{W}$. It is isomorphic to $\mathbb{C}[x_1, \ldots, x_n]$. We can reduce it further, since the symmetric polynomials $p(x_1, \ldots, x_n)$ ($\in \mathbb{C}[x_1, \ldots, x_n]^W$) are central in $\mathcal{H}'_0$.

We fix $\lambda = (\lambda_1, \ldots, \lambda_n) \in \mathbb{C}^n$ and define AKZ_λ as the AKZ equation considered in the representation J^o_λ dual to $J_\lambda = \mathbb{C}[x_1, \ldots, x_n]/((p(x) - p(\lambda))\ \forall p)$. Here the p are symmetric. By dual, we mean $\mathrm{Hom}(J, \mathbb{C})$ with the action of $\mathcal{H}'_0$ via the anti-involution preserving the generators s_i, x_b. It is similar to the considerations of Section 1.2.4. Because of the definition of J^o_λ, there is a natural $\widetilde{W}$–invariant map $\mathrm{tr} : J^o_\lambda \to \mathbb{C}$ dual to the embedding $\mathbb{C} \to \mathbb{C}[x_1, \ldots, x_n]$.

Theorem 1.5.13. *The map $\Phi \mapsto \phi = tr(\Phi)$ is an isomorphism of the space of all (local) solutions of $DAKZ_\lambda$ and the space of solutions of the eigenvalue problem $L_p\phi = p(\lambda)\phi$ for $p \in \mathbb{C}[x_1, \ldots, x_n]^W$.*

The proof follows the same lines as in the nonaffine case. We note that the double affine KZ can be defined for the central extension $\mathcal{H}'$ (nonzero level) of $\mathcal{H}'_0$ as well as the infinite Dunkl operators. The algebra $\mathcal{H}'$ is much more interesting. However, the $\widehat{W}$–invariant L–operators can be constructed only for $\mathcal{H}'_0$, which eventually leads to the relation generalizing the critical level condition in the Kac–Moody theory (see Chapter 0). We will consider the an interpretation and integration of (trigonometric) KZ equations via the Kac–Moody algebras in the next sections. Now we discuss the application to the Harish-Chandra transform.

1.6 Harish-Chandra inversion

In this section, we use the degenerate double Hecke algebras and Dunkl operators to calculate the images of the operators of multiplication by symmetric Laurent polynomials with respect to the Harish-Chandra transform. These formulas were known only in some cases, although they are directly connected with the important problem of making the convolution on the symmetric spaces as explicit as possible. We first solve a more general problem of calculating the transforms of the coordinates for the nonsymmetric Harish-Chandra transform defined by Opdam; this problem has a complete solution. Then we employ the symmetrization. The resulting formulas readily lead to a new simple proof of the Harish-Chandra inversion theorem (see [HC, He3]) and the corresponding theorem from [O3].

We will assume that $k > 0$ and restrict ourselves to compactly supported functions. In this case, we can borrow the growth estimates from [O3]. Hopefully this approach can be extended to any k and to the elliptic radial parts, which were discussed in the previous section. It is also expected to be helpful for developing the analytic theory of the direct and inverse transforms.

For a maximal real split torus A of a semisimple Lie group G, the **Harish-Chandra transform** is the integration of symmetric (W–invariant) functions in terms of $X \in A$ multiplied by the spherical zonal function $\phi(X, \lambda)$, where $\lambda \in (\mathrm{Lie}A \otimes_{\mathbb{R}} \mathbb{C})^*$. The measure is the restriction of the invariant measure on G to the space of double cosets $K\backslash G/K \subset A/W$ for the maximal compact subgroup $K \subset G$ and the restricted Weyl group W. The function ϕ, the kernel of the transform, is a W–invariant eigenfunction of the radial parts of the G–invariant differential operators on G/K; λ determines the set of eigenvalues. The parameter k is given by the root multiplicities ($k = 1$ in the group case).

There is a generalization to arbitrary k due to Calogero, Sutherland, Koornwinder, Moser, Olshanetsky, Perelomov, Heckman, and Opdam; see the papers [HO1, H1, O1] for a systematic ("trigonometric") theory. In the nonsymmetric variant due to Opdam [O3], the Dunkl-type operators from [C10] replace the radial parts of G–invariant operators and their k–generalizations.

The analytic part of the theory is in extending the direct and inverse transforms to various classes of functions. Not much is known here. Hopefully the difference theory will be more promising analytically.

In the papers [C28, C21], a difference counterpart of the Harish-Chandra transform was introduced, which is also a deformation of the Fourier transform in the p–adic theory of spherical functions. Its kernel (a q generalization of ϕ) is defined as an eigenfunction of the q–difference "radial parts" (Macdonald's "minuscule-weight" operators and their generalizations). There are "algebraic" applications in combinatorics (the Macdonald polynomials), representation theory (for instance, for quantum groups at roots of unity), and mathematical physics (see Chapter 0 and the previous sections). This section is toward analytic applications.

The q–Fourier transform is self-dual, i.e., its kernel is $x \leftrightarrow \lambda$ symmetric for $X = q^x$. This holds in the differential theory only for either the so-called rational degeneration with the tangent space $T_e(G/K)$ instead of G/K (see [He3, Du2, Je]) or for a special $k = 0, 1$. For such a special k, the classical differential and the new difference transforms coincide up to a normalization.

The differential case corresponds to the limit $q \to 1$. At the moment, the analytic methods are not mature enough to manage the limiting procedure in detail, which would give a complete solution of the inversion problem. So we develop the corresponding technique without any reference to the q–Fourier transform and the general double affine Hecke algebra. We use only the theory of the degenerate one, which generalizes Lusztig's graded affine Hecke algebra [Lus1] (the GL_n–case is due to Drinfeld). Mainly we need the intertwining operators from [C23] (see [C13] and [KnS] for GL_n).

The key result of the section is that the Opdam transforms of the operators of multiplication by the coordinates coincide with the operators from [C12] (see (1.6.17) below). Respectively, the Harish-Chandra transform sends the operators of multiplications by W–invariant Laurent polynomials to the *difference* operators from (1.6.20).

It is not very surprising that the images of these important operators haven't been found before. The calculation involves the following ingredients, which are new in the harmonic analysis on symmetric spaces:

(a) the differential Dunkl-type operators from [C10, C11];

(b) the Opdam transform [O3];

(c) the difference Dunkl operators [C12, C13].

1.6.1 Affine Weyl groups

Recall the definition of the affine roots and the extended affine Weyl group. Let $\Sigma = \{\alpha\} \subset \mathbb{R}^n$ be the root system of type $A, B, ..., F, G$, normalized by the standard condition $(\alpha,\alpha) = 2$ for long α; $\alpha_1, ..., \alpha_n$ simple roots; $a_1 = \alpha_1^\vee, ..., a_n = \alpha_n^\vee$ simple coroots, where $\alpha^\vee = 2\alpha/(\alpha,\alpha)$; $b_1, ..., b_n$ the dual fundamental weights determined by the relations $(b_i, \alpha_j) = \delta_i^j$ for the Kronecker delta. We will also use

$$Q^\vee = \oplus_{i=1}^n \mathbb{Z}a_i \subset P^\vee = \oplus_{i=1}^n \mathbb{Z}b_i,\ P_+^\vee = \oplus_{i=1}^n \mathbb{Z}_+ b_i \ \text{ for } \ \mathbb{Z}_+ = \{m \geq 0\}.$$

The vectors $\tilde{\alpha} = [\alpha, j] \in \mathbb{R}^n \times \mathbb{R} = \mathbb{R}^{n+1}$ for $\alpha \in \Sigma, j \in \mathbb{Z}$ form the **affine root system** $\Sigma^a \supset \Sigma$ ($z \in \mathbb{R}^n$ are identified with $[z, 0]$). We add $\alpha_0 \overset{\mathbf{def}}{=\!=} [-\theta, 1]$ to the simple roots for the **maximal root** $\theta \in \Sigma$. The corresponding set Σ_+^a of positive roots coincides with $\Sigma_+ \cup \{[\alpha, j],\ \alpha \in \Sigma,\ j > 0\}$.

The set of indices of the orbit of the zero vertex in the affine Dynkin diagram by its automorphisms will be denoted by O ($O = \{0\}$ for E_8, F_4, G_2). Let $O' = \{r \in O, r \neq 0\}$. We identify the indices with the corresponding simple affine roots.

The elements b_r for $r \in O'$ are the so-called minuscule coweights, namely, $(b_r, \alpha) \leq 1$ for $\alpha \in \Sigma_+$.

Given $\tilde{\alpha} = [\alpha, j] \in \Sigma^a,\ b \in P^\vee$, let

$$s_{\tilde{\alpha}}(\tilde{z}) \ = \ \tilde{z} - (z, \alpha^\vee)\tilde{\alpha}, \quad b'(\tilde{z}) \ = \ [z, \zeta - (z, b)] \tag{1.6.1}$$

for $\tilde{z} = [z, \zeta] \in \mathbb{R}^{n+1}$.

The **affine Weyl group** $\widetilde{W}$ is generated by all $s_{\tilde{\alpha}}$ (simple reflections $s_i = s_{\alpha_i}$ for $0 \leq i \leq n$ are enough). It is the semidirect product $W \ltimes Q^\vee$, where the nonaffine Weyl group W is the span of $s_\alpha, \alpha \in \Sigma_+$. We will identify $b \in P^\vee$ with the corresponding translations. For instance,

$$\alpha^\vee \ = \ s_\alpha s_{[\alpha,1]} \ = \ s_{[-\alpha,1]} s_\alpha \ \text{ for } \ \alpha \in R. \tag{1.6.2}$$

The **extended Weyl group** $\widehat{W}$ generated by W and $P^\vee$ is isomorphic to $W \ltimes P^\vee$:

$$(wb)([z, \zeta]) \ = \ [w(z), \zeta - (z, b)] \ \text{ for } \ w \in W, b \in P^\vee.$$

For $b_+ \in P_+^\vee$, let

$$u_{b_+} = w_0 w_0^+ \in W,\ \pi_{b_+} = b_+(u_{b_+})^{-1} \in \widehat{W},\ u_i = u_{b_i}, \pi_i = \pi_{b_i}, \tag{1.6.3}$$

where w_0 (respectively, w_0^+) is the longest element in W (respectively, in W_{b_+} generated by s_i preserving b_+) relative to the set of generators $\{s_i\}$ for $i > 0$.

The elements $\pi_r \stackrel{\text{def}}{=} \pi_{b_r}, r \in O'$, and $\pi_0 = \text{id}$ leave $\{\alpha_i, i \geq 0\}$ invariant and form a group denoted by Π, which is isomorphic to $P^\vee/Q^\vee$ by the natural projection $b_r \mapsto \pi_r$. As with $\{u_r\}$, they preserve the set $\{-\theta, \alpha_i, i > 0\}$. The relations $\pi_r(\alpha_0) = \alpha_r = (u_r)^{-1}(-\theta)$ distinguish the indices $r \in O'$. Moreover (see, e.g., [C12]):

$$\widehat{W} = \Pi \ltimes \widetilde{W}, \quad \text{where} \quad \pi_r s_i \pi_r^{-1} = s_j \ \text{ if } \ \pi_r(\alpha_i) = \alpha_j,\ 0 \leq i, j \leq n.$$

The **length** $l(\widehat{w})$ of $\widehat{w} = \pi_r \widetilde{w} \in \widehat{W}$ is the length of any of the reduced decompositions of $\widetilde{w} \in \widetilde{W}$ with respect to $\{s_i, 0 \leq i \leq n\}$:

$$\begin{aligned} l(\widehat{w}) = |\lambda(\widehat{w})| \ \text{ for } \ \lambda(\widehat{w}) &\stackrel{\text{def}}{=} \Sigma_+^a \cap \widehat{w}^{-1}(-\Sigma_+^a) \\ &= \{\tilde{\alpha} \in \Sigma_+^a,\ l(\widehat{w} s_{\tilde{\alpha}}) < l(\widehat{w})\}. \end{aligned} \tag{1.6.4}$$

For instance, given $b_+ \in P_+^\vee$,

$$\begin{aligned} \lambda(b_+) &= \{\tilde{\alpha} = [\alpha, j],\ \alpha \in \Sigma_+ \text{ and } (b_+, \alpha) > j \geq 0\}, \\ l(b_+) &= 2(b_+, \rho), \ \text{ where } \ \rho = (1/2) \sum_{\alpha \in \Sigma_+} \alpha. \end{aligned} \tag{1.6.5}$$

We will later need the dominant **affine Weyl chamber**

$$\mathfrak{C}^a \ = \ \{z \in \mathbb{R}^n \ \text{ such that } (z, \alpha_i) > 0 \ \text{ for } \ i > 0,\ (z, \theta) < 1\}. \tag{1.6.6}$$

1.6.2 Degenerate DAHA

Let us fix a collection $k = \{k_\alpha\} \subset \mathbb{C}$ such that $k_{w(\alpha)} = k_\alpha$ for $w \in W$, i.e., k_α depends only on the (α, α). We set $k_i = k_{\alpha_i}$ and $k_0 = k_\theta$. Let $\rho_k = (1/2) \sum_{\alpha \in \Sigma_+} k_\alpha \alpha$. One then has: $(\rho_k, \alpha_i^\vee) = k_i$ for $1 \leq i \leq n$.

It will be convenient in this section to switch to the parameters

$$\kappa_\alpha \stackrel{\text{def}}{=} \nu_\alpha k_\alpha \ \text{ for } \ \nu_\alpha \stackrel{\text{def}}{=} (\alpha, \alpha)/2;\ \kappa_i = k_{\alpha_i},\ \kappa_0 = k_0.$$

For instance, $\rho_k = (1/2) \sum_{\alpha \in \Sigma_+} \kappa_\alpha \alpha^\vee$ and $(\rho_k, \alpha_i) = \kappa_i$ for $1 \leq i \leq n$. Also, notice that we use the generators $\{y\}$ in the definition of the degenerate affine Hecke algebra that satisfy the "cross-relation" with the opposite sign, in contrast to the $\{x\}$ used above.

Definition 1.6.1. *The* **degenerate double affine Hecke algebra** $\mathcal{H}'$ *is formed by the group algebra* $\mathbb{C}\widehat{W}$ *and the pairwise commutative* y_b *for* $b \in P^\vee$ *or in* $\mathbb{R}^n$ *satisfying the following relations:*

$$s_i y_b - y_{s_i(b)} s_i = -\kappa_i (b, \alpha_i^\vee), \ \pi_r y_{\tilde b} \ = \ y_{\pi_r(\tilde b)} \pi_r,$$
$$\textit{for } (b, \alpha_0) = -(b, \theta), \ y_{[b,u]} \stackrel{\mathbf{def}}{=\!=} y_b + u, \ 0 \le i \le n, \ r \in O. \tag{1.6.7}$$

❑

Without s_0 and π_r, these are the defining relations of the graded affine Hecke algebra from [Lus1]. This algebra is a degeneration as $q \to 1$, $t = q^\kappa$ of the general double affine Hecke algebra, which will be studied in the following chapters. Note that relations (1.6.7) are dual to those in terms of x_b in this chapter. See, e.g., Definition 1.5.11. The trigonometric Dunkl operators will be changed correspondingly.

We will need the **intertwiners** of $\mathcal{H}'$:

$$\Psi'_i \ = \ s_i + \frac{\kappa_i}{y_{\alpha_i}}, \ \Psi'_0 \ = \ X_\theta s_\theta + \frac{\kappa_0}{1 - y_\theta},$$
$$P'_r \ = \ X_r u_r^{-1}, \ \text{ for } 1 \le i \le n, \ r \in O'. \tag{1.6.8}$$

They belong to the extension of $\mathcal{H}'$ by the field $\mathbb{C}(y)$ of rational functions in $\{y_b\}$. The elements $\{\Psi'_i, \ P'_r\}$ satisfy the homogeneous Coxeter relations for $\{s_i, \pi_r\}$. So the elements

$$\Psi'_{\widehat{w}} \ = \ P'_r \Psi'_{i_l} \cdots \Psi'_{i_1}, \ \text{ where } \ \widehat{w} = \pi_r s_{i_l} \cdots s_{i_1} \in \widehat{W}, \tag{1.6.9}$$

are well defined for reduced decompositions of $\widehat{w}$, and $\Psi'_{\widehat{w}}\Psi'_{\widehat{u}} = \Psi'_{\widehat{w}\widehat{u}}$ whenever $l(\widehat{w}\widehat{u}) = l(\widehat{w}) + l(\widehat{u})$.

The following key property of $\{\Psi'\}$ fixes them uniquely up to left or right multiplications by functions of y:

$$\Psi'_{\widehat{w}} y_b \ = \ y_{\widehat{w}(b)} \Psi'_{\widehat{w}}, \ \widehat{w} \in \widehat{W}, \tag{1.6.10}$$

where $y_{[b,j]} \stackrel{\mathbf{def}}{=\!=} y_b + j$, for instance, $y_{\alpha_0} = 1 - y_\theta$.

The operator P'_1 in the case of GL_n (it becomes of infinite order) plays the key role in [KnS]. The formulas for nonaffine Ψ'_i when $1 \le i \le n$ are well known in the theory of degenerate (graded) affine Hecke algebras. See [Lus1, C10] and [O3], Definition 8.2. However, the applications to Jack polynomials and the Harish-Chandra theory do require the affine intertwiners Φ'_0 and P'_r.

1.6.3 Differential representation

Introducing the variables $X_b = e^{x_b}$ for $b \in P^\vee$, we extend the following formulas to the derivations of $\mathbb{C}[X] = \mathbb{C}[X_b]$:

$$\partial_a(X_b) \ = \ (a, b) X_b, \ a, b \in P^\vee.$$

Note that $w(\partial_b) = \partial_{w(b)},\ w \in W$.

The key measure-function in the Harish-Chandra theory of spherical functions and its k–generalizations is as follows:

$$\tau \overset{\mathbf{def}}{=\!=} \tau(x;k) = \prod_{\alpha\in\Sigma_+} |2\sinh(x_\alpha/2)|^{2\kappa_\alpha}. \tag{1.6.11}$$

Proposition 1.6.2. *(a) The following* **trigonometric Dunkl operators** *acting on the Laurent polynomials* $f \in \mathbb{C}[X] = \mathbb{C}[X_b]$,

$$D_b \overset{\mathbf{def}}{=\!=} \partial_b + \sum_{\alpha\in\Sigma_+} \frac{\kappa_\alpha(b,\alpha^\vee)}{(1-X_\alpha^{-1})}(1-s_\alpha) - (\rho_k, b) \tag{1.6.12}$$

are pairwise commutative, and $y_{[b,u]} = D_b + u$ *satisfy (1.6.7) for the following action of the group* $\widehat{W}$:

$$w^x(f) = w(f) \ \text{ for } \ w\in W,\ b^x(f) = X_b f \ \text{ for } \ b \in P^\vee.$$

For instance, $s_0^x(f) = X_\theta s_\theta(f),\ \pi_r^x(f) = X_r u_r^{-1}(f)$.

(b) The operators D_b *are formally self-adjoint with respect to the inner product*

$$\{f,g\}_\tau \overset{\mathbf{def}}{=\!=} \int f(x)g(-x)\tau dx, \tag{1.6.13}$$

i.e., $\tau^{-1}D_b^+\tau = D_b$ *for the anti-involution* $^+$ *sending* $\widehat{w}^x \mapsto (\widehat{w}^x)^{-1}$ *and* $\partial_b \mapsto -\partial_b$, *where* $\widehat{w} \in \widehat{W}$. ❑

We skip the proof. Concerning the algebraic properties of the Dunkl operators, here and further see the previous sections or [C23, C11].

Comment. (i) We mention that Opdam uses a somewhat different pairing and a different involution in [O3]. We will discuss a q,t-generalization of the involution he considered, "with w_0", in Chapter 2 and, in full generality, in Chapter 3.

(ii) Note that the trigonometric Dunkl operators used in the previous sections are somewhat different; see, e.g., Proposition 1.5.8. Namely, the term $(1-s_\alpha)$ replaces s_α, respectively, the ρ–term appears to ensure the invariance. This renormalization is the most convenient in the Harish-Chandra theory.

1.6.4 Difference-rational case

We introduce the variables

$$\lambda_{a+b} = \lambda_a + \lambda_b,\ \lambda_{[b,u]} = \lambda_b + u,\ b \in P^\vee, u \in \mathbb{C}, \tag{1.6.14}$$

and define the **rational Demazure–Lusztig operators** from [C12] as follows:

$$S_i \;=\; s_i^\lambda \;+\; \frac{\kappa_i}{\lambda_{\alpha_i}}(s_i^\lambda - 1), \quad 0 \le i \le n, \tag{1.6.15}$$

where by $\widehat{w}^\lambda$ we mean the action on $\{\lambda_b\}$:

$$\widehat{w}^\lambda(\lambda_b) = \lambda_{\widehat{w}(b)},\; b^\lambda(\lambda_c) = \lambda_c - (b,c) \;\text{ for }\; b,c \in P^\vee.$$

For instance, $S_0 = s_0^\lambda + \frac{\kappa_0}{1-\lambda_\theta}(s_0^\lambda - 1)$.

We set

$$S_{\widehat{w}} \overset{\mathbf{def}}{=\!=} \pi_r^\lambda S_{i_1} \ldots S_{i_l} \;\text{ for }\; \widehat{w} = \pi_r s_{i_1} \ldots s_{i_l}, \tag{1.6.16}$$

$$\Delta_b \overset{\mathbf{def}}{=\!=} S_b \;\text{ for }\; b \in P^\vee,\; \Delta_i = \Delta_{b_i}. \tag{1.6.17}$$

The definition does not depend on the particular choice of the decomposition of $\widehat{w} \in \widehat{W}$, and the map $\widehat{w} \mapsto S_{\widehat{w}}$ is a homomorphism. The operators Δ_b are pairwise commutative. They are called **difference Dunkl operators.**

The counterpart of τ is the **asymmetric Harish–Chandra** $c\bar{c}$**–function**:

$$\sigma \overset{\mathbf{def}}{=\!=} \sigma(\lambda; k) = \prod_{a \in \Sigma_+} \frac{\Gamma(\lambda_\alpha + \kappa_a)\Gamma(-\lambda_\alpha + \kappa_\alpha + 1)}{\Gamma(\lambda_\alpha)\Gamma(-\lambda_\alpha + 1)}, \tag{1.6.18}$$

where Γ is the classical Γ–function.

Proposition 1.6.3. *(a) The operators $S_{\widehat{w}}$ are well defined and preserve the space of polynomials $\mathbb{C}[\lambda]$ in terms of λ_b. They form the degenerate double Hecke algebra, namely, the map $\widehat{w} \mapsto S_{\widehat{w}}$, $y_b \mapsto \lambda_b$ is a faithful representation of $\mathcal{H}'$.*

(b) The operators $S_{\widehat{w}}$ are formally unitary with respect to the inner product

$$\{f,g\}_\sigma \overset{\mathbf{def}}{=\!=} \int f(\lambda)g(\lambda)\sigma d\lambda, \tag{1.6.19}$$

i.e., $\sigma^{-1}S_{\widehat{w}}^+\sigma = S_{\widehat{w}}^{-1}$ for the anti-involution $^+$ sending $\lambda_b \mapsto \lambda_b$ and $\widehat{w} \mapsto \widehat{w}^{-1}$, where $\widehat{w} \in \widehat{W}$. ❑

The operators

$$\Lambda_p = p(\Delta_1, \ldots, \Delta_n) \;\text{ for }\; p \in \mathbb{C}[X_1^{\pm 1}, \ldots, X_n^{\pm 1}]^W \tag{1.6.20}$$

are W–invariant and preserve $\mathbb{C}[\lambda]^W$ (usual symmetric polynomials in λ). So if we restrict them to W–invariant functions, we will get W–invariant difference operators. However, the formulas for the restrictions of Λ_p are simple enough in special cases only.

Proposition 1.6.4. *Given $r \in O'$, let $m_r = \sum_{w \in W/W_r} X_{w(-b_r)}$, where W_r is the stabilizer of b_r in W. Then $\lambda(b_r)$ belongs to Σ_+ and the restriction of $\Lambda_{m_r}(\Delta_b)$ onto $\mathbb{C}[\lambda]^W$ is as follows:*

$$\Lambda_r = \sum_{w \in W/W_r} \prod_{\alpha \in \lambda(b_r)} \frac{\lambda_{w(\alpha)} + \kappa_\alpha}{\lambda_{w(\alpha)}} w(-b_r), \tag{1.6.21}$$

where $w(-b_r) = -w(b_r) \in P^\vee$ is considered as a difference operator:

$$w(-b_r)(\lambda_c) = \lambda_c + (w(b_r), c).$$

❑

The formula for Λ_r is very close to the corresponding formulas for the general double Hecke algebra from [C16]. Note that in the differential case, the formulas for the invariant operators are much more complicated than in the difference theory. It was (and still is) a convincing argument in favor of the difference theory.

1.6.5 Opdam transform

From now on, we assume that $\mathbb{R} \ni \kappa_\alpha > 0$ for all α and keep the notation from the previous section; i is the imaginary unit; $\Re, \Im$ are the real and imaginary parts.

Theorem 1.6.5. *There exists a solution $G(x, \lambda)$ of the eigenvalue problem*

$$D_b(G(x, \lambda)) = \lambda_b G(x, \lambda), \; b \in P^\vee, \; G(0, \lambda) = 1, \tag{1.6.22}$$

holomorphic for all λ and for x in $\mathbb{R}^n + iU$ for a neighbourhood $U \subset \mathbb{R}^n$ of zero. If $x \in \mathbb{R}^n$, then $|G(x, \lambda)| \le |W|^{1/2} \exp(\max_w (w(x), \Re\lambda))$, so G is bounded for $x \in \mathbb{R}^n$ when $\lambda \in i\mathbb{R}^n$. The solution of (1.6.22) is unique in the class of continuously differentiable functions on $\mathbb{R}^n$ (for a given λ).

This theorem is from [O3] (Theorem 3.15 and Proposition 6.1). Opdam uses the fact that

$$F \overset{\text{def}}{=\!=} |W|^{-1} \sum_{w \in W} G(w(x), \lambda) \tag{1.6.23}$$

is a **generalized hypergeometric function**, i.e., a W-symmetric eigenfunction of the QMBP operators or **Heckman–Opdam operators**, which are the restrictions L'_p of the operators $p(D_{b_1}, \ldots, D_{b_n})$ to symmetric functions:

$$L'_p F(x, \lambda) \; = \; p(\lambda_1, \ldots, \lambda_n) F(x, \lambda),$$

where p is any W–invariant polynomial of $\lambda_i = \lambda_{b_i}$. The operators L'_p generalize the radial parts of Laplace operators on the corresponding symmetric space. The normalization is the same: $F(0, \lambda) = 1$. It fixes F uniquely, so it is W–invariant with respect to λ as well.

A systematic algebraic and analytic theory of F–functions is due to Heckman and Opdam (see [HO1, H1, O1, HS]). There is a formula for G in terms of F (at least for generic λ) via the operators D_b from (1.6.12). The positivity of k implies that it holds for all $\lambda \in \mathbb{C}^n$. See [O3] for a nice and simple argument (Lemma 3.14). This formula and the relation to the affine Knizhnik–Zamolodchikov equation [C10, Mat, C11, O3] are applied to establish the growth estimates for G. Actually, it gives more than was formulated in the theorem (see Corollary 6.5, [O3]).

We introduce the **Opdam transform** (the first component of what he called "Cherednik's transform") as follows:

$$\mathcal{F}(f)(\lambda) \stackrel{\mathbf{def}}{=\!=} \int_{\mathbb{R}^n} f(x)G(-x, \lambda)\tau dx \tag{1.6.24}$$

for the standard measure dx on $\mathbb{R}^n$.

Proposition 1.6.6. *(a) Let us assume that the $f(x)$ are taken from the space $\mathbb{C}_c^\infty(\mathbb{R}^n)$ of $\mathbb{C}$–valued compactly supported ∞–differentiable functions on $\mathbb{R}^n$. The inner product*

$$\{f, f'\}_\tau \stackrel{\mathbf{def}}{=\!=} \int_{\mathbb{R}^n} f(x)f'(-x)\tau dx \tag{1.6.25}$$

satisfies the conditions of part (b), Proposition 1.6.2. Namely, $\{D_b\}$ preserve $\mathbb{C}_c^\infty(\mathbb{R}^n)$ and are self-adjoint with respect to the pairing (1.6.25).

(b) The Opdam transforms of such functions are analytic in λ on the whole $\mathbb{C}^n$ and satisfy the Paley-Wiener condition. A function $g(\lambda)$ is of the PW-type ($g \in PW(\mathbb{C}^n)$) if there exists a constant $A = A(g) > 0$ such that, for any $N > 0$,

$$g(\lambda) \leq C(1 + |\lambda|)^{-N} \exp(A|\Re\lambda|) \tag{1.6.26}$$

for a proper constant $C = C(N; g)$.

Proof. The first claim is obvious. The Paley-Wiener condition follows from Theorem 8.6 [O3]. The transform under consideration is actually the first component of Opdam's transform from Definition 7.9 (ibid.) without the complex conjugation and for the opposite sign of x (instead of λ). Opdam's estimates remain valid in this case. ❑

1.6.6 Inverse transform

In this section we discuss the **inversion** (for positive k). The **inverse Opdam transform** is defined for Paley-Wiener functions $g(\lambda)$ on $\mathbb{C}^n$ by the formula

$$\mathcal{G}(g)(x) \stackrel{\mathbf{def}}{=\!=} \int_{i\mathbb{R}^n} g(\lambda)G(x,\lambda)\sigma d\lambda \tag{1.6.27}$$

for the standard measure $d\lambda$. The transforms of such g belong to $\mathbb{C}_c^\infty(\mathbb{R}^n)$.

The existense readily follows from (1.6.26) and the known properties of the Harish-Chandra c–function (see below). The embedding $\mathcal{G}(PW(\mathbb{C}^n)) \subset \mathbb{C}_c^\infty(\mathbb{R}^n)$ is due to Opdam; it is similar to the classical one from [He3] (see also [GV]).

Let us discuss the shift of the integration contour in (1.6.27). There exists an open neighborhood $U_+^a \subset \mathbb{R}^n$ of the **closure** $\bar{\mathfrak{C}}_+^a \in \mathbb{R}^n$ of the affine Weyl chamber $\mathfrak{C}^a$ from (1.6.6) such that

$$\mathcal{G}(g)(x) \;=\; \int_{\xi+i\mathbb{R}^n} g(\lambda)G(x,\lambda)\sigma d\lambda \tag{1.6.28}$$

for $\xi \in U_+^a$. Indeed, $\kappa_\alpha > 0$ and σ has no singularities in $U_+^a + i\mathbb{R}^n$. Then we use the classical formulas for $|\Gamma(x+iy)/\Gamma(x)|$ for real $x, y, x > 0$. It gives (cf. [He3, O3]):

$$|\sigma(\lambda)| \;\leq\; C(1+|\lambda|)^K, \text{ where } K = 2\sum_{\alpha>0}\kappa_\alpha,\ \lambda \in U_+^a + i\mathbb{R}^n, \tag{1.6.29}$$

for sufficiently large $C > 0$. Thus the products of PW-functions by σ tend to 0 for $|\lambda| \mapsto \infty$, and we can switch to ξ. Actually, we can do this for any integrand analytic in $U_+^a + i\mathbb{R}^n$ and approaching 0 at ∞. We come to the following.

Proposition 1.6.7. *The conditions of part (b), Proposition 1.6.3 are satisfied for*

$$\{g, g'\}_\sigma \stackrel{\mathbf{def}}{=\!=} \int_{i\mathbb{R}^n} g(\lambda)g'(\lambda)\sigma d\lambda \tag{1.6.30}$$

in the class of PW-functions, i.e., the operators $S_{\widehat{w}}$ are well defined on such functions and unitary.

Proof. It is sufficient to check the unitarity for the generators $S_i = S_{s_i}$ $(0 \leq i \leq n)$ and π_r^λ $(r \in O')$. For instance, let us consider s_0. We will integrate over $\xi + i\mathbb{R}$, assuming that $\xi' \stackrel{\mathbf{def}}{=\!=} s_\theta(\xi) + \theta \in U_+^a$ and avoiding the wall $(\theta, \xi) = 1$.

We will apply formula (1.6.19), the formula

$$\int_{\xi+i\mathbb{R}^n} s_0(g(\lambda)\sigma)d\lambda \;=\; \int_{\xi'+i\mathbb{R}^n} g(\lambda)\sigma d\lambda \;=\; \int_{\xi+i\mathbb{R}^n} g(\lambda)\sigma d\lambda,$$

and a similar formula for $(1-\lambda_\theta)^{-1}g\sigma$, where g is of the PW-type in $U_+^a+i\mathbb{R}^n$. Note that $(1-\lambda_\theta)^{-1}\sigma$ is regular in this domain. One then has:

$$\begin{aligned}
&\int_{\xi+i\mathbb{R}^n} S_0(g(\lambda))\ g'(\lambda)\sigma d\lambda \\
&= \int_{\xi+i\mathbb{R}^n} \big(s_0+\kappa_0(1-\lambda_\theta)^{-1}(s_0-1)\big)(g(\lambda))\ g'(\lambda)\sigma d\lambda \\
&= \int_{\xi+i\mathbb{R}^n} g(\lambda)\ \big(s_0+\kappa_0(s_0-1)(1-\lambda_\theta)^{-1}\big)(g'(\lambda)\sigma)d\lambda \\
&= \int_{\xi+i\mathbb{R}^n} g(\lambda)\ S_0^{-1}(g'(\lambda))\sigma d\lambda. \qquad (1.6.31)
\end{aligned}$$

Since (1.6.31) holds for one ξ, it is valid for all of them in U_+^a, including 0. The consideration of the other generators is the same. ❑

Theorem 1.6.8. *Given* $\widehat{w}\in\widehat{W}$, $b\in P^\vee$, $f(x)\in\mathbb{C}_c^\infty(\mathbb{R}^n)$, $g(\lambda)\in PW(\mathbb{C}^n)$,

$$\begin{aligned}
\widehat{w}^x(G(x,\lambda)) &= S_{\widehat{w}}^{-1}G(x,\lambda),\ \textit{e.g.},\ X_bG = \Delta_b^{-1}(G), & (1.6.32)\\
\mathcal{F}(\widehat{w}^x(f(x))) &= S_{\widehat{w}}\mathcal{F}(f(x)),\ \mathcal{G}(S_{\widehat{w}}(g(\lambda))) = \widehat{w}^x(\mathcal{F}(f(x))), & (1.6.33)\\
\mathcal{F}(X_bf(x)) &= \Delta_b(\mathcal{F}(f(x))),\ \mathcal{G}(\Delta_b(g(\lambda))) = X_b\mathcal{G}(g(\lambda)), & (1.6.34)\\
\mathcal{F}(D_b(f(x))) &= \lambda_b\mathcal{F}(f(x)),\ \mathcal{G}(\lambda_bg(\lambda)) = D_b(\mathcal{G}(g(\lambda))). & (1.6.35)
\end{aligned}$$

Proof. Applying the intertwiners from (1.6.8) to $G(x,\lambda)$, we see that

$$\begin{aligned}
(1+\frac{\kappa_i}{\lambda_{\alpha_i}})^{-1}(s_i+\frac{\kappa_i}{\lambda_{\alpha_i}})(G) &= s_i^\lambda(G),\ \text{ for }\ 1\le i\le n, \qquad (1.6.36)\\
(1+\frac{\kappa_0}{1-\lambda_\theta})^{-1}(X_\theta s_\theta+\frac{\kappa_0}{1-\lambda_\theta})(G) &= s_0^\lambda(G),\\
P_r'(G) &= X_r u_r^{-1}(G) = (\pi_r^\lambda)^{-1}(G)\ \text{ for }\ r\in O'.
\end{aligned}$$

Due to the main property of the intertwiners (1.6.10), we obtain these equalities up to λ–multipliers. The scalar factors on the left are necessary to preserve the normalization $G(0,\lambda)=1$, so we can use the uniqueness of $G(x,\lambda)$ from Theorem 1.6.5. Expressing s_j^x in terms of s_j^λ (when applied to G!), we obtain (1.6.32) for $\widehat{w}=s_j$. It is obvious for $\widehat{w}=\pi_r$. Using the commutativity of $\widehat{w}^x$ and $S_{\hat{u}}$ for $\widehat{w},\hat{u}\in\widehat{W}$, we establish this relation in the general case. For $w\in W$ it is due to Opdam.

Formula (1.6.33) results directly from (1.6.32) because we already know that the $\widehat{w}^x$ are unitary for $\{f,f'\}_\tau$ and the $S_{\widehat{w}}$ are unitary with respect to $\{g,g'\}_\sigma$ for the considered classes of functions. See Theorem 1.6.2, Theorem 1.6.3, and formulas (1.6.25) and (1.6.30).

For instance, let us check (1.6.34) for $\mathcal{F}$, which is a particular case of (1.6.33):

$$\mathcal{F}(X_b f(x)) = \int_{\mathbb{R}^n} f(-x) X_b^{-1} G(x,\lambda)\tau dx$$
$$= \int_{\mathbb{R}^n} f(-x)\Delta_b(G(x,\lambda))\tau dx = \Delta_b(\mathcal{F}(f(x))). \tag{1.6.37}$$

Thus the Opdam transforms of the operators of multiplication by X_b are the operators Δ_b.

Since D_b and λ_b are self-adjoint for the corresponding inner products (see Theorem 1.6.2), we obtain (1.6.35), which is in fact the defining property of $\mathcal{F}$ and $\mathcal{G}$. ❑

Corollary 1.6.9. *The compositions $\mathcal{GF} : C_c^\infty(\mathbb{R}^n) \to C_c^\infty(\mathbb{R}^n)$, and $\mathcal{FG} : PW(\mathbb{C}^n) \to PW(\mathbb{C}^n)$ are multiplications by nonzero constants. The transforms $\mathcal{F}, \mathcal{G}$ establish isomorphisms between the corresponding space identifying $\{\ ,\ \}_\tau$ and $\{\ ,\ \}_\sigma$ up to proportionality.*

Proof. The first statement readily follows from Theorem 1.6.8. The composition $\mathcal{GF}$ sends $\mathbb{C}_c^\infty(\mathbb{R}^n)$ into itself, is continuous (due to Opdam), and commutes with the operators X_b (multiplications by X_b). Thus it is a multiplication by a function $u(x)$ of $\mathbb{C}^\infty$–type. It must also commute with D_b. Hence $G(x,\lambda)u(x)$ is another solution of the eigenvalue problem (1.6.22), and we conclude that $u(x)$ has to be a constant.

Let us check that $\mathcal{FG}$, which is a continuous operator on $PW(\mathbb{C}^n)$ (for any fixed A) commuting with multiplications by any λ_b, has to be a multiplication by an analytic function $v(\lambda)$. Indeed, the image of $\mathcal{FG}$ with respect to the standard Fourier transform ($k = 0$) is a continuous operator on $C_c^\infty(\mathbb{R}^n)$ commuting with the derivatives $\partial/\partial x_i$. Thus it is a convolution with some function, and its inverse Fourier transform is the multiplication by a certain $v(\lambda)$. Since $\mathcal{FG}(g) = \text{Const}\ g$ for any $g(\lambda)$ from $\mathcal{F}(C_c^\infty(\mathbb{R}^n))$, v is constant.

The claim about the inner products is obvious because $\{\mathcal{F}(f), g\}_\sigma = \{f, \mathcal{G}(g)\}_\tau$ for $f \in C_c^\infty(\mathbb{R}^n)$, $g \in PW(\mathbb{C}^n)$. ❑

The corollary is due to Opdam ([O3],Theorem 9.13 (1)). His proof is different. He uses Peetre's characterization of differential operators (similar to what Van Den Ban and Schlichtkrull did in [BS]). In his approach, a certain nontrivial analytic argument is needed to check that $\mathcal{GF}$ is multiplication by a constant. The proof above is analytically elementary.

The symmetric case. The transforms can be readily reduced to the symmetric level. If f is W–invariant, then

$$\mathcal{F}(f(x)) = \int_{\mathbb{R}^n} f(x) F(-x,\lambda)\tau dx \tag{1.6.38}$$

for F from (1.6.23). Here we have applied the W–symmetrization to the integrand in (1.6.24) and used that τ is W–invariant. So $\mathcal{F}$ *coincides* with the k–deformation of the Harish-Chandra transform on W–invariant functions up to a minor technical detail. The W–invariance of F in λ results in the W–invariance of $\mathcal{F}(f)$.

As for $\mathcal{G}$, we W–symmetrize the integrand in the definition with respect to x and λ. Namely, we replace G by F and σ by its W–symmetrization. The latter is the genuine Harish-Chandra "$c\bar{c}$– function"

$$\sigma' = \prod_{a\in\Sigma_+} \frac{\Gamma(\lambda_\alpha+\kappa_a)\Gamma(-\lambda_\alpha+\kappa_\alpha)}{\Gamma(\lambda_\alpha)\Gamma(-\lambda_\alpha)} \tag{1.6.39}$$

up to a coefficient of proportionality.

Finally, given a W–invariant function $f \in \mathbb{C}_c^\infty(\mathbb{R}^n)$,

$$f(x) = \text{Const}\int_{i\mathbb{R}^n} g(\lambda)F(x,\lambda)\sigma' d\lambda \text{ for } g=\mathcal{F}(f). \tag{1.6.40}$$

A similar formula holds for $\mathcal{G}$. See [HC, He1] and [GV](Ch.6) for the classical theory.

As an application, we are able to calculate the Fourier transforms of the operators $p(X) \in \mathbb{C}[X_b]^W$ (symmetric Laurent polynomials acting by multiplication) in the Harish-Chandra theory and its k–deformation. They are exactly the operators Λ_p above. In the minuscule case, we obtain formulas (1.6.21). We mention that in [O3] and other papers the pairings serving the Fourier transforms are hermitian. Complex conjugations can be added to ours.

I hope that the method we used can be generalized to negative k, to other classes of functions, and to the p–adic theory (cf. [BS, HO2]). The relations (1.6.32) considered as difference equations for $G(x,\lambda)$ with respect to λ may also help with the growth estimates via the theory of difference equations and the equivalence with the difference Knizhnik–Zamolodchikov equations. However, I believe that the main advantage here will be connected with the q–Fourier transform.

1.7 Factorization and r–matrices

We will define the "factorizable" Kac–Moody algebras following [C1] and introduce at the end of the section the corresponding τ–function as an infinite wedge product. It will be generalized in the next section using the notion of the coinvariant. The r–matrix KZ of trigonometric type will be defined, to be interpreted as the equation for the coinvariant in the next section.

1.7.1 Basic trigonometric r–matrix

Let us start with the simplest example. We set $V = \mathbb{C}_N^{\otimes n} = \overbrace{\mathbb{C}^N \otimes \cdots \otimes \mathbb{C}^N}^{n}$ and

$$r(x) = \frac{1}{2}\coth\left(\frac{x}{2}\right) P + \frac{1}{2} \sum_{1\le l<m\le N} (e_{lm} \otimes e_{ml} - e_{ml} \otimes e_{lm}) + D. \qquad (1.7.1)$$

Here $\coth(x) = (e^x+e^{-x})/(e^x-e^{-x})$, $P = \sum_{1\le l,m\le N} e_{lm}\otimes e_{ml}$ for the standard generators $\{e_{ab}\}$ of $\mathrm{End}(\mathbb{C}^N)$ with the entries $\delta_{li}\delta_{mj}$, and D is any diagonal matrix ($D = \sum c_{lm} e_{ll} \otimes e_{mm}$). So the values of r belong to $\mathrm{End}(\mathbb{C}^N \otimes \mathbb{C}^N)$. Note that P is the permutation matrix: $P(v_1 \otimes v_2) = v_2 \otimes v_1$.

We put $r^{ij} = r(u_i - u_j)^{(i,j)}$, where

$$C^{(i,j)} = \sum 1 \otimes \cdots \otimes \overset{i}{a} \otimes \cdots \otimes \overset{j}{b} \otimes \ldots 1 \text{ if } C = \sum a \otimes b \qquad (1.7.2)$$

for any matrices $\{a, b\}$. Note that $r^{21} = (r(u_2 - u_1))^{(2,1)}$. The following relations can be verified directly:

$$[r^{ij}, r^{ik} + r^{jk}] - [r^{ik}, r^{kj}] = 0, \qquad (1.7.3)$$
$$[r^{ij}, r^{kl}] = 0, \quad r^{ij} + r^{ji} = D^{(i,j)} + D^{(j,i)} \qquad (1.7.4)$$

for pairwise distinct indices i, j, k, l (r^{kj} is not a misprint!).

When $D^{(i,j)} + D^{(j,i)} = 0$ we obtain the relations

$$[r^{ij}, r^{ik} + r^{jk}] + [r^{ik}, r^{jk}] = 0, \qquad (1.7.5)$$

which are nothing but the assumptions of Proposition 1.5.3 for the root system A_{n-1} (the pairs $\{ij\}$ can be identified with the roots). This r–matrix is invariant and unitary. So we can introduce the corresponding KZ equation.

Identifying $P^{(i,j)}$ with the transpositions s_{ij}, we can represent it in the form (1.1.23) from Section 1.1.2. Note that in this section we use u instead of v. The formulas for the operators x_i are straightforward. They involve D; so we arrive at a representation of the degenerate affine Hecke algebra $\mathcal{H}'_n$ of type GL_n in V. Any weight subspaces of V with respect to the standard action of $\mathfrak{gl}_N$ are $\mathcal{H}'_n$–submodules.

In this example, it is not necessary to impose the relation $D^{(i,j)}+D^{(j,i)} = 0$. One can define KZ for any D.

Arbitrary $\mathfrak{g}$. Let us generalize this example to an arbitrary simple finite dimensional Lie algebra $\mathfrak{g}$ of rank$= l$ over $\mathbb{C}$. Let $\Sigma = \{\alpha\} \subset \mathbb{R}^l$ be the root system associated with $\mathfrak{g}$ and $(\ ,\ )$ the W–invariant inner product on $\mathbb{R}^l$ normalized by the condition $(\theta, \theta) = 2$ for the maximal root θ with respect to the simple roots $\alpha_1, \ldots, \alpha_l$.

We choose nonzero

$$e_\alpha \in \mathfrak{g}_\alpha,\ f_\alpha \in \mathfrak{g}_{-\alpha},\ \alpha \in \Sigma_+, \tag{1.7.6}$$

provided we have the relations

$$[h_\alpha, e_\beta] = (\alpha,\beta)e_\beta,\ [h_\alpha, f_\beta] = -(\alpha,\beta)f_\beta, \tag{1.7.7}$$

$$\text{for } h_\alpha = \frac{(\alpha,\alpha)}{2}[e_\alpha, f_\beta]. \tag{1.7.8}$$

The elements $\{e_\alpha\}, \{f_\alpha\}$ are linearly independent and generate the Borel subalgebras $\mathfrak{b}_\pm$, respectively. The elements $h_m = h_{\alpha_m}$ $(m = 1,\ldots,l)$ form a basis in the Cartan subalgebra $\mathfrak{h}$; $h_\alpha = \sum_{m=1}^{l}(\alpha, b_m)h_m$ for the fundamental coweights b_m $((b_m, \alpha_n) = \delta_{mn})$.

Let us connect the above form on $\mathbb{R}^l$ with the the **standard invariant form** (f, f') on $f \in \mathfrak{g} \ni f'$. By the invariant form, we mean that $([f,g],h) = (f,[g,h])$. All invariant forms are proportional to each other. The standard normalization is as follows:

$$(h_\alpha, h_\beta) = (\alpha,\beta), \tag{1.7.9}$$

$$(e_\alpha, f_\beta) = \frac{2}{(\alpha,\alpha)}\delta_{\alpha\beta}, \tag{1.7.10}$$

$$(e_\alpha, e_\beta) = 0 = (f_\alpha, f_\beta). \tag{1.7.11}$$

See [Ka]. In terms of this form, the definition of $\{h_\alpha\}$ does not depend on the particular choice of e_α, f_α in the corresponding weight spaces:

$$h_\alpha = \frac{[e_\alpha, f_\alpha]}{(e_\alpha, f_\alpha)}. \tag{1.7.12}$$

Let us connect the standard form with the Killing form $(f,f')_K \overset{\mathbf{def}}{=\!=} \mathrm{Tr}(\mathrm{ad}f\mathrm{ad}f')$ for $f, f' \in \mathfrak{g}$:

$$(f,f')_K = (2g)(f,f'),\ g = 1 + (\rho,\theta) \text{ (the dual Coxeter number)}, \tag{1.7.13}$$

where $\rho = \frac{1}{2}\sum_{\alpha\in\Sigma_+}\alpha$.

We set

$$\Omega = \sum_a I_a \otimes I_a^* \in \mathfrak{g}\otimes\mathfrak{g} \tag{1.7.14}$$

for a basis $\{I_a\ (a = 1,\ldots,\dim\mathfrak{g})\}$ of $\mathfrak{g}$ and its dual $\{I_a^*\}$: $(I_a, I_{a'}^*) = \delta_{a,a'}$. The definition of Ω does not depend on the choice of the basis $\{I_a\}$. Using the canonical generators e, f, h,

$$\Omega = \sum_{m=1}^{l} h_m \otimes h_m^* + \sum_{\alpha\in R^+}\frac{1}{(e_\alpha, f_\alpha)}(e_\alpha\otimes f_\alpha + f_\alpha\otimes e_\alpha), \tag{1.7.15}$$

where $\{h_m^* = h_{b_m}\}$ are dual to $\{h_m\}$.

Let $A : \mathfrak{h} \to \mathfrak{h}$ be a linear map. The **basic trigonometric r–matrix** is given by the formula:

$$r(x) = \frac{1}{2}\coth\left(\frac{x}{2}\right)\Omega + \frac{1}{2}\sum_{\alpha \in R^+}\frac{f_\alpha \otimes e_\alpha - e_\alpha \otimes f_\alpha}{(e_\alpha, f_\alpha)} + \frac{1}{2}\sum_{m=1}^{l} A(h_i) \otimes h_i^*. \quad (1.7.16)$$

It takes values in $\mathfrak{g} \otimes \mathfrak{g}$. We use the same notation as for vector tensor products:

$$r^{ij} = r(u_i - u_j)^{(i,j)}, (a \otimes b)(i,j) = 1 \otimes \cdots \otimes \overset{m}{a} \otimes \cdots \otimes \overset{j}{b} \otimes \ldots 1$$

for $U(\mathfrak{g})^{\otimes n}$; $U(\mathfrak{g})$ is the universal enveloping algebra of $\mathfrak{g}$. Generalizing formula (1.7.1), we arrive at

Proposition 1.7.1.

$$[r^{ij}, r^{ik} + r^{jk}] - [r^{ik}, r^{kj}] = 0, [r^{ij}, r^{ks}] = 0, \quad (1.7.17)$$

$$r^{ij} + r^{ji} = \Theta^{ij},$$

$$\Theta = (1/2)\textstyle\sum_{m=1}^{l}(A(h_m) \otimes h_m^* + h_m^* \otimes A(h_m)). \quad (1.7.18)$$

Here the indices $1 \le i, j, k, s \le n$ are pairwise distinct.

1.7.2 Factorization and r–matrices

We will establish the one-to-one correspondence between the r–matrices and the Lie subalgebras complementary to the standard holomorphic subalgebra in a Kac–Moody algebra.

In this subsection, we fix a simple Lie algebra $\mathfrak{g}$, an open neighborhood $U_0(\subset \mathbb{C})$ of 0, and the set of pairwise distinct points $u_1, \ldots, u_n \in U_0$ such that $u_i - u_j \in U_0$ $(1 \le i \ne j \le n)$.

The following is a generalization of the classical Yang–Baxter equation (YBE). For unitary r, it is *exactly* YBE.

Definition 1.7.2. *Let $r(x)$ be a holomorphic $\mathfrak{g} \otimes \mathfrak{g}$–valued function of $x \in U_0 \setminus \{0\}$. We call $r(x)$ a* **quasi-unitary** r**–matrix** *if:*
(α) $r(x) - \frac{\Omega}{x}$ is holomorphic at 0,
(β) $[r^{12}, r^{13} + r^{23}] = [r^{12}, r^{32}]$,
(γ) $r^{12} + r^{21} = \Theta$, $d(\Theta)/dx = 0$.
It is called **unitary** *if $r^{12} + r^{21} = 0$.* ❑

Here $r^{ij} = r(x_i - x_j)^{(i,j)}$ for independent variables $x_1, x_2, x_3, \dots$, $r = r^{12}$, $x = x_1 - x_2$. Note that $r^{21} = (r(-x))^{(2,1)}$. The relations (α, β, γ) must be fulfilled whenever $x_i - x_j \in U_0 \setminus \{0\}$. The element Ω is from (1.7.14).

From now on, we set $x_i \stackrel{\mathbf{def}}{=} x - u_i$,

$$\widetilde{\mathfrak{g}}^i = \mathfrak{g}((x_i)) = \{\sum_{k \geq p} g_k x_i^k \mid p \in \mathbb{Z},\ g_k \in \mathfrak{g}\}, \tag{1.7.19}$$

$$\widetilde{\mathfrak{g}}_0^i = \mathfrak{g}[[x_i]] = \{\sum_{k \geq 0} g_k x_i^k \mid g_k \in \mathfrak{g}\}, \tag{1.7.20}$$

$$\widetilde{\mathfrak{g}} = \prod_{i=1}^n \widetilde{\mathfrak{g}}^i, \qquad \widetilde{\mathfrak{g}}_0 = \prod_{i=1}^n \widetilde{\mathfrak{g}}_0^i. \tag{1.7.21}$$

All these are Lie algebras:

$$\begin{aligned} [f,g]_{\widetilde{\mathfrak{g}}} &= ([f^1,g^1],\dots,[f^n,g^n]) \\ &\text{for } f = (f^1,\dots,f^n), g = (g^1,\dots,g^n) \in \widetilde{\mathfrak{g}}, \end{aligned} \tag{1.7.22}$$

where the components f^i, g^i of f, g are formal series in x_i; their commutators are understood coefficient-wise.

The Lie algebras $\widetilde{\mathfrak{g}}$, $\widetilde{\mathfrak{g}}_0$ are direct counterparts of the groups of all adèles and integral adèles in arithmetic.

The **central extension** $\widehat{\mathfrak{g}} = \widetilde{\mathfrak{g}} \oplus \mathbb{C}c$ (c is the center element) of $\widetilde{\mathfrak{g}}$ is introduced as follows:

$$[f + \xi c, g + \zeta c]_{\widehat{\mathfrak{g}}} = [f,g]_{\widetilde{\mathfrak{g}}} + \operatorname{Res}\left(\frac{df}{dx}, g\right) c, \tag{1.7.23}$$

$$\operatorname{Res}\left(\frac{df}{dx}, g\right) = \sum_{i=1}^n \operatorname{Res}_{x_i}\left(\frac{df^i}{dx_i}, g^i\right) dx_i, \tag{1.7.24}$$

where $\operatorname{Res}_{x_i}(\sum_k f_k x_i^k) dx_i = f_{-1}$.

Definition 1.7.3. *The Kac–Moody algebra* $\widehat{\mathfrak{g}}$ *is called* **factorized** *if it is endowed with a subspace, a* **factorizing subalgebra**, $\widetilde{\mathfrak{g}}_r \subset \widetilde{\mathfrak{g}}$ *such that*
(a) $\widetilde{\mathfrak{g}}_r \oplus \widetilde{\mathfrak{g}}_0 = \widetilde{\mathfrak{g}}$,
(b) $\widetilde{\mathfrak{g}}_r$ *is a Lie subalgebra of* $\widetilde{\mathfrak{g}}$,
(c) $\widetilde{\mathfrak{g}}_r$ *is a Lie subalgebra of* $\widehat{\mathfrak{g}}$. ❑

Given an r–matrix $r(x)$ with the values in $\mathfrak{g} \otimes \mathfrak{g}$, let us construct the corresponding factorizing subalgebra $\widetilde{\mathfrak{g}}_r$. For $f = (f^1, \dots, f^n) \in \widetilde{\mathfrak{g}}$, we define a function of x in a certain neighborhood of zero $\widetilde{U}_0 \subset U_0$:

$$\overline{f}(x) = \operatorname{Res}(r(x-y), 1 \otimes f) dy = \sum_{i=1}^n \operatorname{Res}_{y_i}(r(x_i - y_i), 1 \otimes f^i(y_i)) dy_i, \tag{1.7.25}$$

where $(a \otimes b, 1 \otimes c) = (b, c)a \ (a, b, c, \in \mathfrak{g})$. Let $f_r^i = f_r^i(x_i)$ be the expansion of $\overline{f}(x)$ at u_i $(x_i = x - u_i)$. We set

$$\overline{\mathfrak{g}} = \{\overline{f}\},\ \widetilde{\mathfrak{g}}_r = \{f_r \stackrel{\text{def}}{=\!=} (f_r^1(x_1), \ldots, f_r^n(x_n) \mid f \in \widetilde{\mathfrak{g}}\}. \tag{1.7.26}$$

Theorem 1.7.4. *If $r(x)$ is a quasi-unitary r–matrix, then $\overline{\mathfrak{g}}$ is a Lie algebra and $\widetilde{\mathfrak{g}}_r$ is a factorizing subalgebra of $\widetilde{\mathfrak{g}}$. Conversely, every factorizing subalgebra $\widetilde{\mathfrak{g}}_r$ that is invariant with respect to the differentiation d/dx is associated to a quasi-unitary r–matrix defined in some neighborhood $\tilde{U}_0$ of 0. The corresponding r is unique:*

$$r(x - u_i) = \sum_a \overline{\left(\frac{I_a}{x_i}\right)}(x) \otimes I_a^*, \tag{1.7.27}$$

where i can be arbitrary; $\{I_a\}$ is a basis of $\mathfrak{g}$. The expansion f_r of $\overline{f}$ coincides with the first component of the factorization $f = f_r + f_0$ with respect to the decomposition $\widetilde{\mathfrak{g}} = \widetilde{\mathfrak{g}}_r \oplus \widetilde{\mathfrak{g}}_0$.

The proof is based on the relation

$$[f_r, g_r] + [f, g]_r = [f_r, g]_r + [f, g_r]_r,\ \ f, g \in \widetilde{\mathfrak{g}}, \tag{1.7.28}$$

directly following from the condition (b) of Definition 1.7.3 by considering the principal parts of all four terms of the equality. The coincidence of the principal parts is sufficient because of the condition (a). Relation (1.7.28) results in the (equivalent) equality $[f_0, g_0] + [f, g]_0 = [f_0, g]_0 + [f, g_0]_0$. Expressing f_r, g_r in terms of r, we obtain the relation (β) from Definition 1.7.2. ❑

Comment. Here it is not necessary to assume that r depends on the difference of the arguments (see [C1, C3]). The interpretation of r as a projection was formalized by Manin (see [Dr2]) in the definition of the so-called Manin triple, and by Semenov- Tjan-Shanskii [Se]. They considered abstract ("constant") r–matrices.

In [C1], it was done for r depending on the parameter, which resulted in a fruitful direction, "the adèlic approach to soliton theory", with many examples. The paper [C1] was stimulated by [BD], where the unitary case and the corresponding classification of r–matrices were considered. Note that the definition of the classical r–matrix appeared for the first time in paper [KuS].

Let me also mention the book [ES], where one can find some related information and new developments, including the classification due to Belavin and Drinfeld, and a recent theorem of Etingof and Kazhdan about the Drinfeld conjecture concerning the lift of the classical r–matrices to their quantum counterparts. ❑

Arbitrary $\widetilde{\mathfrak{g}}_r$ satisfying condition (a) of Definition 1.7.3 can be obtained by the construction from (1.7.25) and (1.7.26) for a certain series $r(x, y)$

defined on the square of a formal neighborhood of the set $\{u_i\}$. If $r(x, y)$ depends on the difference, i.e., $r(x, y) = r(x - y)$, then condition (α) from Definition 1.7.2 is fulfilled. For such r, the relations (β) and (b) are equivalent. Moreover, imposing (α, β) (or (a,b)), the remaining relations (γ) and (c) are also equivalent. See [C3, C9].

Examples. The Yang r–matrix is given by the formula $r(x) = \frac{\Omega}{x}$ for Ω defined in (1.7.14). It is unitary: $r^{12} + r^{21} = 0$. The corresponding $\widetilde{\mathfrak{g}}_r$ consists of the sets $(f_r^1(x_1), \ldots, f_r^n(x_n)) \in \widetilde{\mathfrak{g}}$ of expansions of $\bar{f}$ at $\{u_i\}$ from the Lie algebra $\bar{\mathfrak{g}}$ of rational $\mathfrak{g}$–valued functions on $\mathbb{P}^1$ with poles at $\{u_1 \ldots u_n\}$ and the normalization condition $f(\infty) = 0$. We then have

$$\overline{\left(\frac{g}{x_i}\right)} = \frac{g}{x - u_i} \qquad (1.7.29)$$

for any i, $g \in \mathfrak{g}$, where $\overline{(\cdot)}$ is defined in (1.7.25).

Given a linear map $A : \mathfrak{h} \to \mathfrak{h}$, let us describe the factorizing subalgebra associated with r from (1.7.16):

$$r(x) = \frac{1}{2}\coth\left(\frac{x}{2}\right)\Omega + \frac{1}{2}\sum_{\alpha \in R^+} \frac{f_\alpha \otimes e_\alpha - e_\alpha \otimes f_\alpha}{(e_\alpha, f_\alpha)} + \sum_{n=1}^{l} A(h_i) \otimes h_i^*. \qquad (1.7.30)$$

First,

$$\overline{\left(\frac{f_\alpha}{x_i}\right)} = \frac{e^{x_i/2}}{e^{x_i/2} - e^{-x_i/2}} f_\alpha, \quad \overline{\left(\frac{e_\alpha}{x_i}\right)} = \frac{e^{-x_i/2}}{e^{x_i/2} - e^{-x_i/2}} e_\alpha, \qquad (1.7.31)$$

$$\overline{\left(\frac{h}{x_i}\right)} = \frac{1}{2}\coth\left(\frac{x_i}{2}\right) h + A(h) \quad (h \in \mathfrak{h}). \qquad (1.7.32)$$

The Lie algebra $\widetilde{\mathfrak{g}}_r$ consists of $(f_r^1(x_1), \ldots, f_r^n(x_n)) \in \widetilde{\mathfrak{g}}$, where $f_r^i(x_i) \in \widetilde{\mathfrak{g}}^i$ are the expansions of $\mathfrak{g}$–valued functions $\bar{f}(x)$ such that
(1) they are rational in terms of $v = e^x$ on $v \in \mathbb{P}^1$;
(2) they have poles at $v_1 = e^{u_1}, \ldots, v_n = e^{u_n}$ only;
(3) $\bar{f}(v = 0) \in \mathfrak{b}_+$, $\bar{f}(v = \infty) \in \mathfrak{b}_-$;
(4)$(\bar{f}(0) + \bar{f}(\infty))|_{\mathfrak{h}} = A((\bar{f}(0) - \bar{f}(\infty))|_{\mathfrak{h}})$.
Here $\mathfrak{b}_+ = \langle e_\alpha, \alpha \in \Sigma_- \rangle$, $\mathfrak{b}_- = \langle f_\alpha, \alpha \in \Sigma_+ \rangle$, and $\mathfrak{h}$ is the Cartan subalgebra.

The simplest way to establish that r is a quasi-unitary solution of the classical YBE in the sense of Definition 1.7.2 is definitely by checking (1–4). Indeed, the functions satisfying these conditions *obviously* form a Lie algebra. So we arrive at (1.7.28), which readily results in (β).

The r–matrix KZ. Given a quasi-unitary r, let

$$\varrho = \sum_a \rho_a I_a \in U(\mathfrak{g}) \text{ for } \{\Omega/x - r(x)\}(x = 0) = \sum_a \rho_a \otimes I_a, \qquad (1.7.33)$$

where $U(\mathfrak{g})$ is the enveloping algebra of $\mathfrak{g}$.

Theorem 1.7.5. *Setting*

$$R_i = \varrho^i - \sum_{j(\neq i)} r^{ji}, \ 1 \le j \le n, \ for \ 1 \le i \le n, \tag{1.7.34}$$

*the following system of differential equations, the r–**matrix KZ**, for a $U(\mathfrak{g})$–valued function $\Phi(u)$ is self-consistent:*

$$\kappa\partial(\Phi)/\partial u_i = R_i\Phi, \ 1 \le i \le n. \tag{1.7.35}$$

for all $\kappa = k^{-1}$.

The theorem can be checked by a straightforward calculation using the relation

$$[r^{12}, \varrho^1] + [\varrho^2, r^{21}] = [r^{12}, r^{21}]. \tag{1.7.36}$$

Recall that $\varrho^i = \varrho^{(i)}$ is ϱ considered in the i–th component of $U(\mathfrak{g})^{\otimes n}$. ❑

The commutation relations are more transparent for the unitary r. The last relation becomes $[r^{12}, \varrho^1 + \varrho^2] = 0$ in this case.

Considering the basic trigonometric r from (1.7.16),

$$\varrho = h_\rho - \frac{1}{2}\sum_{m=1}^{l} A(h_m)h_m^*, \ \rho = \frac{1}{2}\sum_{\alpha\in\Sigma_+}\alpha. \tag{1.7.37}$$

So $\varrho = (1/2)\sum_{\alpha>0} h_\alpha \in \mathfrak{h}$ when $A = 0$.

1.7.3 Two conjectures

If $\widetilde{\mathfrak{g}}_r$ satisfies conditions (a) and (b), then (c) is equivalent to the relation $\mathrm{Res}(\frac{df}{dx}, g) = 0$ for all $f, g \in \widetilde{\mathfrak{g}}_r$. The suffix r stays for "r–matrix" or "rational." The latter does not mean that this subalgebra (a counterpart of the group of principal adèles in arithmetic) must be associated with an algebraic curve. However, in all known examples it is so. One can always find a Lie algebra of the rational $\mathfrak{g}$–valued functions on $\mathbb{P}^1$ or an elliptic curve (containing U_0 as an open subset) such that $\widetilde{\mathfrak{g}}_r$ is formed by expansions of these functions at $u_1, \ldots, u_n$. We conjecture that this is true for all r.

More generally, let us suppose that the spaces

$$\widetilde{\mathfrak{g}}/(\widetilde{\mathfrak{g}}_r + \widetilde{\mathfrak{g}}_0), \ \widetilde{\mathfrak{g}}_r \cap \widetilde{\mathfrak{g}}_0 \tag{1.7.38}$$

are finite dimensional. Such $\widetilde{\mathfrak{g}}_r$ is a Kac–Moody variant of a discrete subgroup of an adèle group in arithmetic with the quotient space of finite volume or compact.

Conjecture 1.7.6. *There exists a complete algebraic curve C over $\mathbb{C}$ containing U_0 such that for the Lie algebra $\widetilde{\mathfrak{g}}_C$ of the expansions at $u_1, \ldots, u_n$ of all $\mathfrak{g}$-valued rational functions on C with poles apart from this set both spaces*

$$(\widetilde{\mathfrak{g}}_r + \widetilde{\mathfrak{g}}_C)/\widetilde{\mathfrak{g}}_r, \ (\widetilde{\mathfrak{g}}_r + \widetilde{\mathfrak{g}}_C)/\widetilde{\mathfrak{g}}_C \tag{1.7.39}$$

are finite dimensional.

Its counterpart in arithmetic was proved by Margulis. It holds when either the number of points u or the rank of the arithmetic group is larger than 1. It seems that we do not need this restriction in the Kac–Moody setup.

A special unitary r–matrix case of the above general conjecture is proved in the paper [BD]. If r is not unitary and does not depend on the difference of the arguments, the conjecture does not look too difficult, but remains an open question. See [C1, C3]. I think that almost nothing is known about the general case (any curves).

We note that the unitary r–matrices ($r^{12} + r^{21} = 0$) have the following interpretation in terms of $\widetilde{\mathfrak{g}}_r$:

$$\{f \in \widetilde{\mathfrak{g}} \mid \mathrm{Res}(f, \widetilde{\mathfrak{g}}_r)dx = 0\} \ = \ \widetilde{\mathfrak{g}}_r.$$

It looks similar to condition (c) from Definition 1.7.3, but of course does not coincide with it. In (c), f must be replaced by df/dx.

An interesting problem concerns the independence of condition (c) in the definition of the factorizing subalgebra.

Conjecture 1.7.7. *The subspace $\widetilde{\mathfrak{g}}_r$ satisfying conditions (a) and (b) from Definition 1.7.3 is a Lie subalgebra of $\widehat{\mathfrak{g}}$ for a proper choice of the central extension.*

All central extensions are described by 2–cocycles on $\widetilde{\mathfrak{g}}$ (see [Ka]). I expect that the conjecture is true even in the general setting of Conjecture 1.7.6. We will outline the proof in the r–matrix case in the following subsection.

1.7.4 Tau function

This subsection is based on the paper [C3] devoted to a generalization of the τ–function introduced by Date, Jimbo, Kashiwara, and Miwa [DJKM1] and a construction due to Kac, Peterson from [KP].

The definition of the **r–matrix τ–function** we give uses the fact that the group $\tilde{\mathcal{G}}_0 = \exp(\widetilde{\mathfrak{g}}_0)$ has a natural structure of an infinite dimensional algebraic variety. Given $f \in \widetilde{\mathfrak{g}}$, let us define the corresponding vector field D_f on this variety. Practically, we need a formula for $\Phi^{-1}D_f(\Phi) \in \widetilde{\mathfrak{g}}_0$ for the generic element $\Phi \in \tilde{\mathcal{G}}_0$.

The coefficients of Φ in the expansions with respect to $x_1, \ldots, x_n$ are precisely the coordinates of $\tilde{\mathcal{G}}_0$. Thus, equating both sides of the following relation, we obtain the complete list of D_f–derivatives of these coefficients (= coordinates) and therefore determine uniquely the corresponding differentiation of Funct$(\tilde{\mathcal{G}}_0)$:

$$\Phi^{-1}D_f(\Phi) = (\Phi f \Phi^{-1})_0, \text{ where } \tilde{\mathfrak{g}} \ni f = f_0 + f_r, \; f_0 \in \tilde{\mathfrak{g}}_0, f_r \in \tilde{\mathfrak{g}}_r. \quad (1.7.40)$$

The decomposition $f = f_0 + f_r$ is the factorization with respect to $\tilde{\mathfrak{g}}_r$; by $\Phi f \Phi^{-1}$, we mean the adjoint action.

Note that $\Phi^{-1}D_g(\Phi) = g$ for $g \in \tilde{\mathfrak{g}}_0$, so the corresponding D_g are left-invariant fields on the group $\tilde{\mathfrak{g}}_0$.

We will assume that r is quasi-unitary. Then $\varrho \in \prod_{i=1}^n \mathfrak{g}$. Actually, $\varrho = 0$ for the most interesting examples. See [C3].

A simple straightforward calculation based directly on the definition of the factorization (see (1.7.28)) gives that

$$D_{[f,f']} = [D_f, D_{f'}] \text{ on Funct}(\tilde{\mathcal{G}}_0). \quad (1.7.41)$$

We introduce the τ–function as the infinite wedge product $\tau = \wedge_g D_g$ of all vector fields D_g, where g runs over the natural basis

$$\{I_A^K = ((I_{a_1}x_1^{k_1}), \ldots, (I_{a_n}x_n^{k_n}))\} \text{ of } \tilde{\mathfrak{g}}_0.$$

Here $\{I_a\}$ is a certain fixed basis of $\mathfrak{g}$, $A = (a_1, \ldots, a_n)$, and $K = (k_1, \ldots, k_n)$, $k_i \geq 0$. Thus τ is defined as a *section of the* $\wedge^{top}\mathcal{T}$ *for the tangent bundle* $\mathcal{T}$ *of* $\tilde{\mathcal{G}}_0$.

Theorem 1.7.8. *The commutators* $\hat{D}_f(\tau) \stackrel{\text{def}}{=} [D_f, \tau]$ *are well defined for* $f \in \tilde{\mathfrak{g}}$ *and*

$$\hat{D}_f(\tau) = \left(\text{Res}(\Phi^{-1}d\Phi, f)_K - \text{Res}(\Phi f \Phi^{-1}, \varrho)_K dx\right)\tau. \quad (1.7.42)$$

Moreover, in the linear space Funct $(\tilde{\mathcal{G}}_0)\,\tau$,

$$[\hat{D}_f, \hat{D}_{f'}] = \hat{D}_{[f,f']} + (2g)Res\left(\frac{df}{dx}, f'\right) \quad (1.7.43)$$

for the dual Coxeter number g. ❑

Thus the central extension of $\tilde{\mathfrak{g}}$ emerges naturally and automatically for the adjoint action of $\{D\}$ in $\wedge^{top}\mathcal{T}$. Here the coefficient $2g$ appears because it is the ratio of the Killing form and the standard invariant form.

From the viewpoint of the next section, τ is the coinvariant for the basic representation of $\hat{\mathfrak{g}}$ with the central charge $2g$. The basic representations are irreducible quotients of the Weyl modules for the zero starting representations of $\mathfrak{g}$.

Returning to a justification of (c) from Definition 1.7.3, this condition is *not* necessary in Theorem 1.7.8. Moreover, an arbitrary central extension of $\widetilde{\mathfrak{g}}$ can be obtained for a proper choice of the basis in $\widetilde{\mathfrak{g}}_0$. We have arrived at the standard cocycle because our basis $\{I_A^K\}$ was invariant with respect to the differentiation d/dx. So, given $\widetilde{\mathfrak{g}}_r$, (c) is equivalent to the existence of a basis in $\widetilde{\mathfrak{g}}_0$ that is "compatible" with the adjoint action of this Lie subalgebra. Then the commutators of the differentiations D_f for the "rational" f will contain no central additions. I think such a basis always exists.

As for the theory of KZ, such a construction would give a definition of the r–matrix KZ for the most general class of r–matrices, which are not unitary and are not supposed to depend on the differences.

1.8 Coinvariant, integral formulas

The main object of this section is the r–matrix coinvariant. We establish that it satisfies the r–matrix KZ when differentiated with respect to the Sugawara elements L_{-1}, and use this to justify the integral formulas for the basic trigonometric KZ. We also discuss KZB. The notation is from the previous section.

1.8.1 Coinvariant

Given $\mathfrak{g}$–modules $V_1, \ldots, V_n$, let $V \stackrel{\text{def}}{=\!=} V_1 \otimes \cdots \otimes V_n$. It has a natural structure of a $\mathfrak{g}$–module and a $\widetilde{\mathfrak{g}}_0$–module:

$$\begin{aligned} &g(v_1 \otimes v_2 \otimes \cdots \otimes v_n) = g^1(0)v_1 \otimes v_2 \otimes \cdots \otimes v_n + \\ &v_1 \otimes g^2(0)v_2 \otimes \cdots \otimes v_n + \cdots + v_1 \otimes v_2 \otimes \cdots \otimes g^n(0)v_n, \\ &\text{where } g = (g^1, \ldots, g^n),\ g^i = \sum_{m \geq 0} g^i_m x_i^m,\ g^i(0) = g^i_0. \end{aligned} \tag{1.8.1}$$

Setting $\widetilde{\mathfrak{g}}_0 \oplus \mathbb{C}c = \widehat{\mathfrak{g}}_0$, we define the induced module, the **Weyl module**,

$$M_V^\sigma = \mathrm{Ind}_{\widehat{\mathfrak{g}}_0}^{\widehat{\mathfrak{g}}} V, \tag{1.8.2}$$

where the central element c acts as $\sigma \in \mathbb{C}$, i.e., $c \cdot v = \sigma v\ \forall v$. Because of the decomposition $\widetilde{\mathfrak{g}}_r \oplus \widetilde{\mathfrak{g}}_0 = \widetilde{\mathfrak{g}}$, given any $m \in M_V^\sigma$, there exists a unique element $\pi(m) \in V$ such that $m - \pi(m) \in \widetilde{\mathfrak{g}}_r M_V^\sigma$. The linear map $\pi : M_V^\sigma \to V$ is called the **coinvariant**. Its defining property is

$$\pi(g_r m) = \pi(m)\ (\forall g \in \widetilde{\mathfrak{g}}_r, m \in M_V^\sigma). \tag{1.8.3}$$

Let $\{I_a\}$ be a basis of $\mathfrak{g}$ and $\{I_a^*\}$ the dual basis. We put $I_{a,k}^i = (I_a)^i x_i^k \in \widetilde{\mathfrak{g}}^i$ and $I_{a,k}^{*i} = (I_a^*)^i x_i^k$. The **Sugawara element** of degree -1 at u_i is given by the series

$$L_{-1}^i = \sum_{m\geq 0}\sum_a I_{a,-1-k}^i I_{a,m}^{*i}, \tag{1.8.4}$$

which belongs to a completion of $U(\widehat{\mathfrak{g}})$ (to a completion of $U(\widehat{\mathfrak{g}}^i)$, to be exact).

The definition does not depend on the choice of the basis $\{I_a\}$, and the action of the Sugawara elements in M_V^σ is well defined. They commute with each other, $[L_{-1}^i, L_{-1}^j] = 0$, because different components of $U(\widehat{\mathfrak{g}})$ are pairwise commutative.

We want to determine the dependence of the coinvariant on the positions of the points $u_1, \ldots, u_n$. So we need to enlarge the algebras and modules under consideration assuming that the elements are functions on u. Let $\mathcal{U}$ be the algebra of $\mathbb{C}$-functions of $u_1, \ldots, u_n \in U_0$, $\widehat{\mathfrak{g}}(U_0) = \mathcal{U}\otimes\widehat{\mathfrak{g}}$, $M_V^\sigma(U_0) = \mathcal{U}\otimes M_V^\sigma$. The functions that we will consider will have singularities along the diagonals $\{u_i = u_j\}$.

The definition of the coinvariant remains unchanged, but now the values of π on $M_V^\sigma(U_0)$ will belong to $\mathcal{U}\otimes V$. When calculating with the coinvariant, we always assume that the u_i are pairwise distinct. Otherwise π is not well defined.

We extend the derivatives $\partial/\partial u_i$ from $\mathcal{U}$ to $g(u) \in \widehat{\mathfrak{g}}(U_0)$ and $m(u) \in M_V^\sigma(U_0)$, setting

$$\frac{\partial}{\partial u_i} I_{a,k}^j = 0, \ \frac{\partial}{\partial u_i}(I_{a,k}^j v) = 0, \tag{1.8.5}$$

for all $v \in V$. Here we mean the original V, without replacing $\mathbb{C}$ by $\mathcal{U}$.

We note that the derivatives may be highly nontrivial at the elements in the form $f_r v_{,}$, where f_r is the "rational" component of $f \in \widetilde{\mathfrak{g}}$, i.e., the projection of f onto $\widetilde{\mathfrak{g}}_r$.

For instance, let $m(u) = (g^i/x_i)_r v$ for $v \in V,\ g \in \mathfrak{g},\ g^i = g^{(i)} = 1\otimes\cdots\otimes \overset{i}{g} \otimes\cdots\otimes 1$. Then

$$\frac{\partial m(u)}{\partial u_k} = \frac{\partial}{\partial u_k}\{\prod_{j=1}^n (r(x_j+u_j-u_i), g)\}v = \tag{1.8.6}$$

$$=\{\prod_{j=1}^n (\delta_{jk}\frac{\partial}{\partial x_k} - \delta_{ik}\frac{\partial}{\partial x_j})(r(x_j+u_j-u_i), g)\}v, \tag{1.8.7}$$

where $(a\otimes b, c) = (b,c)a$ for the standard form on $\mathfrak{g}$.

This example demonstrates that the dependence of the coinvariant of u may be complicated, since the calculation of π is based on the factorization. It also shows that one can express the u-derivatives in terms of x-derivatives, which is important in the next theorem.

We will use the notation from (1.7.35) and (1.7.33):

$$\varrho = \sum_a \rho_a I_a \in U(\mathfrak{g}) \text{ for}$$
$$(\Omega/x - r(x))(x=0) = \sum_a \rho_a \otimes I_a, R_i \ = \ \varrho^i - \sum_{j(\neq i)} r^{ji}. \tag{1.8.8}$$

Here $\varrho^i = \varrho^{(i)}$ and the values of $r^{ij} = r(u_i - u_j)^{(i,j)}$ are considered as endomorphisms of V. Recall that $-r^{ji} = r^{ij}$ for unitary r. It is called **quasi-unitary** if $d(\Theta)/dx = 0$ for $\Theta = r^{12} + r^{21}$.

Theorem 1.8.1. *Let $r(x)$ be a quasi-unitary r–matrix, $\widetilde{\mathfrak{g}}_r$ the factorizing subalgebra of $\widehat{\mathfrak{g}}$ corresponding to $r(x)$. For V, M_V^σ, π, and L_{-1}^i defined above,*

$$\pi\left(\kappa\frac{\partial}{\partial u_i} m(u) + L_{-1}^i m(u)\right)$$
$$= \left(\kappa\frac{\partial}{\partial u_i} + \sum_{j\neq i} r^{i,j}(u_i - u_j) + \rho^i\right)\pi(m(u)), \tag{1.8.9}$$

where $\kappa = \sigma + g$, g is the dual Coxeter number of $\mathfrak{g}$, $m(u) \in M_V^\sigma(U_0)$.

Proof. We will consider here the case when $\kappa = 0$. Only this degeneration will be applied later to the integral formulas. See [C9] for the general case.

One has

$$L_{-1}^i v = \sum_a (I_a^i/x_i) I_a^{*i} v = \sum_a ((I_a^i/x_i)_r + \rho_a^i) I_a^{*i} v \tag{1.8.10}$$

for $v \in V$. The $\widetilde{\mathfrak{g}}_r$–invariance of π (see (1.8.3)) gives:

$$\pi(L_{-1}^i v) = \sum_a \rho_a^i I_a^{*i} v - \sum_{a,j\neq i} (I_a^i/x_i)_r^j I_a^{*i} v, \tag{1.8.11}$$

where f_r^j here and later is the expansion of f_r ($\in \widetilde{\mathfrak{g}}_r$) at u_j with respect to $x_j = x - u_j$. Since $\sum_a (I_a^i/x_i)_r^j I_a^{*i} v = r^{ji} v = r(u_j - u_i)^{(i,j)}$, thanks to relation (1.7.27), we obtain (1.8.9) for a constant $m(u) = v$.

Any element of M_V^σ can be represented as $v + \sum f_r m$ for proper f_r and $m \in M_V^\sigma$. Since the level is critical ($\kappa = 0$) the Sugawara elements commute with any $f = (f^1, \ldots, f^n) \in \widetilde{\mathfrak{g}}$. Indeed, the general relation is

$$[L_{-1}^i, f] = -\kappa \partial f^i / \partial x_i. \tag{1.8.12}$$

See [Ka]. So $\pi(L_{-1}^i(f_r m)) = \pi(f_r(L_{-1}^i m)) = 0$, and (1.8.9) results from the above calculation with v. ❑

As a by-product, we can establish the self-consistency of the r–matrix KZ from (1.7.35):

$$\kappa\partial(\Phi)/\partial u_i = R_i\Phi,\ 1 \le i \le n. \tag{1.8.13}$$

Moreover, we actually have a *generic formula* for solutions of this equation. Namely,

$$\Phi = \exp\Big(-\sum_{i=1}^{n}\big((x_i/\kappa)L_{-1}^i\big)\Big)m \quad \text{for arbitrary constant} \quad m \in M_V^\sigma$$

satisfies the KZ equation.

Comment. This is analogous to the general claim that a τ–function is a universal solution of soliton equations. It is formal and cannot be used for constructing explicit solutions without special analytic or algebraic methods. In soliton theory, the main "constructive" applications of this general claim are

(a) the Backlund–Darboux transformations of the solutions, and

(b) the solutions in terms of θ–functions and degenerations.

The integral formulas for KZ, which will be discussed next, have many points in common with the theory of the BD transforms as well as with the θ–solutions. This is not surprising, since the τ–function in soliton theory (see the previous section) and the coinvariant π *coincide*. The soliton equations are different from KZ because they correspond to different flows. In soliton theory, we consider mainly the flows for the vector fields D_f ($f \in \widetilde{\mathfrak{g}}$) from (1.7.40). As for KZ, the flows correspond to the Sugawara elements.

1.8.2 Integral formulas

We keep the same notation as in the previous subsection. However, we assume now that every V_i $(1 \le i \le n)$ is a highest weight module relative to $\mathfrak{b}_+$. So V_i is generated by the **vacuum vector** ν_i associated to a weight $\lambda_i \in \mathbb{C}^n$ (the highest weight): $h_\alpha(vac) = (\alpha, \lambda)vac$.

The above consideration will be applied to $V' = V \otimes V_{n+1} \otimes \cdots V_{n+m}$, where $V = V_1 \otimes \cdots \otimes V_n$ is from the previous section, and the new $\widehat{\mathfrak{g}}$–modules are one-dimensional:

$$V_i = \mathbb{C} = \mathbb{C}\nu_0 \ (n+1 \le i \le n+m) \ \text{ with zero action of } \mathfrak{g}.$$

The elements ν_0 are the basic vectors in these modules.

Respectively, we introduce the **induced or Weyl $\widehat{\mathfrak{g}}$–module** $M_{V'}^\sigma$, where the central element c acts as the scalar σ, and with the coinvariant $\pi : M_{V'}^\sigma \to V'$. Since the vacuum vectors ν_0 in V_{n+i} $(i > 0)$ are fixed, we can identify V and V'. Thus the values of the coinvariant will actually be in the same old space V. Note that the Weyl modules defined for zero $\mathfrak{g}$–modules are very nontrivial, and there are no obvious connections between M_V^σ and $M_{V'}^\sigma$.

The points $\{u_1, \ldots, u_n\}$ will be called **old points**, and $\{u_{n+1}, \ldots, u_{n+m}\}$ **new points**. The same names will be used for the corresponding indices. Recall that $l \stackrel{\mathrm{def}}{=} \operatorname{rank} \mathfrak{g}$.

We fix a sequence of numbers $(n+1)', \ldots, (n+m)' \in \{1, \ldots, l\}$, *which may coincide*, and define

$$w = \pi(\hat{m}),\ \hat{m} = \nu_1 \otimes \cdots \otimes \nu_n \otimes \left(\frac{f_{(n+1)'}}{x_{n+1}}\right)\nu_0 \otimes \cdots \otimes \left(\frac{f_{(n+m)'}}{x_{n+m}}\right)\nu_0, \quad (1.8.14)$$

$\hat{m} \in M^{\sigma}_{V'}$, $w \in V = V'$.

The definition of w is applicable to any r–matrices. However, w gets sufficiently simple and can be used for the integral formulas only when r are sufficiently special. Therefore we need to reduce the generality at this point.

From now on, we will consider the basic trigonometric r–matrix from (1.7.16):

$$r(x) = \frac{1}{2}\coth\left(\frac{x}{2}\right)\Omega + \frac{1}{2}\sum_{\alpha \in R^+} \frac{f_\alpha \otimes e_\alpha - e_\alpha \otimes f_\alpha}{(e_\alpha, f_\alpha)} + \frac{1}{2}\sum_{m=1}^{l} A(h_i) \otimes h_i^*, \quad (1.8.15)$$

depending on an arbitrarily given endomorphism $A : \mathfrak{h} \to \mathfrak{h}$. Recall that

$$\varrho = h_\rho - \frac{1}{2}\sum_{k=1}^{l} A(h_k) h_k^*, \quad \rho = \frac{1}{2}\sum_{\alpha \in \Sigma_+} \alpha. \quad (1.8.16)$$

We identify the elements from $\mathfrak{h}$ with the corresponding weights and the corresponding linear forms on the complexification $\mathbb{C}^l$ of $\mathfrak{h}$:

$$h_b = \sum_{k=1}^{l} (b, b_k) h_k, \ h_k = h_{\alpha_k}, \ (h_b, \lambda) = (b, \lambda) \text{ for } b, \lambda \in \mathbb{C}^l \quad (1.8.17)$$

Here the invariant bilinear form on $\mathfrak{h}$ is extended to the complexification.

Using the above sequence of numbers $\{i'\}$ for $i > n$, let

$$\Lambda_i = \lambda_i \text{ for } 1 \le i \le n, \ \Lambda_i = -\alpha_{i'} \text{ for } i > n, \ \Lambda = \sum_{i=1}^{n+m} \Lambda_i. \quad (1.8.18)$$

Note that the Λ_i for new $i > n$ have nothing to do with the corresponding weights, which are all zero. Such a uniform notation is convenient in the following definition:

$$\omega_i = \sum_{1 \le k \le n+m, k \ne i} (\Lambda_i, \Lambda_k)\frac{1}{2}\coth\left(\frac{u_i - u_k}{2}\right) + \left(\varrho - \frac{A^*(\Lambda)}{2}, \Lambda_i\right), \quad (1.8.19)$$

where $(Aa, b) = (a, A^*b)$ on $a, b \in \mathfrak{h}$. These functions are logarithmic derivatives. Namely,

$$\omega_i = \partial\omega/\partial u_i \text{ for } \omega = \tag{1.8.20}$$
$$= \prod_{1\le i<j\le n+m} (e^{(u_i-u_j)/2} - e^{(u_j-u_i)/2})^{(\Lambda_i,\Lambda_j)/\kappa} \prod_{1\le i\le n+m} \exp\Big((\varrho - \frac{A^*(\Lambda)}{2}, \Lambda_i)\frac{u_i}{\kappa}\Big).$$

We will also use

$$\omega_J = \sum_{j\in J} \omega_j \tag{1.8.21}$$

for subsets of new points(indices) $J \subset \{n+1, \ldots, n+m\}$.

Definition 1.8.2. *(i) An ordered sequence $c = (j_1, \ldots, j_s\,;\, i)$ for the pairwise distinct new indices $j_1, \ldots, j_s \in \{n+1, \ldots, n+m\}$ and one old index $1 \le i \le n$ is called a* **chain**.

(ii) An ordered sequence $d = (c_1, \ldots, c_r)$ of chains is called a **diagram** *if every new $n+1 \le j \le n+m$ belongs to one and only one chain (no restrictions for the old indices).*

(iii) Given a linear ordering on the new indices $\succ$, a diagram

$$d = \Big(c_1 = (j_1^{(1)}, j_2^{(1)}, \ldots), c_2 = (j_1^{(2)}, j_2^{(2)}, \ldots), \ldots, c_r = (j_1^{(r)}, j_2^{(r)}, \ldots)\Big)$$

is called **well-ordered** *if $j_1^{(1)} \succ j_1^{(2)} \succ \cdots \succ j_1^{(r)}$.*

(iv) For a diagram d, the **component** *of an old i, denoted by $comp_i(d)$, is a union of all new indices connected with i by a chain from d.*

Definition 1.8.3. *(i) Given a chain $c = (j_1, \ldots, j_s; i)$, we set*

$$F_c = \frac{[\cdot\cdot[f_{j'_1}, f_{j'_2}], f_{j'_3}], \ldots, f_{j'_{s-1}}], f_{j'_s}]^i}{(e^{u_{j_1}-u_{j_2}} - 1)(e^{u_{j_2}-u_{j_3}} - 1)\cdots(e^{u_{j_s}-u_i} - 1)}. \tag{1.8.22}$$

(ii) Given a diagram $d = (d_1, \ldots, d_r)$, we set

$$F_d = F_r \cdots F_2 F_1 \in U(\mathfrak{g})^{\otimes n}, \text{ where } F_j = F_{c_j},\ F_{\{\emptyset;i\}} = 1. \tag{1.8.23}$$

Theorem 1.8.4. *We fix an ordering $\succ$. Then $w = \pi(\hat{m}) = \sum_d F_d(vac)$, where $vac = \nu_1 \otimes \cdots \otimes \nu_n \otimes \nu_0 \otimes \cdots \otimes \nu_0$ and the summation is over all well-ordered diagrams. In particular, w does not depend on the central charge σ. Conversely, $\sum_d F_d(vac)$ does not depend on the ordering. The following sums do not depend on the particular choice of the ordering either:*

$$w_i[J] = \sum_d F_d(vac),\ w_i(J) = \sum_{d'} F_{d'}(vac), \tag{1.8.24}$$

where the summation is over all well-ordered diagrams such that $comp_i(d) = J$ (in the first sum) and $J \subset comp_i(d')$ (in the second sum) for any set of new points J.

Theorem 1.8.5. *For any old i $(1 \le i \le n)$ and $R_i = \varrho^i - \sum_{j(\neq i)} r^{ji}$,*

$$R_i w = \omega_i w + \sum_{j=n+1}^{n+m} \omega_j w_i(j) = \omega_i w + \sum_J \omega_J w_i[J], \tag{1.8.25}$$

where J runs over all new subsets ($\subset \{n+1, \ldots, n+m\}$).

Theorem 1.8.6. *For any old i,*

$$(R_i - \kappa\partial/\partial u_i)W = \kappa \sum_{j=n+1}^{n+m} \partial W_i(j)/\partial u_j = \kappa \sum_{new\ J} \partial W_i[J]/\partial u_j, \tag{1.8.26}$$

where $W = w\omega$ and $W_i(j) = w_i(j)\omega$, $W_i[J] = w_i[J]\omega$ for ω from (1.8.20).

Proof. We will establish the first two theorems in the next subsection. Let us deduce the third one from them now. One has:

$$\kappa\partial w/\partial u_i = -\kappa \sum_{j=n+1}^{n+m} \partial w_i(j)/\partial u_j, \tag{1.8.27}$$

since $\partial F_c/\partial u_i = \partial F_{j_1}/\partial u_{j_1} + \ldots + \partial F_{j_s}/\partial u_{j_s}$ for any chain $c = (j_1, \ldots, j_s; i)$. Indeed, the formula for F_c depends only on the differences $u_{j_p} - u_{j_q}$ for $1 \le p, q \le s+1$, $j_{s+1} = i$. Hence, the same holds for any F_d with $comp_i(d) = J$. The relations $\sum_{j=n+1}^{n+m} \omega_j w_i(j) = \sum_{\{new\ J\}} \omega_J w_i[J]$ are obvious because $w_i(j) = \sum_{J \ni j} w_i[J]$. ❑

To get the integral formulas for KZ, we set $\Phi = \int W du_{n+1} \cdots du_{n+m}$, where the integration contours are taken to ensure that

$$\int (\partial W_i(j)/\partial u_j) du_{n+1} \cdots du_{n+m} = 0$$

for all new j. Then $(\kappa\partial/\partial u_i - R_i)\Phi = 0$. The proper choice of contours and the description of the spaces of corresponding solutions Φ can be a difficult problem, especially if the κ are not assumed to be generic (see [Va]). We discuss in this work the algebraic machinery only: the integrands, but not the integrals.

1.8.3 Proof

For the sake of simplicity we will consider here the Yang r–matrix only. We refer the reader to [C9] for the general case (somewhat more general than that considered above). Let us degenerate the main formulas, replacing the trigonometric formulas (hyperbolic, to be exact) by rational ones.

From now on, $r = \Omega/x$ and

$$\omega_i = \sum_{1\le k\le n+m, k\neq i} \frac{(\Lambda_i, \Lambda_k)}{u_i - u_k}, \ \omega_i = \partial\omega/\partial u_i,$$
$$\omega = \prod_{1\le i<j\le n+m} (u_i - u_j)^{(\Lambda_i,\Lambda_j)/\kappa}. \tag{1.8.28}$$

Since r is unitary and $\varrho = 0$, we can simplify R_i:

$$R_i = \sum_{j=1}^{n} r^{ij} = \sum_{j=1}^{n} \frac{\Omega^{ij}}{u_i - u_j}, \ j \neq i, \ 1 \le i \le n. \tag{1.8.29}$$

Given a chain $c = (j_1, \ldots, j_s; i)$, we set

$$F_c = \frac{[\cdot\cdot[[f_{j'_1}, f_{j'_2}], f_{j'_3}], \ldots, f_{j'_{s-1}}], f_{j'_s}]^i}{(u_{j_1} - u_{j_2})(u_{j_2} - u_{j_3}) \cdots (u_{j_s} - u_i)}. \tag{1.8.30}$$

The definition of F_d for a diagram $d = (c_1, \ldots, c_r)$ is the same:

$$F_d = F_r \cdots F_2 F_1 \in U(\mathfrak{g})^{\otimes n} \text{ for } F_j = F_{c_j}. \tag{1.8.31}$$

Let us fix the ordering and check that $w = \pi(\hat{m}) = \sum_d F_d(vac)$, where $vac = \nu_1 \otimes \cdots \otimes \nu_n \otimes \nu_0 \otimes \cdots \otimes \nu_0$ and the summation is over all well-ordered diagrams. It is a straightforward calculation based directly on the defining property of the coinvariant, that is, $\pi(f_r\tilde{v}) = 0$ for $f \in \tilde{\mathfrak{g}}_r$ and $\tilde{v} \in M^{\sigma}_{V'}$.

Recall that

$$\overline{\left(\frac{g}{x_i}\right)} = \frac{g}{x - u_i}, \ \left(\frac{g}{x_i}\right)_r$$
$$= \left(\frac{g}{x_1 + u_1 - u_i}, \ldots, \frac{g}{x_{n+m} + u_{n+m} - u_i}\right), \tag{1.8.32}$$

where $x_i = x - u_i$, $g \in \mathfrak{g}$, and $1 \le i \le n+m$. Here g/x_i is considered as an element of the i–th component of $\tilde{\mathfrak{g}}$ (defined in a formal neighborhood of u_i only) and $g/(x - u_i)$ is a rational function of x (defined globally). So

$$\pi\left(\left(\frac{g}{x_i}\right)^i \tilde{v}\right) = \pi\left(\left(\frac{g}{x_i}\right)^i_r \tilde{v}\right)$$
$$= -\pi\left(\sum_{1\le j\le n+m, j\neq i} \left(\frac{g}{x_j + u_j - u_i}\right)^j \tilde{v}\right). \tag{1.8.33}$$

If $\tilde{v} = \tilde{v}_1 \otimes \cdots \otimes \tilde{v}_{n+m}$ and $\tilde{v}_j \in V_j$ for some $j \neq i$, then further simplification is possible:

$$\left(\frac{g}{x_j + u_j - u_i}\right)^j \tilde{v} = \tilde{v}_1 \otimes \cdots \left(\frac{g}{u_j - u_i}\right)^j \tilde{v}_j \cdots \otimes \tilde{v}_{n+m}. \tag{1.8.34}$$

Moreover, if j is new, then $(g/(u_j - u_i))^j \tilde{v}_j = 0$.

If the point j is new ($j > n$) and $\tilde{v}_j = (f^j/x_j)\nu_0$ for $f \in \mathfrak{g}$, then we can proceed as follows:

$$\left(\frac{g}{x_j + u_j - u_i}\right)^j \tilde{v}_j = \left(\frac{g}{x_j + u_j - u_i}\right)^j \frac{f^j}{x_j}\nu_0 =$$
$$= \left[\left(\frac{g}{x_j + u_j - u_i}\right)^j, \frac{f^j}{x_j}\right]\nu_0 = \frac{1}{u_i - u_j}\frac{[f,g]^j}{x_j}\nu_0. \tag{1.8.35}$$

Formula for the integrand. Now let $\hat{m}$ be from (1.8.14), $w = \pi(\hat{m})$,

$$\hat{m} = \nu_1 \otimes \cdots \otimes \nu_n \otimes \left(\frac{f_{(n+1)'}}{x_{n+1}}\right)\nu_0 \otimes \cdots \otimes \left(\frac{f_{(n+m)'}}{x_{n+m}}\right)\nu_0. \tag{1.8.36}$$

We start with the new index i which is maximal with respect to the ordering $\succ$, and then use (1.8.33) in combination with (1.8.35) to clear up the i–th component, by replacing $(f^i_{i'}/x_i)\nu_0$ by the sum over all remaining components. Applying (1.8.33) again and again, we will eventually come to the element from $V \otimes \nu_0 \otimes \cdots \otimes \nu_0$ that contains pure ν_0 at all new points and coincides with its coinvariant. Note that we have to apply this procedure to *all* terms of the sum obtained after the previous step.

In this calculation, the terms are in one-to-one correspondence with the chains $c = (j_1, \ldots, j_s; i)$ from Definition 1.8.2. The indices $j_1, \ldots, j_s$ must be pairwise distinct because $g\nu_j = 0$ for any $g \in \mathfrak{g}$. There can be only one old i in the chain due to (1.8.34). The old point is always at the end of the above **simplification process**. If all the chains have reached their end points, we start the next chain, picking the maximal new point among the untouched ones and follow the same procedure. The new points that have already been cleared of the f–terms will not participate. We come to the above definition of the diagram and also establish the formula for w, which is the first claim of Theorem 1.8.4.

We can take any ordering $\succ$ in this calculation. The result will of course be the same. Note that if we did not have an interpretation of w in terms of the coinvariant, establishing the fact that different orderings result in the same formula is a mundane combinatorial reasoning (cf. [SV]).

Let us prove that

$$w_i[J] = \sum_d F_d(vac), \; w_i(J) = \sum_{d'} F_{d'}(vac) \tag{1.8.37}$$

do not depend on the ordering as well. Here the summation is over all well-ordered diagrams such that $comp_i(d) = J$ or $comp_i(d') \subset J$. It is sufficient to examine $w_i[J]$ only, because the quantities $\{w_i(J)\}$ can be expressed in terms of $\{w_i[J]\}$.

We will deform the positions of the points $u_1, \ldots, u_{n+m}$. Let

$$\tilde{u}_j = u_j + \delta \text{ for } j \in \{J \cup i\}, \ \tilde{u}_j = u_j \text{ for } j \notin \{J \cup i\}. \tag{1.8.38}$$

The terms F_d will remain unchanged iff they appear in the sum for $w_i[J]$. Moreover,

$$w_i[J] = \lim_{\delta \to \infty} \tilde{w}, \text{ where } \tilde{w} = w(\tilde{u}_1, \ldots, \tilde{u}_{n+m}). \tag{1.8.39}$$

Indeed, all F_d with $comp_i(d) \neq J$ will contain at least one δ in the denominator. This representation does not depend on the ordering.

Thus we proved Theorem 1.8.4 and established that $w = \sum_J w_i[J]$ for any old i.

Calculating $R_i w$. Let us establish the formulas

$$R_i w = \omega_i w + \sum_{newJ} \omega_J w_i[J].$$

See Theorem 1.8.5. To simplify the index notation, we set $i = n$. So we need to check that

$$\left(\sum_{1 \leq j < n} r^{nj}(u_n - u_j) \right) \pi(\hat{m}) = \sum_{J \subset \{n+1, \ldots, n+m\}} \omega_J w_n[J] + \omega_n \pi(\hat{m}). \tag{1.8.40}$$

Since F_d, $w = \pi(\hat{m})$, and $w_i[J]$, all of them, do not depend on the central charge σ, one may put $\kappa = \sigma + g = 0$ and apply Theorem 1.8.1:

$$\left(\sum_{1 \leq j < n} r^{ij}(u_n - u_j) \right) \pi(w) = \pi\left(L^n_{-1} w\right). \tag{1.8.41}$$

Explicitly,

$$\pi\left(L^n_{-1} w\right) = \pi\left(\left(\sum_{\alpha > 0} \left(\frac{e_\alpha}{x_n}\right) \frac{f_\alpha}{(e_\alpha, f_\alpha)} + \sum_{1 \leq p \leq l} \left(\frac{h_p}{x_n}\right) h_p^* \right)^n w \right). \tag{1.8.42}$$

Later on, we will drop the upper right indices, which indicate the component, if confusion is impossible. Once an expression contains x_j, it already means that it acts on the j–th component.

Making use of (1.8.33),(1.8.34), and (1.8.35), we have:

$$\pi\left(\left(\sum_{1 \leq p \leq l} \left(\frac{h_p}{x_n}\right) h_p^* \right) \hat{m} \right) = \omega_n \pi\left(\hat{m}\right). \tag{1.8.43}$$

Indeed, the left-hand side is $\pi(\hat{m})$ multiplied by

$$\sum_{j<n} \frac{1}{u_n - u_j} \sum_p (h_p, \Lambda_j)(h_p^*, \Lambda_n) = \sum_{j<n} \frac{(\Lambda_j, \Lambda_n)}{u_n - u_j}, \tag{1.8.44}$$

that is, the contribution of the old points, plus that of the new points:

$$\sum_{j>n} \frac{1}{u_n - u_j} \sum_p (h_p, \Lambda_j)(h_p^*, \Lambda_n) = \sum_{j>n} \frac{(\Lambda_j, \Lambda_n)}{u_n - u_j}. \tag{1.8.45}$$

The formulas for $j < n$ and $j > n$ are the same, but their meaning is different. In the second formula, we used that $[h_p, f_{j'}] = -(h_p, \alpha_{j'}) = (h_p, \Lambda_j)$ due to the definition of Λ_j at the new points.

Let us now get rid of (e_α/x_n). The calculation is more involved than that with h, because e_α may "interact" with $f_{j'}$ at the new points. However it is still not too complicated (in the case of the basic trigonometric r–matrix). Since ν_j $(1 \le j < n)$ are the highest weight vectors, we move e_α only to the right (to the new points):

$$\begin{aligned} &\pi\left(\left(\left(\frac{e_\alpha}{x_n}\right)\frac{f_\alpha}{(e_\alpha, f_\alpha)}\right)^n \hat{m}\right) = \\ =&\pi\left(\sum_{n+1\le j\le n+m}\left(\frac{[e_\alpha, f_{j'}]}{(u_n-u_j)x_j}\right)^j\left(\frac{f_\alpha}{(e_\alpha, f_\alpha)}\right)^n \hat{m}\{j\}\right). \end{aligned} \tag{1.8.46}$$

Here $\hat{m}\{j\}$ denotes $\hat{m}$ without $f_{j'}/x_j$. Namely (see (1.8.36)):

$$\hat{m}\{j\} = \nu_1 \otimes \cdots \otimes \nu_n \otimes \cdots \left(\frac{f_{(j-1)'}}{x_{j-1}}\right)\nu_0 \otimes \nu_0 \otimes \left(\frac{f_{(j+1)'}}{x_{j+1}}\right)\nu_0 \otimes \cdots. \tag{1.8.47}$$

Next we move every $([e_\alpha, f_{j'}]/x_j)$ to all components $\tilde{j} \ne j$ to clear the j–th component of $[e, f]$. Then we will get the triple commutators $[[e, f], f]$ at the new $\tilde{j}$.

Next we will go to the components $\hat{j} \ne j, \tilde{j}$ and produce the four-term commutators. Then the five-term commutators will be produced, and so on. Note that the poles will always be simple after every step.

We stop this process when the s–fold commutator is 0 or when we come to an old point (including the point u_n) for the first time.

Let us examine the resulting commutators. Since $f_{j'}$ are simple, any commutator given by this procedure is proportional to

$$\text{(a) } (e_\beta/x_k)^k, \quad \text{(b) } (h_\beta/x_k)^k, \text{ or } \quad \text{(c) the central element } c \tag{1.8.48}$$

for a certain $\beta > 0$ and a new index k.

In case (a), we may continue and get another new point.

In case (b), there will be one more (final) step when we replace $(h_\beta/x_k)^k$ by the corresponding sum over all points (old and new). After this we will stop.

In case (c), we simply plug in $c = \sigma = -g$.

Note that since one never obtains f–terms in this procedure, the only way to reach the old points is via (b), but the point u_n must be excluded here.

Thus the result is given in terms of the *chains* again. Namely, we arrive at the chains

$$j_s = j, j_{s-1} = \tilde{j}, j_{s-2} = \hat{j}, \ldots, j_1 = k.$$

However, now we read $\{j\}$ in the opposite order and the chain describe a different process, that is, the process of eliminating e_α/x_n. Note that $\beta = \alpha_{k'}$ for the last index $k = j_1$.

Once we know all the indices, it is not difficult to determine the exact formula up to the contribution of the central element (the case (c)) and the chains that go back to the starting point u_n. The calculation is based on the identities:

$$\begin{aligned}
&\text{if } [\cdot\cdot[f_{j_1'}, f_{j_2'}], \ldots, f_{j_s'}] = c f_\alpha, \text{ then} \\
&[\cdot\cdot[e_\alpha, f_{j_s'}], f_{j_{s-1}'}], \ldots, f_{j_1'}]) = (-1)^{s-1} c(f_\alpha, e_\alpha) h_{j_1'}.
\end{aligned} \tag{1.8.49}$$

Setting $\phi = (u_n - u_{j_1})(u_{j_1} - u_{j_2}) \cdots (u_{j_{s-1}} - u_{j_s})$, we obtain

$$\begin{aligned}
&\pi\left(\left(\left(\frac{e_\alpha}{x_n}\right) \frac{f_\alpha}{(e_\alpha, f_\alpha)}\right) \hat{m}\right) = \\
&= \phi^{-1} \pi\Big(\big((f_\alpha)^n \otimes 1 \otimes \cdots (h_{\alpha_{k'}}/x_k) \cdots \otimes 1\big)\, \hat{m}\{J\}\Big).
\end{aligned} \tag{1.8.50}$$

Here $k = j_1$, $\hat{m}\{J\}$ denotes $\hat{m}$ without $(f_{j'}/x_j)$ for all $j \in J$, namely,

$$\begin{aligned}
\hat{m}\{J\} &= \nu_1 \otimes \cdots \otimes \nu_n \otimes \cdots \left(\frac{f_{1'}}{x_1}\right) \nu_0 \otimes \cdots \otimes \overbrace{\nu_0}^{j_1} \otimes \\
&\cdots \otimes \left(\frac{f_{()'}}{x_{()}}\right) \nu_0 \cdots \overbrace{\nu_0}^{j_2} \cdots \left(\frac{f_{()'}}{x_{()}}\right) \nu_0 \cdots\cdots \overbrace{\nu_0}^{j_s} \cdots \left(\frac{f_{(n+m)'}}{x_{n+m}}\right).
\end{aligned} \tag{1.8.51}$$

Here the set $\{j\}$ is not ordered. For instance, one can get $j_2, \ldots, j_3, \ldots, j_1, \ldots$.

Comparing (1.8.49) with the definition of F_c from (1.8.30) for the chain $c = (j_1, \cdots, j_n; n)$, represented in the form

$$F_c = \phi^{-1} \pi\left(\left((f_\alpha)^n \otimes 1 \otimes \cdots \overbrace{1}^{k} \cdots \otimes 1\right) \hat{m}\{J\}\right), \tag{1.8.52}$$

we see that these two expressions are different only due to the appearance of $h_{\alpha_{k'}}/x_k$ in (1.8.49). These h–terms readily result in a sum over all, old and new, indices $i \neq k$.

Note the cancelation of $(-1)^{s-1}$ from the denominator of the first formula with $(-1)^{s-1}$ from (1.8.49).

This procedure gives the first step. We still need to clear the new components $i \notin J$ that contain $\{f\}$. We will follow exactly the process of calculating w and numerate the terms by the diagrams. The final formula will be the sum of F_d over all diagrams d multiplied by proper ω coming from $h_{\alpha_{k'}}/x_k$ at the first step.

Here we need to use that $w_n[J]$ do not depend on the ordering, because when eliminating e_α we cannot control the endpoint $k = j_1$ of the (first) chain. So we may need to change the ordering of the indices to make k the maximal element.

We arrive at the formula

$$R_n w - \omega_n w - \sum_{new J} \omega_J w_n[J] = S, \tag{1.8.53}$$

where S is a sum of the terms with denominators that may not contain the "chain products" in the form $(u_n - u_{j_1})(u_{j_1} - u_{j_2}) \cdots (u_{j_p} - u_i)$ for any new j and old $i < n$. It is the contribution of the terms of type (c) and the chains ending at the n–th component. The latter can come from h (the case (b)) or from certain e if we go back and fuse them with $(f_\alpha)\nu_n$.

Let us check that $S = 0$. It can be represented as $S = \sum_J S_n[J]$ exactly in the same manner as $w = \sum_J w_n[J]$. We will prove that $S_n[J] = 0$ for any new set J. Consider the deformation $\tilde{u}$ from (1.8.38) for $i = n$ and tend δ to ∞. Then $w \to w_n[J]$. In the sum $\omega_n + \omega_J = \sum_{i,j} \frac{(\Lambda_i, \Lambda_n)}{u_i - u_j}$, exactly one index from any pair $\{i, j\}$ belongs to $\{J \cup n\}$. So it goes to 0 as δ approaches ∞. Since all the terms of R_n contain the differences $u_i - u_n$ for some $i < n$ in the denominators, the left-hand side of (1.8.53) identically equals zero. The right-hand side tends to $S_n[J]$. Thus the latter is zero. The proof of the theorem is completed.

Comment. The direct (combinatorial) definition of the w and W in the rational case is due to Schechtman and Varchenko (Preprint MPI/89-51, 1989). They established that W satisfies KZ up to exact derivatives in terms of the new $\{u_j, j > n\}$. This theorem generalized the paper by Date, Jimbo, Matsuo, and Miwa [DJMM] (the SL_2–case) and that by Matsuo for SL_n. There were also results of Dotsenko, Fateev, Aomoto, Christe, and Flume in this direction. An extended version of the MPI-preprint is [SV].

Paper [C9] (first published as preprint RIMS–699, 1990) contained the interpretation of the Schechtman–Varchenko via the coinvariant, the trigonometric generalization, and Theorem 1.8.5 with new explicit formulas for the exact derivatives. The approach via the Kac–Moody algebras dramatically simplified the formulas from [SV] and their justification.

Theorem 1.8.5 has applications not only to KZ. Actually, it is a pure algebraic statement and must have algebraic corollaries. Indeed, let us choose the parameters $\{u_{n+1}, \ldots, u_{n+m}\}$ to ensure the relations $\omega_j = 0$ for $j > n$. Then w is an eigenvector of the *pairwise commutative* matrices R_i $(1 \le i \le n)$

with the eigenvalues $\{\omega_i\}$. This eigenvalue problem is called the **Gaudin model**. The first results on the diagonalization of the "Hamiltonians" $\{R_i\}$ were obtained by Babujan and Flume in [BF]. See also [FFR]. We will not discuss this direction here.

1.8.4 Comment on KZB

In the elliptic KZ theory, we must first examine the **Baxter–Belavin r–matrix.** It is unique among unitary elliptic classical r–matrices. There are also some other examples of non-unitary type. An important variant of Belavin's r in infinite matrices was introduced by Shibukawa–Ueno.

Theorem 1.8.1 holds for *all* such r. However, there are problems with the integral formulas. Our method can be generalized, but we need to assume that the rational extensions $\bar{f}$ of the elements $f = f_\alpha/x_i$ and $f = e_\alpha/x_i$ are proportional to f_α and e_α, respectively, with certain scalar meromorphic functions as coefficients of proportionality.

As was demonstrated in [FW1], the KZB equation due to Bernard [Be] leads to a Lie algebra of elliptic functions satisfying this very property (see also [Fe]). In this subsection, we will comment on it. The above approach to the integral formulas also can be extended to this case, but we are not going to discuss it here.

Let E be an algebraic elliptic curve, 0 its zero point, and x a local parameter in a neighborhood $U_0 \subset E$ of 0. We fix pairwise distinct $u_1, \ldots, u_n \in U_0$. Given $\mu \in \mathbb{C}$, a **Baker function** φ is a function on E, that is regular apart from U_0 and such that $e^{-\mu x^{-1}}\varphi(x)$ is meromorphic with poles at points $u_1, \ldots, u_n$. Thus Baker functions have an "exponential" singularity at 0. We denote the space of such functions by B_μ.

Baker functions are completely determined by their principal parts, that may be arbitrary. More precisely, given $p(x) = \sum_{i=1}^{n} \sum_{1\le j<\infty} c_{ij}(x-u_i)^{-j}$, there exists a unique function $\tilde{p}(x) \in B_\mu$ such that $\tilde{p}(x) - p(x)$ is holomorphic on U_0. Here μ is generic. It is important because if $\mu = 0$ and there is no essential singularity at 0, then one can define $\tilde{p}$ only when the sum of residues of p equals zero.

Let $\widetilde{p_0}$ be a rational function on E with the principal part

$$q(x) = p(x) - \left(\sum_{i=1}^{n} \mathrm{Res}_{u_i} p(x)dx\right) x^{-1} \tag{1.8.54}$$

normalized by the condition $\widetilde{p_0} - q(x) = x(\cdot)$ (no constant term at 0).

We keep the notation of the previous section: $\mathfrak{g}$ is a simple Lie algebra and $\Sigma = \{\alpha\} \subset \mathbb{R}^l$ is the corresponding root system. Let us fix a vector $\lambda \in \mathbb{C}^n$ and set $\lambda_\alpha = (\lambda, \alpha)$ for the standard invariant form on $\mathbb{C}^l$.

Given a principal part

$$p(x) = \sum_{i=1}^{k}\sum_{j=1}^{m_i} \frac{c_{i,j}}{(x-u_i)^j},$$

let $\widetilde{p_\alpha}$ be the function $\tilde{p}$ from B_{λ_α} with the principal part p. Using this definition we set:

$$\overline{p(x)e_\alpha} = \widetilde{p_{-\alpha}}(x)e_\alpha,\ \overline{p(x)f_\alpha} = \widetilde{p_\alpha}(x)f_\alpha, \tag{1.8.55}$$

$$\overline{p(x)h_\alpha} = \widetilde{p_0}(x)h_\alpha - \left(\sum_{i=1}^{k} \mathrm{Res}_{u_i} p(x)dx\right)\partial_{\lambda_\alpha}, \tag{1.8.56}$$

where $\partial_{\lambda_\alpha}$ is the differentiation $\partial_{\lambda_\alpha}(\lambda_\beta) = (\alpha, \beta)$.

Let us extend $\mathbb{C}$ to the noncommutative algebra $\mathcal{L}$ of differential operators of λ with meromorphic coefficients (we will not specify which and where). We set $\mathfrak{G} = \mathfrak{G} \otimes \mathcal{L}$.

Proposition 1.8.7. *The space $\bar{\mathfrak{G}}$ linearly generated by*

$$\{\overline{p(x)e_\alpha},\ \overline{p(x)f_\alpha},\ \overline{p(x)h_\alpha}, h_\alpha\}$$

for all $\alpha \in \Sigma_+$ and for all principal parts p at $\{u_1, \ldots, u_n\}$ is a Lie algebra.

Proof. The commutator $[\overline{p(x)f_\alpha}, \overline{q(x)f_\beta}]$ is proportional to $[f_\alpha, f_\beta]$ and the coefficient of proportionality is from $B_{\lambda_{\alpha+\beta}}$. So it belongs to $\bar{\mathfrak{G}}$, as do the commutators $[f, e]$, $[e, e]$. The commutators $[\overline{p(x)h_\alpha}, \overline{q(x)h_\beta}]$ are zero, since $\widetilde{p_0}$ do not depend on λ. Let us calculate $[\overline{q(x)h_\alpha}, \overline{p(x)f_\beta}]$. The commutator $[\widetilde{q_0}h_\alpha, \overline{p(x)f_\beta}]$ belongs to $\bar{\mathfrak{g}}$. As to the ∂_λ–term, we need to check that

$$[c\partial_{\lambda_\alpha} + c\frac{h_\alpha}{x},\ \overline{q(x)f_\beta}] \tag{1.8.57}$$

does not contain a pole at $x = 0$. It is easy, since

$$[\partial_{\lambda_\alpha} + \frac{h_\alpha}{x},\ e^{\lambda_\beta x^{-1}} f_\beta] = \left((\alpha, \beta)\frac{f_\beta}{x} - \frac{[h_\alpha, f_\beta]}{x}\right) e^{\lambda_\beta x^{-1}} = 0. \tag{1.8.58}$$

❑

We note that if one of u_i is 0, say, $u_1 = 0$, then the definition still works well. For instance, $\overline{x^{-1}h_\alpha} = -\partial_{\lambda_\alpha}$ at $u_1 = 0$.

One may introduce $\widetilde{\mathfrak{G}}$ and the Kac–Moody algebra $\widehat{\mathfrak{G}}$ using the same definitions. The coefficients of all series are taken from $\mathcal{L}$. Expanding the elements from $\bar{\mathfrak{G}}$, we define $\widetilde{\mathfrak{G}}_r$. We see that

$$\widetilde{\mathfrak{G}} = \widetilde{\mathfrak{G}}_r + \widetilde{\mathfrak{G}}_0,\ \widetilde{\mathfrak{G}}_r \cap \widetilde{\mathfrak{G}}_0 = \mathfrak{H} = \mathfrak{h} \otimes \mathcal{L}. \tag{1.8.59}$$

Thus $\widetilde{\mathfrak{G}}_r$ is somewhat larger than we required in the definition of the *factorizing* Lie subalgebra. The latter must have the zero intersection with the "holomorphic" Lie subalgebra.

Nevertheless, we can introduce the r–matrix describing the projection of $\widetilde{\mathfrak{G}}$ onto $\widetilde{\mathfrak{G}}_r$ "up to $\mathfrak{h}$." It will satisfy the non-unitary Yang–Baxter equation up to $\mathfrak{h}$ and will not depend on the differences, in contrast to the Baxter–Belavin r–matrix. The notion of the coinvariant must be properly changed as well.

Let V_i be $\mathfrak{g}$–modules, $V = \otimes_{i=1}^{n} V_i$, and M_V^σ the corresponding Weyl module. We set $\mathfrak{M}_V^\sigma = M_V^\sigma \otimes \mathcal{L}$ and define the **coinvariant**

$$\pi : \mathfrak{M}_V^\sigma \to \mathfrak{V}/(\mathfrak{h}\mathfrak{V}) \text{ for } \mathfrak{V} = V \otimes \mathcal{L}. \tag{1.8.60}$$

The abstract relation between π and the Sugawara elements and the integral formulas can be extended to this setup.

To obtain KZB in the form [FW1], we need to add the derivative with respect to the τ–parameter of the elliptic curve, the so-called parabolic equation. After this we can transform the r–matrix and make it dependent on the difference. The latter means that the rational continuation $\overline{(\cdot)}$ will be invariant with respect to the shifts in x. The Baker functions will be replaced by the multivalued functions on E; the translations by the periods will produce the multiplicators. The definitions of $\pi, \hat{m}$, and w follow along the same lines.

Eventually we come to the integral formulas due to Felder–Varchenko [FV1]. The justification of their integral formulas is not far from [C9], with an important reservation. The parabolic equation (a special feature of KZB) did not appear in [C9].

Note that the interpretation of KZB via Baker functions is equivalent to the Felder interpretation. Using the Baker functions could be convenient for extending the integral formulas to any (some) algebraic curves (there are certain results in this direction). However, the general theory over algebraic curves is not "integrable," so a connection with reasonable special functions, like the hypergeometric function, cannot be expected. Such a connection is an important motivation of the integral formulas considered in this section.

Chapter 2

One-dimensional DAHA

2.0 Overview

This chapter is mainly based on series of lectures on the one-dimensional double Hecke algebra $\mathcal{HH}$ ("double H") delivered by the author at Harvard University in 2001 and on the paper [CO] written with V. Ostrik.

Sections 0.4–0.5 from Chapter 0 give a comprehensive introduction to the A_1–case. We would like to give some references concerning the relations to classical theory of functions in this section, and comment on the material and the structure of this chapter.

2.0.1 Classical origins

There are deep relations to the theory of special functions including the q–functions and the classical Fourier analysis. We will not try to reconstruct systematically the history of the subject and review the connections. There are many comments in this chapter, but they are fragmentary. Chapter 0 contains additional information, especially Section 0.3; see also [M5, O3, C25, C28, EG].

As for the classical Fourier analysis, we recommend the book [Ed], although it is not directly related to the topics of this chapter. The theory of the Riemann zeta function is one of the major achievements of the classical theory of functions. Riemann was always referred to as the greatest master of the Fourier analysis.

The papers [AI], [AW] are a good introduction to the basic hypergeometric function and the Rogers polynomials; see also [GR]. The book [An] can be definitely recommended to those who want to understand the theory of q–functions. The paper [Ba] remains the best on the analytic theory of the q–Gamma function and the multiple Gamma functions.

We discuss in detail how Bessel functions, Hankel transform, and their nonsymmetric counterparts (a new development of this classical field!) appear in the theory of DAHA. The exposition is from scratch, however, in my opinion, the paper [Kos] (the Bessel functions via $SL(2,\mathbb{R})$) and [HO1, He3] (spherical functions and generalizations) are a good addition to this chapter. See also [DV] for information about Harish-Chandra, the creator of the harmonic analysis on symmetric spaces.

The focus in this chapter is on the q–Fourier transform and the corresponding representation theory of $\mathcal{H}$. This transform was introduced in [C19] as a q–generalization of the classical Hankel and Harish-Chandra transforms. It also generalizes the Macdonald–Matsumoto p–adic transform. The new q–theory has deep relations to the Macdonald orthogonal polynomials, algebraic combinatorics, the Gaussian sums, conformal field theory, and the Verlinde algebras.

The objective of the Harvard lectures was to convince the audience that the q–theory provides *unification* as well as *simplification.* The same approach was followed in [CO] and this chapter. The exposition is "modern:" it is entirely based on the recent theory of the nonsymmetric Macdonald polynomials, although we consider the symmetric (even) case in full detail too. See [Du1, O3, M5, KnS, C23, C20].

This chapter is devoted to the A_1–case, but the last two sections, a review of the general theory including the relations to the p–adic theory and the degenerate DAHAs. A systematic exposition and proofs of the claims of these two sections are in Chapter 3.

Concerning the A_1 theory, this chapter is self-contained and with complete proofs. We follow the principle *the more proof the better.* Quite a few theorems are proved twice using different tools.

The A_1–proofs we give are mainly of a general nature and can be smoothly transferred to the case of arbitrary root systems. However, there is one important formula that has no reasonable multidimensional counterpart, namely, the explicit expansion of the one-dimensional Macdonald measure, the truncated theta function. This formula is a variant of the q–binomial theorem. The expansions of multidimensional Macdonald's measures in higher ranks generally do not have "nice" coefficients, with a reservation about the constant term and some other "small" coefficients. In dimension one, all coefficients are simple q–products. Fortunately the role of this special formula is limited in the theory of DAHA. See [C21, C28] and Chapter 3.

2.0.2 Main results

Let us mention the most interesting ones.

(a) A detailed proof of the main formula for the q–Mellin transform, found by V. Ostrik. It helps in controlling the spaces of analytic functions involved in the formula and is closely related to the analytic theory of the shift operator from [C26].

(b) Integrating the Gaussian over $\mathbb{R}$ with respect to the q–measure, performed by P. Etingof, which led to the Appell functions. This chapter is mainly devoted to the imaginary integration and the Jackson-type summations. The real integration completes the picture.

(c) A theory of the q–Fourier transform for the inner product without the conjugation $q \mapsto q^{-1}$, generalizing the one from [O3] and necessary in the harmonic analysis for real q. In a way, Opdam's w_0 is replaced by T_{w_0}. There are relations to [O4, O5].

(d) The classification of irreducible finite dimensional representations of the double Hecke algebra of type A_1 (with V. Ostrik and A. Oblomkov), including the Fourier–invariant ones and the theory at the roots of unity; cf. [C27, C28].

(e) The limit of the multidimensional generalization of the q–Fourier transform to the p–adic Macdonald–Matsumoto theory [Ma] continues the analysis from [C23] and demonstrates that DAHA can play an important role in the classical p–adic theory.

(f) Applications to the Gaussian sums and the nonsymmetric Verlinde algebras ("perfect representations"), including formulas for the *non-cyclotomic* Gaussian sums, the deformations of the Verlinde algebra, and its degenerations. The main references are [C27, C29].

2.0.3 Other directions

In spite of a simple definition, the one-dimensional DAHA has an impressive spectrum of applications, making it a serious "rival" of the celebrated $\mathfrak{sl}_2$ and $\widehat{\mathfrak{sl}}_2$. Let me mention the relations and applications that appeared beyond the scope of the chapter, partially or completely.

(a) DAHA has direct relations to $\mathfrak{sl}_2$, to the super Lie algebra $\mathfrak{osp}(2|1)$, and their quantum counterparts. The rational degeneration of $\mathcal{H}$ is a quotient of $U(\mathfrak{osp}(2|1))$.

(b) The Weyl and Heisenberg algebras are its special cases. The nonsymmetric Verlinde algebra is a counterpart of the N–dimensional representation of the one-dimensional Weyl algebra as q is a primitive N–th root of unity.

(c) The one-dimensional DAHA for $C^\vee C_1$ governs major families of one-dimensional orthogonal polynomials and the basic hypergeometric function. It has applications in the theory of dimers (Okounkov, Reshetikhin and others).

(d) It also provides quantization of the cubic surfaces (Oblomkov) through the center at the roots of unity; this method was used by Etingof–Oblomkov–Rains to quantize Del Pezzo surfaces.

(e) Through the Vasserot-Varagnolo duality, $\mathcal{H}$ is connected with the toroidal (double q–Kac–Moody) algebras of type A. Also, $\mathcal{H}$ naturally appears in the K–theory of the affine flag varieties.

(f) The definition of the one-dimensional DAHA is based on the topology of an elliptic curve with one puncture, which is an interesting independent topic; connections to algebraic geometry, the theory of elliptic functions, and number theory are expected.

(g) Recently Gaitsgory and Kazhdan continued Kapranov's results and defined the "double p–adic Lie group" which is associated with $\mathcal{HH}$ in a way similar to the classical construction of the p–adic affine Hecke algebra.

(h) Last but not least, the q–Fourier transform of $\mathcal{HH}$ is a *self-dual* unification of the classical (real) spherical transform and the p–adic one. There are first results toward the analytic q–theory (Koeling and Stokman).

Actually, this chapter is more about the general double Hecke algebra technique than about the special one-dimensional applications and the results themselves. The key claim of the chapter is very general and simple to formulate:

q–Fourier transform is self-dual.

The real goal of this chapter is to justify and clarify this fact, and to show its origins in the mathematics and physics.

Acknowledgments. *The participants of the Harvard lectures helped a lot to achieve this goal. My special thanks go to P. Etingof and D. Kazhdan, who organized the lectures, and to A. Braverman, D. Nikshych, and A. Polyshchuk. I am very grateful to V. Ostrik, the coauthor of [CO], and to P. Etingof for permission to include his theorem. In the classification of the finite dimensional representations, some unpublished results by A. Oblomkov were used.*

2.1 Euler's integral and Gaussian sum

One of the main objectives of this chapter is in connecting the Gaussian sums with the Gauss integrals in one theory via the theory of q–functions.

The key points of our approach will be explained in the following three subsections. We will begin with the Euler–Gauss integral, extend it by q, make q a root of unity, and finally arrive at the classical Gaussian sums, following [C26, C27].

The technique is actually quite elementary. The key step is in calculating two q–integrals, similar to those considered by Ramanujan (see Askey's paper in [An]). There is nothing here beyond the classical calculus of the nineteenth century, however, it is not surprising that the unification of the Euler–Gauss integral with the Gaussian sums was achieved only recently.

In my opinion, the main reason of such a delay was that the q–functions and q–integrals did not play any significant role in mathematics until relatively recent breakthrough in mathematical physics, the theory of integrable models and quantum groups. Only now are we beginning to understand the true role of q–functions, the roots of unity, and the p–adic methods in mathematics.

2.1.1 Euler's integral, Riemann's zeta

We begin with one of the most famous classical formulas:

$$2\int_0^\infty e^{-x^2}x^{2k}dx = \Gamma(k+\frac{1}{2}), \tag{2.1.1}$$

where k is a complex number with $\Re k > -\frac{1}{2}$. Actually, it is the best way to introduce the Γ–function, so it is more of a definition than a formula. Indeed, it readily results in

(i) the functional equation $\Gamma(x+1) = x\Gamma(x)$,

(ii) the meromorphic continuation of $\Gamma(x)$ to all complex x,

(iii) the infinite Weierstrass product formula for Γ.

For $k = 0$, this formula reduces to the Poisson integral

$$2\int_0^\infty e^{-x^2}dx = \sqrt{\pi}. \tag{2.1.2}$$

Other remarkable special values are

$$\Gamma(n+1) = n! \ \text{ and } \Gamma(n+\frac{1}{2}) = \frac{(2n)!}{2^{2n}(n)!}\sqrt{\pi},\ n \in \mathbb{Z}_+.$$

Formula (2.1.1) has tremendous applications in both mathematics and physics. In the first place, it gave birth to analytic number theory. The following "perturbation" of (2.1.1) leads to the analytic continuation of Riemann's zeta–function $\zeta(s)$ and its functional equation:

$$\int_{-\infty}^\infty \frac{|x|^{2k}}{e^{x^2}-1}dx = \zeta(k+\frac{1}{2})\Gamma(k+\frac{1}{2}), \Re k > 0.$$

The functional equation is due to Euler and Riemann. The approach through the above formula is the so-called Riemann's second proof of the functional equation.

It is interesting that the integral in the left-hand side did appear in physics (Landau and Lifshitz) as a kind of perturbation of (2.1.1).

There are three celebrated, closely connected, and entirely open problems concerning the behavior of $\zeta(k+1/2)$ in the critical strip $\{-1/2 < \Re k < 1/2\}$:

(i) the Riemann hypothesis: all zeros are k–imaginary,

(ii) the distribution of the zeros beyond the $T\log T$ formula,

(iii) the growth estimates in this strip as $\Im k \to \infty$.

Intensive computer experiments confirm (i) and strongly indicate that the imaginary parts of the zeros are distributed randomly subject to the approximate formula $T\log T$ for the number of zeros in the strip between 0 and iT. Roughly speaking, there is nothing to expect beyond the $T\log T$ formula (and the Riemann hypothesis). The statistical approach to the problem, including connections with the matrix models, is relatively recent. It leads to conjectural formulas for the higher correlators (Bogomolny et.al.), which are in full harmony with the computer simulations (Odlyzko).

It is really important to know that (ii) is "empty." There is the classical connection of the small zeros of the ζ at the critical line with the so-called Gramm points. It gave birth to several qualitative "laws" for the distribution of the zeros. If (ii) is empty, then these laws cannot be true. Indeed, *all* of them were rejected by computers.

Quite a few specialists in the nineteenth century and at the beginning of the twentieth centuries suspected that the zeros of the zeta function are far from being regular. Now we are certain that they are completely irregular, modulo Riemann's hypothesis and the $T\log T$ formula.

As for (iii), nothing positive is known toward the celebrated Lindelöf conjecture, that is, $\zeta(\sigma+it)=O(t^{\epsilon})$ for every positive ϵ and $\sigma\geq 1/2$.

M.C. Gutzwiller writes: "The Riemann zeta function displays the essence of chaos in quantum mechanics, analytically smooth, and yet seemingly unpredictable."

A much more "predictable" extension of (2.1.1) is the Fourier analysis. Instead of perturbing the Gaussian toward the zeta, we multiply it by the Bessel function. The corresponding integral, the **Hankel transform** of the Gaussian, can be calculated without difficulties. The theory is plane and square. All standard facts about the usual Fourier transform take place for the Hankel transform.

2.1.2 Extension by q

Another approach to zeta is via the so-called p–adic integration. Its origin is in the theory of Gaussian sums. A cyclotomic counterpart of (2.1.1) at $k=0$ is the Gauss formula

$$\sum_{m=0}^{2N-1} e^{\frac{\pi m^2}{2N}i} = (1+i)\sqrt{N}, \quad N\in\mathbb{N}. \tag{2.1.3}$$

The study of the p–adic properties of the Gaussian sums, Bernoulli numbers, and similar objects is one of the most challenging directions in modern arithmetic. The p–adic Gamma function and even a counterpart of Riemann's zeta, Kubota–Leopold's p–adic zeta, appeared in this way.

The most direct approach could be by adding k to (2.1.3). The so called modular Gauss–Selberg sums are an example of such a theory; see [Ev]. Our approach is based on the q–Mehta–Macdonald integrals from [C21].

We are going *to connect the classical Hankel transform and the Fourier transform on the group $\mathbb{Z}_N$ in one theory.*

We will extend (2.1.3) by adding q and its variant with Bessel functions. Dependent on the integration path, there are three choices, namely, the imaginary axis (compact theory), the real axis (noncompact theory), and the loop from zero to $+\infty$ (the Jackson theory). The latter theory leads to infinite q–sums; the generalized Gaussian sums are limits of the such sums as q approaches a root of unity.

Using the Jackson sums is an interesting feature of the difference setting. For instance, the Jackson-type counterpart of the q–zeta function is an *infinite series,* that converges to the classical Riemann's zeta for *all* complex k as $q \to 1$ [C26].

Actually, the most natural way of reaching the Gaussian sums is in replacing the Hankel transform by its trigonometric variant, the **Harish-Chandra spherical transform.** Unfortunately the latter is significantly more difficult to deal with than the Hankel transform, both algebraically and analytically. Quite a few remarkable properties of the classical Fourier transform and Hankel transform are missing for the Harish-Chandra transform, in spite of intensive research in this direction.

The main basic facts about the Harish-Chandra transform are as follows:

(i) the exact analytic description of the Fourier–image of the space of compactly supported functions,

(ii) the exact inversion formula, generalizing the self-duality of the Hankel transform.

Technically, passage to the Harish-Chandra theory is by replacing the measure $|x|^{2k}dx$, which makes the Bessel functions pairwise orthogonal, by $|\sinh(x)|^{2k}$, which governs the orthogonality of the hypergeometric and spherical functions.

The first serious problem is that the integral

$$2\int_0^{\infty} e^{-x^2}|\sinh(x)|^{2k}dx$$

becomes transcendental apart from $k \in \mathbb{Z}_+/2$. When we multiply the Gaussian by the spherical or, more generally, the hypergeometric function, the integral gets even worse. It is essentially the so-called heat kernel problem.

Such a product can be exactly integrated for the Harish-Chandra measure only when $k = 0$ and in the group case $k = 1$. Unfortunately we need this integral because it is nothing but the Harish-Chandra transform of the Gaussian, which is important to know at almost all levels of harmonic analysis on symmetric spaces. Generally, only approximate methods can be used.

A bypass was suggested in [C19, C21]. We go from the Bessel function to the *next but one* level, the **basic hypergeometric function**, a q–generalization of the classical hypergeometric function. The resulting transform is self-dual and has all the other good properties of the Hankel transform. This chapter contains a complete algebraic theory of this q–transform and some elements of the analytic theory.

To explain this development we will begin with the Hankel transform, including the elementary properties of the Bessel functions "from scratch." These functions were a must for quite a few generations of mathematicians, but not anymore.

To put things in perspective, let us start with a multidimensional counterpart of formula (2.1.1).

2.1.3 Mehta–Macdonald formula

Let $\mathbb{R}^n$ be a Euclidean vector space with scalar product $(\cdot,\cdot)$ and $R \subset \mathbb{R}^n$ a reduced irreducible root system. For any root α, we set

$$\alpha^\vee = \frac{2\alpha}{(\alpha,\alpha)}, \ \tilde{\alpha} = \frac{\sqrt{2}\alpha}{|\alpha|}.$$

Note that $(\tilde{\alpha},\tilde{\alpha}) = 2$. We will need $\rho = \frac{1}{2}\sum_{\alpha>0}\alpha$. Given $\alpha \in R$ and $x \in \mathbb{R}^n$, let $\tilde{x}_\alpha = (x,\tilde{\alpha})$. By dx we mean the standard measure on the Euclidean space $\mathbb{R}^n$.

Theorem 2.1.1. *(Mehta–Macdonald integral) For any complex number k such that $\Re k > -\frac{1}{h^\vee}$, where $h^\vee$ is the dual Coxeter number of R, we have the identity*

$$\int_{\mathbb{R}^n} \prod_{\alpha\in R} |\tilde{x}_\alpha|^k e^{-(x,x)/2} dx = (2\pi)^{n/2} \prod_{\alpha>0} \frac{\Gamma(k(\rho,\alpha^\vee)+k+1)}{\Gamma(k(\rho,\alpha^\vee)+1)}. \tag{2.1.4}$$

This formula was conjectured by Mehta for the root system of type A_n. Bombieri readily deduced it from Selberg's integral. Then Macdonald extended it in [M2] to all root systems and proved his formula for the classical systems B, C, D. Opdam found a uniform proof using the technique of shift operators [O1].

Example. Let us check the Mehta–Macdonald formula for the root system of type A_1. In this case $n = 1$ and there is only one positive root $\alpha :\ (\alpha,\alpha) = 2$. Thus $\tilde{\alpha} = \alpha^\vee = \alpha,\ \rho = \frac{\alpha}{2}$, and $(\rho,\alpha^\vee) = 1$. Let $u = \tilde{x}_\alpha$. Then formula (2.1.4) reads as

$$\int_{-\infty}^{\infty} |u|^{2k} e^{-u^2/4} \frac{du}{\sqrt{2}} = (2\pi)^{1/2} \frac{\Gamma(2k+1)}{\Gamma(k+1)}.$$

Changing the variables and using the doubling formula for the Γ–function

$$2^{2k}\Gamma(k+\frac{1}{2}) = \sqrt{\pi}\frac{\Gamma(2k+1)}{\Gamma(k+1)},$$

we obtain (2.1.1). Thus (2.1.4) is a generalization of formula (2.1.1) to higher rank root systems.

2.1.4 Hankel transform

For any complex number $k \notin -\frac{1}{2} - \mathbb{Z}_+$, we define the function

$$\phi^{(k)}(t) = \sum_{m=0}^{\infty} \frac{t^{2m}\Gamma(k+\frac{1}{2})}{m!\Gamma(k+m+\frac{1}{2})}.$$

The convergence is for all t and is very fast. It is clear that $\phi^{(k)}(0) = 1$. The function $\phi^{(k)}(t)$ is related to the classical Bessel function $J_k(t)$ by the formula

$$\phi^{(k)}(t) = \Gamma(k+\frac{1}{2})\, t^{-k+1/2}\, J_{k-1/2}(2it). \tag{2.1.5}$$

Theorem 2.1.2. *For any complex k with $\Re k > -\frac{1}{2}$,*

$$\int_{-\infty}^{\infty} e^{-x^2}\phi^{(k)}(\lambda x)\phi^{(k)}(\mu x)|x|^{2k}dx = \phi^{(k)}(\lambda\mu)e^{\lambda^2+\mu^2}\Gamma(k+\frac{1}{2}), \tag{2.1.6}$$

which reduces to (2.1.1) as $\lambda = \mu = 0$.

Definition 2.1.3. *The* **real Hankel transform** *of an even function $f(x)$ is given by the formula*

$$\mathbf{H}^k_{re}(f)(\lambda) = \frac{1}{\Gamma(k+\frac{1}{2})}\int_{-\infty}^{\infty} \phi^{(k)}(x\lambda)f(x)|x|^{2k}dx.$$

The Hankel transforms are even functions of λ; the transforms of odd $f(x)$ are zero. Formula (2.1.6) states that

$$\mathbf{H}^k_{re}(\phi^{(k)}(\mu x)e^{-x^2}) = e^{\mu^2}(\phi^{(k)}(\mu\lambda)e^{\lambda^2}). \tag{2.1.7}$$

Definition 2.1.4. *The* **imaginary Hankel transform** *of an even function $f(x)$ is*

$$\mathbf{H}^k_{im}(f)(\lambda) = \frac{1}{i\Gamma(k+\frac{1}{2})}\int_{i\mathbb{R}} \phi^{(k)}(x\lambda)f(x)|x|^{2k}dx.$$

Let V^k_{re} be the linear span of the functions $\phi^{(k)}(\lambda x)e^{-x^2}$ treated as functions of $x \in \mathbb{R}$. Similarly, V^k_{im} is the span of $\phi^{(k)}(\lambda x)e^{\lambda^2}$ considered as functions of $\lambda \in i\mathbb{R}$. It is clear that $\mathbf{H}^k_{re}$ maps V^k_{re} to V^k_{im} and $\mathbf{H}^k_{im}$ maps V^k_{im} to V^k_{re}. As an immediate consequence of (2.1.6), we obtain the **inversion** formula.

Corollary 2.1.5. *The maps* $\mathbf{H}^k_{re}$ *and* $\mathbf{H}^k_{im}$ *are inverse to each other.*

Here we can replace the spaces V^k_{re} and V^k_{im} by their suitable completions.

Corollary 2.1.6. *Let* U_{re} *(respectively,* U_{im}*) be the linear span of functions in the form* $p(x)e^{-x^2}$ *(respectively,* $p(\lambda)e^{\lambda^2}$*), where* p *runs through even polynomials. Then* $\mathbf{H}_{re}$ *maps* U_{re} *to* U_{im}*,* $\mathbf{H}_{im}$ *maps* U_{im} *to* U_{re}*,* $\mathbf{H}_{im}\mathbf{H}_{re} = id$*, and* $\mathbf{H}_{re}\mathbf{H}_{im} = id$*. Moreover,*

$$\mathbf{H}_{re} : U^{\leq n}_{re} \mapsto U^{\leq n}_{im}, \ \mathbf{H}_{im} : U^{\leq n}_{im} \mapsto U^{\leq n}_{re},$$

where $U^{\leq n}_{re} \subset U_{re}$ *(respectively,* $U^{\leq n}_{im} \subset U_{im}$*) are the subspaces generated by* $p(x)e^{-x^2}$ *(respectively,* $p(\lambda)e^{\lambda^2}$*) for all even polynomials* p *of degree* $\leq n$*.*

Proof. Differentiating formula (2.1.7) $2n$ times in terms of the variable μ and setting $\mu = 0$, we obtain that $\mathbf{H}_{re}(x^{2n})$ is e^{λ^2} times a polynomial in λ. A better and more constructive proof will be presented below.

2.1.5 Gaussian sums

Let N be a natural number and q a primitive N–th root of 1. We will also need $q^{1/4}$, which will be picked in primitive $4N$–th roots of 1. We will consider the so-called **generalized Gaussian sums**,

$$\tau = \sum_{j=0}^{2N-1} q^{j^2/4}.$$

Note that $q^{j^2/4}$ depends only on the residue of j modulo $2N$, so we can assume that the summation index runs through $\mathbb{Z} \bmod 2N$. A natural choice for $q^{1/4}$ is $e^{\pi\imath/2N}$. In this case, we have the celebrated formula of Gauss:

$$\tau = (1+\imath)\sqrt{N}. \tag{2.1.8}$$

Comment. The standard definition of the Gaussian sum is $\tau' = \sum_{j=0}^{n-1} q^{j^2}$. Gauss proved that for $q = e^{2\pi i/n}$,

$$\tau' = \begin{cases} \sqrt{n} & \text{if } n \bmod 4 = 1, \\ i\sqrt{n} & \text{if } n \bmod 4 = 3, \\ (1+i)\sqrt{n} & \text{if } n \bmod 4 = 0, \\ 0 & \text{if } n \bmod 4 = 2. \end{cases}$$

Formula (2.1.8) corresponds to $n = 4N$. It somewhat resembles Fresnel's integral

$$\int_{-\infty}^{\infty} e^{ix^2} dx = (1+i)\sqrt{\frac{\pi}{2}},$$

which can be obtained from (2.1.2) by changing the variable $x \mapsto \frac{1+i}{\sqrt{2}}x$ and shifting the contour of integration. In a sense, $\sqrt{2N}$ is a substitute for $\sqrt{\pi}$ when we switch from the real theory to roots of unity. ❑

There are two levels in the theory of Gaussian sums:

(a) checking that $\tau\bar{\tau} = 2N$, which is simple and entirely conceptual,

(b) finding $\arg(\tau)$, which is not.

Let us recall the proof of (a).

We consider the space V of complex-valued functions on $\mathbb{Z} \bmod 2N$, pick a primitive $2N$–th root $q^{1/2}$ of 1, and introduce the Fourier transforms $\mathbf{F}_+$ and $\mathbf{F}_-$ on V:

$$\mathbf{F}_\pm(f)(\lambda) = \sum_{x=0}^{2N-1} q^{\pm x\lambda/2} f(x).$$

Then

(i) $\mathbf{F}_-\mathbf{F}_+ \ = \ \mathbf{F}_+\mathbf{F}_- \ = \$ multiplication by $2N$,

(ii) $\mathbf{F}_+(\gamma) = \tau\gamma^{-1}$ and $\mathbf{F}_-(\gamma^{-1}) = \bar{\tau}\gamma$,

where $\gamma(x) = q^{x^2/4}$ for a primitive $4N$–th root of unity $q^{1/4}$.
Indeed,

$$\sum_{\lambda\in\mathbb{Z} \bmod 2N} q^{(x-y)\lambda/2} = 2N$$

for $x = y$ and zero otherwise. We obtain (i). As for (ii), it suffices to examine

$$\begin{aligned}\mathbf{F}_+(\gamma) \ &= \sum_{x\in\mathbb{Z} \bmod 2N} q^{x\lambda/2} q^{x^2/4} \\ &= \sum_{x\in\mathbb{Z} \bmod 2N} q^{(x+\lambda)^2/4} q^{-\lambda^2/4} \ = \ \tau q^{-\lambda^2/4}.\end{aligned}$$

The relation $\tau\bar{\tau} = 2N$ is immediate from (i) and (ii). By the way, it is easy to show that $\bar{\tau} = Tr(\mathbf{F}_+)$, but it is not helpful for calculating $\arg(\tau)$.

We are going to deduce (2.1.8) from a q–variant of (2.1.1). There will be two interesting corollaries:

(1) a formula for τ for any primitive root q,

(2) a generalization of (2.1.8) to an arbitrary integral k.

The former is closely connected with known formulas, but the latter is new.

2.2 Imaginary integration

In this section, we will discuss the imaginary integration, which is called the **compact case** in the context of symmetric spaces.

Let us first switch from $\mathbb{R}$ to $i\mathbb{R}$. Recall our main formula:

$$\int_{-\infty}^{\infty} e^{-x^2}|x|^{2k}dx = \Gamma(k+\frac{1}{2}),\ \Re k > -1/2.$$

Changing the variable $x \mapsto ix$, we obtain

$$\frac{1}{i}\int_{i\mathbb{R}} e^{x^2}|x|^{2k}dx = \Gamma(k+\frac{1}{2}),\ \Re k > -1/2. \tag{2.2.1}$$

These two formulas are of course equivalent, but their q–counterparts are not. It is similar to the Harish-Chandra theory; real integration corresponds to the **noncompact case**, which is very different from the compact case (imaginary integration).

Comment. (i) There is an improvement of (2.2.1):

$$\int_{-\varepsilon+i\mathbb{R}} e^{x^2}(-x^2)^k dx = \Gamma(k+\frac{1}{2})\cos(\pi k), \tag{2.2.2}$$

which holds for *all* complex k provided that $\varepsilon > 0$. It can be readily deduced from the well-known classical definition of the Γ–function:

$$\frac{1}{2}\int_C \frac{(-z)^k e^{-z}}{i\sqrt{-z}}dz = \Gamma(k+\frac{1}{2})\cos(\pi k),$$

by change the variable $z \mapsto -x^2$. Here the path of the integration C begins at $z = -\varepsilon i + \infty$, moves to the left down the positive real axis to $-\varepsilon\imath$, then circles the origin, and returns along the positive real axis to $\varepsilon i + \infty$. The branches of the log in the expression $(-z)^k = e^{k\log(-z)}$ and $\sqrt{-z}$ are standard, with the cutoff at -$\mathbb{R}_+$.

(ii) The use of the Gaussian e^{x^2} makes the classical formula in terms of e^{-z} somewhat more elegant, but of course it is not significant. It is different in the q–theory, which does require the formula in terms of the Gaussian. This is directly connected with the appearance of the double Hecke algebra. The (formal) conjugation by the Gaussian is an automorphism of the latter. The substitution $z = -x^2$ does not seem meaningful from the viewpoint of this algebra.

2.2.1 Macdonald's measure

Let q be a real number, $0 < q < 1$. We set $q = e^{-1/a}$ for a real number $a > 0$. The q–counterpart of x^{2k} is the following function of a complex variable x:

$$\delta_k(x) = \prod_{j=0}^{\infty}\frac{(1-q^{2x+j})(1-q^{-2x+j})}{(1-q^{2x+j+k})(1-q^{-2x+j+k})}. \tag{2.2.3}$$

If k is a non-negative integer, formula (2.2.3) reads

$$\delta_k(x) = \prod_{j=0}^{k-1}(1-q^{2x+j})(1-q^{-2x+j}). \tag{2.2.4}$$

Theorem 2.2.1. *Assume that $\Re k > 0$. Then $\delta_k(x)$ is regular for $x \in i\mathbb{R}$ and*

$$\frac{1}{i}\int_{i\mathbb{R}} q^{-x^2}\delta_k(x)dx = 2\sqrt{\pi a}\prod_{j=0}^{\infty}\frac{1-q^{k+j}}{1-q^{2k+j}}. \tag{2.2.5}$$

Comment. The right- and left-hand sides of formula (2.2.5) are well defined for any number k such that $\Re k \notin -\mathbb{Z}_+$. However, formula (2.2.5) *is not* true when $\Re k < 0$. ❑

Before proving the theorem, let us get (2.2.1) as the limit of (2.2.5) when $q \to 1$.

Proposition 2.2.2. *As $q \to 1$, the leading term of the left-hand side of formula (2.2.5) is $\left(\frac{a}{4}\right)^{-k}\sqrt{a}\int_{i\mathbb{R}} e^{z^2}(-z^2)^k dz$, and the leading term of the right-hand side of this formula is $2a^{-k}\sqrt{\pi a}\frac{\Gamma(2k)}{\Gamma(k)}$. Hence (2.2.5) implies (2.2.1).*

Proof. We introduce the q–counterpart of the Γ–function:

$$\Gamma_q(x) = (1-q)^{1-x}\prod_{j=0}^{\infty}\frac{1-q^{x+j}}{1-q^{1+j}}.$$

It is well known (and not difficult to prove) that $\lim_{q\to 1}\Gamma_q(x) = \Gamma(x)$. Hence the right-hand side of formula (2.2.5) as $q \to 1$ is $2a^{-k}\sqrt{\pi a}\frac{\Gamma(2k)}{\Gamma(k)}$ asymptotically.

Now consider the left-hand side of formula (2.2.5). Changing the variables $x \mapsto \sqrt{a}z$,

$$\frac{1}{i}\int_{i\mathbb{R}} q^{-x^2}\delta_k(x)dx = \frac{1}{i}\int_{i\mathbb{R}} e^{z^2}\delta_k(\sqrt{a}z)\sqrt{a}dz.$$

Lemma 2.2.3. *(Stirling–Moak) One has*

$$\lim_{a\to\infty}\delta_k(\sqrt{a}z)\left(\frac{a}{4}\right)^k = (-z^2)^k,$$

where the standard branch of the logarithm is taken for $(-z^2)^k$ and k is an arbitrary complex number.

It is obvious in the case $k \in \mathbb{Z}_+$. Indeed, asymptotically,

$$\delta_k(\sqrt{a}z) = \prod_{j=0}^{k-1}(1-e^{j/a}e^{2z/\sqrt{a}})(1-e^{j/a}e^{-2z/\sqrt{a}}) \approx \left(-\frac{4z^2}{a}\right)^k.$$

We omit the proof for general k. Actually, this formula can be applied only if one can estimate the remainder. The most convenient formula for the latter is due to Moak. We will not discuss his formula in more detail, because the integrand above contains the Gaussians, that makes the convergence estimates simple. ❑

It follows from the lemma that the left-hand side of formula (2.2.5) is asymptotically $\left(\frac{a}{4}\right)^{-k}\sqrt{a}\int_{i\mathbb{R}} e^{z^2}(-z^2)^k dz$. Comparing it with the asymptotical behavior of the right-hand side and using the doubling formula

$$2^{2k-1}\Gamma(k+1/2) = \sqrt{\pi}\Gamma(2k)/\Gamma(k),$$

we obtain (2.2.1) from (2.2.5). ❑

2.2.2 Meromorphic continuations

It is instructional to calculate the difference between the right-hand side and the left-hand side of formula (2.2.5) as $-1 < k < 0$. It is based on the general technique of meromorphic continuation of the integrals depending on parameters "to the left" and can be applied to many functions in place of the Gaussian.

Comment. A good example is the integral $\int_{i\mathbb{R}}(q^{x^2}-1)^{-1}\delta_k(x)dx$, i.e., a certain q–variant of Riemann's zeta, up to q–Gamma factors. It has a meromorphic continuation in k to the left (and other interesting properties). See [C26]. Surprisingly, it *does not* converge to $\zeta(k+1/2)\Gamma(k+1/2)$ for $\Re k < 0$, as one may expect. Its limit as $a \to \infty$ is a combination of Γ–factors (the ζ does not appear at all). We can obtain the desired limiting behavior by switching from $(q^{x^2}-1)^{-1}$ to $(q^{x^2}+1)^{-1}$. In this case, the integral does converge to what would be expected. Calculating these and similar limits requires the technique of analytic continuation based on the shift operator, which will be considered later in this section. ❑

We will involve the intermediate function

$$f_\varepsilon(k) = \frac{1}{i}\int_{\varepsilon+i\mathbb{R}} q^{-x^2}\delta_k(x)dx,$$

where $\varepsilon > 0$. Let $f(k)$ be the left-hand side of (2.2.5) and $g(k)$ the right-hand side. The difference $f_\varepsilon - f$ is $2\pi i$ times the sum of the residues of $\frac{1}{i}q^{-x^2}\delta_k(x)$ over its poles in the strip $0 < \Re x < \varepsilon$. The calculation depends on the domain of k. For instance, f_ε and $f(k)$ coincide for $\Re k > 2\varepsilon$ because the strip contains no x–poles for such k.

Let $0 < \Re k < 2\varepsilon$. Then the poles are $\frac{k}{2} + n\pi i a$, $n \in \mathbb{Z}$. So

$$f_\varepsilon = f + \Pi_k,\ \Pi_k = \frac{2\pi i}{i}\mathrm{Res}_{\,x=k/2}\delta_k(x)\sum_{n=-\infty}^{\infty} q^{-(k/2+n\pi i a)^2}.$$

Since $\lim_{\varepsilon\to 0}^{x=k/2+\varepsilon}(1-q^{-2x+k}) = -a/2$,

$$\mathrm{Res}_{\,x=k/2} = -(a/2)\prod_{j=0}^{\infty}\frac{(1-q^{k+j})(1-q^{-k+j})}{(1-q^{2k+j})(1-q^{j+1})}.$$

On the other hand,

$$\sum_{n=-\infty}^{\infty} q^{-(k/2+n\pi ia)^2} = (2\sqrt{\pi a})^{-1} \sum_{m=-\infty}^{\infty} q^{m^2} q^{mk}, \tag{2.2.6}$$

thanks to the functional equation for the theta function. Finally,

$$\Pi_k = -\frac{\sqrt{\pi a}}{2}\left(\sum_{m=-\infty}^{\infty} q^{m^2} q^{mk}\right)\prod_{j=0}^{\infty}\frac{(1-q^{k+j})(1-q^{-k+j})}{(1-q^{2k+j})(1-q^{j+1})}.$$

If $-2\varepsilon < \Re k < 0$, the poles are $-\frac{k}{2} + n\pi ia$, $n \in \mathbb{Z}$, and $f_\varepsilon = f - \Pi_{-k}$. The minus of $-\Pi_{-k}$ reflects the change $a/2 \mapsto -a/2$ in the calculation of the residues.

The function $f_\varepsilon(k)$ is well defined for $|\Re k| < 2\varepsilon$ and equals $g(k) + \Pi_k$ in this strip. Recall that $g(k)$ is given by the product formula meromorphic for all complex k and coinciding with $f(k)$ as $\Re k > 0$. Therefore

$$f(k) = f_\varepsilon + \Pi_{-k} = g(k) + \Pi_k + \Pi_{-k} \text{ for } -1 < \Re k < 0. \tag{2.2.7}$$

2.2.3 Using the constant term

We can reformulate Theorem 2.2.1 in a purely algebraic way. Indeed, the function q^x is periodic with period $\omega = 2\pi ia$; so is $\delta_k(x)$. Actually, the period of $\delta_k(x)$ is $\omega/2$, but we will use ω in the following calculation:

$$\frac{1}{i}\int_{i\mathbb{R}} q^{-x^2}\delta_k(x)dx = \frac{1}{i}\int_0^{\omega}\sum_{n\in\mathbb{Z}} q^{-(x+n\omega)^2}\delta_k(x)dx$$

$$= \frac{1}{i}\int_0^{\omega}\sum_{n\in\mathbb{Z}} q^{n^2/4+nx}\frac{\delta_k(x)}{2\sqrt{\pi a}}dx.$$

The second equality results from the functional equation for the theta function, in a form slightly different from (2.2.6).

Since

$$\int_0^{\omega} q^{nx}dx = \begin{cases} 0 & \text{if } n \neq 0, \\ 2\pi ia & \text{if } n = 0, \end{cases}$$

we need to, first, represent $\delta_k(x)$ as a Laurent series and, second, calculate the **constant term** CT of $\widehat{\gamma_-}(x)\widehat{\delta}_k(x)$, where

$$\widehat{\gamma_-}(x) \stackrel{\mathbf{def}}{=\!=} \sum_{n\in\mathbb{Z}} q^{n^2/4+nx}$$

and $\widehat{\delta}_k(x)$ is the expansion of $\delta_k(x)$ as a Laurent series of q^x. To be more exact, we will take the expansion with the coefficients that are power series in terms of the variables q, q^k. Then it is determined uniquely.

Generally speaking, a product of two Laurent series is not well defined, but here we have perfect convergence, thanks to the $q^{n^2/4}$ in $\widehat{\gamma_-}(x)$. Finally,

$$\frac{1}{i}\int_{i\mathbb{R}} q^{-x^2}\delta_k(x)dx = \frac{2\pi i a}{i2\sqrt{\pi a}} CT(\widehat{\gamma_-}(x)\widehat{\delta}_k(x)),$$

and Theorem 2.2.1 is equivalent to

Theorem 2.2.4. *We have the equality:*

$$CT\left(\widehat{\gamma_-}(x)\widehat{\delta}_k(x)\right) = 2\prod_{j=0}^{\infty}\frac{1-q^{k+j}}{1-q^{2k+j}}.$$

The expansion of δ_k can be calculated explicitly. Let us introduce the function

$$\mu_k(x) = \prod_{j=0}^{\infty}\frac{(1-q^{2x+j})(1-q^{-2x+j+1})}{(1-q^{2x+k+j})(1-q^{-2x+k+j+1})}. \tag{2.2.8}$$

It is closely connected with δ_k:

$$\delta_k(x) = \frac{\mu_k(x)+\mu_k(-x)}{1+q^k}. \tag{2.2.9}$$

Following Macdonald, we call it the **truncated theta function**, since for $k \in \mathbb{Z}_+$,

$$\mu_k(x) = \prod_{j=0}^{k-1}(1-q^{2x+j})(1-q^{-2x+j+1}).$$

As $\mathbb{Z}_+ \ni k \to \infty$, it becomes the classical theta function.

Theorem 2.2.5. *(Ramanujan's ${}_1\Psi_1$-summation)*

$$\mu_k(x)/CT(\mu_k(x)) = 1 + \frac{q^k-1}{1-q^{k+1}}(q^{2x}+q^{1-2x}) + \cdots$$

$$+\frac{(q^k-1)(q^k-q)(q^k-q^2)\cdots(q^k-q^m)}{(1-q^{k+1})(1-q^{k+2})(1-q^{k+3})\cdots(1-q^{k+m})}(q^{2mx}+q^{m-2mx}) + \cdots.$$

Proof. Use $\mu_k(x+\frac{1}{2}) = \frac{q^{2x+k}-1}{q^{2x}-q^k}\mu_k(x)$. ❑

Theorem 2.2.6. *(Constant Term Conjecture)*

$$CT(\mu_k(x)) = \frac{(1-q^{k+1})^2(1-q^{k+2})^2\cdots}{(1-q^{2k+1})(1-q^{2k+2})\cdots(1-q)(1-q^2)\cdots}. \tag{2.2.10}$$

Proof. First,

$$CT(\mu_{k+1}(x)) = \frac{(1-q^{2k+1})(1-q^{2k+2})}{(1-q^{k+1})^2} CT(\mu_k(x)). \tag{2.2.11}$$

Indeed, $\mu_{k+1}(x) = (1-q^{2x+k})(1-q^{-2x+k+1})\mu_k(x)$ and Theorem 2.2.5 results in

$$CT(\mu_{k+1}(x)) = CT((1-q^{2x+k}-q^{-2x+k+1}+q^{2k+1})\mu_k(x))$$

$$= (1+q^{2k+1})CT(\mu_k(x)) - 2q^{k+1}\frac{q^k-1}{1-q^{k+1}}CT(\mu_k(x))$$

$$= \frac{CT(\mu_k(x))}{1-q^{k+1}}((1+q^{2k+1}-q^{k+1}-q^{3k+2}-2q^{2k+1}+2q^{k+1})$$

$$= CT(\mu_k(x))\frac{(1+q^{k+1})(1-q^{2k+1})}{1-q^{k+1}}.$$

Let us denote the right-hand side of (2.2.10) by c_k. Then $CT(\mu_k)/c_k$ is a periodic function in terms of k with period 1. Hence the expansion of c_k in terms of q, q^k is invariant under the substitution $q^k \mapsto qq^k$. So it does not depend on k and is a series in terms of q. It must be 1 because $\mu_0 = 1$. ❑

Comment. The above calculation is well known (see [An, AI]). The celebrated constant term conjecture due to Macdonald is for arbitrary root systems. It was proved first for A_n in [BZ], then for BC_n in [Kad], and then for G_2, F_4 using computers. The case of E_6 appeared to be beyond the capacity of modern computers. The proof from [C16] is based on the shift operators, generalizing those introduced by Opdam in the differential setup. Let us formulate the statement for an arbitrary reduced irreducible root system $R = \{\alpha\} \subset \mathbb{R}^n$.

In terms of the one-dimensional μ above:

$$\mu_k^R(x) = \prod_{\alpha>0} \mu_k(x_\alpha/2).$$

The Macdonald conjecture then reads

$$CT(\mu_k^R(x)) = \prod_{j=1}^{\infty}\prod_{\alpha>0} \frac{(1-q^{k(\rho,\alpha^\vee)+j})^2}{(1-q^{k(\rho,\alpha^\vee)+j+k})(1-q^{k(\rho,\alpha^\vee)+j-k})}.$$

There are no good formulas for other coefficients of μ_k^R. Generally speaking, they are not q-products. This was the main problem with calculating $CT(\mu^R)$. If Theorem 2.2.5 actually existed, it would be straightforward.

2.2.4 Shift operator

We are going to prepare the main tool for proving Theorem 2.2.4. The **shift operator** is

$$S(f(x)) \stackrel{\text{def}}{=\!=} \frac{f(x-\frac{1}{2}) - f(x+\frac{1}{2})}{q^x - q^{-x}}.$$

It is a q–variant of the differentiation $x^{-1}d/dx$. It plays an important role in the theory of the so-called basic hypergeometric function. See, e.g., [AI].

Let $g_k = \prod_{j=0}^{\infty} \frac{1-q^{k+j}}{1-q^{2k+j}}$. Given a function $f(x)$ defined on the imaginary line, its **q–Mellin transform** is introduced by the formula

$$\Psi_k(f) = \frac{1}{ig_k}\int_{i\mathbb{R}} f(x)\delta_k(x)dx.$$

The function Ψ_k is an analytic function of the variable k in the half-plane $\Re k > 0$, provided we have integrability.

Let us examine the behavior of the q–Mellin transform under the q–variant of the integration by parts.

Theorem 2.2.7. *Let the function $f(x)$ be analytic in an open neighborhood of the strip $|\Re x| \le 1$. Provided that the integrals below are well defined,*

$$\Psi_k(f) = (1-q^{k+1})\Psi_{k+1}(f) + q^{k+3/2}\Psi_{k+2}(S^2(f)). \tag{2.2.12}$$

Proof is direct. Later we will give a better one, with modest calculations. However, the theorem will be applied to various classes of functions, so an explicit proof is helpful for controlling the analytic matters.

First, we check that $S^2(f)$

$$= \frac{f(x-1)}{(q^x-q^{-x})(q^{x-1/2}-q^{-x+1/2})} + \frac{f(x+1)}{(q^x-q^{-x})(q^{x+1/2}-q^{-x-1/2})}$$
$$-\frac{f(x)}{q^x-q^{-x}}\left(\frac{1}{q^{x-1/2}-q^{-x+1/2}} + \frac{1}{q^{x+1/2}-q^{-x-1/2}}\right).$$

Second, we will change the variables and move the contour of integration. The resulting formulas are

$$\begin{aligned}
&\int_{i\mathbb{R}} S^2(f)\delta_k(x)dx \\
&= \int_{i\mathbb{R}} \frac{f(x)\delta_k(x+1)dx}{(q^{x+1}-q^{-x-1})(q^{x+1/2}-q^{-x-1/2})} - \\
&- \int_{i\mathbb{R}} \frac{f(x)\delta_k(x)dx}{(q^x-q^{-x})(q^{x+1/2}-q^{-x-1/2})} + \\
&+ \int_{i\mathbb{R}} \frac{f(x)\delta_k(x-1)dx}{(q^{x-1}-q^{-x+1})(q^{x-1/2}-q^{-x+1/2})} - \\
&- \int_{i\mathbb{R}} \frac{f(x)\delta_k(x)dx}{(q^x-q^{-x})(q^{x-1/2}-q^{-x+1/2})}.
\end{aligned} \tag{2.2.13}$$

Let us denote the first two terms in the right-hand side by A and the second two terms (lines) by B. So the integral is $A + B$.

Third,

$$\delta_k(x+1) = \frac{(1-q^{2x+k})(1-q^{2x+1+k})(1-q^{-2x-2})(1-q^{-2x-1})}{(1-q^{-2x-2+k})(1-q^{-2x-1+k})(1-q^{2x})(1-q^{2x+1})}\delta_k(x),$$

and

$$\delta_k(x-1) = \frac{(1-q^{-2x+k})(1-q^{-2x+1+k})(1-q^{2x-2})(1-q^{2x-1})}{(1-q^{2x-2+k})(1-q^{2x-1+k})(1-q^{-2x})(1-q^{-2x+1})}\delta_k(x).$$

Fourth,

$$A = -\int_{i\mathbb{R}} \frac{(q^{x+1/2}+q^{-x-1/2})(1-q^{2k-1})f(x)\delta_k(x)dx}{(1-q^{-2x-2+k})(1-q^{-2x-1+k})(q^x-q^{-x})q^{2x+1}},$$

and

$$B = \int_{i\mathbb{R}} \frac{(q^{x-1/2}+q^{-x+1/2})(1-q^{2k-1})f(x)\delta_k(x)dx}{(1-q^{2x-2+k})(1-q^{2x-1+k})(q^x-q^{-x})q^{-2x+1}}.$$

Now,

$$\frac{A+B}{(1-q^{2k-1})} = \int_{i\mathbb{R}} Gf(x)\delta_k(x)dx,$$

where

$$G = \frac{1+q+q^k+q^{2x}+q^{-2x}-q^{k-2}-q^{2k-3}-q^{2k-2}-q^{2x+2k-2}-q^{-2x+2k-2}}{(1-q^{2x-2+k})(1-q^{2x-1+k})(1-q^{-2x-2+k})(1-q^{-2x-1+k})q^{3/2}}.$$

Regrouping the terms:

$$\frac{A+B}{(1-q^{2k-1})} = (q^{1-k}-q^{k-1})\int_{i\mathbb{R}} \frac{q^{-1/2}f(x)\delta_k(x)dx}{(1-q^{2x-1+k})(1-q^{-2x-1+k})}$$
$$+\int_{i\mathbb{R}} \frac{q^{-3/2}(1-q^{2k-3})(1+q^{k-1})(1+q^{2-k})f(x)\delta_k(x)dx}{(1-q^{2x-2+k})(1-q^{2x-1+k})(1-q^{-2x-2+k})(1-q^{-2x-1+k})}.$$

Taking into account that

$$\delta_{k-1}(x) = \frac{\delta_k(x)}{(1-q^{2x-1+k})(1-q^{-2x-1+k})},\ \delta_{k-2}(x)$$
$$= \frac{\delta_k(x)}{(1-q^{2x-2+k})(1-q^{2x-1+k})(1-q^{-2x-2+k})(1-q^{-2x-1+k})},$$

and that

$$\frac{g_k}{g_{k-1}} = (1-q^{2k-1})(1+q^{k-1}),$$
$$\frac{g_k}{g_{k-2}} = (1-q^{2k-3})(1+q^{k-2})(1-q^{2k-1})(1+q^{k-1}),$$

we come to the formula

$$q^{k-1/2}\Psi_k(S^2(f)) = \Psi_{k-2}(f) - (1-q^{k-1})\Psi_{k-1}(f),$$

which is equivalent to the statement of the theorem. ❑

2.2.5 Applications

The simplest example is $f(x) = 1$. We obtain the formula

$$CT(\delta_k(x)) = \frac{(1-q^k)(1-q^{2k+1})}{(1-q^{k+1})(1-q^{2k})}(CT(\delta_{k+1}(x)),$$

which is equivalent to the formula (2.2.11) above.

Now let $f(x) = q^{-x^2}$. It is easy to check that $S(q^{-x^2}) = q^{-1/4}q^{-x^2}$. Therefore

$$\Psi_k = (1-q^{k+1})\Psi_{k+1} + q^{k+1}\Psi_{k+2}$$

for $\Psi_k = \Psi_k(q^{-x^2})$ or, equivalently,

$$\Psi_k - \Psi_{k+1} = -q^{k+1}(\Psi_{k+1} - \Psi_{k+2}).$$

Introducing the function

$$\phi_k = (-1)^k q^{(k+1)k/2}(\Psi_k - \Psi_{k+1}),$$

we see that it is periodic with period 1 in the variable k. In particular, it can be extended to the whole complex plane, where it is analytic. The function ϕ_k is also quasi-periodic with period ω because Ψ_k is ω-periodic. Namely:

$$\phi_{k+\omega} = q^{k\omega+\omega^2/2}\phi_k \;=\; e^{-2\pi ik}e^{-\pi i\omega}\phi_k.$$

Recall that $\omega = 2\pi ia, q = e^{-1/a}$. These are the defining properties of the classical

$$\vartheta_3(k) = \sum_{m=-\infty}^{\infty} e^{2\pi imk}e^{\pi im^2\omega}.$$

For instance, ϕ_k has only one zero in the fundamental parallelogram if it is not zero identically. The zero is known to be $(\omega+1)/2$.

However, it is straightforward to check that the ϕ_k vanish at $k=0$. Indeed,

$$\Psi_0 = \frac{2}{i}\int_{i\mathbb{R}} q^{-x^2}dx = 2\sqrt{\pi a},$$

$$\Psi_1 = \frac{1}{i(1-q)}\int_{i\mathbb{R}} q^{-x^2}(1-q^{2x})(1-q^{-2x})dx$$

$$= \frac{1}{i(1-q)}\left(2\int_{i\mathbb{R}} q^{-x^2}dx - \int_{i\mathbb{R}} q^{-x^2+2x}dx - \int_{i\mathbb{R}} q^{-x^2-2x}dx\right)$$

$$= \frac{2}{i}\int_{i\mathbb{R}} q^{-x^2}dx = \Psi_0.$$

Therefore, $\phi = 0$ and Ψ_k is a constant, which is the value of $\Psi = \Psi_k(q^{-x^2})$ at zero, that is, $2\sqrt{\pi a}$. ❑

Comment. The above argument can be simplified using a reformulation of Theorem 2.2.4 in terms of the Laurent series. First, we check a variant of Theorem 2.2.7 for ω–periodic even functions $f(x)$ and the integration over the imaginary period. Second, we replace integrating over the period by taking the constant term and switch entirely to the Laurent expansions. Third, we apply the shift formula to $f(x) = \widehat{\gamma_-}(x)$. In this setting, ϕ_k is not considered as an analytic function, but becomes a formal series in terms of the variables q and q^k. For such series, the equality $\phi_{k+1} = -q^{k+1}\phi_k$ immediately implies that $\phi_k = 0$.

2.3 Jackson and Gaussian sums

We are going to change the imaginary integration in the main formula of the previous section to the Jackson summation.

Recall that the imaginary q–Mellin transform is

$$\Psi_k(f) = \frac{1}{\imath g_k}\int_{\imath\mathbb{R}} f(x)\delta_k(x)dx, \; g_k = \prod_{j=0}^{\infty}\frac{1-q^{k+j}}{1-q^{2k+j}},$$

$$\delta_k(x) = \prod_{j=0}^{\infty}\frac{(1-q^{2x+j})(1-q^{-2x+j})}{(1-q^{2x+k+j})(1-q^{-2x+k+j})}.$$

Using this transform, Theorem 2.2.1 reads

$$\Psi_k(q^{-x^2}) = 2\sqrt{\pi a}, \text{ where } q = e^{-1/a},\ \Re k > 0.$$

The next theorem will be very close to Theorem 2.2.1. There was one place in its proof where we used special features of the imaginary integration more than absolutely necessary. Let us first somewhat improve it.

We introduced $\phi_k = \Psi_k(q^{-x^2}) - \Psi_{k+1}(q^{-x^2})$, checked that $\phi_k = -q^{k+1}\phi_{k+1}$, and found that $\phi_0 = 0$ and therefore $\phi_k = 0$ for all $k \in \mathbb{Z}_+$. This part remains unchanged. Now it is easy to see that the function $\Psi_k(q^{-x^2})$ is analytic in terms of the $K = q^k$ considered as a new variable in a neighborhood of $K = 0$; so is ϕ_k. However, the latter has infinitely many zeros in a neighborhood of $K = 0$, which results in $\phi_k = 0$. The rest of the proof is the same as in the previous section.

2.3.1 Sharp integration

For an analytic function $f(x)$ in a neighborhood of the positive real axis, we define its **sharp q–Mellin transform**

$$\Psi_k^\sharp(f) = \frac{1}{\imath g_k}\int_C f(x)\delta_k(x)dx, \tag{2.3.1}$$

where the path of the integration C begins at $z = -\varepsilon\imath + \infty$, moves to the left down the positive real axis to $-\varepsilon\imath$, then moves up to $\varepsilon\imath$ and returns along the positive real axis to $\varepsilon\imath + \infty$.

The behavior of C near 0 is not important in the q–theory. Recall that the classical paths of this type used for Γ and ζ *must* go around zero. Our C is like a pencil aimed at zero from $+\infty$; that is why we call it sharp.

Theorem 2.3.1. *Provided that $\Re k > 0$ and $|\Im k| < 2\varepsilon$ (equivalently, $k/2$ sits inside C),*

$$\Psi_k^\sharp(q^{x^2}) = (-a\pi)\prod_{j=1}^{\infty}\frac{(1-q^{j+k})(1-q^{j-k-1})}{(1-q^j)^2}\sum_{j=-\infty}^{\infty}q^{(k-j)^2/4}. \tag{2.3.2}$$

Comment. The sum $\sum_{j=-\infty}^{\infty} q^{(k-j)^2/4}$ can be expressed as the infinite product

$$q^{k^2/4}\prod_{j=1}^{\infty}(1-q^{j/2})(1+q^{j/2-1/4+k/2})(1+q^{j/2-1/4-k/2})$$

by Jacobi's triple product formula. ❑

Proof. We begin with the following lemma. Recall that $q = e^{-1/a}$ and $\omega = 2\pi\imath a$.

Lemma 2.3.2. *Assume that $f(x)$ is analytic for all $x \in \mathbb{C}$ and $\lim_{\xi\to\infty} f(x+\xi)e^{c\xi} = 0$ for arbitrary $c > 0$ and $x \in \mathbb{C}$. Then the function $\Psi_k^\sharp(f)$ has an analytic continuation to all $k \in \mathbb{C}$ and, moreover, vanishes at the points from $\mathbb{Z}_+ + \mathbb{Z}\omega$.*

Proof. By Cauchy's theorem, the integral (2.3.2) is the following sum of residues:

$$\Psi_k^\sharp(f) = (-a\pi)\prod_{j=1}^{\infty}\frac{(1-q^{j-k-1})}{(1-q^j)}\times$$

$$\times\sum_{j=0}^{\infty}\frac{1-q^{j+k}}{1-q^k}\prod_{l=1}^{j}\frac{1-q^{l+2k-1}}{1-q^l}f\left(\frac{k+j}{2}\right)q^{-kj}.$$

The right-hand side gives the desired analytic continuation. ❑

Using the above formula we can reformulate Theorem 2.3.1 in a purely algebraic way:

$$\sum_{j=0}^{\infty}q^{(k-j)^2/4}\frac{1-q^{j+k}}{1-q^k}\prod_{l=1}^{j}\frac{1-q^{l+2k-1}}{1-q^l} \tag{2.3.3}$$
$$=\prod_{j=1}^{\infty}\frac{1-q^{j+k}}{1-q^j}\left(\sum_{j=-\infty}^{\infty}q^{(k-j)^2/4}\right).$$

The left-hand side is a **Jackson sum**, i.e., the summation of the values of a given function (here $q^{(k-j)^2/4}$) over $\mathbb{Z}$ with some weights. There is a proof of the theorem directly in terms of the Jackson summation, without using the sharp integration. It will not be discussed here (see [C21]).

Comment. (i) Formula (2.3.3) obviously holds for $k = 0$, to be more exact, as $k \to 0$ (we have $1 - q^k$ in the denominator). Indeed, we obtain

$$1 + 2\sum_{j=1}^{\infty} q^{j^2/4} = \sum_{j=-\infty}^{\infty} q^{j^2/4}.$$

Hence, the difference between the left-hand and the right-hand sides of formula (2.3.3) has a zero of the *second* order at $k = 0$.

(ii) For $k = 1$, formula (2.3.3) reads

$$\sum_{j=0}^{\infty} q^{(1-j)^2/4}\frac{(1-q^{j+1})^2}{1-q} = \sum_{j=-\infty}^{\infty} q^{j^2/4}$$

and can be checked by a simple calculation.

(iii) It holds for $k = -1/2$. Indeed, in this case the left-hand side is $2q^{1/16}$ and the right-hand side can be transformed using Jacobi's triple product formula as follows:

$$q^{1/16}\prod_{j=1}^{\infty}(1-q^{j/2})(1+q^{j/2})(1+q^{j/2-1/2})\prod_{j=1}^{\infty}\frac{1-q^{j-1/2}}{1-q^j}$$
$$= q^{1/16}\prod_{j=1}^{\infty}(1+q^{j/2-1/2})(1-q^{j-1/2})$$
$$= 2q^{1/16}\prod_{j=1}^{\infty}\frac{(1-q^j)(1-q^{j-1/2})}{(1-q^{j/2})} = 2q^{1/16}.$$

❑

Let us denote the right-hand side of formula (2.3.2) by Π_k. We then have the following properties:

(a) $\Pi_k = -q^{k+1}\Pi_{k+1}$; (b) $\Pi_{k+2\omega} = q^{\omega^2+k\omega}\Pi_k$;

(c) the zeros of Π_k are $\{0, \omega, 1/2+\omega\} \mod \mathbb{Z} + 2\omega\mathbb{Z}$.

In (c), all zeros are simple.

These properties determine Π_k uniquely. The only property which is not immediate is (c). It can be proved using Jacobi's triple product or deduced from (a) and (b). Indeed, the latter give that Π_k has three zeros inside the parallelogram of periods with the sum 1/2. Two of them are obvious: $\{0, \omega\}$.

Now we are going to deduce Theorem 2.3.1 from the shift formula.

2.3.2 Sharp shift formula

Let $\Psi^\sharp_k \stackrel{\mathbf{def}}{=} \Psi^\sharp_k(q^{x^2})$ be the left-hand side of formula (2.3.2). It is clear that

(b') $\Psi^\sharp_{k+2\omega} = q^{\omega^2+k\omega}\Psi^\sharp_k$,

(c') $\Psi^\sharp_k$ has zeros at 0 and ω.

Thus, if we also prove

(a') $\Psi^\sharp_k = -q^{k+1}\Psi^\sharp_{k+1}$,

then we will obtain $\Pi_k = c\Psi^\sharp_k$ for a constant c, which has to be 1 (use the normalization). So we need to employ a variant of the shift formula (2.2.12) for the sharp integration.

Lemma 2.3.3.

$$\Psi^\sharp_k(f) = (1-q^{k+1})\Psi^\sharp_{k+1}(f) + q^{k+3/2}\Psi^\sharp_{k+2}(S^2(f)), \qquad (2.3.4)$$

provided we have the existence of the integrals. Here

$$S(f)(x) = \frac{f(x-1/2)-f(x+1/2)}{q^x - q^{-x}};$$

$f(x)$ is an even function continuous on the sharp integration path C, and, moreover, analytic in the domain

$$\{x \mid -1-\delta < \Re x < 1+\delta,\ -\epsilon < \Im x < \epsilon\} \text{ for } \delta > 0.$$

Proof is a straightforward adjustment of that in the imaginary case. The analyticity is necessary to ensure the invariance of the integration with respect to the shifts by ± 1. ❑

Using the formula $S(q^{x^2}) = q^{1/4}q^{x^2}$ (note the change of sign, compared with the previous section),

$$\Psi^\sharp_k = (1-q^{k+1})\Psi^\sharp_{k+1} + q^{k+2}\Psi^\sharp_{k+2}.$$

So the function $\phi_k = \Psi^\sharp_k + q^{k+1}\Psi^\sharp_{k+1}$ is periodic: $\phi_{k+1} = \phi_k$. On the other hand, (b') results in $\phi_{k+2\omega} = q^{\omega^2+k\omega}\phi_k$. Combining this with the 1–periodicity, we conclude that ϕ_k has only one zero in the parallelogram of periods. However, we already know that ϕ_k has a zero of order 2 at $k=0$. Hence

$$\phi_k = 0 \Rightarrow \Psi^\sharp_k = -q^{k+1}\Psi^\sharp_{k+1} \Rightarrow \Psi^\sharp_k = \Pi_k, \qquad (2.3.5)$$

and Theorem 2.3.1 is proved. ❑

2.3.3 Roots of unity

We have demonstrated that

imaginary integration $\longleftrightarrow$ *sharp integration* $\Longrightarrow$ *Jackson summation,*

where the first arrow is a certain relation, the second is a straight implication. The next step toward a unification of the Gauss integrals and the Gaussian sums will be the limiting procedure

Jackson summation $\longrightarrow$ *Gaussian sums.*

Switching in (2.3.3) to the roots of unity, we come to the following theorem.

Theorem 2.3.4. *Let $q^{1/4}$ be a primitive $4N$–th root of unity and k be an integer such that $0 < k \le N/2$. Then*

$$\sum_{j=0}^{N-2k} q^{\frac{(k-j)^2}{4}} \frac{1-q^{j+k}}{1-q^k} \prod_{l=1}^{j} \frac{1-q^{l+2k-1}}{1-q^l} = \prod_{j=1}^{k} \frac{1}{1-q^j} \sum_{j=0}^{2N-1} q^{\frac{(k-j)^2}{4}}. \tag{2.3.6}$$

Proof. By Galois theory, it is sufficient to pick any primitive $q^{1/4}$, so we may assume that $q^{1/4} = e^{\pi\imath/2N}$. Let us substitute $\widehat{q}^{1/4} = e^{\pi\imath/2N - \pi/a}$ for q in formula (2.3.3).

Lemma 2.3.5. *(Siegel) Let $j_m = h + 2Nm$ where $h, m \in \mathbb{Z}$. Then asymptotically, as $a \to +\infty$,*

$$\sum_{m=0}^{\infty} \widehat{q}^{(j_m-k)^2/4} \approx \frac{\sqrt{a}}{2N} q^{(h-k)^2/4}.$$

Proof. First,

$$\sum_{m=-\infty}^{\infty} \widehat{q}^{(j_m-k)^2/4} = q^{(h-k)^2/4} \sum_{m=-\infty}^{\infty} e^{-\pi(h+2Nm-k)^2/4a}.$$

Second, the sum

$$\Sigma = \sum_{m=-\infty}^{\infty} e^{-\pi(h+2Nm-k)^2/4a} = \sum_{m=-\infty}^{\infty} e^{-\pi N^2(m+(h-k)/2N)^2/a}$$

approximates the integral

$$S = \int_{-\infty}^{\infty} e^{-\pi N^2(x+(h-k)/2N)^2/a} dx$$

with the difference $S - \Sigma$ bounded as $a \to +\infty$. Third,

$$S \approx \int_{-\infty}^{\infty} e^{-\pi N^2 x^2/a} dx = \frac{\sqrt{a}}{N}$$

asymptotically. ❑

Now we compare the asymptotics of the left-hand side and the right-hand side of formula (2.3.3) as $\widehat{q} \to q$, or, equivalently, $a \to +\infty$. We represent either side of this formula as the sum of $2N$ subsums corresponding to all possible values $j \bmod 2N$.

Due to the lemma, the right-hand side approaches the right-hand side of formula (2.3.6) multiplied by $\sqrt{a}/N$. To manage the left-hand side of formula (2.3.3), we note that the product $\prod_{l=1}^{j} \frac{1-q^{l+2k-1}}{1-q^l}$ is a periodic function of j of period N because $q^N = 1$. Moreover, this product vanishes for $N - 2k + 1 \le j < N$. Therefore, the left-hand side of formula (2.3.3) tends to the left-hand side of formula (2.3.6) times $\sqrt{a}/N$. Theorem 2.3.4 is proved. ❑

2.3.4 Gaussian sums

We come to the following corollary.

Corollary 2.3.6. *Taking* $q^{1/4} = e^{\pi\imath/2N}$,

$$\sum_{j=0}^{2N-1} q^{j^2/4} = (1+\imath)\sqrt{N}.$$

Proof. Letting $k = [N/2]$ in formula (2.3.6),

$$\sum_{j=0}^{2N-1} q^{j^2/4} = \begin{cases} q^{n^2/4}\prod_{j=1}^{n}(1-q^j) & \text{if } N = 2n, \\ q^{n^2/4}(1+q^{(2n-1)/4})\prod_{j=1}^{n}(1-q^j) & \text{if } N = 2n+1. \end{cases}$$

Only even $N = 2n$ will be considered. Setting $\Pi = \prod_{j=1}^{n}(1-q^j)$, we obtain $\Pi\overline{\Pi} = 2N$, since it is the value of $(X^N-1)(X+1)(X-1)^{-1}$ at $X = 1$. Here the bar is the complex conjugation. On the other hand,

$$\arg(1-e^{\imath\phi}) = \phi/2 - \pi/2 \text{ for angles } 0 < \phi < 2\pi, \text{ and}$$

$$\arg \Pi = \frac{\pi}{N}\frac{n(n+1)}{2} - \frac{\pi n}{2} = \frac{\pi(1-n)}{4}.$$

❑We now have the following reduction of Theorem 2.3.4, which is important from the viewpoint of applications to Gaussian sums. It has a direct relation to the "little" double affine Hecke algebra (see [C27] and Section 2.10).

Theorem 2.3.7. *Let* $n = [N/2]$. *Then, for* $0 < k \le n$,

$$\sum_{j=0}^{n-k} q^{j^2-kj}\frac{1-q^{2j+k}}{1-q^k}\prod_{l=1}^{2j}\frac{1-q^{l+2k-1}}{1-q^l} = \prod_{j=1}^{k}\frac{1}{1-q^j}\sum_{j=0}^{N-1} q^{j^2-kj}. \tag{2.3.7}$$

Proof. Considering the variant of (2.3.6) for $q^{1/4} \mapsto -q^{1/4}$, check that (2.3.7) equals either the half-sum of these two formulas for even k or the half-difference for odd k. ❑

Corollary 2.3.8. *For even $N = 2n$ and $k = n$,*

$$\sum_{j=0}^{N-1}(-1)^j q^{j^2} = \prod_{j=1}^{n}(1-q^j), \tag{2.3.8}$$

which can be rewritten as

$$\sum_{j=0}^{n-1}(-1)^j q^{j^2} = \prod_{j=1}^{n-1}(1-q^j), \tag{2.3.9}$$

due to the substitution $j \mapsto j+n$.

We are going to use these formulas for $q = \exp(\pi\imath m/n)$ to calculate the corresponding Legendre symbol in terms of the integer parts $[mj/(2n)]$. Let $n > 0$ be any integer and $m > 0$ an odd integer, assuming that they are relatively prime: $(m,n) = 1$. We set

$$\{\frac{m}{n}\} \stackrel{\mathbf{def}}{=\!=} \imath^{(n-1)(m-1)/2}(-1)^{\sum_{j=1}^{n-1}[\frac{mj}{2n}]}, \tag{2.3.10}$$

$$G(m,n) \stackrel{\mathbf{def}}{=\!=} \sum_{j=1}^{n} e^{\pi\imath j^2\frac{m}{n}+\pi\imath jm}. \tag{2.3.11}$$

Formula (2.3.11) is the classical definition of the generalized Gaussian sum (see, e.g., [Ch]). As we will see in the next theorem, the first definition extends the Legendre symbol $(\frac{m}{n})$. The latter is ± 1 as m is a quadratic residue (nonresidue) modulo n, where n is an odd prime number or its power.

Theorem 2.3.9. *(a) For odd m coprime with n,*

$$G(m,n) = \sqrt{n}\left(\frac{1+\imath}{\sqrt{2}}\right)^{1-n}\{\frac{m}{n}\}. \tag{2.3.12}$$

(b) Taking $n = p^a$ for odd prime p and odd a, $\{\frac{m}{n}\} = (\frac{m}{n})$.

Proof. The $G(m,n)$ is the sum from (2.3.9) for $q = e^{\pi\imath/n}$. The product on the right-hand side of this equality was calculated in Corollary 2.3.6 as $m = 1$. Indeed, $G(1,n)$ is given by (2.3.12).

(a) The equality $|G(m,n)| = \sqrt{n}$ is immediate:

$$G(m,n)\overline{G(m,n)} = \frac{1}{2}\prod_{j=1}^{N-1}(1-q^j) = \frac{N}{2} = n.$$

Here $G(1,n)$ is sufficient to examine (apply the Galois automorphisms). The formula for the argument of $G(m,n)$ is direct from formula (2.3.9). Using $\arg(1-e^{i\phi}) = \phi/2 - \pi/2$ for $0 < \phi < 2\pi$:

$$\arg(G(m,n)) = \sum_{j=1}^{n-1}(\frac{\pi mj}{2n} - \left[\frac{mj}{2n}\right]\pi - \frac{\pi}{2})$$

$$= \frac{\pi m}{2n}\frac{n(n-1)}{2} - (n-1)\frac{\pi}{2} - \pi\sum_{j=1}^{n-1}\left[\frac{mj}{2n}\right] =$$

$$= \frac{\pi}{4}(m-1)(n-1) - \frac{\pi}{4}(n-1) + \sum_{j=1}^{n-1}\left[\frac{mj}{2n}\right] \bmod 2\pi.$$

(b) For $m = 1$, the coincidence is evident. Generally,

$$G(m,n) = \pm G(1,n) = \{\frac{m}{n}\}G(1,n).$$

Using the left-hand side of (2.3.9), the sign is plus if m is a quadratic residue modulo n and is constant on all nonresidues m. We use that $\mathbb{Z}_n^*$ is cyclic. It cannot always be plus because it would give the invariance of $G(1,n)$ under the Galois automorphisms $q \mapsto q^m$ and would result in $G(1,n) \in \mathbb{Q}$. Therefore the sign is minus at the nonresidues. ❑

2.3.5 Etingof's theorem

So far we have considered the imaginary integration, the Jackson summation, and its variant at roots of unity. However, the most natural choice is of course $\int_{i\varepsilon+\mathbb{R}} q^{x^2}\delta_k(x)dx$. The calculation of the latter integral was performed by P. Etingof. I am grateful for his permission to include his note. Here we use i (instead of $\imath$) for the imaginary unit.

Recall that $q = \exp(-1/a)$, $\omega = 2\pi i a$. Let k be a positive real number, and $\epsilon > 0$ a small positive number. We will use the function Π_k, the right-hand side of (2.3.2), and its properties, including information about the zeros. Let

$$\Psi_k = \frac{1}{g_k}\int_{i\epsilon+\mathbb{R}} q^{x^2}\delta_k(x)dx.$$

It is clear that Ψ_k is well defined.

Our goal is to calculate Ψ_k, which will be done in Theorem 2.7.4 below. To formulate the theorem, we need some definitions.

First, the theta function is

$$\theta(k,\tau) = \sum_{n\in\mathbb{Z}}(-1)^n e^{2\pi ink + \pi in(n-1)\tau}, \ \Im(\tau) > 0.$$

It is a periodic entire function with period 1 that satisfies the equation

$$\theta(k+\tau,\tau) = -e^{-2\pi i k}\theta(k,\tau).$$

It is defined by this equation uniquely up to scaling. The zeros of $\theta(k,\tau)$ with respect to k are $m+n\tau$, $m,n\in\mathbb{Z}$ (all of them are simple).

Define the following (degenerate) **Appell function**:

$$A(k,\tau) = 2\pi i \sum_{n\in\mathbb{Z}\setminus 0} \frac{e^{2\pi i n^2\tau + 2\pi i n k}}{e^{2\pi i n\tau} - 1}.$$

It is not expressed via theta functions.

Theorem 2.3.10. *The function Ψ_k extends to an entire function, and one then has*

$$\Psi_k = 2\sqrt{\pi a}\cdot e^{-\pi i k} q^{-k(k+1)/2} \frac{\theta(k,\omega)}{\theta'(0,\omega)} \times$$
$$\times\left(\frac{\theta'}{\theta}(k,\omega) - \frac{\theta'}{\theta}(1/2,\omega) - A(k,\omega) + A(1/2,\omega)\right).$$

We present the proof as a chain of lemmas.

Lemma 2.3.11. *The function Ψ_k has an analytic continuation to the region $\Re(k) > 0$, $|\Im(k)| < |\omega|$.*

Proof. The contour of integration can be replaced by

$$(-\infty + i\epsilon, i\epsilon] \cup [i\epsilon, \frac{1}{2}\omega - i\epsilon] \cup [\frac{1}{2}\omega - i\epsilon, \frac{1}{2}\omega - i\epsilon + \infty),$$

or by

$$(\frac{1}{2}\omega - i\epsilon + \infty, \frac{1}{2}\omega - i\epsilon] \cup [\frac{1}{2}\omega - i\epsilon, i\epsilon] \cup [i\epsilon, \infty + i\epsilon).$$

This implies the statement. ❑

Lemma 2.3.12.

$$\Psi_{k+\omega} - \Psi_k = -2i\Psi_k^\sharp = -2i\Pi_k.$$

In particular, Ψ_k extends to a holomorphic function in the half-plane $\Re(k) > 0$.

Proof. This follows directly from Cauchy's residue formula. ❑

Lemma 2.3.13. *One has*

$$\Psi_k = -q^{k+1}\Psi_{k+1}.$$

Proof. This is proved by using the shift formula in the same way as relation (2.3.5). Actually, the sharp integration is a variant of the real integration, so they result in the same multiplier. However, their behavior in the imaginary direction is different. ❑

Lemma 2.3.14. *Ψ_k extends to an entire function.*

Proof. This is immediate from Lemma 2.3.13. ❑

Let us now define the entire function

$$F_k = q^{k(k+1)/2} e^{\pi i k} \Psi_k.$$

Lemma 2.3.15. *F_k is periodic with period 1 and*

$$e^{2\pi i k} F_{k+\omega} + F_k = 2i e^{\pi i k} q^{k(k+1)/2} \Pi_k, \tag{2.3.13}$$

where Π_k is the right-hand side of (2.3.2).

Proof. The real periodicity is direct from Lemma 2.3.13. Equation (2.3.13) is checked following the same lines as formula (2.2.7). ❑

Now, we are going to construct at least one holomorphic solution of equation (2.3.13) and then correct it by adding a solution of the homogeneous equation, i.e., a multiple of $\theta(k, \omega)$.

To construct such a solution, let us express the right-hand side of (2.3.13) in terms of the theta function.

Lemma 2.3.16. *One has*

$$i e^{\pi i k} q^{k(k+1)/2} \Pi_k = C\, \theta(k + \omega + 1/2, 2\omega)\, \frac{\theta(k, \omega)}{\theta'(0, \omega)},$$

where C is a constant.

Proof is a combination of the translational properties of Π_k and the information about its zeros. ❑

Now consider the following two equations:

$$G_{k+\omega} - G_k = 1, \text{ and} \tag{2.3.14}$$

$$H_{k+\omega} - H_k = \theta(k + \omega + \frac{1}{2}, 2\omega) - 1. \tag{2.3.15}$$

It is clear from Lemma 2.3.16 that if G_k solves (2.3.14) and H_k solves (2.3.15), then $-2C\frac{\theta(k,\omega)}{\theta'(0,\omega)}(G_k + H_k)$ solves (2.3.13). Concerning (2.3.14), it is satisfied by

$$G_k = -\frac{1}{2\pi i} \frac{\theta'}{\theta}(k, \omega).$$

On the other hand,

$$H_k = \frac{1}{2\pi i} A(k, \omega)$$

solves (2.3.15). This implies that

$$F_k = -\frac{C}{\pi i} \frac{\theta(k, \omega)}{\theta'(0, \omega)} (A(k, \omega) - \frac{\theta'}{\theta}(k, \omega) - \beta)$$

for constants C, β (note that the right-hand side is an entire function). So we need to find these constants.

Lemma 2.3.17. *The function F_k satisfies the conditions $F_0 = 2\sqrt{\pi a}$ and $F_{1/2} = 0$.*

Proof. The first statement is clear (use the Gauss integral); the second one follows from the fact that g_k has a pole at $k = -1/2$ and F_k is 1–periodic. ❑

Now we see that $\beta = A(1/2, \omega) - \frac{\theta'}{\theta}(1/2, \omega)$ and $C = 2\pi i \sqrt{\pi a}$. This completes the proof of the theorem.

Comment. The function $A(k, \omega)$ has the following connection to Appell functions (see, e.g., [Po] and references therein). The Appell function $\kappa(u, k, \tau)$ is defined as a unique holomorphic 1–periodic in k solution of the equation

$$\kappa(u, k + \tau, \tau) = e^{2\pi i u} \kappa(u, k, \tau) + \theta(k + \frac{1+\tau}{2}, \tau).$$

It has the Fourier series expansion

$$\kappa(u, k, \tau) = \sum_{n \in \mathbb{Z}} \frac{e^{\pi i n^2 \tau + 2\pi i n k}}{e^{2\pi i n \tau} - e^{2\pi i u}}.$$

The Appell function has a pole at $u = 0$, so one can introduce its regular part:

$$\kappa_0(k, \tau) = \sum_{n \in \mathbb{Z} \backslash 0} \frac{e^{\pi i n^2 \tau + 2\pi i n k}}{e^{2\pi i n \tau} - 1}.$$

Then

$$A(k, \omega) = 2\pi i (\kappa_0(k, 2\omega) + \kappa_0(k + \omega, 2\omega)).$$

2.4 Nonsymmetric Hankel transform

Recall that we want *to connect the Hankel transform and the Fourier transform on $\mathbb{Z}_N$ in one theory.*

We will interpret the integral

$$\int_{\mathbb{R}} e^{-x^2}|x|^{2k}dx = \Gamma(k+1/2), \ \Re k > -1/2,$$

as the structural constant of the classical Hankel transform, and prove Theorem 2.1.2. Our approach is different from the usual treatment of similar integrals in classical works on Bessel functions. Then we will switch to the nonsymmetric theory and demonstrate that it provides significant simplifications.

2.4.1 Operator approach

Recall that $\phi_\lambda^{(k)}(x) \stackrel{\mathbf{def}}{=\!=} \phi^{(k)}(\lambda x)$ is given as follows:

$$\phi^{(k)}(t) = \sum_{n=0}^{\infty} \frac{t^{2n}\Gamma(k+1/2)}{n!\Gamma(k+n+1/2)}, \ k \notin -1/2 - \mathbb{Z}_+.$$

It is even in both x and λ: $\phi^{(k)}(t) = \phi^{(k)}(-t)$. The connection with the classical Bessel function is established in (2.1.5).

Our first step is to reprove the classical formula (see, e.g., [Lu]) for the Gauss integral in the presence of the Bessel functions, using the operator interpretation of the latter.

Theorem 2.4.1. *For arbitrary complex λ, μ and $\Re(k) > -1/2$,*

$$\int_{\mathbb{R}} \phi_\lambda^{(k)}(x)\phi_\mu^{(k)}(x)e^{-x^2}|x|^{2k}dx = \Gamma(k+1/2)\phi_\lambda^{(k)}(\mu)e^{\lambda^2+\mu^2}. \tag{2.4.1}$$

Proof is based on the theory of the differential operator

$$L = \frac{d^2}{dx^2} + \frac{2k}{x}\frac{d}{dx}.$$

Lemma 2.4.2. *(i) The function $\phi_\lambda^{(k)}(x)$ is a unique even solution of the eigenvalue problem $L\phi(x) = 4\lambda^2\phi(x)$ with the normalization $\phi(0) = 1$.*

(ii) The operator L is self-adjoint with respect to the scalar product

$$\langle f(x), g(x)\rangle = \int_{\mathbb{R}} f(x)g(x)|x|^{2k}dx.$$

Proof. Both statements are straightforward. Concerning (ii), we check that $L = |x|^{-k} \circ H \circ |x|^k$, where $H = \frac{d^2}{dx^2} + \frac{k(1-k)}{x^2}$, and use that H is self-adjoint for the scalar product $\int_{\mathbb{R}} f(x)g(x)dx$.

❑

Corollary 2.4.3. *Asymptotically as* $|x| \to \infty$,

$$\phi_\lambda^{(k)}(x) \sim C(\lambda)(e^{2\lambda x} + e^{-2\lambda x})|x|^{-k}$$

for a constant $C(\lambda)$.

Proof. The eigenfunctions of H with the eigenvalue $4\lambda^2$ are asymptotically $e^{2\lambda x}$ and $e^{-2\lambda x}$. ❑

Let $f(x)$ be an even function of $x \in \mathbb{R}$ such that $f(x)e^{cx} \to 0$ as $x \to \infty$ for any c. The **Hankel transform** is defined as follows:

$$\mathbf{F}(f)(\lambda) \stackrel{\text{def}}{=} \frac{1}{\Gamma(k+1/2)} \int_{\mathbb{R}} f(x)\phi_\lambda^{(k)}(x)|x|^{2k}dx, \ \Re k > -1/2. \tag{2.4.2}$$

Let $\mathbf{F}^{op}$ be the corresponding transform on the operators: $\mathbf{F}^{op}(A) = \mathbf{F}\,A\,\mathbf{F}^{-1}$.

Lemma 2.4.4. *(a)* $\mathbf{F}^{op}(L) = 4\lambda^2$,
(b) $\mathbf{F}^{op}(4x^2) = L_\lambda \stackrel{\text{def}}{=} \frac{d^2}{d\lambda^2} + \frac{2k}{\lambda}\frac{d}{d\lambda}$,
(c) $\mathbf{F}^{op}(4x\frac{d}{dx}) = -4\lambda\frac{d}{d\lambda} - 4 - 8k$.

Proof. Formulas (a) and (b) hold because the operator L is self-adjoint and $\phi_\lambda^{(k)}(x)$ is its eigenfunction. Formula (c) formally follows from (a), (b), and the identity

$$[L, x^2] = 4x\frac{d}{dx} + 2 + 4k.$$

❑

We set $\phi_{\pm,\lambda}^{(k)} = \phi_\lambda^{(k)}(x)\, e^{\pm x^2}$ and

$$L^+ = e^{x^2} \circ L \circ e^{-x^2},\ L^- = e^{-x^2} \circ L \circ e^{x^2}.$$

Explicitly,

$$L^+ = L + 4x\frac{d}{dx} + 2 + 4k + 4x^2,\ L^- = L - 4x\frac{d}{dx} - 2 - 4k + 4x^2.$$

Using Lemma 2.4.4,

$$\mathbf{F}^{op}(L^-) = L^+.$$

It is immediate from Lemma 2.4.2(i) that $L^\pm\phi_{\pm,\lambda}^{(k)} = 4\lambda^2\phi_{\pm,\lambda}^{(k)}$ and

$$\mathbf{F}(\phi_{-,\mu}^{(k)}) = C_\mu e^{\mu^2}\phi_{+,\mu}^{(k)}(\lambda),$$

where C_μ does not depend on λ. Therefore

$$\int_{\mathbb{R}} \phi_\lambda^{(k)}(x)\phi_\mu^{(k)}(x)e^{-x^2}|x|^{2k}dx = \Gamma(k+1/2)C_\mu\phi_\mu^{(k)}(\lambda)e^{\lambda^2+\mu^2}.$$

The left-hand side of this equality is invariant under the change $\lambda \leftrightarrow \mu$, and so is the right-hand side. Hence $C_\mu = C_0 = 1$. The theorem is proved. ❑

The key step in this proof is the calculation of the Hankel transform for the algebra of symmetric (even) differential operators. Thanks to the formula for $[L, x^2]$, which gives the answer for the Euler operator $x\frac{d}{dx}$, it is is simple. For arbitrary root systems, the L–operators and the symmetric (W–invariant) functions in terms of the coordinates denerate the algebra of all symmetric differential operators. However, this statement becomes much more involved technically.

The symmetry $x \leftrightarrow \lambda$ of the function $\phi_\lambda^{(k)}(x)$, which was used in the proof, holds in general, but the justification is not straightforward.

Also note that the calculation we performed was simple, but not completely sufficient to clarify why the final formula (2.4.1) is so nice. Certainly other methods are needed for the multidimensional counterparts of 2.4.1.

The operator method we have used gives a really simple prove of formula (2.4.1) and its multidimensional generalizations in the *nonsymmetric setting*, which is based on the Dunkl operators instead of the L–operators.

2.4.2 Nonsymmetric theory

The Hankel transform sends even functions to even functions (and is zero when applied to odd functions), by construction. In this subsection we consider its nonsymmetric version. The reflection $f(x) \mapsto f(-x)$ will be denoted by s.

Key definition. (C. Dunkl) $\mathcal{D} = \frac{d}{dx} - \frac{k}{x}(s-1)$. ❑

The operator $\mathcal{D}$ is obviously odd, i.e., $s\mathcal{D}s = -\mathcal{D}$. The restriction of the even operator $\mathcal{D}^2$ to the space of even functions coincides with $L = \frac{d^2}{dx^2} + \frac{2k}{x}\frac{d}{dx}$:

$$\mathcal{D}^2(f) = \mathcal{D}(f') = f'' + \frac{2k}{x} f' \text{ for even } f(x).$$

The following two lemmas are counterparts of Lemmas 2.4.2 and 2.4.4.

Lemma 2.4.5. *(i) If $\lambda \neq 0$ or $\lambda = 0$ and $k \notin -1/2 - \mathbb{Z}_+$, then the eigenvalue problem $\mathcal{D}\psi = 2\lambda\psi$ has a unique analytic at 0 solution $\psi_\lambda^{(k)}(x)$ with the normalization $\psi_\lambda^{(k)}(0) = 1$. Moreover, $\psi_\lambda^{(k)}(x) = \psi^{(k)}(\lambda x)$ and*

$$\psi^{(k)}(t) = \phi^{(k)}(t) + \frac{(\phi^{(k)}(t))'}{2} \text{ for } \phi^{(k)}(t) = \frac{\psi^{(k)}(t) + \psi^{(k)}(-t)}{2}.$$

If $\lambda = 0$ and $k = -1/2 - n$, $n \in \mathbb{Z}_+$, then $\psi = 1 + Cx^{2n+1}$ for $C \in \mathbb{C}$.

(ii) Let $\mathcal{D}^*$ be the adjoint of $\mathcal{D}$ with respect to the scalar product

$$\langle f(x), g(x) \rangle = \int_{\mathbb{R}} f(x) g(x) |x|^{2k} dx.$$

Then $\mathcal{D}^* = -\mathcal{D}$.

Comment. For any k, there is an extra solution $\psi = x/|x|^{2k+1}$ of the equation $\mathcal{D}\psi = 0$, but it is analytic at 0 only as $k \in -\frac{1}{2} - \mathbb{Z}_+$.

Proof. (i) We set $\psi = \psi_0 + \psi_1$, where ψ_0 is even and ψ_1 is odd. Then $\mathcal{D}\psi = 2\lambda\psi$ is equivalent to the system of equations

$$\psi_0' = 2\lambda\,\psi_1,\ \psi_1' + \frac{2k}{x}\psi_1 = 2\lambda\psi_0.$$

We obtain that ψ_0 satisfies Lemma 2.4.2(i).

(ii) The operator $|x|^k \circ \mathcal{D} \circ |x|^{-k} = \frac{d}{dx} - \frac{k}{x}s$ is anti-self-adjoint with respect to the scalar product $(f(x), g(x)) = \int_{\mathbb{R}} f(x)g(x)dx$. ❑

Following Dunkl, we introduce the **nonsymmetric Hankel transform**:

$$\mathbf{F}(f)(\lambda) = \frac{1}{\Gamma(k+1/2)} \int_{\mathbb{R}} f(x)\psi_\lambda^{(k)}(x)|x|^{2k}dx;$$

$\mathbf{F}^{op}$ is its action on operators.

Lemma 2.4.6. *(a)* $\mathbf{F}^{op}(\mathcal{D}) = -2\lambda$,

(b) $\mathbf{F}^{op}(s) = s_\lambda,\ s_\lambda f(\lambda) \mapsto f(-\lambda)$,

(c) $\mathbf{F}^{op}(2x) = \mathcal{D}_\lambda \stackrel{\text{def}}{=\!=} \frac{d}{d\lambda} - \frac{k}{\lambda}(s_\lambda - 1)$. ❑

The main result of this subsection is the following theorem.

Theorem 2.4.7. *(Master formula)*

$$\int_{\mathbb{R}} \psi_\lambda^{(k)}(x)\psi_\mu^{(k)}(x)e^{-x^2}|x|^{2k}dx = \Gamma(k+1/2)\psi_\lambda^{(k)}(\mu)e^{\lambda^2+\mu^2}. \tag{2.4.3}$$

Proof is parallel to the proof of Theorem 2.4.1. However, the necessary calculations are simpler than in the symmetric case (even in the rank one case). We introduce the functions $\psi_{\pm,\lambda}^{(k)}(x) = \psi_\lambda^{(k)}(x)e^{\pm x^2}$ and operators $\mathcal{D}^+ = e^{x^2} \circ \mathcal{D} \circ e^{-x^2}$ and $\mathcal{D}^- = e^{-x^2} \circ \mathcal{D} \circ e^{x^2}$. Easy calculations show that

$$\mathcal{D}^\pm \psi_{\pm,\lambda}^{(k)} = 2\lambda\psi_{\pm,\lambda}^{(k)},\ \mathcal{D}^\pm = \mathcal{D} \mp 2x,\ \mathbf{F}^{op}(\mathcal{D}^-) = \mathcal{D}_\lambda^+.$$

Hence

$$\mathbf{F}(\psi_{-,\mu}^{(k)}) = C_\mu e^{\mu^2}\psi_{+,\mu}^{(k)}(\lambda), \tag{2.4.4}$$

where C_μ does not depend on λ. Finally, $C_\mu = C_0 = 1$, thanks to the symmetry $\lambda \leftrightarrow \mu$. ❑

Comment. Theorem 2.4.7 is equivalent to Theorem 2.4.1, which is a special feature of the one-dimensional setup. Generally, the nonsymmetric formula results in the symmetric one, but not the other way round.

In the first place, Theorem 2.4.7 implies Theorem 2.4.1 since

$$\phi_\lambda^{(k)}(x) = (\psi_\lambda^{(k)})_0(x) = (\psi_\lambda^{(k)}(x) + \psi_{-\lambda}^{(k)}(x))/2, \text{ and for } \mu.$$

To deduce Theorem 2.4.7 from Theorem 2.4.1 we need to show that

$$\int_{\mathbb{R}} (\psi_\lambda^{(k)})_1(x)\,(\psi_\mu^{(k)})_1(x)\,e^{-x^2}|x|^{2k}dx = \Gamma(k+1/2)(\psi_\lambda^{(k)})_1(\mu)\,e^{\lambda^2+\mu^2},$$

where $(\psi_\lambda^{(k)})_1(x)$ is the odd component of $(\psi_\lambda^{(k)})(x)$. We may ignore the "cross-terms" $(\psi_\lambda^{(k)})_1(x)(\psi_\mu^{(k)})_0(x)$, since they are odd and their integrals are zero. To calculate the integral above, we can use the same Theorem 2.4.1 because of the following **shift–formula**:

$$\frac{1}{x}(\psi_\lambda^{(k)})_1 = \frac{2\lambda}{1+2k}\phi_\lambda^{(k+1)}.$$

This equivalence somewhat clarifies why the nonsymmetric Hankel transform did not appear (as far as we know) in classical works on Bessel functions. It adds nothing new to the symmetric (even) one. ❑

Using the operator $e^{-\mathcal{D}^2/4}$, we can rewrite formula (2.4.3) in the following form:

$$\int_{\mathbb{R}} \psi_\lambda^{(k)}(x)e^{-\mathcal{D}^2/4}(f(x))e^{-x^2}|x|^{2k}dx = \Gamma(k+1/2)f(\lambda)e^{\lambda^2}, \tag{2.4.5}$$

where $f(x)$ is a function from a suitable completion of the space linearly generated by the functions $\psi_\mu^{(k)}(x)$. Indeed, Theorem 2.4.7 shows that (2.4.5) holds for $f(x) = \psi_\mu^{(k)}(x)$. Formula (2.4.5) leads to the following entirely algebraic definition of the Hankel transform:

$$\mathbf{F} = e^{x^2} \circ e^{\mathcal{D}^2/4} \circ e^{x^2}.$$

We will use it later to introduce the truncated Hankel transform.

Since $\mathcal{D}$ is nilpotent, (2.4.5) results in the following important corollaries.

Corollary 2.4.8. *(cf. Corollary 2.1.6) The nonsymmetric Hankel transform restricts to a map* $\mathbf{F} : \mathbb{C}[x]e^{-x^2} \to \mathbb{C}[\lambda]e^{\lambda^2}$ *and, moreover, preserves the filtration by degrees of polynomials.* ❑

We set $\mathcal{B}_x = \sum_\mu \mathbb{C}\,\psi_\mu^{(k)}(x)e^{-x^2}$, $\mathcal{B}_\lambda = \sum_\mu \mathbb{C}\,\psi_\mu^{(k)}(\lambda)e^{+\lambda^2}$.

Corollary 2.4.9. *(***Inversion***) Introducing the imaginary Hankel transform*

$$\mathbf{F}_{im}(g)(x) \stackrel{\text{def}}{=\!=} \frac{1}{i\Gamma(k+1/2)}\int_{i\mathbb{R}} g(\lambda)\psi_x^{(k)}(-\lambda)|\lambda|^{2k}d\lambda,$$

$\mathbf{F}\circ\mathbf{F}_{im} = id$ *and* $\mathbf{F}_{im}\circ\mathbf{F} = id$ *in the spaces* $\mathcal{B}_x, \mathcal{B}_\lambda$ *or* $\mathbb{C}[x]e^{-x^2}$, $\mathbb{C}[\lambda]e^{\lambda^2}$ *or their suitable completions.*

Proof. This follows from the formula

$$\frac{1}{i}\int_{i\mathbb{R}} \psi_\lambda^{(k)}(-x)\psi_\mu^{(k)}(x)e^{x^2}|x|^{2k}dx = \Gamma(k+1/2)\psi_\lambda^{(k)}(\mu)e^{-\lambda^2-\mu^2}, \qquad (2.4.6)$$

which results from (2.4.3) upon $x \mapsto ix, \lambda \mapsto i\lambda, \mu \mapsto -i\mu$. The passage to $\mathbb{C}[x]e^{-x^2}$ and $\mathbb{C}[\lambda]e^{\lambda^2}$ is either by means of (2.4.5) or via the completion. ❑

Corollary 2.4.10. (**Plancherel formula**) *For functions $f(x), g(x)$ in $\mathcal{B}_x$ or those in $\mathbb{C}[x]e^{-x^2}$, we set $\widehat{f} = \mathbf{F}(f)$, $\widehat{g} = \mathbf{F}(g)$. Then*

$$\int_{\mathbb{R}} f(x)g(x)|x|^{2k}dx = \frac{1}{i}\int_{i\mathbb{R}} \widehat{f}(-\lambda)\widehat{g}(\lambda)|\lambda|^{2k}d\lambda. \qquad (2.4.7)$$

Proof. The left-hand side of (2.4.7) can be readily calculated for $f(x) = \psi_{\mu_1}^{(k)}(x)e^{-x^2}$ and $g(x) = \psi_{\mu_2}^{(k)}(x)e^{-x^2}$, thanks to formula (2.4.4). We substitute $\sqrt{2}x \mapsto x$ and use that $\psi_\mu^{(k)}(x)$ depends on the product μx. Similarly, we calculate the right-hand side using (2.4.6). Completing or using (2.4.5), we switch to the space $\mathbb{C}[x]e^{-x^2}$. ❑

Concerning completions, when $k \in \mathbb{R}$, the last corollary determines the nonsymmetric Hankel transform on the space $L^2(\mathbb{R}, |x|^{2k})$. Indeed, the latter space is the L^2–completion of the space $\mathbb{C}[x]e^{-x^2}$ with respect to the scalar product $\int_{\mathbb{R}} f(x)\overline{g(x)}|x|^{2k}dx$.

To be more exact, we first use this corollary for real-valued functions. The image of the corresponding L^2 will be the $\mathbb{R}$–subspace of $L^2(\imath\mathbb{R}, |\lambda|^{2k})$ formed by functions satisfying $\overline{f(\lambda)} = f(-\lambda)$. Then we extend the coefficients from $\mathbb{R}$ to $\mathbb{C}$.

2.4.3 Rational DAHA

Concerning the notation, the rational DAHA is denoted by $\mathcal{H}''$. The notation $\mathcal{H}'$ is reserved for the trigonometric limit which leads to the Harish-Chandra theory. In a sense, $\mathcal{H}''$ is a double degeneration of DAHA.

This algebra is defined as follows:

$$\mathcal{H}'' = \langle \mathcal{D}, x, s\rangle / \{s\mathcal{D}s = -\mathcal{D}, sxs = -x, [\mathcal{D}, x] = 1 + 2ks, s^2 = 1\};$$

$\mathcal{H}''$ acts on the vector space $\mathbb{C}[x]$ via the **Dunkl operator**:

$$s(f(x)) = f(-x),\ x(f(x)) = xf(x),\ \mathcal{D}(f) = (\frac{d}{dx} - \frac{k}{x}(s-1))f.$$

We will call this module the **polynomial representation**.

Theorem 2.4.11. *(a) The $\mathcal{H}''$–module $\mathbb{C}[x]$ is faithful.*

(b) (PBW property) The elements $x^n\mathcal{D}^m s^\varepsilon$, $n, m \in \mathbb{Z}_+$, $\varepsilon = 0, 1$, form a basis of $\mathcal{H}''$.

Proof. It is clear that any element of $\mathcal{H}''$ can be expressed in the following form:

$$H = \sum_{n,m,\varepsilon} c_{n,m,\varepsilon} x^n \mathcal{D}^m s^\varepsilon, \; c_{n,m,\varepsilon} \in \mathbb{C}.$$

If $c_{n,m,\varepsilon} \neq 0$ at least once, then the image of H is nonzero in the space of either even or odd functions. This gives both (a) and (b). ❑

Because the polynomial representation is always faithful, we will identify $\mathcal{D}$ considered as a generator of the rational DAHA with its image, the Dunkl operator.

The algebra $\mathcal{H}''$ has an automorphism ω defined by

$$\omega(\mathcal{D}) = -2x, \; \omega(2x) = \mathcal{D}, \; \omega(s) = s. \tag{2.4.8}$$

It can be represented as follows:

$$\omega = e^{x^2} \circ e^{\mathcal{D}^2/4} \circ e^{x^2} = e^{\mathcal{D}^2/4} \circ e^{x^2} \circ e^{\mathcal{D}^2/4}. \tag{2.4.9}$$

Here we extend $\mathcal{H}''$ by adding e^{x^2}, $e^{\mathcal{D}^2/4}$, and treat the latter as inner automorphisms in the resulting greater algebra $\widehat{\mathcal{H}}''$. It is simple to see that both preserve $\mathcal{H}''$. The automorphism ω is nothing but an algebraic version of the Hankel transform $\mathbf{F}$.

The coincidence of the two representations of ω can be deduced from the defining relations or (much simpler) checked in the polynomial representation. It defines the action of the *projective* $PSL(2,\mathbb{Z})$ (due to Steinberg) on $\mathcal{H}''$.

The polynomial representation is always faithful, but not always irreducible. The reducibility may occur only for special, *singular*, values of k.

Theorem 2.4.12. *The $\mathcal{H}''$-module $\mathbb{C}[x]$ is irreducible if and only if $k \notin -1/2 - \mathbb{Z}_+$.*

Proof. If $k \notin -1/2 - \mathbb{Z}_+$, then the equation $\mathcal{D}\psi = 0$ has a unique solution $\psi = 1$ in $\mathbb{C}[x]$. Since the operator $\mathcal{D}$ is nilpotent in $\mathbb{C}[x]$, any submodule should contain 1 and therefore the whole $\mathbb{C}[x]$. If $k = -1/2 - n$, then x^{2n+1} generates a nontrivial submodule of $\mathbb{C}[x]$, since $\mathcal{D}(x^{2n+1}) = 0$. ❑

We see that $\mathcal{H}''$ can have finite dimensional irreducible representations for some k. It is not difficult to describe them all. Generally, the theory of finite dimensional representations of double Hecke algebras associated with root systems is far from being complete.

2.4.4 Finite dimensional modules

We will use that

$$[h, x] = x, \; [h, \mathcal{D}] = -\mathcal{D} \;\; \text{for } h = (x\mathcal{D} + \mathcal{D}x)/2. \tag{2.4.10}$$

Note that h is $x\frac{d}{dx}+k+1/2$ in the polynomial representation. Since it is faithful, (2.4.10) is the claim that $x, \mathcal{D}$ are homogeneous operators of degrees ± 1, which is obvious.

These relations are the defining relations of $\mathfrak{osp}(2|1)$, but we will not rely on the theory of this super Lie algebra. For our purpose, a reduction to $\mathfrak{sl}_2$ is sufficient.

Namely, we will use that the elements $e = x^2$, $f = -\mathcal{D}^2/4$, and h satisfy the defining relations of $\mathfrak{sl}_2(\mathbb{C})$. Indeed, $[e, f] = h$ because

$$[\mathcal{D}^2, x^2] = [\mathcal{D}^2, x]x + x[\mathcal{D}^2, x] = 2\mathcal{D}x + x(2\mathcal{D}),$$

and the relations $[h, e] = 2e, [h, f] = -2f$ readily result from (2.4.10). The Casimir operator $C = h^2 - 2h + 4ef$ becomes

$$(h^2 - 2h)(1) = h(h-2)(1) = (k+1/2)(k-3/2) \text{ in } \mathbb{C}[x].$$

The module $\mathbb{C}[x]$ is the Verma module with the h–lowest weight $1/2+k$ over $U(\mathfrak{osp}(2|1))$. It is the direct sum of the two Verma modules with the h–lowest weights $1/2+k$ and $3/2+k$, formed by even and odd functions respectively, with respect to the action of $U(\mathfrak{sl}_2)$.

Let $k = -n - \frac{1}{2}$ for $n \in \mathbb{Z}_+$. Then $V_{2n+1} = \mathbb{C}[x]/(x^{2n+1})$ is an irreducible representation of $\mathcal{H}''$. The elements of V_{2n+1} can be identified with polynomials of degree smaller than $2n+1$.

Theorem 2.4.13. *Finite dimensional representations of $\mathcal{H}''$ exist only as $k = -n - 1/2$ or $k = n + 1/2$ for $n \in \mathbb{Z}_+$. Given such k, the algebra $\mathcal{H}''$ has a unique finite dimensional irreducible representation up to isomorphisms. It is either V_{2n+1} for negative k or its image under the $\mathcal{H}''$-automorphism:*

$$x \mapsto x, \mathcal{D} \mapsto \mathcal{D}, \ s \mapsto -s, k \mapsto -k, \tag{2.4.11}$$

in the case of positive k.

Proof. Let V be an irreducible finite dimensional representation of $\mathcal{H}''$. Then the subspaces V^0, V^1 of V formed, respectively, either by s–invariant or s–anti-invariant vectors are preserved by e, f, and h. Indeed, $sxs = -x$, $s\mathcal{D}s = -\mathcal{D}$, and $sh = hs$. Note that s leaves all h–eigenspaces invariant. For instance, it commutes with the $\mathfrak{sl}_2$–action. One obtains

$$\mathcal{D}x = h + k + 1/2, \ x\mathcal{D} = h - k - 1/2 \ \text{ in } V^0,$$

and the other way round in V^1. Note that x and D send V^0 to V^1.

Let us check that $\pm k \in -1/2 - \mathbb{Z}_+$. All h–eigenvalues in V are integers thanks to the general theory of finite dimensional representations of $\mathfrak{sl}_2(\mathbb{C})$.

We pick a highest vector v, i.e., a nonzero h–eigenvector $v \in V$ with the maximal possible eigenvalue m. Using the automorphism (2.4.11), we may assume (till the end of the proof) that it belongs to V^0. Then $m \in \mathbb{Z}_+$ (the theory of $\mathfrak{sl}_2$) and $x(v) = 0$ because the latter is an h–eigenvector with the eigenvalue $m+1$. Hence $\mathcal{D}x(v) = 0$, $m + k + 1/2 = 0$, and $k = -1/2 - m$.

Let U^0 be a nonzero irreducible $\mathfrak{sl}_2(\mathbb{C})$–submodule of V^0. The spectrum of h in U^0 is $\{-n, -n+2, \ldots, n-2, n\}$ for an integer $n \geq 0$. Let $v_l \neq 0$ be an h–eigenvector with the eigenvalue l. If $e(v) = 0$, then $v = cv_n$ for a constant c, and if $f(v) = 0$, then $v = cv_{-n}$.

Let us check that $\mathcal{D}x(v_n) = 0$, $x\mathcal{D}(v_{-n}) = 0$, and

$$\mathcal{D}x(v_l) \neq 0 \text{ for } l \neq n, \quad x\mathcal{D}(v_l) \neq 0 \text{ for } l \neq -n.$$

Both operators $\mathcal{D}x$ and $x\mathcal{D}$ obviously preserve U^0:

$$\mathcal{D}x(v_l) = (l + k + 1/2)v_l, \quad x\mathcal{D}(v_l) = (l - k - 1/2)v_l.$$

Hence,

$$\mathcal{D}^2x^2(v_l) = ((\mathcal{D}x)^2 + (1-2k)(\mathcal{D}x))(v_l) = (l + k + 1/2)(l - k + 3/2)v_l.$$

Setting $l = n$, we obtain that $(n + k + 1/2)(n - k + 3/2) = 0$, and therefore we come to the formula $k = -1/2 - n$, since $k < 0$ and $n - k + 3/2 > 0$. Thus $\mathcal{D}x(v_n) = 0$. The case of $x\mathcal{D}$ is analogous.

The next claim is that $x(v_n) = 0$, $\mathcal{D}(v_{-n}) = 0$. Indeed, $x(v') = 0$ and $\mathcal{D}(v') = 0$ for $v' = x(v_n)$. Therefore $v' \in V^1$, $s(v') = -v'$, and

$$0 = [\mathcal{D}, x](v') = (1 + 2ks)(v') = (1 + 1 + 2n)v' = (2 + 2n)v'.$$

Hence $v' = x(v_n) = 0$. A similar argument shows that $\mathcal{D}(v_{-n}) = 0$.

Now we use the formula

$$\mathcal{D}(x^2(v_l)) = x(2 + x\mathcal{D})(v_l) = (2 + l - k - 1/2)x(v_l) = (2 + l + n)x(v_l)$$

and obtain that $x(v_l) \in \mathcal{D}(U^0)$ for any $-n \leq l \leq n$. Hence $U = U^0 + \mathcal{D}(U^0)$ is x–invariant. It is obviously $\mathcal{D}$–invariant and s–invariant ($\Leftarrow \mathcal{D}(V^0) \subset V^1$). Also, the sum is direct.

Finally, U is an $\mathcal{H}''$–module and has to coincide with V because the latter was assumed to be irreducible. The above formulas make the $\mathcal{H}''$–isomorphism $U \simeq V_{2n+1}$ explicit: the h–eigenvectors $x^i(v_{-n}) \in U$ go to the monomials $x^i \in V_{2n+1}$. ❑

2.4.5 Truncated Hankel transform

Let us consider the representation V_{2n+1} closely. We endow it with the following two scalar products:

$$\langle f, g \rangle_+ \stackrel{\text{def}}{=} \text{Res}\,(f(x)g(x)x^{-2n-1})$$

and

$$\langle f, g\rangle_- \stackrel{\text{def}}{=\!=} \operatorname{Res}\,(f(x)g(-x)x^{-2n-1}),$$

analogous to the standard scalar products considered above. Here by Res , we mean the coefficient of x^{-1}. One has: $\langle sf, g\rangle_\pm = \langle f, sg\rangle_\pm$, and

$$\langle xf, g\rangle_\pm = \pm\langle f, xg\rangle_\pm,\ \langle \mathcal{D}f, g\rangle_\pm = \mp\langle f, \mathcal{D}g\rangle_\pm. \tag{2.4.12}$$

Note that if $f(x) = \sum_{i=0}^{2n} a_i x^i$, then

$$\langle f, f\rangle_\pm = \sum_{i=0}^{2n} (\pm)^i a_i a_{2n-i},$$

so these scalar products are far from being positive definite.

In contrast to $\mathbb{C}[x]$, the Gaussians are well defined in V_{2n+1}:

$$e^{\pm x^2} = \sum_{m=0}^{2n} (\pm x^2)^m/m!.$$

We may introduce the **truncated Hankel transform** on this space by the old formula $\mathbf{F}_+ = e^{x^2}e^{\mathcal{D}^2/4}e^{x^2}$. Multiplication by e^{x^2} and $e^{\mathcal{D}^2/4}$ is well defined in the space V_{2n+1} because x and $\mathcal{D}$ are nilpotent. Similarly, $\mathbf{F}_-(f(x)) = \mathbf{F}_+(f(-x))$. Another (equivalent) approach requires the truncated nonsymmetric Bessel functions, which will not be discussed here. See [CM].

Theorem 2.4.14. *The truncated* **inversion** *reads as*

$$\mathbf{F}_-\circ\mathbf{F}_+ \;=\; (-1)^n\, id \;=\; \mathbf{F}_+\circ\mathbf{F}_-.$$

Let $\widehat{f} = \mathbf{F}_+(f)$ *and* $\widehat{g} = \mathbf{F}_+(g)$. *We have the* **Plancherel formula**

$$\langle f, g\rangle_+ = (-1)^n\langle \widehat{f}, \widehat{g}\rangle_-.$$

Proof. Using (2.4.12), we conclude that F_+, inducing the automorphism ω on $\mathcal{H}''$, sends the anti-involution corresponding to $\langle\cdot,\cdot\rangle_+$ to that of $\langle\cdot,\cdot\rangle_-$. We need to examine the generators $s, x, \mathcal{D}$. The representation V_{2n+1} is irreducible, so the scalar products $\langle f, g\rangle_+$ and $\langle \hat{f}, \hat{g}\rangle_-$ have to be proportional.

To find the proportionality coefficient, let us calculate the Hankel transform explicitly. First, $\mathbf{F}_+(1)$ has to be proportional to x^{2n}, since 1 is a unique eigenvector of $\mathcal{D}$ and ω sends $\mathcal{D}$ to $-2x$. Hence $\mathbf{F}_+(x^m)$ is proportional to $\mathcal{D}^m\mathbf{F}_+(1)$, that is, x^{2n-m}. Second,

$$\mathbf{F}_+(e^{-x^2}) = (e^{x^2}e^{\mathcal{D}^2/4}e^{x^2})e^{-x^2} = e^{x^2};\ \text{therefore}$$

$$\mathbf{F}_+(x^{2m}) = (-1)^m\frac{m!}{(n-m)!}x^{2n-2m},\ \text{and}$$

$$\mathbf{F}_+(x^{2m+1}) = (\mathcal{D}/2)\mathbf{F}_+(x^{2m}) = (-1)^m\frac{m!}{(n-m-1)!}x^{2n-2m-1}.$$

In particular, $\mathbf{F}_+(1) = \frac{x^{2n}}{n!}$ and $\mathbf{F}_+(x^{2n}) = (-1)^n n!$. We obtain the inversion formula. The **Plancherel formula** holds, since

$$\langle 1, x^{2n}\rangle_+ = 1 \text{ and } \langle \mathbf{F}_+(1), \mathbf{F}_+(x^{2n})\rangle_- = (-1)^n \langle x^{2n}, 1\rangle_- = (-1)^n.$$

❑

2.5 Polynomial representation

Recall Theorem 2.2.1:

$$\frac{1}{i}\int_{i\mathbb{R}} q^{-x^2}\delta_k(x)dx = 2\sqrt{\pi a}\prod_{j=0}^{\infty}\frac{1-q^{k+j}}{1-q^{2k+j}},\ \Re k > 0, \tag{2.5.1}$$

where

$$\delta_k(x) = \prod_{j=0}^{\infty}\frac{(1-q^{j+2x})(1-q^{j-2x})}{(1-q^{k+j+2x})(1-q^{k+j-2x})}$$

for $q = e^{-1/a}$.

For a Laurent series $f(x) = \sum_{n=-\infty}^{\infty} c_n q^{nx}$ in the variable q^x, $\langle f(x)\rangle \stackrel{\textbf{def}}{=\!=} c_0$, the constant term. We expand $\delta_k(x)$ in a Taylor series in terms of q^k. It becomes a Laurent series in terms of q^x with the coefficients from $\mathbb{C}[q^k][[q]]$. Let

$$\widehat{\gamma_-} \stackrel{\textbf{def}}{=\!=} \sum_{n=-\infty}^{\infty} q^{nx+n^2/4}.$$

It is a definition, but the right-hand side does coincide with q^{-x^2} in the space of distributions on periodic functions in the variable x with the period $\omega = 2\pi ia$. Formula (2.5.1) is equivalent to

$$\langle \widehat{\gamma_-}\delta_k(x)\rangle = 2\prod_{j=0}^{\infty}\frac{1-q^{k+j}}{1-q^{2k+j}}. \tag{2.5.2}$$

2.5.1 Rogers' polynomials

We are going to interpret the latter formula as a calculation of the structural constant of the difference spherical Fourier transform in the A_1–case. Generally speaking, the spherical transform is an integration with the spherical (or hypergeometric) functions. In this setup, the spherical functions are Rogers' polynomials and the integration is the constant term functional.

Definition 2.5.1. *The* **Rogers polynomials** $p_n(x) \in \mathbb{C}(q, q^k)[q^x + q^{-x}]$ *are uniquely defined by the properties:*

(a) $p_n(x) = q^{nx} + q^{-nx} + \sum_{|m|<n} c_m q^{mx},\ n > 0,\ p_0 = 1;$

(b) $\langle p_n(x) q^{mx}\delta_k(x)\rangle = 0$ *for all* m *with* $|m| < n$.

The first two nontrivial Rogers' polynomials are:

$$p_1 = q^x + q^{-x},\ p_2 = q^{2x} + q^{-2x} + \frac{(1-q^k)(1+q)}{1-q^{k+1}}.$$

The formula for p_1 is immediate from the following simple lemma.

Lemma 2.5.2. *Let* $p_n(x) = \sum_m c_m q^{mx}$. *Then* $c_m \neq 0$ *only for even* $n - m$.

Proof. The Laurent series $\delta_k(x)$ involves only even powers of q^x. ❑

To calculate p_2, we use that $\delta_k(x)$ can be replaced in (b) by

$$\delta_k^0(x) \stackrel{\text{def}}{=\!=} \frac{\delta_k(x)}{\langle \delta_k(x) \rangle}.$$

We calculated it in Theorem 2.2.5 in terms of $(q^{2nx} + q^{-2nx})$:

$$\delta_k^0(x) = 1 + \frac{(1+q)(q^k-1)}{2(1-q^{k+1})}(q^{2x} + q^{-2x}) + \dots .$$

The constant term of p_2 is the first coefficient of this expansion multiplied by -2.

More generally, this argument gives that the coefficients of all $p_n(x)$ lie in $\mathbb{Q}(q, q^k)$, since so do the coefficients of $\delta_k^0(x)$.

Rogers' polynomials $p_n(x)$ play the role of the Bessel functions in the following theorem

Theorem 2.5.3.

$$\begin{aligned} &\langle p_n(x) p_m(x) \widehat{\gamma_-} \delta_k^0(x) \rangle = \\ &\qquad = p_n((m+k)/2) p_m(k/2) q^{(m^2+n^2+2k(m+n))/4} \langle \widehat{\gamma_-} \delta_k^0(x) \rangle. \end{aligned} \tag{2.5.3}$$

Proof will be given later. ❑

Obviously (2.5.3) results in the following $m \leftrightarrow n$–invariance:

$$p_n((m+k)/2) p_m(k/2) = p_m((n+k)/2) p_n(k/2).$$

Conversely, this formula can be readily deduced from this symmetry, as we will see.

The most natural proof of this theorem goes via the q–counterparts of the *nonsymmetric* Bessel functions.

2.5.2 Nonsymmetric polynomials

Recall formulas (2.2.8) and (2.2.9):

$$\delta_k^0(x) = \frac{\mu_k^0(x) + \mu_k^0(-x)}{2},$$

where

$$\mu_k(x) = \prod_{j=0}^{\infty} \frac{(1 - q^{j+2x})(1 - q^{j+1-2x})}{(1 - q^{k+j+2x})(1 - q^{k+j+1-2x})},$$

$$\mu_k^0(x) = \mu_k(x)/\langle\mu_k(x)\rangle = 1 + \frac{q^k - 1}{1 - q^{k+1}}(q^{2x} + q^{1-2x}) + \cdots .$$

We use the following linear order on the set of monomials q^{nx} : $q^{nx} \succ q^{mx}$ if either $|m| < |n|$ or $n = -m$ and n is negative (so $q^{-x} \succ q^x$).

Definition 2.5.4. *The* **nonsymmetric polynomials** $e_n(x) \in \mathbb{C}(q, q^k)[q^x, q^{-x}]$ *for $n \in \mathbb{Z}$ are uniquely defined by the properties:*
(a) $e_n(x) = q^{nx}$ + lower terms with respect to $\succ$,
(b) $\langle e_n(x)q^{-mx}\mu_k^0(x)\rangle = 0$ if $q^{mx} \prec q^{nx}$ (note the minus sign!).

The definition and the beginning of the expansion of μ_k^0 give that

$$e_0 = 1,\ e_1 = q^x,\ e_{-1} = q^{-x} + \frac{1 - q^k}{1 - q^{k+1}}q^x,\ e_2 = q^{2x} + q\frac{1 - q^k}{1 - q^{k+1}}.$$

Similar to p_n, each $e_n(x)$ involves only monomials q^{mx} with even $n - m$ because $\mu_k^0(x)$ contains only even powers of q^x. Similar to $p_n(x)$,

$$e_n(x) \in \mathbb{Q}(q, q^k)[q^x, q^{-x}],\ \text{since } \mu_k^0(x) \in \mathbb{Q}(q, q^k)[[q^x, q^{-x}]].$$

The main technical advantage of the theory of e–polynomials versus the p–polynomials is that we can construct them using the intertwining operators of the double Hecke algebra. The following **creation operator** is due to Knop and Sahi in the case of A_n. See [KnS] and [C16, C23] for the general theory.

Proposition 2.5.5. *Denoting $\pi(f(x)) = f(1/2 - x)$,*

$$q^{(1-n)/2}e_n = q^x\pi(e_{1-n}(x)) \text{ for } n > 0.$$

Proof. It is clear that $q^{(n-1)/2}q^x\pi(e_{1-n}(x)) = q^{nx}$ + lower terms. So it suffices to check that

$$\langle q^x\pi(e_{1-n}(x))q^{-mx}\mu_k^0(x)\rangle = 0,\ m = n-1, n-2, \ldots, 1-n. \tag{2.5.4}$$

The following two properties of π are evident:

$$(a)\ \langle\pi(f)\rangle = \langle f\rangle,\quad (b)\ \pi(\mu_k^0(x)) = \mu_k^0(x).$$

Using them, we see that (2.5.4) is true for $m = n-1, \ldots, 2-n$. The only remaining case $m = 1-n$ is trivial: $\pi(e_{1-n}(x))q^{nx}\mu_k^0(x)$ involves only odd powers of q^x. ❑

We can deduce (2.5.3) from its nonsymmetric analog. Setting

$$\varepsilon_n(x) \;=\; e_n(x)/e_n(-k/2),$$

the nonsymmetric variant of Theorem 2.5.3 reads as follows:

$$\langle \varepsilon_n(x)\varepsilon_m(x)\widehat{\gamma_-}\mu_k^0(x)\rangle = \langle \widehat{\gamma_-}\mu_k^0(x)\rangle\, \varepsilon_n(m_\sharp)\, q^{(m^2+n^2+2k(|m|+|n|))/4}, \qquad (2.5.5)$$

where $n_\sharp \overset{\textbf{def}}{=\!=} (n + \mathrm{sgn}(n)k)/2$ and we set $\mathrm{sgn}(0) = -1$.

The nonsymmetric (and more general) formula is easier to prove than the symmetric one, similar to the case of the Hankel transform considered above.

The coefficient of proportionality in formula (2.5.5) is a combination of formula (2.5.2) and the constant term conjecture (2.2.10):

$$\langle \widehat{\gamma_-}\mu_k^0(x)\rangle = \prod_{j=0}^{\infty}\frac{1-q^j}{1-q^{j+k}}.$$

Formula (2.5.5) implies that $\varepsilon_n(m_\sharp)$ is invariant under the change $m \leftrightarrow n$. This of course can be established directly. Let us do it.

2.5.3 Double affine Hecke algebra

We denote the double affine Hecke algebra by "double H" $\mathcal{HH}$. It depends on the two parameters $q^{1/2}$ and $t^{1/2}$. The generators are $X^{\pm 1}, Y^{\pm 1}, T$; the relations are

$$\left\{\begin{array}{c} TXT = X^{-1},\ TY^{-1}T = Y, \\ Y^{-1}X^{-1}YXT^2q^{1/2} = 1, \\ (T - t^{1/2})(T + t^{-1/2}) = 0 \end{array}\right\}.$$

Comment. If $t = 1$, then $\mathcal{HH}$ is the extension of the Weyl algebra by the reflection S, i.e.,

$$\langle X, Y, S\rangle/\{Y^{-1}X^{-1}YX = q^{-1/2}, S^2 = 1, SXS = X^{-1}, SYS = Y^{-1}\}.$$

❑

Theorem 2.5.6. *(a) (PBW property) The elements* $X^nT^\varepsilon Y^m$, $n, m \in \mathbb{Z}$, $\varepsilon = 0, 1$, *form a basis of* $\mathcal{HH}$.

(b) Using $s(f(x)) = f(-x)$ *and* $\pi(f(x)) = f(1/2 - x)$, *the formulas*

$$T \mapsto t^{1/2}s + \frac{t^{1/2} - t^{-1/2}}{q^{2x} - 1}(s - 1),\ X \mapsto q^x,\ Y \mapsto \pi T \qquad (2.5.6)$$

define a representation of $\mathcal{HH}$ *in the space* $\mathbb{C}[q^x, q^{-x}]$. *It is faithful for* q *apart from roots of unity.*

Proof. We will start with the following lemma, which is simple to check.

Lemma 2.5.7. *Setting* $\pi \stackrel{\mathbf{def}}{=\!=} YT^{-1}$, *the algebra* $\mathcal{H}$ *can be alternatively described as follows:*

$$\mathcal{H} = \langle T, X^{\pm 1}, \pi\rangle / \left\{ \begin{array}{c} TXT = X^{-1},\ \pi^2 = 1,\ \pi X \pi^{-1} = q^{1/2}X^{-1}, \\ (T - t^{1/2})(T + t^{-1/2}) = 0 \end{array} \right\}.$$

❑

The lemma gives that the formulas from part (b) of Theorem 2.5.6 define a representation of $\mathcal{H}$. Indeed, the formula for T is well known in the theory of the affine Hecke algebra of type A_1. This operator does satisfy $TXT = X^{-1}$. So we need to check only the relations involving Y or, equivalently, π. The relations with π are simple.

(a) Any element of $\mathcal{H}$ is a linear combination of $X^n T^{\varepsilon} Y^m$. These monomials are linearly independent in $\mathcal{H}$ if they are independent for at least one pair of special values of q, t. Let us take q that is not a root of unity. Then the images of these monomials in $\mathbb{C}[q^x, q^{-x}]$ are linearly independent. It is immediate for $t^{1/2} = 1$, i.e., for generic t. It is also true for arbitrary t, and not difficult to check.

We obtain (a), and the remaining part of (b). ❑

Comment. In fact, the theorem can be reformulated as follows:

$$\mathbb{C}[q^x, q^{-x}] = \mathrm{Ind}_{<T,Y>}^{\mathcal{H}}(\mathbb{C}),$$

where the subalgebra $\langle T, Y\rangle \subset \mathcal{H}$ acts on the one-dimensional space $\mathbb{C}$ via the character $T \mapsto t^{1/2}$, $Y \mapsto t^{1/2}$. Indeed, this relation includes the PBW theorem and readily results in the formulas from part (b). However, this coincidence does not include the fact that the polynomial representation is faithful for generic q. ❑

We introduce the **conjugation** $f \mapsto \bar{f}$ on the polynomial representation:

$$\overline{q^x} = q^{-x},\ \overline{q^{1/2}} = q^{-1/2},\ \overline{t^{1/2}} = t^{-1/2},$$

and define the scalar product:

$$\langle f, g\rangle \stackrel{\mathbf{def}}{=\!=} \langle f\bar{g}\mu_k^0(x)\rangle, \quad \text{where } q^k = t.$$

Here it is necessary to change the field of coefficients $\mathbb{C}$. It will be replaced by the ring $\mathbb{C}[q^{\pm 1/2}, t^{\pm 1/2}]$ or its field of rationals. We will do the same for $\mathcal{H}$. From now on $q^{1/2}$ *and* $t^{1/2}$ *will be considered as formal parameters,* unless stated otherwise. Note that the operators T and π from (2.5.6) preserve the space $\mathbb{C}[q^{\pm 1/2}, t^{\pm 1/2}, q^{\pm x}]$.

Theorem 2.5.8. *The operators* T, X, Y, π, q, t *are unitary with respect to the scalar product* $\langle\ ,\ \rangle$.

Proof. It is evident for X, q, t. It is also clear that $\pi^* = \pi = \pi^{-1}$. For T, we use the formula $\mu_k^0(-x) = \frac{1-q^{2x+k}}{q^k - q^{2x}}\mu_k^0(x)$. It gives that the adjoint of s is $(1-q^{2x}t)(t-q^{2x})s$. Concerning Y, use $Y = \pi T$. ❑

Since the coefficints of the polynomials $e_n(x)$ have denominators, the algebra $\mathcal{H}$ and the polynomial representation will be, as a matter of fact, considered over the field $\mathbb{C}(q^{1/2}, t^{1/2})$ in what follows, although we will continue to use the notation $\mathcal{H}$ and $\mathbb{C}[q^{\pm x}]$.

Moreover, it is sufficient to take $\mathbb{Q}$ instead of $\mathbb{C}$. Note that the coefficients of e_n do not require the square roots; they are expressed in terms of q, t. However, the action of T and π does require the square roots of q, t.

Theorem 2.5.9. *Considering the polynomials $e_n(x)$ as elements of the $\mathcal{H}$–module $\mathbb{C}[q^{\pm x}]$,*

$$Y e_n(x) = q^{-n_\sharp}\, e_n(x),\ n_\sharp \stackrel{\mathbf{def}}{=\!=} (n + sgn(n)k)/2,\ sgn(0) = -1,\ n \in \mathbb{Z}.$$

Proof. The operator Y is unitary and preserves the filtration induced by $\prec$, which is easy to check. Note that T does not preserve this filtration. It readily gives that $e_n(x)$ are eigenvectors of Y. Use the definition. Examining the leading term q^{nx} of $e_n(x)$, one obtains the corresponding eigenvalue. ❑

Theorem 2.5.10. *Setting $\varepsilon_n(x) \stackrel{\mathbf{def}}{=\!=} e_n(x)/e_n(-k/2)$, we obtain the* **duality formula**

$$\varepsilon_n(m_\sharp) = \varepsilon_m(n_\sharp).$$

Equivalently, the pairing $\{f\,,\,g\} \stackrel{\mathbf{def}}{=\!=} f(Y^{-1})(g)(-k/2)$ is symmetric on $\mathbb{C}[q^{\pm x}]$, where $f(Y^{-1}) = f(q^x \mapsto Y^{-1})$.

Proof. First, we recall the definition of the *anti-involution* ϕ of $\mathcal{H}$:

$$\phi(X) = Y^{-1},\ \phi(Y) = X^{-1},\ \phi(T) = T,\ \phi(q) = q,\ \phi(t) = t. \tag{2.5.7}$$

Later it will be associated with the Fourier transform on the space of the generated functions.

Second, we need the **evaluation map** : $\mathcal{H} \to \mathbb{C}(q^{1/2}, t^{1/2})$ defined by

$$\{X^n T^\varepsilon Y^m\} = t^{-n/2} t^{\varepsilon/2} t^{m/2}.$$

It is obviously ϕ–invariant. Explicitly,

$$\{H\} \text{ is the evaluation of } H(1) \in \mathbb{C}[q^{\pm x}]$$

for $H \in \mathcal{H}$. The evaluation of a polynomial in $q^{\pm x}$ is its value at the point $x = -k/2$.

Third, the scalar product $\{A, B\} = \{\phi(A)B\}$ on $\mathcal{H}$ is symmetric, i.e., $\{A, B\} = \{B, A\}$. It follows from $\{\phi(H)\} = \{H\}$ for $H \in \mathcal{H}$. It coincides with $\{f\,,\, g\}$ on Laurent polynomials upon the substitution $X = q^x$.

Finally,

$$\{e_n(X), e_m(X)\} = \{e_n(Y^{-1})e_m(X)\} = e_n(m_\sharp)e_m(-k/2), \tag{2.5.8}$$

where we used that $Y^{-1}e_n(x) = q^{n_\sharp}e_n(x)$ in the polynomial representation.

❑

2.5.4 Back to Rogers' polynomials

The relation of the Macdonald polynomials to the double affine Hecke algebras was the first obvious confirmation of their importance. These algebras were designed for a somewhat different purpose: to connect the Knizhnik–Zamolodchikov equation with the quantum many-body (eigenvalue) problem in the q–case. The paper [C22] is devoted to it. It was an important step in the new theory of generalized q–hypergeometric functions, which appear as solutions of the QMBP, that is equivalent to the proper QKZ. However, the application to the celebrated Macdonald constant term conjecture and the norm conjecture was the first recognized success of the double Hecke algebras.

In the A_1–case, symmetric means even. We will constantly use $X = q^x$ instead of x. Then the reflection s sends $X \mapsto X^{-1}$. Concerning operators, we call them symmetric if they commute with s. Symmetric operators on the space $\mathbb{C}[X^{\pm 1}]$ preserve the subspace $\mathbb{C}[X + X^{-1}]$ of symmetric polynomials. For the operators we consider it is necessary and sufficient.

Recall that $p_0 = 1$ and the other p_n are defined from the relations

$$p_n(x) = X + X^{-1} + \text{lower terms}, \ \langle p_n(x)p_m(x)\delta_k^0(x)\rangle = \delta_{mn}C_n,$$

where $\delta_k^0(x) = (\mu_k^0(x) + \mu_k^0(-x))/2$ and C_n are certain constants.

Theorem 2.5.11. *(a) The operator $Y + Y^{-1}$ commutes with T and is symmetric. Setting $\varpi(f)(x) \stackrel{\text{def}}{=} f(x + 1/2)$, $\mathcal{L} \stackrel{\text{def}}{=} Y + Y^{-1} \mid {}_{\mathbb{C}[X+X^{-1}]}$,*

$$\mathcal{L} = \frac{t^{1/2}X - t^{-1/2}X^{-1}}{X - X^{-1}}\varpi + \frac{t^{1/2}X^{-1} - t^{-1/2}X}{X - X^{-1}}\varpi^{-1}.$$

(b) For any $n > 0$,

$$p_n = (1 + t^{1/2}T)e_n = (1 + s)\left(\frac{t - X^2}{1 - X^2}e_n\right), \tag{2.5.9}$$

$$\mathcal{L}p_n = (q^{n/2}t^{1/2} + q^{-n/2}t^{-1/2})p_n. \tag{2.5.10}$$

Proof. (a) The commutativity of T and $Y+Y^{-1}$ immediately results from the defining relations. Generally, it is due to Bernstein and Zelevinsky. The symmetric polynomials $f \in \mathbb{C}[X+X^{-1}]$ are exactly those satisfying $Tf = t^{1/2}f$. Therefore $Y+Y^{-1}$ is symmetric. The calculation of the restriction of $Y+Y^{-1}$ to $\mathbb{C}[X+X^{-1}]$ is simple, especially if one uses that it is symmetric.

(b) Let us use (2.5.9) to introduce the polynomials $\tilde{p}_n$. We must check that they coincide with p_n. First,

$$\mathcal{L}\tilde{p}_n = (q^{n/2}t^{1/2} + q^{-n/2}t^{-1/2})\tilde{p}_n,$$

since T commutes with $Y+Y^{-1}$. Second, $(Y+Y^{-1})^* = Y+Y^{-1}$, where $*$ is with respect to the scalar product $\langle f\bar{g}\mu_k^0(x)\rangle$ because the operator Y is unitary. It readily results in $\mathcal{L}^* = \mathcal{L}$, where $\mathcal{L}^*$ is now defined with respect to the scalar product $\langle f\bar{g}\delta_k^0(x)\rangle$, which is the symmetrization of the above pairing with μ.

Note that here we can drop the conjugation and use $\langle fg\delta_k^0(x)\rangle$, because $\mathcal{L}$ is bar-invariant. Respectively, $\mathcal{L}$ has $*$–invariant eigenfunctions.

For generic q and t, the eigenvalues of $\mathcal{L}$ in the space of even polynomials are pairwise distinct, and therefore the eigenvectors are pairwise orthogonal. We obtain the desired coincidence with p. ❑

Comment. When $k=1$, we have $\delta_1(x) = (1-q^{2x})(1-q^{-2x})$ and it is easy to show that $p_n^{(k=1)}(x) = \frac{q^{(n+1)x}-q^{-(n+1)x}}{q^x-q^{-x}}$. Thus $p_n^{(k=1)}(x)$ is the character of the irreducible representation of $\mathfrak{sl}_2$ of dimension $n+1$. Is there any reasonable interpretation of the coefficients of the polynomilas $e_n^{(k=1)}$ or general e_n^k considered as rational functions or series in terms of q,t? ❑

2.5.5 Conjugated polynomials

Let us examine the action of the bar-involution

$$\bar{X} = X^{-1},\ q^{1/2} \mapsto q^{-1/2},\ t^{1/2} \mapsto t^{-1/2}$$

on the ε-polynomials. It preserves p_n and the renormalized polynomials $p_n/p_n(-k/2)$, thanks to formula (2.5.10) and the relation $\mathcal{L}^* = \mathcal{L}$. One can also see it using that δ_k^0 is bar-invariant.

We will need the following extension to $\mathcal{H}$ of the **Kazhdan–Lusztig involution** (cf. [KL1]):

$$\eta(T) = T^{-1},\ \eta(\pi) = \pi,\ \eta(X) = X^{-1},\ q^{\frac{1}{2}} \mapsto q^{-\frac{1}{2}},\ t^{\frac{1}{2}} \mapsto t^{-\frac{1}{2}}. \tag{2.5.11}$$

Introducing $T_0 = \pi T \pi$, we obtain $\eta(T_0) = T_0^{-1}$, so it coincides with the Kahdan-Lusztig involution on the affine Hecke algebra genearated by $T_1 = T,\ T_0$.

The importance of η in the theory of double affine Hecke algebras is due to the following proposition.

Proposition 2.5.12. *For arbitrary* $f \in \mathbb{C}(q^{1/2}, t^{1/2})[q^{\pm x}]$,

$$H(\bar{f}) = \overline{\eta(H)f}, \;\; H \in \mathcal{H}.$$

For $m \in \mathbb{Z}$, *the polynomial* $\overline{\varepsilon_m}(x)$ *is an eigenvector of* $\eta(Y)$ *with the eigenvalue* $q^{m_\sharp}$.

Proof. It suffices to check that $\overline{T}$ is T^{-1}, which is straightforward. ❑

Proposition 2.5.13. *For all* $m \in \mathbb{Z}$,
(a) $\overline{\varepsilon_m}(x) = t^{-1/2}T(\varepsilon_m(x))$,
(b) $\overline{\varepsilon_m}(x) = t^{-1/2}X^{-1}\varepsilon_{1-m}(x)$,
(c) $XT(\varepsilon_m(x)) = \varepsilon_{1-m}(x)$.

Proof. We will work backwards and prove part (c) first. Using Proposition 2.5.5,

$$\begin{aligned} XT(e_m) &= X\pi Y(e_m) = q^{-m_\sharp}X\pi(e_m) \\ &= q^{-m_\sharp+m/2}e_{1-m} = q^{-k/2}e_{1-m} = t^{-1/2}e_{1-m} \end{aligned}$$

for $m > 0$. Thus (c) results from $e_{1-m}(-k/2)/e_m(-k/2) = t^{1/2}$, which is a simple corollary of the evaluation formula in the next section.

To check it, one can also proceed as follows. In the relation $X\pi(\varepsilon_m) = C\varepsilon_{1-m}$, we need to find C. The right-hand side is C at $-k/2$. So we need to evaluate the left-hand side at $-k/2$. It suffices to know $\varepsilon_m(1/2 + k/2)$. Using the duality:

$$\varepsilon_m(1/2 + k/2) = \varepsilon_1(m_\sharp) = q^{m_\sharp}q^{k/2}.$$

(b) Thanks to the previous proposition, $\overline{\varepsilon_m}(x)$ is an eigenvector of $\eta(Y)$ with eigenvalue $q^{m_\sharp}$. Since $\eta(Y) = YT^{-2} = \pi Y^{-1}\pi$, we obtain that $\overline{\varepsilon_m}(x)$ is proportional to $\pi\varepsilon_m(x)$ and to $X^{-1}\varepsilon_{1-m}(x)$. Thus

$$\overline{\varepsilon_m}(x) = CX^{-1}\varepsilon_{1-m}(x) \text{ for a constant } C,$$

which has to be $t^{-1/2}$ due to $\varepsilon_n(-k/2) = 1$.

(a) It is a combination of (b) and (c). ❑

2.6 Four corollaries

Let us emphasize the main points of the previous section. So far we have five key definitions and five key properties of the double Hecke algebra and the e–polynomials.

2.6.1 Basic definitions

(i) Double affine Hecke algebra:

$$\mathcal{H} = \langle X, Y, T\rangle / \left\{ \begin{array}{cc} (T - t^{1/2})(T + t^{-1/2}) = 0, & TXT = X^{-1}, \\ Y^{-1}X^{-1}YXT^2q^{1/2} = 1, & TY^{-1}T = Y \end{array} \right\}.$$

(ii) There is also an alternative description in terms of the generators $T, X, \pi = YT^{-1}$:

$$\mathcal{H} = \langle X, T, \pi\rangle / \left\{ \begin{array}{cc} TXT = X^{-1}, & \pi X \pi^{-1} = q^{1/2}X^{-1}, \\ \pi^2 = 1, & (T - t^{1/2})(T + t^{-1/2}) = 0 \end{array} \right\}.$$

(iii) The μ–function $\mu_k^0(x)$ is uniquely defined by the properties

$$\mu_k^0(x + 1/2) = \frac{q^{2x+k} - 1}{q^{2x} - q^k}\mu_k^0(x), \text{ and } \langle \mu_k^0(x)\rangle = 1,$$

where $\langle\,\rangle$ is the constant term.
(iv) The formal conjugation is the automorphism of $\mathbb{C}[X^{\pm1}]$ defined by $\bar{X} = X^{-1}, \bar{q} = q^{-1}, \bar{t} = t^{-1}$. We will constantly identify X with q^x.
(v) The linear order $\prec$ on the set of monomials X^n:

$$X^n \prec X^m \text{ if either } |n| < |m| \text{ or } n = -m > 0.$$

Let us summarize what has been proved.
(a) (PBW property) The elements $X^nT^\varepsilon Y^m$, $n, m \in \mathbb{Z}$, $\varepsilon = 0, 1$, form a basis of $\mathcal{H}$.
(b) (Polynomial representation) For $s(X) = X^{-1}$, $\varpi(X) = q^{1/2}X$, the formulas

$$T \mapsto t^{1/2}s + \frac{t^{1/2} - t^{-1/2}}{q^{2x} - 1}(s - 1), \ \pi \mapsto s\varpi, \ X \mapsto X, \ Y \mapsto s\varpi T$$

define a representation of $\mathcal{H}$ in the space $\mathcal{P} \stackrel{\mathbf{def}}{=\!=} \mathbb{C}[X^{\pm1}]$.
(c) (Unitary structure) The operators X, q, Y, π, T are unitary with respect to the scalar product. $\langle f, g\rangle = \langle f\bar{g}\mu_k^0(x)\rangle$ on the space $\mathcal{P}$.
(d) (Nonsymmetric polynomials) They can be defined as follows:

$$\begin{aligned} &e_n(x) = X^n \mod \{X^m \prec X^n\}, \\ &Ye_n = q^{-n_\sharp}e_n, \ n_\sharp = (n + \operatorname{sgn}(n)k)/2, \ 0_\sharp = -k/2. \end{aligned} \tag{2.6.1}$$

(e) (Duality) $\varepsilon_n(m_\sharp) = \varepsilon_m(n_\sharp)$, where $\varepsilon_n(x) \stackrel{\mathbf{def}}{=\!=} e_n(x)/e_n(-k/2)$.

Let us discuss some applications. Our main instrument will be the duality and, in particular, the *anti-involution* $\phi : \mathcal{H} \to \mathcal{H}$ from (2.5.7) defined on the generators by

$$\phi(X) = Y^{-1}, \ \phi(Y) = X^{-1}, \ \phi(T) = T, \ \phi(q) = q, \ \phi(t) = t.$$

2.6.2 Creation operators

In this subsection we will consider q and $t = q^k$ as numbers (not as formal variables). We will assume that q is not a root of unity. The coefficients of the polynomials e_n are rational functions of q and q^k, so the polynomials $e_n(x)$ are not well defined for some particular values of q and k. The theory of e–polynomials is (relatively) simple because they can be produced using the **intertwining operators**. The latter play the role of the creation (raising) operators in Lie theory.

Corollary 2.6.1. *The polynomials $e_n(x)$, $e_{1-n}(x)$ for $n > 1$ are well defined for $k \notin -\{[n/2], \ldots, n-1\}$. Explicitly, they can be obtained using the following operations:*

(A) Introducing $\Pi \stackrel{\mathbf{def}}{=\!=} X\pi$, *i.e.,* $\Pi(X^m) = q^{m/2}X^{1-m}$,

$$e_{1-n} = q^{-n/2}\Pi e_n \text{ for } n \in \mathbb{Z}.$$

(B) Assuming that $q^{2n_\sharp} \neq 1$,

$$e_{-n} = q^{k/2}(T + \frac{t^{1/2} - t^{-1/2}}{q^{2n_\sharp} - 1})e_n.$$

Proof. Recall that Y preserves the filtration on $\mathcal{P} = \mathbb{C}[X^{\pm 1}]$ induced by $\prec$, as well as the subspaces $\mathbb{C}[X^{\pm 2}]$ and $X\mathbb{C}[X^{\pm 2}]$. The eigenvector of Y with an eigenvalue $q^{-n_\sharp}$ is clearly well defined if $q^{-n_\sharp} \neq q^{-m_\sharp}$ for all m such that $X^m \prec X^n$ and $n - m$ is even. Since q is not a root of unity, this condition is equivalent to $n_\sharp \neq m_\sharp$. It always holds when $n, m > 0$ or $n, m \leq 0$. If $n > 0$, $m \leq 0$, then the conditions $(n + k)/2 \neq (m - k)/2$ or $k \neq (n - m)/2$ are sufficient.

(A) Let us check that $\phi(\Pi) = \Pi$:

$$\phi(\Pi) = \phi(X\pi) = \phi(X\pi^{-1}) = \phi(XTY^{-1}) = XTY^{-1} = \Pi.$$

Using $\Pi X \Pi^{-1} = q^{1/2}X^{-1}$, we obtain that $\Pi Y^{-1}\Pi^{-1} = q^{1/2}Y$. So we only need to calculate the coefficient of proportionality. Compare the proof of Proposition 2.5.5, which is based directly on the definition.

(B) The relation $TXT = X^{-1}$ implies

$$\left(T + \frac{t^{1/2} - t^{-1/2}}{X^2 - 1}\right) X = X^{-1} \left(T + \frac{t^{1/2} - t^{-1/2}}{X^2 - 1}\right).$$

Applying ϕ we come to

$$\left(T + \frac{t^{1/2} - t^{-1/2}}{Y^{-2} - 1}\right) Y = Y^{-1} \left(T + \frac{t^{1/2} - t^{-1/2}}{Y^{-2} - 1}\right),$$

which gives the desired result up to a coefficient of proportionality. We obtain the latter coefficient from consideration of the action of T on the leading term of e_n. ❑

Comment. (i) We will see later that the polynomials $e_n, e_{1-n}, n > 1$, are not well defined for $k \in -\{[n/2], \ldots, n-1\}$. The polynomial e_n is well defined if and only if e_{1-n} is well defined, thanks to (A).

(ii) The corollary gives an inductive procedure for calculating the non-symmetric polynomials, which will be used a great deal for the classification of the finite dimensional representations:

$$1 = e_0 \xrightarrow{A_0} e_1 \xrightarrow{B_1} e_{-1} \xrightarrow{A_{-1}} e_2 \xrightarrow{B_2} e_{-2} \xrightarrow{A_{-2}} \cdots.$$

2.6.3 Standard identities

Recall that $\varepsilon_n(x) = e_n(x)/e_n(-k/2)$. We continue using X instead of q^x in the formulas.

Corollary 2.6.2. *(a) (**Pieri formula**) Setting $\nu = 1$ for $m \le 0$ and $\nu = -1$ for $m > 0$,*

$$X^{-1}\varepsilon_m = \frac{t^{1/2+\nu}q^{-m+1} - t^{-1/2}}{t^\nu q^{-m+1} - 1}\varepsilon_{m-1} - \frac{t^{1/2} - t^{-1/2}}{t^\nu q^{-m+1} - 1}\varepsilon_{1-m}, \tag{2.6.2}$$

$$X\varepsilon_m = \frac{t^{-1/2+\nu}q^{-m} - t^{1/2}}{t^\nu q^{-m} - 1}\varepsilon_{m+1} - \frac{t^{-1/2} - t^{1/2}}{t^\nu q^{-m} - 1}\varepsilon_{1-m}. \tag{2.6.3}$$

*(b) (**evaluation formula (A_1)**) We have the equalities*

$$e_m(-k/2) = t^{-|m|/2} \prod_{0<j<|m|'} \frac{1 - q^j t^2}{1 - q^j t}, \tag{2.6.4}$$

where $|m|' = m$ if $m > 0$ and $|m|' = 1 - m$ if $m \le 0$.

*(c) (**norm formula**) For $m, n \in \mathbb{Z}$,*

$$\langle \varepsilon_m, \varepsilon_n \rangle = \delta_{mn}\langle \varepsilon_m \overline{\varepsilon_m} \mu_k^0(x)\rangle = \delta_{mn} \prod_{0<j<|m|'} \frac{1 - q^j}{t^{-1} - q^j t}, \tag{2.6.5}$$

$$\langle e_m, e_n \rangle = \delta_{mn} \prod_{0<j<|m|'} \frac{(1 - q^j)(1 - q^j t^2)}{(1 - q^j t)(1 - q^j t)}. \tag{2.6.6}$$

Proof. (a) By Theorem 2.5.9, $Y\varepsilon_n = q^{-n_\sharp}\varepsilon_n$. Evaluating this equality at points $X = q^{m_\sharp}$, we obtain $(Y\varepsilon_n)(m_\sharp) = q^{-n_\sharp}\varepsilon_n(m_\sharp)$. Let $m \le 0$. Using the formula for the action of Y in $\mathcal{P}$,

$$t^{1/2}\varepsilon_n((m-1)_\sharp) + \frac{t^{1/2} - t^{-1/2}}{q^{1-2m_\sharp} - 1}(\varepsilon_n((m-1)_\sharp) - \varepsilon_n((1-m)_\sharp)) =$$

$$= q^{-n_\sharp}\varepsilon_n(m_\sharp).$$

Now we apply Theorem 2.5.10:

$$t^{1/2}\varepsilon_{m-1}(n_\sharp) + \frac{t^{1/2}-t^{-1/2}}{q^{1-m}t-1}(\varepsilon_{m-1}(n_\sharp) - \varepsilon_{1-m}(n_\sharp)) \doteq q^{-n_\sharp}\varepsilon_m(n_\sharp).$$

This gives formula (2.6.2)) at points $x = n_\sharp$. Since there are infinitely many such points we obtain the first formula in (a) for $m \le 0$. Positive m are considered in the same way. The relations

$$\varepsilon_m = q^{(m-1)/2}\Pi\varepsilon_{1-m} \tag{2.6.7}$$

connect the second formula with the first.

(b) Assume that $m \le 0$. The leading terms on the left-hand side and the right-hand sides of (2.6.2) for $\nu = 1$ are $(1/e_m(0_\sharp))X^{m-1}$ and, respectively,

$$\frac{(t^{3/2}q^{-m+1} - t^{-1/2})}{e_{m-1}(0_\sharp)(tq^{-m+1}-1)}X^{m-1}.$$

This implies (b). The case $m > 0$ is analoguos.

(c) Let $m \le 0$. Using (2.6.2):

$$\langle\varepsilon_m, \varepsilon_m\rangle = \langle X^{-1}\varepsilon_m, X^{-1}\varepsilon_m\rangle =$$

$$= \frac{t^{3/2}q^{-m+1} - t^{-1/2}}{tq^{-m+1}-1}\cdot\frac{t^{-3/2}q^{m-1} - t^{1/2}}{t^{-1}q^{m-1}-1}\langle\varepsilon_{m-1},\varepsilon_{m-1}\rangle +$$

$$+\frac{t^{1/2}-t^{-1/2}}{tq^{-m+1}-1}\cdot\frac{t^{-1/2}-t^{1/2}}{t^{-1}q^{m-1}-1}\langle\varepsilon_{1-m},\varepsilon_{1-m}\rangle.$$

Involving Π, which is unitary by Corollary 2.6.1(b),

$$\langle\varepsilon_{1-m},\varepsilon_{1-m}\rangle = \langle\varepsilon_m,\varepsilon_m\rangle \text{ since } \varepsilon_m = q^{(m-1)/2}\Pi\varepsilon_{1-m}, \tag{2.6.8}$$

$$\langle\varepsilon_m,\varepsilon_m\rangle = \frac{t^{-1}-tq^{-m+1}}{1-q^{-m+1}}\langle\varepsilon_{m-1},\varepsilon_{m-1}\rangle,$$

and we obtain (c) by induction. In the case $m > 0$, one uses (2.6.8). ❑

Comment. (i) The existence of the three-term relation in the form of (a) can be seen directly from the orthogonality of polynomials $e_n(x)$. The approach based on the duality is simpler and readily gives the exact coefficients.

(ii) Formula (b) shows that the polynomial e_n is not well defined for values of k from Corollary 2.6.1(a), because otherwise $e_n(-k/2)$ would be defined. So we have the complete list of singular k for each e_n.

(iii) Formula (b) as $k = 1$ is a polynomial in the variable q with positive integral coefficients for either positive odd m or negative even m. These coefficients have some combinatorial interpretation. Note that $p_n^{(k=1)}(1/2)$ is the so-called q–dimension of the representation of $\mathfrak{sl}_2$ of dimension $n+1$. There is a partial interpretation of $e_n(-k/2)$ and $p_n(k/2)$ considered as a series in terms of $q, t^{1/2}$. However, it is for A_1 only and there is no known connection with representation theory.

2.6.4 Changing k to $k+1$

Now we are going to discuss the shift formula in the presence of the polynomials $p_n^{(k)}(x)$. The p–polynomials appear because of the following corollary.

Corollary 2.6.3. *Let $\mathcal{A} = (t^{1/2}X - t^{-1/2}X^{-1})$ and $\mathcal{B} = (t^{1/2}Y^{-1} - t^{-1/2}Y)$. The operator $S \stackrel{\mathbf{def}}{=\!=} t^{-1/2}\mathcal{A}^{-1}\mathcal{B}$ preserves the space $\mathbb{C}[X+X^{-1}]$. Its restriction to this space is $\frac{\varpi^{-1}-\varpi}{X-X^{-1}}$ for $\varpi(f)(x) \stackrel{\mathbf{def}}{=\!=} f(x+1/2)$, and*

$$S(p_n^{(k)}(x)) = (q^{-n/2} - q^{n/2})p_{n-1}^{(k+1)} \text{ for } n > 0. \tag{2.6.9}$$

Proof. It is straightforward to check the following equality in $\mathcal{H}$:

$$(T + t^{-1/2})(t^{1/2}X - t^{-1/2}X^{-1}) = (t^{1/2}X^{-1} - t^{-1/2}X)(T - t^{1/2}).$$

As a corollary, we obtain that the eigenspace of T in $\mathcal{P}$ with the eigenvalue $-t^{-1/2}$ is $\mathcal{A}\mathbb{C}[X + X^{-1}]$. Applying the anti-involution ϕ to this equality we obtain

$$(t^{1/2}Y^{-1} - t^{-1/2}Y)(T + t^{-1/2}) = (T - t^{1/2})(t^{1/2}Y - t^{-1/2}Y^{-1}).$$

Therefore $\mathcal{B}$ maps $\mathbb{C}[X + X^{-1}]$ to the T–eigenspace of $\mathcal{P}$ with eigenvalue $-t^{-1/2}$, that is, $\mathcal{A}\mathbb{C}[X+X^{-1}]$. This shows that S is well defined on $\mathbb{C}[X+X^{-1}]$ and, moreover, preserves this space. The formula for the restriction of S to $\mathbb{C}[X + X^{-1}]$ is simple. Cf. Theorem 2.5.11.

The formula for S (or, directly, the definition) givs that the left-hand side and the right-hand side of (2.6.9) coincide. Therefore it suffices to check

$$\langle S(p_n^{(k)}(x))g(x)\delta_{k+1}^0(x)\rangle = 0$$

for any $g(x) \in \mathbb{C}[q^x + q^{-x}]$ of degree $m < n - 1$. Let us prove that

$$\langle S(p_n^{(k)}(x))g(x)\mu_{k+1}^0(x)\rangle = 0$$

for any $g(x) \in \mathbb{C}[q^{\pm x}]$ of degree $m < n - 1$. Symmetrizing the latter for even g, we obtain the former.

First,

$$\mu_{k+1}^0(x) = \frac{1 - q^{k+1}}{(1 - q^{2k+1})(1 + q^{k+1})}(1 - q^{k+2x})(1 - q^{k+1-2x})\mu_k^0(x) =$$

$$= \mathcal{A}(x)\tilde{\mathcal{A}}(x)\mu_k^0(x),$$

where $\mathcal{A}(x) = t^{1/2}X - t^{-1/2}X^{-1} = q^{(k+2x)/2} - q^{-(k+2x)/2}$ and $\tilde{\mathcal{A}}$ is a polynomial in q^x of the first degree. Second,

$$\langle S(p_n^{(k)}(x))g(x)\mu_{k+1}^0(x)\rangle = \langle \mathcal{B}p_n^{(k)}(x)\tilde{\mathcal{A}}g(x)\mu_k^0(x)\rangle.$$

Here $\tilde{\mathcal{A}}g(x)$ is a polynomial of degree $m' < n$. Third, $p_n^{(k)}(x)$ is a linear combination of $e_n^{(k)}(x)$ and $e_{-n}^{(k)}(x)$ by Theorem 2.5.11(b). The same holds for $\mathcal{B}p_n^{(k)}(x)$. We get the desired result. ❑

2.6.5 Shift formula

Let us discuss an analytic interpretation of this corollary. We need to choose the space of functions $\mathfrak{F}$ and the integration ϱ. The following cases can be considered:

(a) $\mathfrak{F}$ is the space of even Laurent series in terms of q^x with the constant term taken as the functional ϱ;

(b) $\mathfrak{F}$ is the space of even analytic functions on the strip

$$-1/2 - \delta \leq \Re x \leq 1/2 + \delta \text{ for } \delta > 0, \quad \text{and } \varrho = \int_{i\mathbb{R}} (\cdot)\,;$$

(c) $\mathfrak{F}$ is the space of continuous even functions on the real line and $\varrho = \int_{\mathbb{R}}$;

(d) $\mathfrak{F}$ is a space of even functions that are defined and continuous on the two lines $z = \pm\varepsilon\imath + \mathbb{R}$ for small $\varepsilon > 0$, i.e., on $C \cup \{-C\}$ for the the sharp integration path C from (2.3.1), and, moreover, they are analytic in the rectangle

$$\{x \mid -\varepsilon < \Im x < \varepsilon,\ -1/2 - \delta \leq \Re x \leq 1/2 + \delta\}.$$

In this case,

$$\varrho = 2\int_C = \int_{\widehat{C}}, \text{ where } \widehat{C} = C \cup \{-\overline{C}\}$$

for the complex conjugation (preserving C, but changing its orientation).

In all cases, the integration ϱ is invariant under the change of variable $x \mapsto -x$, because the integration paths are s–invariant. Note that, topologically, $\widehat{C}$ is the cross of the diagonals at the origin, both directed upwards. On even functions, $\frac{1}{2}\int_{\widehat{C}}$ coincides with the sharp integration from (2.3.1). Note that all the integrations defined above are zero for odd functions.

The analytic properties of f are necessary to ensure the invariance of ϱ under the change of variables $x \mapsto x \pm 1/2$. The corresponding domains of analyticity allow us to deform the contours after the shift by $\pm 1/2$. The real case is the most relaxed: no analiticity domains are necessary and there is no problem with the real translation. However, the q–Gauss integrals are the most difficult to calculate in this case.

Corollary 2.6.4. *Let $\mathfrak{F}, \varrho$ be one of the pairs above. Then, for any $f \in \mathfrak{F}$ such that either the left-hand or the right-hand side below is well defined, the other is well defined too and*

$$\varrho(S(f)p_{n-1}^{(k+1)}(x)\delta_{k+1}(x)) = q^k(q^{(-k-n)/2} - q^{(k+n)/2})\varrho(fp_n^{(k)}(x)\delta_k(x)).$$

Proof will be given below. ❑

The corallary readily results in the main property of the q–Mellin transform, namely, Theorem 2.2.7. The proof given there was by straightforward integration.

Recall the notation $g_k = \prod_{j=0}^{\infty} \frac{1-q^{k+j}}{1-q^{2k+j}}$. The q–Mellin transform was defined as follows:

$$\Psi_k(f(x)) = \frac{1}{g_k}\varrho(f(x)\delta_k(x)).$$

Let us deduce from the corollary that

$$\Psi_k(f(x)) = (1-q^{k+1})\Psi_{k+1}(f(x)) + q^{k+3/2}\Psi_{k+2}(S^2(f(x))). \tag{2.6.10}$$

First,

$$\delta_{k+1}(x) = (1-tX^2)(1-tX^{-2})\delta_k(x) =$$
$$= \left(1+t^2-t\left(p_2^{(k)}(x) - \frac{(1-t)(1+q)}{1-tq}\right)\right)\delta_k(x) =$$
$$= \left(\frac{(1+t)(1-t^2q)}{1-tq} - tp_2^{(k)}(x)\right)\delta_k(x).$$

Using Corollary 2.6.4 twice,

$$\begin{aligned}
&\varrho(f(x)\delta_{k+1}(x))\\
&= \frac{(1+t)(1-t^2q)}{1-tq}\varrho(f(x)\delta_k(x)) - t\varrho(f(x)p_2^{(k)}(x)\delta_k(x)) =\\
&= \frac{(1+t)(1-t^2q)}{1-tq}\varrho(f(x)\delta_k(x)) -\\
&- \frac{t^{-1}q^{-1}}{(tq-t^{-1}q^{-1})(tq^{3/2}-t^{-1}q^{-3/2})}\varrho(f(x)\delta_{k+2}(x)),
\end{aligned}$$

which is equivalent to formula (2.6.10).

2.6.6 Proof of the shift formula

First, we switch from δ to μ:

$$\delta_{k+1}(x) = (1-q^{-2x})(1-q^{k+2x})\mu_k(x).$$

Let us rewrite it in terms of $\mathcal{A}$, $q^x = X$ and $q^{k/2} = t^{1/2}$:

$$\mathcal{A}^2\mu_k(x) = -t^{-1/2}\frac{\mathcal{A}}{X-X^{-1}}\delta_{k+1}(x).$$

Since $f(x)$ is an even function,

$$\varrho(\mathcal{A}^2 f(x)\mu_k(x)) = \varrho(-t^{-1/2}\frac{\mathcal{A}}{X-X^{-1}}f(x)\delta_{k+1}(x)) =$$
$$= -\frac{1+t^{-1}}{2}\varrho(f(x)\delta_{k+1}(x)). \tag{2.6.11}$$

When replacing $\frac{\mathcal{A}}{X-X^{-1}}f(x)$ by its symmetrization, we used the invariance of ϱ and $\delta_{k+1}(x)$ under the change $x \mapsto -x$.

Second, we take an even function $g(x)$ of the same type as f such that $\overline{g(x)} = g(x)$. For instance, it can be a Laurent polynomial or a convergent Laurent series with the coefficients invariant under $q \mapsto q^{-1}$ and $t \mapsto t^{-1}$. It is not restrictive because later $g(x)$ will be one of $p_n^{(k)}$, which are bar-invariant. One gets

$$\begin{aligned}&\frac{1+t^{-1}}{2}\varrho(S(f(x))S(g(x))\delta_{k+1}(x))\\ &\stackrel{(1)}{=} -\varrho(\mathcal{A}^2S(f(x))S(g(x))\mu_k(x)) \stackrel{(2)}{=} \varrho(\mathcal{A}^2S(f(x))\overline{S(g(x))}\mu_k(x))\\ &\stackrel{(3)}{=} -\varrho(\mathcal{B}(f(x))\overline{\mathcal{B}(g(x))}\mu_k(x)) \stackrel{(4)}{=} \varrho(\mathcal{B}^2(f(x))\overline{g(x)}\mu_k(x)),\end{aligned}$$

where the first equality follows from (2.6.11), the second follows from $\overline{S} = -S$, and the third follows from $\bar{\mathcal{A}} = -\mathcal{A}$. Only the fourth equality requires some comment. In the algebraic variant, i.e., for the constant term integration (a), we can simply use that Y is unitary. See Theorem 2.5.8. The same argument is applied for the other three integrations, since ϱ is invariant under $x \mapsto -x$ and $x \mapsto x \pm 1/2$. Note that the latter symmetry is necessary when collecting $\mathcal{B}$ together.

Third, we apply the t-symmetrizer $\mathbf{P} \stackrel{\text{def}}{=\!=} (1+t)^{-1}(1+t^{1/2}T)$. It projects $\mathbb{C}[X^{\pm 1}]$ onto $\mathbb{C}[X+X^{-1}]$. One then has

$$\mathbf{P}\mathcal{B}^2\mathbf{P} = -\mathcal{B}\mathcal{B}'\mathbf{P}, \text{ where } \mathcal{B}' = t^{1/2}Y - t^{-1/2}Y^{-1}. \tag{2.6.12}$$

Indeed, $\mathcal{B}\mathcal{B}' = -Y^2 - Y^{-2} + t + t^{-1}$ commutes with T and, in particular, is even. So it suffices to check that $\mathbf{P}\mathcal{B}(\mathcal{B}+\mathcal{B}')(f) = 0$ for any $f \in \mathbb{C}[X+X^{-1}]$. However, $\mathcal{B}+\mathcal{B}' = (t^{1/2}-t^{-1/2})(Y+Y^{-1})$ commutes with T. Hence $f' = (\mathcal{B}+\mathcal{B}')(f) \in \mathbb{C}[X+X^{-1}]$ and $T\mathcal{B}(f') = -t^{-1/2}f'$. Cf. the proof of Corollary 2.6.3 above. Formula (2.6.12) is now checked.

Combining the previous formulas, we obtain

$$\begin{aligned}1/2(1+t^{-1})\varrho(S(f(x))S(g(x))\delta_{k+1}(x)) &= \varrho(\mathcal{B}^2(f(x))\overline{g(x)}\mu_k(x)) =\\ = \varrho(\mathcal{B}^2(\mathbf{P}(f(x)))\overline{\mathbf{P}(g(x))}\mu_k(x)) &= \varrho(\mathbf{P}\mathcal{B}^2\mathbf{P}(f(x))\overline{g(x)}\mu_k(x)) =\\ = -\varrho(\mathcal{B}\mathcal{B}'(f(x))\overline{g(x)}\mu_k(x)) &= -\varrho(f(x)\overline{\mathcal{B}\mathcal{B}'(g(x))}\mu_k(x)) =\\ &= -1/2(1+t)\varrho(f(x)\overline{\mathcal{B}\mathcal{B}'(g(x))}\delta_k(x)).\end{aligned}$$

Thus the relation

$$\varrho(S(f(x))S(g(x))\delta_{k+1}(x)) = -t\varrho(f(x)\overline{\mathcal{B}\mathcal{B}'(g(x))}\delta_k(x)) \tag{2.6.13}$$

is proved, provided that $f(x), g(x) \in \mathbb{C}[q^x + q^{-x}]$ and $\overline{g(x)} = g(x)$.

Now we simply take $g(x) = p_n^{(k)}(x)$ in (2.6.13) for $n \geq 1$. Using Theorem 2.5.11,

$$\begin{aligned}\mathcal{BB}'(p_n^{(k)}(x)) &= (-q^n t - q^{-n}t^{-1} + t + t^{-1})p_n^{(k)}(x)\\ &= (tq^{n/2} - t^{-1}q^{-n/2})(q^{-n/2} - q^{n/2})p_n^{(k)}(x).\end{aligned}$$

Using Corollary 2.6.3,

$$\varrho(S(f)p_{n-1}^{(k+1)}\delta_{k+1}) = -t(tq^{n/2} - t^{-1}q^{-n/2})\varrho(f(x)p_n^{(k)}\delta_k).$$

Corollary 2.6.4 is proved.

2.7 DAHA–Fourier transforms

Recall the notation. We set $0_\sharp = -k/2$ and $n_\sharp = (n + \mathrm{sgn}(n)k)/2$ for integers $n \neq 0$, where $t = q^k$. For generic q, t, the Laurent polynomials $\varepsilon_n(x)$ are uniquely defined by the properties $Y\varepsilon_n(x) = q^{-n_\sharp}\varepsilon_n(x)$ and $\varepsilon_n(0_\sharp) = 1$. We permanently use $X = q^x$.

The Gaussian is $\widehat{\gamma_-} = \sum_{n=-\infty}^{\infty} q^{n^2/4}X^n$. We will also use γ, a solution of the corresponding difference equation treated as a formal symbol. When Jackson's summation is considered, $\gamma = q^{x^2}$.

The scalar product $\langle f, g\rangle = \langle f\bar{g}\mu_k^0\rangle$ is given in terms of

$$\mu_k^0(x) = 1 + \frac{q^k - 1}{1 - q^{k+1}}(q^{2x} + q^{1-2x}) + \ldots,$$

the constant term functional $\langle\cdot\rangle$, and the involution $f \mapsto \bar{f}$ of the space of Laurent polynomials $\mathcal{P} = \mathbb{C}(q^{1/2}, t^{1/2})[X^{\pm 1}]$, where

$$\bar{X} = X^{-1},\ \bar{q}^{1/2} = q^{-1/2},\ \bar{t}^{1/2} = t^{-1/2}.$$

Our aim is the following theorem.

Theorem 2.7.1. *(***Master formulas***) For arbitrary $m, n \in \mathbb{Z}$,*

$$\langle\varepsilon_n(x)\varepsilon_m(x)\widehat{\gamma_-}\mu_k^0(x)\rangle = q^{\frac{m^2+n^2+2k(|m|+|n|)}{4}}\varepsilon_m(n_\sharp)\langle\widehat{\gamma_-}\mu_k^0(x)\rangle, \tag{2.7.1}$$

$$\langle\varepsilon_n(x)\overline{\varepsilon_m}(x)\widehat{\gamma_-}\mu_k^0(x)\rangle = q^{\frac{m^2+n^2+2k(|m|+|n|)}{4}}\overline{\varepsilon_m}(n_\sharp)\langle\widehat{\gamma_-}\mu_k^0(x)\rangle. \tag{2.7.2}$$

Before proving the theorem, we need to establish several general facts.

2.7.1 Functional representation

For $\mathbb{Z}_\sharp \stackrel{\text{def}}{=} \{n_\sharp | n \in \mathbb{Z}\} \subset \mathbb{C}$, let $\hat{\mathcal{F}}$ be the space of functions on $\mathbb{Z}_\sharp$ and let $\mathcal{F} \subset \hat{\mathcal{F}}$ be the subspace of functions with compact support. We have an evident **discretization map** $\chi : \mathbb{C}[X^{\pm 1}] \to \hat{\mathcal{F}}$.

Theorem 2.7.2. *The space $\hat{\mathcal{F}}$ has the natural structure of an $\mathcal{H}$–module making the map $\chi : \mathcal{P} \to \hat{\mathcal{F}}$ an $\mathcal{H}$-homomorphism; $\mathcal{F} \subset \hat{\mathcal{F}}$ is an $\mathcal{H}$-submodule.*

Proof. Let $f \in \hat{\mathcal{F}}$. We define $(Xf)(n_\sharp) = q^{n_\sharp} f(n_\sharp)$, $(\pi f)(n_\sharp) = f(1/2 - n_\sharp)$. Note that the set $\mathbb{Z}_\sharp$ is invariant under $\pi : x \mapsto 1/2 - x$. It is not s–invariant. However, the operator

$$(Tf)(n_\sharp) = \frac{t^{1/2} q^{2n_\sharp} - t^{-1/2}}{q^{2n_\sharp} - 1} f(-n_\sharp) - \frac{t^{1/2} - t^{-1/2}}{q^{2n_\sharp} - 1} f(n_\sharp) \tag{2.7.3}$$

is well defined because $-n_\sharp \notin \mathbb{Z}_\sharp$ for $n = 0$, precisely when the first term of (2.7.3) vanishes. ❑

Definition 2.7.3. *(a) The* **Fourier transform S** *acts from $\mathcal{P}$ to $\mathcal{F}$ and is given by the formula*

$$\mathbf{S}(f)(n_\sharp) \stackrel{\text{def}}{=} \langle f \varepsilon_n(x) \mu_k^0(x) \rangle.$$

(b) Its **antilinear counterpart,** *conjugating q and t, is introduced as follows:*

$$\mathbf{E}(f)(n_\sharp) \stackrel{\text{def}}{=} \langle \bar{f} \varepsilon_n(x) \mu_k^0(x) \rangle.$$

❑

The fact that the functions $\mathbf{S}(f)$ and $\mathbf{E}(f)$ have compact supports readily follows from orthogonality relations for the polynomials $\varepsilon_n(x)$; any polynomial is their linear combination.

Theorem 2.7.4. *(a) The formulas*

$$\epsilon(X) = Y,\ \epsilon(Y) = X,\ \epsilon(T) = T^{-1},\ \epsilon(q) = q^{-1}; \epsilon(t) = t^{-1}$$

can be extended to an antilinear automorphism of $\mathcal{H}$. The formulas

$$\sigma(X) = Y^{-1},\ \sigma(T) = T,\ \sigma(Y) = q^{-1/2} Y^{-1} X Y = X T^2$$

define a linear automorphism of $\mathcal{H}$, fixing q, t. The connection with the involution η from (2.5.11) is as follows: $\eta = \epsilon\sigma = \sigma^{-1}\epsilon$.

(b) For any $H \in \mathcal{H}$, $f \in \mathcal{P}$,

$$\mathbf{S}(H(f)) = \sigma(H)(\mathbf{S}(f)),\ \ \mathbf{E}(H(f)) = \epsilon(H)(\mathbf{E}(f)).$$

Proof. Claim (a) is straightforward. Let us check (b) for the operator $\mathbf{E}$. It holds for Y because of the definition of ε_n and, therefore, for X, thanks to the duality (Theorem 2.5.8). Hence it holds for $T^2 = q^{-1/2}X^{-1}Y^{-1}XY$ and for $T = (t^{1/2} - t^{-1/2})^{-1}(T^2 - 1)$. Here we use that X, T, Y are unitary for $\langle f, g \rangle$. The special case $t = 1$ is not a problem because it is sufficient to check (b) for generic t.

The automorphism corresponding to the operator $\mathbf{S}$ is $\epsilon\eta = \sigma$, thanks to Proposition 2.5.12. ❑

We already defined automorphisms $\epsilon, \sigma, \eta = \epsilon\sigma$ of the algebra $\mathcal{H}$ and the anti-involution ϕ. See (2.5.7). Completing the preparation to proving the master formulas, let us introduce two more automorphisms.

Proposition 2.7.5. *(a) The formulas*

$$\tau_+ : X \mapsto X,\ T \mapsto T,\ Y \mapsto q^{-1/4}XY,$$

$$\tau_- : Y \mapsto Y,\ T \mapsto T,\ X \mapsto q^{1/4}YX$$

can be extended to linear automorphisms of $\mathcal{H}$.

(b) We have the following identities in $\mathrm{Aut}(\mathcal{H})$*:*

$$\tau_+\tau_-^{-1}\tau_+ = \sigma = \tau_-^{-1}\tau_+\tau_-^{-1},\ \sigma^2 = T^{-1}(\cdot)T,$$

$$\sigma\tau_+\sigma^{-1} = \tau_-^{-1},\ \epsilon\tau_+\epsilon = \tau_-,\ \phi\tau_+\phi = \tau_-.$$

Proof. Straightforward. ❑

Comment. It is immediate from (b) that σ^2 commutes with τ_+ and τ_-. This also follows from $\sigma^2 = T(\cdot)T^{-1}$. The automorphism τ_- is inner in the representation $\mathcal{P}$. This is not true for τ_+. ❑

The automorphism τ_+ is directly related to the Gaussian $\widehat{\gamma_-}$, namely,

$$H\widehat{\gamma_-} = \widehat{\gamma_-}\tau_+(H)$$

in the polynomial representation. Actually, what we need here is an abstract function γ satisfying the relation:

$$\gamma(x + 1/2) = q^{1/4}q^x\gamma(x),$$

which readily results in

$$\gamma H \gamma^{-1} = \tau_+(H)$$

in any functional space.

2.7.2 Proof of the master formulas

We start with **S**. Treating γ^{-1} as $\widehat{\gamma_-}$ and γ as q^{x^2}, we put

$$\varepsilon_m^-(x) \stackrel{\mathbf{def}}{=\!=} \widehat{\gamma_-}\varepsilon_m(x),\ \varepsilon_m^+(n_\sharp) \stackrel{\mathbf{def}}{=\!=} (\gamma\varepsilon_m)(n_\sharp).$$

Note that they belong to different spaces. Then (2.7.1) is equivalent to the formula

$$\mathbf{S}(\varepsilon_m^-)(n_\sharp) = q^{m_\sharp^2 - 0_\sharp^2}\varepsilon_m(n_\sharp)q^{n_\sharp^2}\langle\widehat{\gamma_-}\mu_k^0(x)\rangle$$

or, equivalently, to

$$\mathbf{S}(\varepsilon_m^-) = q^{m_\sharp^2 - 0_\sharp^2}\varepsilon_m^+\langle\widehat{\gamma_-}\mu_k^0(x)\rangle.$$

Actually, it is sufficent to establish that

$$\mathbf{S}(\varepsilon_m^-) = C_m\varepsilon_m^+ \tag{2.7.4}$$

for a constant C_m, since the left-hand side of (2.7.1) is $m \leftrightarrow n$–symmetric and this constant can only be 1.

Introducing the operators

$$Y^+ \stackrel{\mathbf{def}}{=\!=} \gamma Y\gamma^{-1},\quad Y^- \stackrel{\mathbf{def}}{=\!=} \gamma^{-1}Y\gamma, \tag{2.7.5}$$

we claim that $\sigma(Y^-) = Y^+$. Indeed, $Y^- = \tau_+^{-1}(Y) = q^{1/4}X^{-1}Y$ and

$$\sigma(Y^-) = \sigma(q^{1/4}X^{-1}Y) = q^{-1/4}YY^{-1}XY = q^{-1/4}XY = \tau_+(Y).$$

Obviously $Y^\pm(\varepsilon_m^\pm) = q^{-m_\sharp}\varepsilon_m^\pm$. For generic q and k, the eigenvalues are distinct, the $\mathcal{H}$–module $\mathcal{P}\widehat{\gamma_-}$ is irreducible, and $\{\varepsilon_m^-\}$ is its basis. Similarly, all eigenvectors of Y^+ in $\mathcal{F}\gamma$ are simple. This gives (2.7.4) and therefore (2.7.1). Now we can make q, t arbitrary, provided that (2.7.1) is well defined.

Switching to (2.7.2), its left-hand side is not $m \leftrightarrow n$–symmetric anymore. So we have to proceed in a slightly different way.

First, we set $\bar{\varepsilon}_m^{(-)} \stackrel{\mathbf{def}}{=\!=} \overline{\varepsilon_m}(x)\widehat{\gamma_-}$, $\bar{\varepsilon}_m^{(+)}(n_\sharp) \stackrel{\mathbf{def}}{=\!=} \overline{\varepsilon_m}(n_\sharp)\gamma(n_\sharp)$, and come to the relation

$$\mathbf{S}(\bar{\varepsilon}_m^{(-)})(n_\sharp) = C_m\overline{\varepsilon_m}(n_\sharp)q^{n_\sharp^2}. \tag{2.7.6}$$

Indeed, for $\widetilde{Y} \stackrel{\mathbf{def}}{=\!=} \eta(Y)$,

$$\widetilde{Y}(\overline{\varepsilon_n}) = q^{n_\sharp}\overline{\varepsilon_n}$$

and $\bar{\varepsilon}_m^{(\pm)}$ are eigenvectors of the operators

$$\widetilde{Y}^+ \stackrel{\mathbf{def}}{=\!=} \gamma\widetilde{Y}\gamma^{-1} = \tau_+(\widetilde{Y}),\ \widetilde{Y}^- \stackrel{\mathbf{def}}{=\!=} \gamma^{-1}\widetilde{Y}\gamma = \tau_+^{-1}(\widetilde{Y}).$$

Then we observe that $\sigma(\widetilde{Y}^-) = \widetilde{Y}^+$, which results from the relation $\sigma = \tau_+\tau_-^{-1}\tau_+$:

$$\begin{aligned}\sigma(\widetilde{Y}^-) &= \sigma\tau_+^{-1}\eta(Y) = \sigma\tau_+^{-1}\epsilon\sigma(Y) = \tau_+(\tau_+^{-1}\sigma\tau_+^{-1})\epsilon\sigma(Y)\\ &= \tau_+\tau_-^{-1}\epsilon\sigma(Y) = \tau_+\epsilon\sigma\tau_-^{-1}(Y) = \tau_+\epsilon\sigma(Y) = \widetilde{Y}^+.\end{aligned}$$

Formula (2.7.6) is now checked.

Second, we establish that

$$\mathbf{E}(\varepsilon_n^-)(m_\sharp) = D_n \overline{\varepsilon_n}(m_\sharp) q^{m_\sharp^2}. \tag{2.7.7}$$

This formula follows from the identity $\epsilon(Y^-) = \tau_+\eta(Y)$ for the operator Y^- from (2.7.5). Let us prove it:

$$\begin{aligned}\epsilon(Y^-) &= \epsilon\tau_+(Y) = \tau_+\eta(\sigma^{-1}\epsilon\tau_+^{-1}\epsilon\tau_+)(Y)\\ &= \tau_+\eta(\sigma^{-1}\tau_-^{-1}\tau_+)(Y) = \tau_+\eta\tau_-(Y) = \tau_+\eta(Y).\end{aligned}$$

Third, formulas (2.7.6) and (2.7.7) result in $C_m = q^{m_\sharp^2} C$ and $D_n = q^{n_\sharp^2} C$, where C does not depend on m and n, which concludes the proof of formula (2.7.2). ❑

2.7.3 Topological interpretation

We use that the relations of an $\mathcal{H}$ are mainly of group nature and introduce the **braid group (elliptic)**

$$\mathcal{B}_q = \langle T, X, Y, q^{1/4} \rangle / \left\langle \begin{array}{c} TXT = X^{-1}, TY^{-1}T = Y, \\ Y^{-1}X^{-1}YXT^2 q^{1/2} = 1 \end{array} \right\rangle.$$

Now $T, X, Y, q^{1/4}$ are treated as group generators and $q^{1/4}$ is assumed central. The double affine Hecke algebra $\mathcal{H}$ is the quotient of the group algebra of $\mathcal{B}_q$ by the quadratic Hecke relation. It is easy to see that the change of variables $q^{1/4}T \mapsto T$, $q^{-1/4}X \mapsto X$, $q^{1/4}Y \mapsto Y$ defines an isomorphism $\mathcal{B}_q \cong \mathcal{B}_1 \times \mathbb{Z}$, where the generator of $\mathbb{Z}$ is $q^{1/4}$.

In this subsection, we give a topological interpretation of the group

$$\mathcal{B}_1 = \langle T, X, Y \rangle / \langle TXT = X^{-1}, TY^{-1}T = Y, Y^{-1}X^{-1}YXT^2 = 1 \rangle.$$

Let E be an elliptic curve over $\mathbb{C}$, i.e., $E = \mathbb{C}/\Lambda$, where $\Lambda = \mathbb{Z} + \mathbb{Z}\imath$. Topologically, the lattice can be arbitrary. Let $o \in E$ be the zero point, and -1 the automorphism $x \mapsto -x$ of E. We are going to calculate the fundamental group of the space $(E \setminus o)/\pm 1 = \mathbb{P}^1_{\mathbb{C}} \setminus o$. Since this space is contractible, its usual fundamental group is trivial. We can take the quotient after removing all (four) ramification points of -1. However, it would enlarge the fundamental group dramatically. Thus we need to understand this space in a more refined way.

Let us fix the **base (starting) point** $\star = -\varepsilon - \varepsilon\imath \in \mathbb{C}$ for small $\varepsilon > 0$.

Proposition 2.7.6. *We have an isomorphism* $\mathcal{B}_1 \cong \pi_1^{orb}((E \setminus o)/\pm 1)$, *where* $\pi_1^{orb}(\cdot)$ *is the orbifold fundamental group, which will be defined in the process of proving the proposition.*

Proof. The projection map $E \setminus o \to (E \setminus o)/\pm 1 = \mathbb{P}^1_{\mathbb{C}} \setminus o$ has three branching points, which come from the nonzero points of order 2 on E. So by definition, $\pi_1^{orb}((E \setminus o)/\pm 1)$ is generated by three *involutions* A, B, C, namely, the clockwise loops from $\star$ around the branching points in $\mathbb{P}^1_{\mathbb{C}}$. There are no other relations. We claim that the assignment $A = XT$, $B = T^{-1}Y$, $C = XTY$ defines a homomorphism

$$\pi_1^{orb}((E \setminus o)/\pm 1) \to \mathcal{B}_1.$$

Indeed, A and B are obviously involutive. Concerning C, the image of its square is

$$XTYXTY = T^{-1}X^{-1}YXTY = T^{-1}YT^{-1}Y = 1.$$

This homomorphism is an isomorphism. The inversion is given by the formulas $ACB = T$, $ABCA = X$ and $AC = Y$. ❑

This approach can hardly be generalized to arbitrary root systems; the following (equivalent) constructions can. The definition of the fundamental group is modified as follows. We follow [C13]. See also paper [Io] (in this paper the orbifold group is not involved and the construction due to v.d. Lek is used).

We switch from E to its universal cover $\mathbb{C}$ and define the **paths** as curves $\gamma \in \mathbb{C} \setminus \Lambda$ from $\star$ to $\widehat{w}(\star)$, where $\widehat{w} \in \widehat{W} = \{\pm 1\} \ltimes \Lambda$. The generators T, X, Y are shown as the arrows in Figure 2.1.

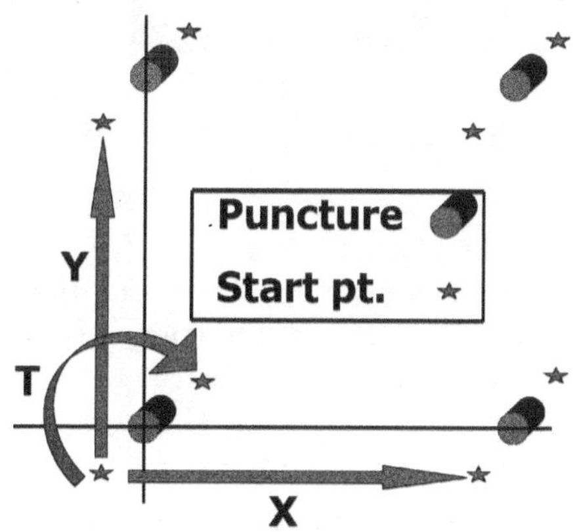

Figure 2.1: Generators of $\mathcal{B}_1$

The **composition of the paths** is via $\widehat{W}$: we add the image of the second path under $\widehat{w}$ to the first path if the latter ends at $\widehat{w}(\star)$. The corresponding variant of Proposition 2.7.6 reads as follows.

Proposition 2.7.7. *The fundamental group of the above paths modulo homotopy is isomorphic to $\mathcal{B}_1$, where T is the half-turn, i.e., the clockwise half-circle from $\star$ to $s(\star)$, and X, Y are 1 and $\imath$ considered as vectors in $\mathbb{R}^2(=\mathbb{C})$ originated at the base point $\star$.*

Proof is in Figures 2.1 and 2.2. Figure 2.1 gives that when we first use X then T and then again X (note that after T the direction of X will be opposit!), and then again T, the loop corresponding to the product $TXTX$ will contain no punctures inside. Thus it equals *id* in the fundamental group. The reasoning for $TY^{-1}TY^{-1}T$ is the same. Concerning the "commutator" relation, see Figure 2.2. ❑

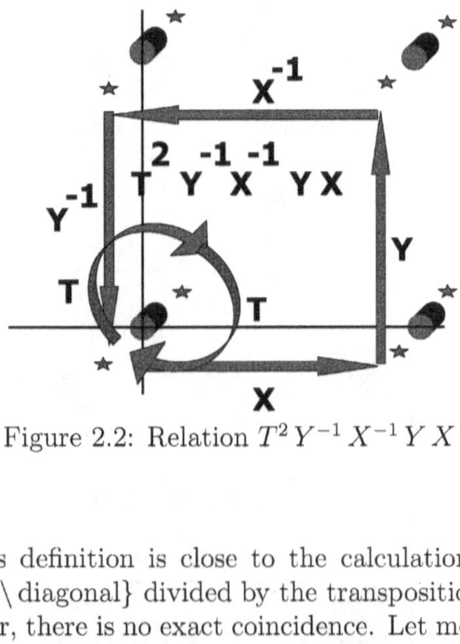

Figure 2.2: Relation $T^2Y^{-1}X^{-1}YX = 1$

Actually, this definition is close to the calculation of the fundamental group of $\{E \times E\backslash \text{diagonal}\}$ divided by the transposition of the components. See [Bi]. However, there is no exact coincidence. Let me mention the relation to the elliptic braid group due to v.d. Lek, although he removes all points of second order and his group is significantly larger.

Noncommutative Kodaira–Spencer map. A topological interpretation is the best way to understand why the group $SL_2(\mathbb{Z})$ acts in $\mathcal{B}_1$ projectively.

Its elements act in $\mathbb{C}$ natuarally, by the corresponding real linear transformations. On E, they commute with the reflection -1, preserve o, and permute the other three points of second order. Given $g \in SL_2(\mathbb{Z})$, we set $g = \exp(h), g_t = \exp(th)$ for the proper $h \in \mathfrak{sl}_2(\mathbb{R})$, $0 \le t \le 1$.

The position of the base point $\star$ will become $g(\star)$, so we need to go back, i.e., connect the image with the base point by a path. To be more exact, the g–image of $\gamma \in \pi_1^{orb}$ will be the union of the paths

$$\{g_t(\star)\} \cup g(\gamma) \cup \{\widehat{w}(g_{-t}(\star))\},$$

where the path for γ goes from $\star$ to the point $\widehat{w}(\star)$.

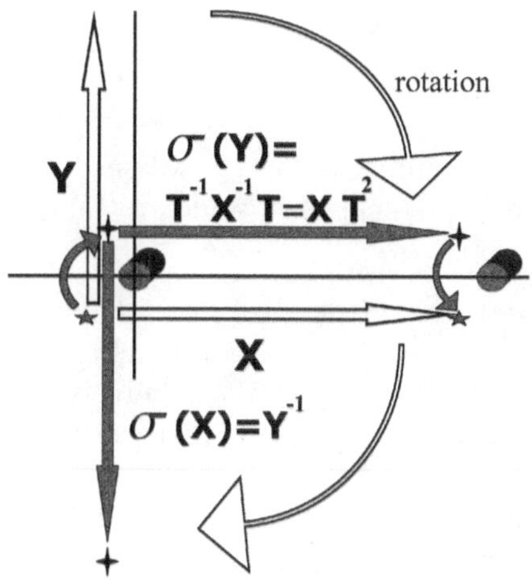

Figure 2.3: Relations $\sigma(X) = Y^{-1}$, $\sigma(Y) = XT^2$

Figure 2.3 shows the action of the automorphism σ corresponding to the rotation of the periods and $\star$ by 90° with the origin taken as the center. Here the dark straight arrows show the images of X, Y (straight white arrows) with respect to this rotation. The quarter of a turn from the point $\star$ is the rotation path $\{g_t(\star)\}$ of this point as $0 \le t \le 1$. The other quarter of a turn is its X–image with the opposit orientation.

We can always choose the base point sufficiently close to 0 and connect it with its g–image in a small neighborhood of zero. This makes the corresponding automorphism of $\mathcal{B}_1$ unique up to powers of T^2. All such automorphisms fix T, because they preserve zero and the orientation.

Thus we have constructed a homomorphism

$$\alpha : SL_2(\mathbb{Z}) \to \mathrm{Aut}_T(\mathcal{B}_1)/T^{2\mathbb{Z}},$$

where $\mathrm{Aut}_T(\mathcal{B}_1)$ is the group of automorphisms of $\mathcal{B}_1$ fixing T. The elements from $T^{2\mathbb{Z}} = \{T^{2n}\}$ are identified with the corresponding inner automorphisms.

Let τ_+, τ_- be the α–images of the matrices $\begin{pmatrix} 1 & 1 \\ 0 & 1 \end{pmatrix}$ and $\begin{pmatrix} 1 & 0 \\ 1 & 1 \end{pmatrix}$. Then $\sigma = \tau_+\tau_-^{-1}\tau_+$ corresponds to $\begin{pmatrix} 0 & 1 \\ -1 & 0 \end{pmatrix}$, and σ^2 has to be the conjugation by T^{2l-1} for some l. Similarly,

$$\tau_+\tau_-^{-1}\tau_+ = T^{2m}\tau_-^{-1}\tau_+\tau_-^{-1}.$$

Using the rescaling $\tau_\pm \mapsto T^{2m_\pm}\tau_\pm$ for $m_- + m_+ = m$, we can eliminate T^{2m} and make $0 \leq l \leq 5$.

Note that, generally, $l \bmod 6$ is the invariant of the action, that is due to Steinberg.

Taking the "simplest" pullbacks for $\tau_\pm$, we easily check that $l = 0$ and calculate the images of the generators under $\tau_\pm$ and σ. We arrive at the relations from Proposition 2.7.5.

Abstract construction. Generalizing, let $\mathcal{E}$ be an algebraic, or complex analytic, or symplectic, or real analytic manifold, or similar. It may be noncompact and singular. We assume that there is a continuous family of *topological isomorphisms* $\mathcal{E} \to \mathcal{E}_t$ for manifolds $\mathcal{E}_t$ as $0 \leq t \leq 1$, and that $\mathcal{E}_1$ is isomorphic to $\mathcal{E}_0 = \mathcal{E}$.

The path $\{\mathcal{E}_t\}$ in the moduli space $\mathcal{M}$ of $\mathcal{E}$ induces an outer automorphism ε of the fundamental group $\pi_1(\mathcal{E}, \star)$ defined as above. The procedure is as follows. We take the image of $\gamma \in \pi_1(\mathcal{E}, \star)$ in $\pi_1(\mathcal{E}_1, \star_1)$ for the image $\star_1$ of the base point $\star \in \mathcal{E}$ and conjugate it by the path from $\star_1$ to $\star$. Thus, *the fundamental group* $\pi_1(\mathcal{M})$ (whatever it is) *acts in* $\pi_1(\mathcal{E})$ *by outer automorphisms modulo inner automorphisms.*

The above considerations correspond to the case when a group G acts in $\mathcal{E}$ preserving a submanifold D. Then $\pi_1(\mathcal{M})$ acts in $\pi_1^{orb}((\mathcal{E} \setminus D)/G)$ by outer automorphisms.

Another variant is with a Galois group taken instead of $\pi_1(\mathcal{M})$ assuming that $\mathcal{E}$ is an algebraic variety over a field that is not algebraically closed.

The action of $\pi_1(\mathcal{M})$ on an *individual* $\pi_1(\mathcal{E})$ generalizes, in a way, the celebrated **Kodaira–Spencer map** and is of obvious importance. However, calculating the fundamental groups of algebraic (or similar) varieties, generally speaking, is difficult. The main examples are the products of algebraic curves and related configuration spaces. Not much can be extracted from the action above without an explicit description of the fundamental group.

2.7.4 Plancherel formulas

Recall that we have two representations of the double affine Hecke algebra $\mathcal{H}$, the polynomial representation $\mathcal{P} = \mathbb{C}[X^{\pm 1}]$ and the representation $\mathcal{F}$ in the space of finitely supported functions on the set $\mathbb{Z}_\sharp$. We constructed two Fourier transforms $\mathbf{S}, \mathbf{E} : \mathcal{P} \to \mathcal{F}$ defined by

$$\mathbf{S}(f)(n_\sharp) = \langle f\varepsilon_n(x)\mu_k^0(x)\rangle_0, \quad \mathbf{E}(f)(n_\sharp) = \langle \overline{f}\varepsilon_n(x)\mu_k^0(x)\rangle_0,$$

where $\langle\cdot\rangle_0$ denotes the constant term.

The space $\mathcal{P}$ is equiped with the scalar product $\langle f, g\rangle_0 \stackrel{\mathbf{def}}{=\!=} \langle f\bar{g}\mu_k^0(x)\rangle_0$. Its $\mathcal{F}$–counterpart is as follows:

$$\begin{aligned} &\langle f, g\rangle_1 \stackrel{\mathbf{def}}{=\!=} \langle f\bar{g}\mu_k^1\rangle_1, \text{ where } \bar{g}(n_\sharp), = \overline{g(n_\sharp)}, \\ &\langle f\rangle_1 \stackrel{\mathbf{def}}{=\!=} \sum_{n\in\mathbb{Z}} f(n_\sharp), \quad \mu_1^k(z) \stackrel{\mathbf{def}}{=\!=} \mu_k(z)/\mu_k(0_\sharp). \end{aligned} \tag{2.7.8}$$

A simple calculation for $n \in \mathbb{Z}_{>0}$ gives that

$$\mu_k^1(n_\sharp) = \mu_k^1((1-n)_\sharp) = q^{-k(n-1)} \prod_{j=1}^{n-1} \frac{1-q^{2k+j}}{1-q^j}. \tag{2.7.9}$$

Note that $(1-n)_\sharp = 1/2 - n_\sharp$ and hence $\pi(\mu_k^1) = \mu_k^1$. Also, $\overline{\mu_k^1} = \mu_k^1$, which makes the form $\langle\cdot,\cdot\rangle_1$ symmetric.

Theorem 2.7.8. (**Plancherel formula I**). *For any $f, g \in \mathcal{P}$,*

$$\langle f, g\rangle_0 = \langle \mathbf{S}(f), \mathbf{S}(g)\rangle_1 = \langle \mathbf{E}(f), \mathbf{E}(g)\rangle_1. \tag{2.7.10}$$

Proof. Let $H \mapsto H^\star$ denote taking the adjoint with respect to the scalar product $\langle\cdot,\cdot\rangle_0$. From Theorem 2.5.8,

$$X^\star = X^{-1},\, Y^\star = Y^{-1} T^\star = T^{-1} q^\star = q^{-1}, t^\star = t^{-1}.$$

It is an anti-automorphism of $\mathcal{H}$.

The theorem follows formally from the following statements.

(i) For automorphisms σ and ϵ from Theorem 2.7.4, $\sigma\star\sigma^{-1} = \star$, $\epsilon\star\epsilon^{-1} = \star$. Use that σ and ϵ are homomorpisms of the group $\mathcal{B}_q$. Any group automorphisms commute with the inversion $g \mapsto g^{-1}$.

(ii) The representations $\mathcal{P}$ and $\mathcal{F}$ of the algebra $\mathcal{H}$ are irreducible. This is true only for sufficiently generic q, t, but we may assume this because it sufices to check (2.7.10) for generic q, t.

(iii) The relations (2.7.8) for $f = g = 1$, that guarantees that the coefficient of proportionality in the Plancherel formula is 1. ❑

We introduce the **characteristic functions** and **delta functions** $\chi_m^\sharp, \delta_m^\sharp \in \mathcal{F}$ by the formulas

$$\chi_m^\sharp(n_\sharp) = \delta_{mn}, \quad \delta_m^\sharp(n_\sharp) = \delta_{mn}/\mu_k^1(n_\sharp), \tag{2.7.11}$$

where δ_{mn} is the Kronecker delta.

Theorem 2.7.9. *(a)* $\mathbf{E}(\varepsilon_m(x)) = \delta^\sharp_m$, *(b)* $\mathbf{S}(\varepsilon_m(x)) = t^{1/2}T^{-1}(\delta^\sharp_m)$.

Proof. (a) By the definition of $\mathbf{E}$ and the norm formulas (2.6.5),

$$\mathbf{E}(\varepsilon_m(x)) = \chi^\sharp_m\langle\varepsilon_m(x), \varepsilon_m(x)\rangle_0 = \delta^\sharp_m.$$

Instead of using the norm formulas we can apply the Plancherel theorem:

$$\langle\varepsilon_m(x), \varepsilon_m(x)\rangle_0 = \langle\chi^\sharp_m, \chi^\sharp_m\rangle_1\langle\varepsilon_m(x), \varepsilon_m(x)\rangle_0^2.$$

Since $\langle\chi^\sharp_m, \chi^\sharp_m\rangle_1 = \mu^1_k(m_\sharp)$, we obtain that $\langle\varepsilon_m(x), \varepsilon_m(x)\rangle_0 = \mu^1_k(m_\sharp)^{-1}$.
(b) From (a), $\mathbf{S}(\overline{\varepsilon_m}(x)) = \delta^\sharp_m$. Then we use part (a) of Proposition 2.5.13. ❑

Corollary 2.7.10. *Let* $\widehat{\varepsilon}(m_\sharp; x) \stackrel{\mathbf{def}}{=\!=} \varepsilon_m(x)$. *For any* $g \in \mathcal{F}$,

$$\mathbf{S}^{-1}(g) = t^{-1/2}\langle g\, T^{(1)}(\widehat{\varepsilon}(m_\sharp; x))\mu^1_k\rangle_1, \tag{2.7.12}$$

where $T^{(1)}$ *acts on the first argument of* $\widehat{\varepsilon}$ *via (2.7.3).*

Proof. Using that $\mathbf{S}(\varepsilon_m(x)) = t^{1/2}T^{-1}\delta^\sharp_m$ due to Theorem 2.7.9,

$$\begin{aligned}\mathbf{S}(\, t^{-1/2}\langle g\, T^{(1)}(\widehat{\varepsilon}(m_\sharp; x))\,) &= t^{-1/2}\langle g\, T(t^{1/2}T^{-1}\delta^\sharp_m)\mu^1_k\rangle_1 \\ &= \langle g\, \delta^\sharp_m\mu^1_k\rangle_1 = \langle g\, \chi^\sharp_m\rangle_1 = g(m_\sharp).\end{aligned} \tag{2.7.13}$$

❑

We are going to drop the bar-conjugation in the scalar products:

$$\langle\!\langle f, g\rangle\!\rangle_0 \stackrel{\mathbf{def}}{=\!=} \langle fg\mu^0_k(x)\rangle_0 \text{ on } \mathcal{P}, \quad \langle\!\langle f, g\rangle\!\rangle_1 \stackrel{\mathbf{def}}{=\!=} \langle fg\mu^1_k\rangle_1 \text{ on } \mathcal{F}.$$

Theorem 2.7.9 and Theorem 2.7.2. result in the following corollary.

Corollary 2.7.11. *For arbitrary* $n, m \in \mathbb{Z}$,

$$\begin{aligned}&\langle\varepsilon_n(x)\varepsilon_m(x)\mu^0_k(x)\rangle_0 = t^{1/2}T^{-1}\delta^\sharp_n(m_\sharp), \text{ where} \\ &T(\delta^\sharp_n) = \frac{t^{1/2}q^{2n_\sharp} - t^{-1/2}}{q^{2n_\sharp} - 1}\delta^\sharp_{-n} - \frac{t^{1/2} - t^{-1/2}}{q^{2n_\sharp} - 1}\delta^\sharp_n.\end{aligned} \tag{2.7.14}$$

In particular, the inequality $\langle\varepsilon_n(x)\varepsilon_m(x)\mu^0_k(x)\rangle_0 \neq 0$ *implies* $m = \pm n$. ❑

Theorem 2.7.12. *(***Plancherel formula II***) For any* $f, g \in \mathcal{P}$, *we have*

$$t^{-1/2}\langle\!\langle\mathbf{S}(f), T\mathbf{S}(g)\rangle\!\rangle_1 = \langle\!\langle f, g\rangle\!\rangle_0 = t^{-1/2}\langle\!\langle\mathbf{E}(f), T\mathbf{E}(g)\rangle\!\rangle_1. \tag{2.7.15}$$

Proof. Let $\diamond$ denote the anti-involution corresponding to the scalar product $\langle\langle f, g\rangle\rangle_0$. It equals $\star \circ \eta$:

$$X^\diamond = X,\ T^\diamond = T,\ \pi^\diamond = \pi,\ Y^\diamond = TYT^{-1}.$$

The same anti-involution serves the scalar product $\langle\langle f, g\rangle\rangle_1$.

Following Theorem 2.7.8 (see (i–iii)), we need to check that

$$\sigma \diamond \sigma^{-1} = \epsilon \diamond \epsilon^{-1} = T^{-1} \diamond T.$$

This is straightforward. ❑

Preparing for the inversion theorems, let us introduce the **conjugated** Fourier transforms $\bar{\mathbf{S}}, \bar{\mathbf{E}}$:

$$\begin{aligned} \bar{\mathbf{S}} &: f \mapsto \langle f\, \bar{\varepsilon}_n(x)\mu_k^0(x)\rangle_0, \\ \bar{\mathbf{E}} &: f \mapsto \langle \overline{f\, \varepsilon_n}(x)\mu_k^0(x)\rangle_0. \end{aligned} \tag{2.7.16}$$

Theorem 2.7.4 states that, for any $H \in \mathcal{H\!H}$, $f \in \mathcal{P}$,

$$\mathbf{S}(H(f)) = \sigma(H)(\mathbf{S}(f)),\ \ \mathbf{E}(H(f)) = \epsilon(H)(\mathbf{E}(f)).$$

It is not difficult to find the automorphisms corresponding to $\bar{\mathbf{S}}$ and $\bar{\mathbf{E}}$:

$$\bar{\mathbf{S}}(H(f)) = \sigma^{-1}(H)(\bar{\mathbf{S}}(f)),\ \ \bar{\mathbf{E}}(H(f)) = \eta\epsilon(H)(\bar{\mathbf{E}}(f)).$$

Following Theorem 2.7.1, we get another **master formula:**

$$\begin{aligned} &\langle \overline{\varepsilon_n}(x)\overline{\varepsilon_m}(x)\widehat{\gamma_-}\mu_k^0(x)\rangle \\ &= q^{m_\sharp^2+n_\sharp^2-2o_\sharp^2}\, t^{-1/2}\, (T\overline{\varepsilon_m}(x))(n_\sharp)\langle\widehat{\gamma_-}\mu_k^0(x)\rangle. \end{aligned} \tag{2.7.17}$$

Algebraically, it is equvalent to the formula

$$\sigma^{-1}\tau_+^{-1}\eta(Y) \ = \ \tau_+(T(\eta(Y)T^{-1}),$$

resulting from $\sigma^{-2} = T(\cdot)T^{-1}$ and $\eta\tau_\pm\eta = \tau_\pm^{-1}$:

$$\sigma^{-1}\tau_+^{-1}\eta = \tau_+\tau_+^{-1}(\sigma^{-2}\sigma)\tau_+^{-1}\eta = \tau_+\sigma^{-2}(\tau_+^{-1}(\sigma)\tau_+^{-1})\eta$$

$$= \tau_+\sigma^{-2}\tau_-^{-1}\eta = \tau_+\sigma^{-2}\tau_-^{-1}\eta = \tau_+\sigma^{-2}\eta\tau_-.$$

2.7.5 Inverse transforms

Let us summarize what we have done so far. We have defined four transforms acting from $\mathcal{P}$ to $\mathcal{F}$:

$$\mathbf{S}_0(f) = \langle f\,\varepsilon_n(x)\mu_k^0(x)\rangle_0,\ \ \mathbf{E}_0(f) = \langle \overline{f}\,\varepsilon_n(x)\mu_k^0(x)\rangle_0,$$

$$\bar{\mathbf{S}}_0(f) = \langle f\,\overline{\varepsilon}_n(x)\mu_k^0(x)\rangle_0,\ \ \bar{\mathbf{E}}_0(f) = \langle \overline{f}\,\overline{\varepsilon}_n(x)\mu_k^0(x)\rangle_0.$$

Let us introduce their counterparts acting in the opposite direction, from the space $\hat{\mathcal{F}}$ of all functions on $\mathbb{Z}_\sharp$ to $\mathcal{P}$. Replacing $0 \rightsquigarrow 1$ in these two formulas,

$$\langle\cdot,\cdot\rangle_0 \rightsquigarrow \langle\cdot,\cdot\rangle_1,\ \ \mu_k^0 \rightsquigarrow \mu_k^1,$$

$$\mathbf{S}_1(f) = \langle f\,\varepsilon_n(x)\mu_k^1\rangle_1,\ \ \mathbf{E}_1(f) = \langle \overline{f}\,\varepsilon_n(x)\mu_k^1\rangle_1,$$
$$\bar{\mathbf{S}}_1(f) = \langle f\,\overline{\varepsilon}_n(x)\mu_k^1\rangle_1,\ \ \bar{\mathbf{E}}_1(f) = \langle \overline{f}\,\overline{\varepsilon}_n(x)\mu_k^1\rangle_1.$$

Theorem 2.7.13. (**Inversion theorem**) *We have the following identities:*

$$\bar{\mathbf{S}}_1\circ\mathbf{S}_0 = \mathrm{id} = \mathbf{S}_1\circ\bar{\mathbf{S}}_0,\ \ \bar{\mathbf{S}}_0\circ\mathbf{S}_1 = \mathrm{id} = \mathbf{S}_0\circ\bar{\mathbf{S}}_1,$$

$$\mathbf{E}_1\circ\mathbf{E}_0 = \mathrm{id} = \bar{\mathbf{E}}_1\circ\bar{\mathbf{E}}_0,\ \ \mathbf{E}_0\circ\mathbf{E}_1 = \mathrm{id} = \bar{\mathbf{E}}_0\circ\bar{\mathbf{E}}_1.$$

Proof. We know that the automorphisms corresponding to $\mathbf{S}$, $\mathbf{E}$, $\bar{\mathbf{S}}$, $\bar{\mathbf{E}}$ are σ, ϵ, σ^{-1}, $\eta\sigma$, respectively. This gives that all maps in the theorem are homomorphisms of $\mathcal{H}$–modules. Now we use that the $\mathcal{H}$–modules $\mathcal{P}$, $\mathcal{F}$ are irreducible and the inversion formulas are true for $f = 1 \in \mathcal{P}$ or $f = \delta_0^\sharp \in \mathcal{F}$. ❑

Jackson master formulas. We can obtain the Jackson master formulas from the old ones by *formal* conjugating performed together with replacing $\widehat{\gamma_-}$ by $\gamma = q^{x^2}$. The proofs of these formulas (including the convergence) are a straightforward copy of those in the compact case.

For example, formula (2.7.17) results in

$$\langle\varepsilon_m(x)\varepsilon_n(x)\gamma\mu_k^1\rangle_1 = q^{2\rho_\sharp^2-m_\sharp^2-n_\sharp^2}\,t^{1/2}T^{-1}(\varepsilon_m(x))(n_\sharp)\langle\gamma\mu_k^1\rangle_1. \tag{2.7.18}$$

Formulas (2.7.1) and (2.7.2) read as follows:

$$\langle\overline{\varepsilon_m}(x)\overline{\varepsilon_n}(x)\gamma\mu_k^1\rangle_1 = q^{2\rho_\sharp^2-m_\sharp^2-n_\sharp^2}\,\overline{\varepsilon_m}(n_\sharp)\langle\gamma\mu_k^1\rangle_1, \tag{2.7.19}$$

where $2\rho_\sharp^2 - m_\sharp^2 - n_\sharp^2 = -\frac{m^2+n^2+2k(|m|+|n|)}{4}$, and, respectively,

$$\langle\varepsilon_m(x)\overline{\varepsilon_n}(x)\gamma\mu_k^1\rangle_1 = q^{2\rho_\sharp^2-m_\sharp^2-n_\sharp^2}\,\varepsilon_m(n_\sharp)\langle\gamma\mu_k^1\rangle_1. \tag{2.7.20}$$

The symmetrization of these formulas (it does not matter which one is taken) gives a direct counterpart of Theorem 2.5.3, namely,

$$\langle p_n(x)p_m(x)\gamma\delta_k^1(x)\rangle_1 = \\ = p_n((m+k)/2)p_m(k/2)q^{-(m^2+n^2+2k(m+n))/4}\,\langle\gamma\delta_k^1(x)\rangle_1. \qquad (2.7.21)$$

The Plancherel formulas can also be transformed to the Jackson case with ease using Theorem 2.7.13. We mention here the papers [KS1, KS2] devoted to a variant of the Fourier transform $\mathbf{S}_1$ and its inversion in the *analytic* setting.

Note that formulas (2.7.18),(2.7.19), and (2.7.20) have counterparts for the integration over the contour $i\epsilon + \mathbb{R}, \epsilon > 0$ instead of the Jackson integration (which is essentially the sharp integration in the A_1–case). Cf. case (c) in Section 2.6.5. The constant $\int_{i\epsilon+R}\gamma\mu_k$ must be replaced by the formula from Section 2.3.5. No other changes are necessary.

Comment. (i) The limit $q \to \infty$ of the transform $\bar{\mathbf{S}}_1 : \mathcal{F} \to \mathcal{P}$ is the p–adic spherical transform due to Matsumoto [Ma]. Its inverse $\mathbf{S}_0$ is also compatible with the limit. In the papers [O4, O5], Opdam developed the Matsumoto theory of "nonsymmetric" spherical functions towards the theory of nonsymmetric polynomials. The operator T_{w_0} appears there in the inverse transform in a way similar to what we are doing ([O4], Proposition 1.12).

(ii) As $q \to 1$, the transform $\mathbf{S}_0$ becomes the Hankel transform. One can also obtain the Harish-Chandra spherical transform under the same limit by switching from the variable x to the variable $X = q^x$. That is, we need to rewrite all formulas in terms of X and then leave X untouched in the limit. Note that the self-duality of the Fourier transform and the Gaussian do not survive in this limit (and in the p–adic case).

(iii) We can transform the norm formula (2.6.5) and its variant without conjugation (2.7.14) to the Jackson case:

$$\langle\varepsilon_m(x)\overline{\varepsilon_n}(x)\mu_k^1\rangle_1 = \mu_k^1(n_\sharp)^{-1}\delta_{mn},$$

$$\langle\varepsilon_m(x)\varepsilon_n(x)\mu_k^1\rangle_1 = t^{1/2}T^{-1}(\delta_n^\sharp)(m_\sharp).$$

However, these formulas hold only for $\Re k << 0$. In the absence of the nonsymmetric polynomials, $\Re k < 0$ is sufficient. It is the A_1–case of the Aomoto conjecture [A2], deduced by Ian Macdonald in general from the constant term conjecture proved in [C16]. See [A2, M6].

2.8 Finite dimensional modules

Finite dimensional representations can appear either for *singular* k, namely, for half-integers, or for special q (roots of unity). In this section we discuss either generic q and singular k or generic k as q is a root of unity. Let us start with the case of generic q.

2.8.1 Generic q, singular k

We assume that q is not a root of unity. When $q^{a/b}$ appear in the claims and/or formulas for $a/b \in \mathbb{Q}$, then we will assume in the next theorem that $q^{1/b}$ is somehow fixed and $q^{a/b} = (q^{1/(b)})^a$. In particular, a product $q^a t^b$ can be 1, but may not be a nontrivial root of unity.

Theorem 2.8.1. *(i) An arbitrary irreducible finite dimensional representation V is a quotient of either the polynomial represenation $\mathcal{P} = \mathbb{C}[X^{\pm 1}]$ or its image under the automorpisms ι, ς_y, $\iota\varsigma_y$, where*

$$\iota : T \mapsto -T,\ X \mapsto X,\ Y \mapsto Y,\ q^{1/2} \mapsto q^{1/2},\ t^{1/2} \mapsto t^{-1/2}, \tag{2.8.1}$$

$$\varsigma_y : T \mapsto T,\ X \mapsto X,\ Y \mapsto -Y,\ q^{1/2} \mapsto q^{1/2},\ t^{1/2} \mapsto t^{1/2}. \tag{2.8.2}$$

The $\mathcal{P}$ is Y-semisimple if and only if k is not a negative integer. It is irreducible if and only if $k \neq -1/2 - n$, $n \in \mathbb{Z}_+$.

(ii) Let $k = -1/2 - n$ for $n \in \mathbb{Z}_+$. The polynomials e_m are well defined for all m and form a basis of $\mathcal{P}$. They are multiplied by $(-1)^m$ under the automorphism of $\mathcal{H}$

$$\varsigma_x : T \mapsto T,\ X \mapsto -X,\ Y \mapsto Y,\ q^{1/2} \mapsto q^{1/2},\ t^{1/2} \mapsto t^{1/2}. \tag{2.8.3}$$

The values $e_m(-k/2)$ are nonzero and therefore the polynomials

$$\varepsilon_m = e_m/e_m(-k/2) \text{ exist for } M = \{-2n \le m \le 2n+1\}.$$

The series μ_k^0 and the pairing $\langle f, g\rangle$ on $\mathcal{P}$ are well defined for $k = -1/2 - n$. The scalar squares of e_m are nonzero precisely at the same set M. The radical Rad_0 of the pairing $\langle f, g\rangle$ is $\oplus_{m \notin M} \mathbb{C} e_m$.

(iii) Continuing (ii), $Rad_0 = (e_{-2n-1})$ as an ideal in $\mathcal{P}$, and the $\mathcal{H}$-module $\mathcal{P}/Rad_0$ is the greatest finite dimensional quotient of $\mathcal{P}$. It is the direct sum of the two non-isomorphic $\mathcal{H}$-submodules of dimension $2n+1$:

$$V^{\pm}_{2n+1} \stackrel{\mathbf{def}}{=\!=} \oplus_{m=1}^{2n+1} \mathbb{C}\varepsilon^{\pm}_m \text{ mod } Rad_0,\ \varepsilon^{\pm}_m = \varepsilon_m \pm \varepsilon_{-2n-1+m}, \tag{2.8.4}$$

$$V^{\pm}_{2n+1} \cong \mathbb{C}[X^{\pm 1}]/(\varepsilon^{\mp}) \text{ for } \varepsilon^{\pm} = \varepsilon_{n+1} \pm \varepsilon_{-n}. \tag{2.8.5}$$

These modules are orthogonal to each other with respect to $\langle\, ,\, \rangle$, and V^{-}_{2n+1} is isomorphic to the ς_x-image of V^{+}_{2n+1}.

iv) The representation V^{+}_{2n+1} is the quotient of $\mathcal{P}$ by the radical Rad of the pairing $\{f\,,\, g\} = f(Y^{-1})(g)(-k/2)$ from Theorem 2.5.10, which is $V^{-}_{2n+1} + Rad_0$. Respectively, V^{-}_{2n+1} corresponds to the paring

$$\{f\,,\, g\}_- \;=\; f(Y^{-1})(g)\,|_{X \mapsto -t^{-1}}\,.$$

The discretization map

$$\chi : f \mapsto f(z),\ z \in \bowtie' \stackrel{\mathbf{def}}{=\!=} \{1/4 + m/2 \mid -n \le m \le n\}$$

identifies V^+_{2n+1} *with* $F_{2n+1} \stackrel{\text{def}}{=\!=}$ *Funct* $(\bowtie')$, *where the action of* $\mathcal{HH}$ *is via formulas (2.7.3) from Theorem 2.7.2. The* $\varepsilon^{\pm}$ *from (ii) are proportional to*

$$e^{\pm} = e_{n+1} \pm t^{-1/2} e_{-n} = X^{-n} \prod_{m=-n}^{n} (X \pm q^{1/4+m/2}).$$

Proof. Let V be an arbitrary $\mathcal{HH}$–module and v a Y–eigenvector of weight λ: $Y(v) = q^{\lambda} v$. Following Corollary 2.6.1, we construct the chain

$$v = v_0 \xrightarrow{A_0} v_1 \xrightarrow{B_1} v_{-1} \xrightarrow{A_{-1}} v_2 \xrightarrow{B_2} \ldots, \tag{2.8.6}$$

where the operation A_m is $q^{-m/2} X\pi$ and the operation B_m is the application of

$$t^{1/2}(T + \frac{t^{1/2} - t^{-1/2}}{q^{-2\lambda_m} - 1}).$$

These come from the intertwining operators, so we will call them intertwiners too. The normalization is adjusted to the case $v = 1 \in \mathbb{C}[X^{\pm 1}]$, when $v_m = e_m$, $\lambda_m = -m_\sharp$. Generally, v_m is a Y–eigenvector of weight

$$\lambda_m = -\lambda - m/2 \ \text{ for } m > 0, \quad \lambda_m = \lambda - m/2 \ \text{ for } m \leq 0.$$

The intertwiners A_m are always invertible. Because q is not a root of unity, the chain is infinite and contains only invertible intertwiners B_m unless it occurs for some $m \in \mathbb{Z}_+$ that

(a) either $q^{2\lambda} = q^{-m}$ (nonexistence),

(b) or $q^{2\lambda} = t^{\pm 1} q^{-m}$ (noninvertibility).

We treat B_m as elements of the nonaffine Hecke algebra $\mathbf{H} = \mathbb{C} + \mathbb{C}T$. Note that B_m is simply s in the degenerate case $k = 0$, and there are no finite dimensional representations in this case.

(i) Let V be a finite dimensional irreducible module. Then $k \neq 0$ and at least one of the conditions (a), (b) must hold for some m. Applying intevertible intertwiners, we can always make

$$\text{either} \ \ (a')\ \lambda = 0 \ \ \text{or} \ \ (b')\ \lambda = k/2.$$

In the second case, we may need to use the automorphisms ι, and ς_y. Let us start with (b').

The corresponding chain of intertwiners results in the weights

$$\lambda_0 = k/2,\ \lambda_1 = -1/2 - k/2,\ \lambda_{-1} = 1/2 + k/2, \tag{2.8.7}$$
$$\ldots,\ \lambda_m = -m/2 - k/2,\ \lambda_{-m} = m/2 + k/2, \ldots,\ m > 0.$$

Here all intertwiners exist and are invertible, respectively, if and only if $0 \neq m/2 + k/2 \neq -k/2$ for $m \in \mathbb{Z}_+$. These inequalities make V infinite dimensional, so we can assume that $k \in -\mathbb{Z}_+/2$. Recall that the B–intertwiners are applied to the v_m with positive indices m.

Let us check that $\tilde{v} \stackrel{\text{def}}{=} (T - t^{1/2})(v) = 0$. Indeed, $\tilde{v}$ is a Y–eigenvector of weight $-k/2$. Taking $v_0 = \tilde{v}$, the corresponding chain (2.8.6) produces the weights

$$\lambda_0 = -k/2,\ \lambda_1 = -1/2 + k/2,\ \lambda_{-1} = 1/2 - k/2,\ -1 + k/2,\ 1 - k/2, \ldots.$$

The existence and invertibility of the intertwiners are satisfied, respectively, as $0 \neq m/2 - k/2 \neq k/2$ for $m \in \mathbb{Z}_+$. Since $k < 0$, the chain is always invertible and $\dim V = \infty$ unless $\tilde{v} = 0$. Thus, under assumption (b'), k has to be in $-\mathbb{Z}_+/2, k \neq 0$, and V is a quotient of $\mathcal{P}$, where the covering map sends $1 \mapsto v$.

Actually, case (b') with integral k is also a part of (a'). Indeed, if $k = -n$ for natural n, then the chain behaves as follows. It loses the existence at $m/2 = n/2$ and then the invertibility at $m/2 = n$. Thus we can reach $\lambda = 0$ using the invertible intertwiners in this case, which leads to (a').

Let us prove that V is always infinite dimensional in case (a'), i.e., when we can find $\lambda = 0$ in V. Instead of (2.8.6), we use its variant with the space $V_0 = \mathbf{H}v$. The latter is an irreducible two-dimensional module over the affine Hecke algebra $\langle T, Y\rangle$. The operator Y is not semisimple there. This module is the space of all solutions of $(Y-1)^2(w) = 0$. The operations A, B transform V_0 to

$$V_m \stackrel{\text{def}}{=} \{w \mid (Y - q^{\lambda_m})^2(w) = 0 \text{ for } \lambda_m = -m/2, m \in \mathbb{Z}_+\}.$$

All intertwiners here are invertible unless $\pm k/2 = n/2$ for a positive integer n. Note that the invertibility of all intertwiners is possible only for infinite dimensional V.

Let us consider the case $\{(a'),\ k \in \mathbb{Z} \setminus \{0\}\}$. In this case, the chain from $\lambda_0 = 0$ loses the invertibility exactly once at $m = n$. The dimension of the next (after V_n) space V_{-n} becomes 1 and then remains unchanged. Therefore V is infinite dimensional too.

We conclude that an irreducible V containing a Y–eigenvector of weight 0 is always infinite dimensional, and, moreover, it is Y–non-semisimple precisely as $k = \pm n$ for a positive integer n.

In the case under consideration, we may assume that $k = -n$ using ι, so it is covered by (b') and V is a quotient of $\mathcal{P}$. However, the polynomial representation is irredicible for such k. Indeed, let us assume that there is a submodule $W \subset \mathcal{P}$. Then it contains at least one Y–eigenvector v. Denoting its weight by λ, W contains the whole space

$$\{v \mid (Y - q^{\lambda})^2(v) = 0\},$$

which is two-dimensional. Using the intertwiners, we can make $\lambda = 0$ here. So 1 belongs to W and we get $W = \mathcal{P}$.

(ii) Corollary 2.6.1 provides the existence of e_m for any half-integral k. The statement about μ_k^0 is direct from Theorem 2.2.5. The other claims follow from formulas (2.6.4) and (2.6.6) from Corollary 2.6.2:

$$e_m(-k/2) = t^{-|m|/2} \prod_{0<j<|m|'} \frac{1-q^j t^2}{1-q^j t}, \tag{2.8.8}$$

where $|m|' = m$ if $m > 0$, $|m|' = 1-m$ if $m \le 0$, and

$$\langle e_l\,,\, e_m\rangle = \delta_{lm} \prod_{0<j<|m|'} \frac{(1-q^j)(1-q^j t^2)}{(1-q^j t)(1-q^j t)}. \tag{2.8.9}$$

(iii) Let us show that $\mathcal{P}$ has a unique maximal finite dimensional quotient and calculate it as $k = -1/2 - n$ for $n \in \mathbb{Z}_+$. We can construct e_m explicitly using the chain of intertwiners from $v_0 = 1$. The intertwiners are well defined. However, one of them, namely,

$$B_{2n+1} = t^{1/2}(T + \frac{t^{1/2}-t^{-1/2}}{tq^{2n+1}-1}) = t^{1/2}(T + \frac{t^{1/2}-t^{-1/2}}{t^{-1}-1}) = t^{1/2}(T - t^{1/2})$$

is not invertible. Nevertheless $e_{-2n-1} = B_{2n+1}(e_{2n+1})$ because the leading monomial of the latter is X^{-2n-1}. We see that this chain does produce all e_m regardless of the noninvertibility of B_{2n+1}. By the way, this consideration makes the above reference to Corollary 2.6.1 (concerning the existence of $\{e_m\}$) unnecessary. See (ii).

The polynomials e_m and $e_{-2n-1+m}$ have coinciding weights for $1 \le m \le 2n+1$, namely, $-m_\sharp = -(-2n-1+m)_\sharp$. We obtain that the image of the space J linearly generated by e_m, e_{1-m} for $m > 2n+1$ is zero in V. Indeed, otherwise $\dim V$ would be infinite. Moreover, all eigenvectors in V are images of $\{e_m, e_{1-m}\}$ as $1 \le m \le 2n+1$ up to proportionality. The multiplicities of the images can be either all equal to 1 or all equal to 2. If the multiplicities are all 2, then V has to be $\mathcal{P}/J$.

Let us check that J is an $\mathcal{H}$–submodule of $\mathcal{P}$. The Y–invariance of J is obvious. It is T–invariant, since the intertwiners are always well defined and

$$e_{-2n-1} = t^{1/2}(T - t^{1/2})(e_{2n+1}) \Rightarrow T(e_{-2n-1}) = -t^{-1/2} e_{-2n-1}.$$

Hence it is π–invariant. The A-operations give the $X\pi$–invariance of J. So it is $X^{\pm 1}$–invariant, which can also be seen from the Pieri rules (Corollary 2.6.2). Obviously $J = (e_{-2n-1})$ as an ideal, because the dimension of the quotient $\mathcal{P}/J$ is $2(2n+1)$.

If the Y–multiplicities are all 1, then the dimension of V becomes $2n+1$ and V has to be irreducible. We may set $V = \mathcal{P}/(e)$ for a polynomial $e = X^{n+1} + \ldots + \text{const}\cdot X^{-n}$ and a nonzero constant. Here e must be a Y–eigenvector. Indeed, there is a linear combination of $e, Y(e)$ that is lower than

e, unless these vectors are proportional. Therefore $e = e_{n+1} + ce_{-n}$ (both have coinciding weights, namely, $-1/4$).

It is easy to find that $c = \pm t^{-1/2}$ using the A, B–invariance of (e), i.e., $e = e^{\pm}$ up to proportionality. Switching to the ε–polynomials, e has to be proportional to $\varepsilon_{n+1} \pm \varepsilon_{-n}$.

However, it is easier to obtain the value of c using (iv), which will be considered next. Indeed, if we know that the ideal $V^-_{2n+1} = (e^-)$ is a nontrivial submodule of $\mathcal{P}/J$, then its orthogonal complement with respect to $\langle\, , \rangle$ is obviously $V^+_{2n+1} = (e^+)$. Their sum is direct and they are non-isomorphic as $\mathcal{H}$–modules because the $e^{\pm}$ are non-proportional. Note that they are transposed by ς_x (acting on polynomials) because $\varsigma_x(e^+) = \pm e^-$. So $c = \pm t^{-1/2}$.

By the way, the modules $V^{\pm}_{2n+1}$ are not isomorphic to each other because otherwise $\mathcal{P}/J$, generated by 1 as an $\mathcal{H}$–module, would contain a proper submodule generated by $\varepsilon_0^+ + \varepsilon_0^- = 2\varepsilon_0 = 2$. The latter is impossible.

(iv) The radical *Rad* of the pairing $\{\, , \}$ contains a Y–eigenvector e' if and only if $e'(-k/2) = 0$. This follows directly from the definition (see Theorem 2.5.10). Since the e–polynomials form a basis in $\mathcal{P}$, a given polynomial f belongs to the radical if and only if $f(m_\sharp) = 0$ for all m such that $e_m(-k/2) \neq 0$.

The quotient $V = \mathcal{P}/Rad$ is finite dimensional. Indeed, it contains all

$$\varepsilon_m - \varepsilon_{-2n-1+m} \text{ for } 1 \le m \le 2n+1.$$

Actually, one of them is sufficient for $\dim V < \infty$ because *Rad* is an ideal.

By the way, it gives that $e_m(-k/2) \neq 0 \neq e_{1-m}(-k/2)$ for the $1 \le m \le 2n+1$ and that this is the complete list of such m, without using the explicit formula (2.6.4).

As an immediate application, we obtain that the set $\{m_\sharp\}$ for e_m with nonzero values at $-k/2$ is precisely $\bowtie'$, and V^+_{2n+1} coincides with $F_{2n+1} =$ Funct $(\bowtie)$. This proves the product formula for e from (iii) and justifies that $e = e_{n+1} - t^{-1/2}e_{-n}$. ❑

The theorem gives that for each negative half-integer $k = -1/2 - n$, there are four irreducible finite dimensional representations, namely, $\varsigma_x^{\pm 1}\varsigma_y^{\pm 1}(V^+_{2n+1})$ for all possible combinations of the signs. This is different from the case of the rational DAHA, which has only one such representation (Theorem 2.4.13) as k is a half-integer. The automorphisms ς do not have counterparts in the rational limit.

We note that a construction connected with the finite dimensional quotients of $\mathcal{P}$ appeared in [DS]. The authors did not consider the double Hecke algebras, but found some finitely supported measures for orthogonal polynomials, which is directly related to the theorem.

2.8.2 Additional series

If it is not supposed that $q^a t^b$ do not represent nontrivial roots of unity for integral a, b, then the **additional series** of finite dimensional representations will appear. We continue to assume that q, t are not roots of unity. Parts (i)–(ii) of the theorem below are due to A. Oblomkov. An example of the additional representation was considered in the Appendix of [C20].

In practical terms, we need to go back to the q, t–notation instead of using the k. The proof of the statement that finite dimensional representations are quotients of the polynomial representation up to the products of the automorphisms ς_y and ι remains unchanged, so we only need to describe the quotients of the polynomial representation.

Following the previous section, we consider the chain (2.8.6):

$$v_0 = 1 \xrightarrow{A_0} v_1 = X \xrightarrow{B_1} v_{-1} \xrightarrow{A_{-1}} v_2 \xrightarrow{B_2} \ldots, \qquad (2.8.10)$$

where the operation A_{-m} is $q^{m/2}X\pi$ and the operation B_m is the application of

$$t^{1/2}(T + \frac{t^{1/2} - t^{-1/2}}{tq^m - 1}), \ m > 0.$$

Because q is not a root of unity, the chain is infinite and contains only invertible intertwiners B_m (A_m are always invertible), unless for some $m \in \mathbb{Z}_+$,

(a) either $tq^m = 1$ (nonexistence),

(b) or $tq^m = t^{-1}$ (noninvertibility).

There is only one case which was not covered by the previous theorem, namely, when

$$t = -q^{-n/2} \text{ for some integral } n > 0. \qquad (2.8.11)$$

Note that the case when $t = -1$ and q is generic is excluded from the consideration; we leave it as an exercise.

Let as impose (2.8.11). Then (b) may hold for some m, but we have no problem with (a) and the polynomials e_m always exist. Indeed, the B–intertwiners always create the polynomials with the desired leading term, even if they are not invertible.

Because e_{-n} is proportional to $B_n(e_n)$, one has:

$$T(e_{-n}) = t^{1/2}e_{-n} \text{ and } s(e_{-n}) = e_{-n}.$$

Thus $e_{-n} = X^{-n}+\text{const}X^{-n+2}\ldots + X^n$. Only the terms with X^{-n+2l} can appear in the decomposition.

Using the argument as above, we conclude that the space

$$\mathbb{C}e_{-n} \oplus \mathbb{C}e_{n+1} \oplus \mathbb{C}e_{-n-1} \oplus \ldots$$

is an ideal (e_{-n}), and, moreover, an $\mathcal{H}$–submodule. The following holds.

Theorem 2.8.2. *(i) Assuming that q, t are not roots of unity, the nonzero finite dimensional representations of $\mathcal{H}$ that are not described in Theorem 2.8.1 exist only for $t = -q^{\pm n/2}$ as $0 \neq n \in \mathbb{Z}_+$. Given q, t, such a representation is unique, has dimension $2n$, and is irreducible. For $t = -q^{-n/2}$, it is isomorphic to $V_{2n} \stackrel{\text{def}}{=\!=} \mathcal{P}/(e_{-n})$. The automorphisms ς_x and ς_y do not change its isomorphism class; $\iota(V_{2n})$ corresponding to $t = -q^{n/2}$ is not a quotient of $\mathcal{P}$ for such t.*

(ii) The series μ_k^0 and the pairing $\langle f, g\rangle$ on $\mathcal{P}$ are well defined for such t. The values $e_m(t^{-1/2})$ of e_m and their scalar squares from (2.8.8) and (2.8.9) are nonzero precisely for $m = 0, 1, -1, \ldots, n$. The radical Rad_0 of the pairing $\langle f, g\rangle$ coincides with (e_{-n}), so its restriction to V_{2n} is well defined.

(iii) The discretization map identifies V_{2n} with the space

$$F_{2n} \stackrel{\text{def}}{=\!=} Funct\Big(t^{-1/2}q^{\frac{1-n}{2}}, \ldots, t^{-1/2}, t^{1/2}q^{1/2}, \ldots, t^{1/2}q^{\frac{n}{2}}\Big). \tag{2.8.12}$$

Its kernel coincides with (e_{-n}) and is the radical Rad of the pairing $\{f, g\} = f(Y^{-1})(g)(X \mapsto t^{-1/2})$ from Theorem 2.5.10.

Proof. Claim (i) has been mainly checked. The invariance of V_{2n} with respect to the substitution $X \mapsto -X$ follows from the structure of e_{-n} described above. The map $Y \mapsto -Y$ acts in V_{2n}, since the Y–spectrum of V, which is the inverse of the set (2.8.12) in the multiplicative notation, that is,

$$\Big\{t^{1/2}q^{\frac{n-1}{2}}, \ldots, t^{1/2}, t^{-1/2}q^{-1/2}, \ldots, t^{-1/2}q^{\frac{-n}{2}}\Big\},$$

is invariant with respect to the multiplication by -1.

Claim (ii) is straightforward. Claim (iii) readily folows from the descrition of the Y–spectrum of V_{2n}. ❑

2.8.3 Fourier transform

To conclude the consideration of the case of generic q, let us describe the action of the projective $PSL(2, \mathbb{Z})$ on V_{2n+1}^+ for $k = -1/2 - n$ and V_{2n} for $t = -q^{-n/2}$.

We will begin with the case $k = -1/2 - n$.

Generally, the σ–invariance (up to proportionality) of an $\mathcal{H}$–module V gives that if $Y(v) = q^{\lambda}v$, then $X(v') = q^{-\lambda}v'$ for some v'. Applying this to the case under consideration, we conclude that V_{2n+1}^+ and $\varsigma_x\varsigma_y(V_{2n+1}^+)$ are σ–invariants, but $\varsigma_x(V_{2n+1}^+)$ and $\varsigma_y(V_{2n+1}^+)$ are not; the latter two are transposed by σ.

We will study V_{2n+1}^+ only. Applying $\varsigma_x\varsigma_y$, one can manage the module $\varsigma_x\varsigma_y(V_{2n+1}^+)$. Actually, the σ–invariant module $\varsigma_x(V_{2n+1}^+) \oplus \varsigma_x(V_{2n+1}^+)$ must be considered too. This case follows along the same lines; we leave it as an exercise.

It is convenient to switch from V^+_{2n+1} to its "functional realization" F_{2n+1} from Theorem 2.8.1(iv).

This space has the following scalar product:

$$\langle f,g\rangle' \stackrel{\mathbf{def}}{=\!=} \langle f\bar{g}\mu^1\rangle',\ \bar{g}(m_\sharp)=\overline{g(m_\sharp)},\ \langle f\rangle' = \sum_{m_\sharp\in\bowtie'} f(m_\sharp), \tag{2.8.13}$$

$$\mu^1(m_\sharp)=\mu^1((1-m)_\sharp)=q^{-k(m-1)}\prod_{i=1}^{m-1}\frac{1-q^{2k+i}}{1-q^i} \text{ as } m>0.$$

Recall that $(1-m)_\sharp = 1/2-m_\sharp$, $\pi(\mu^1)=\mu'$, and $\pi(\bowtie')=\bowtie'$. Also, $\overline{\mu^1}=\mu^1$ and therefore the form $\langle\cdot,\cdot\rangle'$ is symmetric. The operators X,Y,T,q,t are unitary with respect to this scalar product.

We come to the following truncation of Theorem 2.7.2.

Theorem 2.8.3. (**Truncated master formula**) *(i) We introduce the* **Gaussian** *by the formula* $\gamma(m_\sharp)=q^{m_\sharp^2}$. *Then it induces* τ_+ *upon the action* $H\mapsto\gamma H\gamma^{-1}$ *on* $H\in\mathcal{H}$. *Respectively, the automorphisms of* F_{2n+1}

$$\mathbf{S}(f)(m_\sharp)=\langle f\varepsilon_m\mu^1\rangle',\ \mathbf{E}(f)(m_\sharp)=\langle \bar{f}\varepsilon_m\mu^1\rangle' \tag{2.8.14}$$

induce σ *and* ϵ *on* $\mathcal{H}$.

(ii) For $l_\sharp, m_\sharp\in\bowtie'$,

$$\langle\varepsilon_l\overline{\varepsilon_m}\gamma\mu^1\rangle' = q^{-\frac{m^2+n^2+2k(|m|+|n|)}{4}}\varepsilon_l(m_\sharp)\langle\gamma\mu^1\rangle',\quad \langle\gamma\mu^1\rangle' = \tag{2.8.15}$$

$$=\sum_{j=n+1}^{2n+1} q^{\frac{(j-k)^2}{4}}\frac{1-q^{j+k}}{1-q^k}\prod_{i=1}^{j}\frac{1-q^{i+2k-1}}{1-q^i} = q^{1/16}\prod_{i=1}^{n}(1-q^{1/2-i}). \tag{2.8.16}$$

❑

Comment. In the simplest case $k=-1/2$, the module V^+_{2n+1} becomes one-dimensional. Formula (2.8.16) is then a trivial identity. Nevertheless, it is directly related to the A_1–case of the Dedekind–Macdonald η–type identity [M1]. The latter is closely connected with the Kac–Moody algebras [Ka]. The case of arbitrary n seems beyond the Kac–Moody theory. It is definitely more reasonable to discuss the connection with [M1, Ka] for arbitrary root systems, because there are many *incidental* coincidences in the rank one case. See [C28].

❑

Now let us consider the **additional series**, namely, the modules V_{2n} for $t=-q^{-n/2}, n\in\mathbb{N}$. The analysis is straightforward, so we will simply formulate the counterpart of the previous theorem in this case.

We set

$$q=e^a,\ k=-n/2+\pi i a \text{ for } a>0,\ n\in\mathbb{N}. \tag{2.8.17}$$

The formulas below are algebraic identities in terms of q, t and their fractional powers. This special setting is convenient for the continued use of the "exponential" notation.

We use the standard $m_\sharp$ and the functional realization F_{2n} from Theorem 2.8.2 in the set:

$$\bowtie'_{-} \stackrel{\text{def}}{=} \Big\{\frac{1-n-k}{2}, \ldots, -\frac{k}{2}, \frac{k+1}{2}, \ldots, \frac{n+k}{2}\Big\}. \tag{2.8.18}$$

The Gaussian will be $\gamma(m_\sharp) = q^{m_\sharp^2}$. For instance,

$$\gamma(m_\sharp) = \exp\Big(\frac{(m-n/2)^2}{4a} + \frac{(m-n/2)\pi i}{2} - \frac{\pi^2 a}{4}\Big) \ \text{ for } \ m > 0.$$

The ε_m are the spherical polynomials, normalized by $\varepsilon_m(0_\sharp) = 1$; $\mu^1 = \mu/\mu(0_\sharp)$.

Theorem 2.8.4. *(Additional series) Provided (2.8.17), for $l_\sharp, m_\sharp \in \bowtie'_{-}$,*

$$\langle \varepsilon_l \overline{\varepsilon_m} \gamma \mu^1 \rangle'_{-} = q^{-\frac{m^2+n^2+2k(|m|+|n|)}{4}} \varepsilon_l(m_\sharp) \langle \gamma \mu^1 \rangle'_{-},$$

$$\langle \gamma \mu^1 \rangle'_{-} = q^{\frac{(j-k)^2}{4}} + \sum_{j=1}^{n} q^{\frac{(j-k)^2}{4}} \frac{1-q^{j+k}}{1-q^k} \prod_{i=1}^{j} \frac{1-q^{i+2k-1}}{1-q^i},$$

where the Gaussian sum is given by the formulas:

$$\langle \gamma \mu^1 \rangle'_{-} = e^{-\frac{\pi^2 a}{4}} \prod_{i=0}^{l-1} (1 + q^{-i}) \ \ \textit{for} \ \ m = 2l, \ l \in \mathbb{N}, \ \ \textit{and} \tag{2.8.19}$$

$$\langle \gamma \mu^1 \rangle'_{-} = \sqrt{2}\, e^{\frac{1}{16a} - \frac{\pi^2 a}{4}} \prod_{i=0}^{l-1} (1 + q^{-i-1/2}), \ m = 2l+1, l \in \mathbb{Z}_+. \tag{2.8.20}$$

❑

2.8.4 Roots of unity q, generic k

Let us study the case that is, in a sense, opposite to the previous one. We assume that $q^{1/2}$ is a primitive $2N$–th root of unity for $N \geq 1$ and consider generic k. In this subsection, $k \notin \mathbb{Z}/2$. We will continue using the symbol $\mathbb{C}$ for the field of constants, although it now can be made $\mathbb{Q}(q^{1/4}, t^{1/2})$. As above, by generic, we mean that *all fractional powers of q must be defined in terms of the "highest" primitive root of unity.*

Theorem 2.8.5. *Let $k \notin \mathbb{Z}/2$.*

(i) The polynomials $e_m(x)$, $\varepsilon_m(x)$ are well defined and constitute a basis of $\mathcal{P} = \mathbb{C}[X^{\pm 1}]$. In particular, $e_{-N} = X^N + X^{-N}$, $e_{-2N} = X^{2N} + X^{-2N} + 2 = e_{-N}^2$. The form $\langle \cdot, \cdot \rangle_0$ is also well defined:

$$\langle \varepsilon_l(x), \varepsilon_m(x) \rangle_0 = \delta_{lm} (\mu^1(m_\sharp))^{-1},$$

where μ^1 is given by formula (2.8.13).

(ii) The vectors $\varepsilon_m(x)$ $(m \le -N)$ and the vectors $\varepsilon_m(x)$ $(m \ge N+1)$ form a basis of the radical Rad_0 of the pairing $\langle \cdot, \cdot \rangle_0$ on the space $\mathcal{P}$. The $\mathcal{H}$–module $V_{2N} \overset{\mathbf{def}}{=\!=} \mathcal{P}/\operatorname{Rad}_0 = \oplus_{N \ge n \ge -N+1} \mathbb{C}\varepsilon_m(x)$ is irreducible of dimension $2N$ with the simple Y–spectrum:

$$\Big\{\frac{k+N-1}{2}, \ldots, \frac{k}{2}, -\frac{k+1}{2}, \ldots, -\frac{k+N}{2}\Big\}.$$

As an ideal, $Rad_0 = (e_{-N})$, so $V_{2N} = \mathcal{P}/(X^N + X^{-N})$.

(iii) The polynomials $\varepsilon_l - \varepsilon_m$ for $m_\sharp = l_\sharp \bmod N$ linearly generate the radical Rad of the pairing $\{f, g\} = f(Y^{-1})(g)(-k/2)$ from Theorem 2.5.10. The quotient $\mathcal{P}/Rad$ is irreducible and isomorphic to the $\mathcal{H}$–module Funct $(\bowtie_N)$ for the set $\bowtie_N \overset{\mathbf{def}}{=\!=} \mathbb{Z}_\sharp mod N$ of cardinality $4N$, under the action from Theorem 2.7.2. As an ideal,

$$Rad = (X^{2N} + X^{-2N} - t^N - t^{-N}).$$

(iv) Irreducible quotients of $\mathcal{P}$ are either

$$V^C \overset{\mathbf{def}}{=\!=} \mathcal{P}/(X^{2N} + X^{-2N} + C) \ \ for \ \ C \ne 2$$

or $V_{2N} = \mathcal{P}/(X^N + X^{-N})$, which is a quotient of V^2 by the submodule isomorphic to the image of V_{2N} under ς_y. The dimension of V^C is $4N$ and it is Y–semisimple with simple weights constituting the set $-\bowtie_N$. It is also X–semisimple (with simple weights) unless $C = -2$. The module $\mathcal{P}/Rad$ from (iii) equals V^{-C_t} as $C_t = t^N + t^{-N} = q^{kN} + q^{-kN}$. Here $C_t \ne \pm 2$, since $k \notin \mathbb{Z}/2$ by the asumption.

Proof. (i) The existence of the series μ^0, the e–polynomials, and the ε–polynomials readily follows from Theorem 2.2.5, Corollary 2.6.1, and (2.8.8). To check this, one can also involve the Pieri rules from Corollary 2.6.2 or the chain (2.8.6) directly. Let us calculate e_{-N}.

We get $Y(e_{-N}) = q^{N/2+k/2} e_{-N} = -t^{1/2} e_{-N}$, where $q^{N/2} = -1$ because $q^{1/2}$ is a primitive root of unity. Then $e_{-N} = t^{1/2}(T + t^{-1/2}) e_N$, thanks to (2.8.6), so $T(e_{-N}) = t^{1/2} e_{-N}$. The latter immediately gives that e_{-N} is s–invariant (i.e., even) and that

$$\pi(e_{-N}) = YT^{-1}(e_{-N}) = -e_{-N}, \text{ where } \pi(f(x)) = f(1/2 - x);$$

so $e_{-N}(x+1/2) = -e_{-N}$. Then, e_{-N} can contain only the monomials X^N and X^{-N} with some coefficients. Combining this with the s–invariance of e_{-N}, we obtain the formula for e_{-N}. Note the relation $e_{-N}(-k/2) = t^{N/2} + t^{-N/2}$, that matches the evaluation formula (2.6.4).

The calculation of e_{-2N} is similar. It is s–invariant and equals $X^{2N} + X^{-2N} + C$ for a constant C which must be $C = 2$ due to (2.6.4).

(ii) The norm formula (2.8.9) gives a desired description of Rad_0 as a linear space. Since the radical is an ideal, $Rad_0 = (e_{-N})$. As an immediate application, we obtain that (e_{-N}) is an $\mathcal{H}$–submodule.

Note that it is not difficult to check directly the $\mathcal{H}$–invariance of (e_{-N}) using the Pieri formula for Xe_{-N} together with the above formulas for the action of Y, T on e_{-N}.

(iii) The calculation of the radical Rad of the form $\{\,,\,\}$ is similar to that for Rad from the previous theorem. The module $\mathcal{P}/Rad$ (of dimension $4N$) has to be irreducible because of the following argument. If it contains a nontrivial submodule W, then, using the corresponding chain of invertible intertwiners, we can find there a Y–eigenvector either of weight $k/2$ or of weight $N/2 + k/2$, i.e., either 1 or e_{-N}. The former is impossible. The latter is impossible too because the generator $X^{2N} + X^{-2N} - t^N - t^{-N}$ of Rad is not divisible by $e_{-N} = X^N + X^{-N}$:

$$e_{-N}(k/2) = q^{kN/2} + q^{-kN/2} = q^{-kN/2}(q^{kN} + 1) = 0 \Leftrightarrow k \in 1/2 + \mathbb{Z}.$$

(iv) The element $X^{2N} + X^{-2N}$ is central in $\mathcal{H}$. Indeed, it commutes with π and with

$$T_i + \frac{t^{\frac{1}{2}} - t^{-\frac{1}{2}}}{X_i^2 - 1} \text{ for } i = 0, 1,\ T_1 = T,\ X_1 = X,\ T_0 = \pi T_1 \pi,\ X_0 = \pi(X),$$

generating a proper localization of $\mathcal{H}$. Therefore it is central in $\mathcal{H}$ without the localization. Thus V^C is an $\mathcal{H}$–module indeed.

Using the chain of intertwiners (see (iii)), it is irreducible if and only if $X^{2N} + X^{-2N} + C$ is divisible by $X^N + X^{-N}$; so it holds for $C = 2$ only. Otherwise the Y–spectrum of V^C is simple. Concerning the X–spectrum, the polynomial $X^{2N} + X^{-2N} + C$ has simple roots unless $C = \pm 2$. The action of X is not semisimple in V^{-2} (use the affine Hecke algebra $\langle T, X\rangle$). ❑

Comment. The form $\langle\,,\,\rangle$ making the generators X, Y, T, q, t unitary and inducing $\star$ on $\mathcal{H}$ cannot be introduced on V^C for $C \neq 2$. It follows from formula (6.26) (before Proposition 6.3) in [C28]. The form $\{\,,\,\}$ inducing ϕ exists only in $\mathcal{P}/Rad$ from (iii). Note that if V_{2N} had such a form, then $\star\cdot\phi = \epsilon$ would be an inner automorphism of V_{2N} and result in $\mathrm{Spec}_X = -\mathrm{Spec}_Y$, which is impossible.

Conversely, the module $\mathcal{P}/Rad$ cannot have a $\star$–invariant form, because it would lead to $C = 2$.

Similarly, τ_- acts in $\mathcal{P}$, but does not act in $\mathcal{P}/Rad$ for generic k. Technically, it is because τ_- multiplies $e_{-2N} = X^{2N} + X^{-2N} + 2$ by the constant $q^{-Nk} \neq 1$. See the following corollary. ❑

Corollary 2.8.6. *Let q be as above, but k is not supposed generic.*

(i) Assuming that e_n is well defined,

$$\tau_-(e_n) \;=\; q^{-n^2/4-|n|k/2}\, e_n \;\; \text{for} \;\; n \in \mathbb{Z},$$

where the canonical action τ_- in $\mathcal{P}$ is used.

(ii) The module $\tau_-(V^C)$ equals $V^{\tilde{C}}$ for $\tilde{C} - 2 = (C-2)q^{-Nk}$. Therefore it is τ_-–invariant if and only if either $C = 2$ (and k is arbitrary) or $q^{Nk} = 1$. The module V^{-C_t} is τ_-–invariant for $C_t = t^N + t^{-N}$ if and only if $q^{2kN} = 1$, that is, $2k \in \mathbb{Z}$.

(iii) If the radical Rad of $\{\,,\,\}$ is τ_-–invariant, then the projective $PSL_2(\mathbb{Z})$ generated by $\{\tau_\pm\}$ acts in $\mathcal{P}/Rad$ by inner automorphisms.

Proof. The first claim is straightforward. The action of τ_- in $\mathcal{P}$ exists due to the definition of the polynomial representation. The natural normalization is $\tau_-(1) = 1$. Using that $\tau_-(e_{-2N}) = q^{-Nk}e_{-2N}$, we arrive at (ii).

If τ_- preserves *Rad*, then it can be restricted to $\mathcal{P}/Rad$. We introduce the action of τ_+ from the relation $\{\tau_+(v), w\} = \{v, \tau_-(w)\}$, that is necessary and sufficient to match the relation $\phi\tau_-\phi = \tau_+$. See Proposition 2.7.5. ❑

2.9 Classification, Verlinde algebras

In this section we continue to assume that $q^{1/2}$ is a primitive $2N$–th root of unity, $N \geq 1$. However, now k may be arbitrary (possibly special). Let us begin with some preliminary technical remarks.

The next theorem is stated over $\mathbb{C}$ for the sake of simplicty; the actual field of constants is $\mathbb{Q}(q^{1/4}, t^{1/2})$.

Recall that the Y–weights λ are defined as follows: $Yv = q^{\lambda}v$, $v \neq 0$.

The classification below gives the list of all possible pairs $\{\lambda, k\}$. In the process of proving the next theorem, we give a complete description of the defining parameters of the corresponding irreducible representations.

Note that consideration of the *regular* case can be simplified using the calculation of the center of DAHA. It makes the parametrization $X \leftrightarrow Y$–symmetric. Using the center makes sense for the *special* cases too, although the technique of intertwiners seems more convenient.

There is a variant of the general theory for odd N when one picks $q^{1/2}$ in primitive N–th roots of unity. This variant will not be considered in this section, not will we consider the (equivalent) case of the little DAHA generated by $\langle X^2, T, Y^2\rangle \in \mathcal{H}$ (see below).

2.9.1 The classification list

Definition 2.9.1. *(i) A number $\lambda \in \mathbb{C}$ is called* **regular** *if the orbit $O_\lambda \stackrel{\text{def}}{=\!=} \pm\lambda + \mathbb{Z}/2 \mod N$ is simple, i.e., contains $4N$ elements. Equivalently, $2\lambda \notin \mathbb{Z}/2 \mod N$. Otherwise it is called* **half-singular** *as $2\lambda \in 1/2 + \mathbb{Z} \mod N$, and* **singular** *as $2\lambda \in \mathbb{Z} \mod N$.*

*(ii) An irreducible $\mathcal{H}$-module V is called a Y-***principal** *module if its dimension is $4N$ and Y has a simple spectrum in it. Otherwise it is called a Y-***special** *module.*

Note that the orbit O_λ can consist of $4N$ or $2N$ elements. In either case,

$$
(-q^{-2N\lambda}) \prod_{j=0}^{2N-1} (Y - q^{\lambda+j/2})(Y^{-1} - q^{\lambda+j/2})
$$
$$
= Y^{2N} + Y^{-2N} - q^{2N\lambda} - q^{-2N\lambda} \tag{2.9.1}
$$

is a *central* element of $\mathcal{H}$. The Y-roots in this orbi-product are simple if $|O_\lambda| = 4N$ and are of multiplicity two otherwise. This fact is not needed in the proof of the following theorem, but readily gives that Y-*cyclic* $\mathcal{H}$-modules are direct sums of Jordan Y-blocks of dimension no greater than two.

We will use the automorphisms ι and ς_y from (2.8.1,2.8.2) in the following theorem.

Theorem 2.9.2. *Let V be an irreducible finite dimensional $\mathcal{H}$-module with a Y-eigenvector of weight λ. The following list exhausts all possible pairs $\{\lambda, k\}$.*

(A) Let λ be regular. Then V is Y-principal if $k \notin \pm 2\lambda + \mathbb{Z}$. Otherwise, $k \in \pm 2\lambda + \mathbb{Z}$, so $k \notin \mathbb{Z}/2$ and, up to ι, ς_y and their product, V is one of the irreducible quotients of $\mathcal{P}$ described in Theorem 2.8.5. These quotients are Y-principal modules unless $V = \mathcal{P}/(X^N + X^{-N})$.

(B) Let λ be half-singular, i.e., $2\lambda \in 1/2+\mathbb{Z}$, and either $k \notin \mathbb{Z}/2$ or $k \in \mathbb{Z}$. Then V is Y-semisimple. The dimension can be $2N$ (then the Y-spectrum is simple) or $4N$ (then all weights are of multiplicity 2).

(C) Let λ be singular, i.e., $2\lambda \in \mathbb{Z}$, and $k \notin \mathbb{Z}$. Then the dimension of V is $4N$ and it is not Y-semisimple, unless $t = 1$.

(D) Let λ be either half-singular under $k \in 1/2 + \mathbb{Z}$ or λ singular under $k \in \mathbb{Z}$. Then V is a quotient of $\mathcal{P}$ or its image under the automorphisms $\iota, \varsigma_y, \iota\varsigma_y$.

Proof. Given a Y-eigenvector $v = v_0$ of weight $\lambda = \lambda_0$, we will apply the chain of intertwiners from (2.8.6):

$$
v_0 \xrightarrow{A_0} v_1 \xrightarrow{B_1} \cdots \longrightarrow v_{-m+1} \xrightarrow{A_{-m+1}} v_m \xrightarrow{B_m} \cdots v_{2N}, \tag{2.9.2}
$$

to construct v_{1-m}, v_m for $1 \le m \le 2N$. Recall that

$$A_{-m+1} = q^{(m-1)/2} X\pi, \quad B_m = t^{1/2}(T + \frac{t^{1/2} - t^{-1/2}}{q^{-2\lambda_m} - 1}), \quad \text{and}$$
$$\lambda_1 = -\frac{1}{2} - \lambda, \lambda_{-1} = \frac{1}{2}\lambda, \ldots, \lambda_m = -\frac{m}{2} - \lambda = \lambda_{-m} \quad \text{for} \quad m > 0.$$

We can also start here with the space $V_0 = V(\lambda) \subset V$ of all λ–eigenvectors. Then the composition

$$K \stackrel{\text{def}}{=\!=} B_{2N} A_{1-2N} \ldots B_2 A_{-1} B_1 A_0$$

preserves $V_0 = V(\lambda)$ if K is well defined. Similarly, the *chain* can be applied to the space $\widehat{V}_0 = \widehat{V}(\lambda)$ of the *generalized eigenvectors*, satisfying $(Y - \lambda)^M(u) = 0$ for sufficiently large M. In this case, however, we need to go back to the formula

$$B_m = t^{1/2}(T + \frac{t^{1/2} - t^{-1/2}}{Y^{-2} - 1}).$$

(A) All intertwiners exist and are invertible if $k/2 \notin O_\lambda$. We can set $K(v_0) = q^\kappa v_0$ for arbitrary $\kappa \in \mathbb{C}$ and uniquely extend it to the action of $\mathcal{H\!H}$ on $\oplus_{m=1-2N}^{2N} \mathbb{C} v_m$. It is a simple exercise. This representation is principal. If $k/2$ belongs to the orbit O_λ, then we use the intertwiners to find a Y–eigenvector v_0 of weight $\pm k/2 \mod N/2$ such that the B–intertwiner is zero on it. Indeed, if $B(v_0) \neq 0$, then we take it as v_0. Using ι and ς_y, we can assume that this weight is $k/2$. This means that V is a quotient of $\mathcal{P}$. Now we can apply Theorem 2.8.5.

(B) All intertwiners exist and are invertible in this case. We can make $\lambda = -1/4$. Then $S = q^{-1/4} X\pi$ creates the weight $-1/2 - \lambda = -1/2$ and therefore preserves the space V_0 of all eigenvectors of weight $-1/4$. The operators K, S act in V_0 and satisfy the relation $S^2 = 1, SKS = K^{-1}$. The irreduciblity of V_0 as a $\{S, K\}$–module is necessary and sufficient for the irreduciblity of V. Similar to (A), we can uniquely extend an arbitrary action of S, K on V_0 to a structure of an $\mathcal{H\!H}$–module on V. Thus V_0 is either one-dimensional ($K = \pm 1$, $S = \pm 1$) or two-dimensional. We obtain the desired result.

(C) The intertwiners are invertible provided we have their existence. We can find v_0 with $\lambda = 0$. Then $\mathcal{H}_y v_0$ is a two-dimensional irreducible representation of the affine Hecke algebra $\mathcal{H}_Y = \langle T, Y \rangle$, which coincides with

$$\widehat{V}_0 = \widehat{V}(0) = \{u \mid (Y - 1)^2(u) = 0\}.$$

The action of Y here is not semisimple. We take invertible

$$L \stackrel{\text{def}}{=\!=} A_{1-N} \ldots B_2 A_{-1} B_1 A_0$$

instead of K. It sends $\lambda = 0$ to $-N/2$ and $\widehat{V}_0$ to $\widehat{V}_N$. Recall that $q^{N/2} = -1$, so the intertwiner B_N acting in $\widehat{V}_N$ is singular.

We obtain that $(Y+1)^2 = 0$ on $\widehat{V}_N$ and T preserves this space. If the action of the operator T in this space is known, then this information is sufficient to reconstruct uniquely and completely the $\mathcal{H}$–action in V. Indeed, we know how Y acts on $\widehat{V}_N$ and know the action of $X\pi$, that sends $\widehat{V}_N$ back to $\widehat{V}_{1-N}$.

The T–action depends on one parameter, because we can conjugate T by the matrices from the centralizer of Y (the Y–action is known). Any choice of this parameter results in the corresponding action of $\mathcal{H}$ on V. Here we need to exclude the case $k = 0$, which requires a somewhat different consideration.

(D) Similar to the second part of (A), we can find a Y–eigenvector v_0 of weight $\pm k/2 \mod N/2$ such that the B–intertwiner is zero on it. Using ι and ς_y if necessary, we obtain a surjection $\mathcal{P} \to V$. ❑

Actually, we have proved more than what was stated in the theorem. In cases (A–C), we gave the following complete description of *all* parameters that determine the irreducible representations:

(1) Given a Y–weight λ, we need to know either the action of K for (A) or K and S for (B) in the corresponding Y–eigenspace.

(2) In case (C), we need to know the action of T in $\{v \mid (Y+1)^2(v) = 0\}$ as $k \neq 0$; this action can be arbitrary compatible with the action of Y.

(3) An explicit description of the irreducible modules in case (D) requires an extension of Theorem 2.8.5 to the integral and half-integral k, which we are going to discuss now.

2.9.2 Special spherical representations

First of all, the substitution

$$T \mapsto -T,\ t^{1/2} \mapsto -t^{1/2}$$

identifies the polynomial representations for $t^{1/2}$ and $-t^{1/2}$. It does not act on monomials X^m, so the nonsymmetric polynomials e_m remain unchanged, as long as they are well defined. The spherical polynomials ε_m for even m do not change either; $\varepsilon_m \mapsto -\varepsilon_m$ for odd m.

Thus it is sufficient to decompose $\mathcal{P}$ upon the transformation $k \mapsto k + N$, and we can assume that $N/2 \leq k < N/2$. We will also use the outer involutions ι, ς of $\mathcal{H}$ from (2.8.1) and (2.8.2).

Theorem 2.9.3. *(i) Let $k \in \mathbb{Z}/2 \mod 2N$ and $-N/2 \leq k < N/2$. Then the representations $V^C = \mathcal{P}/(X^{2N} + X^{-2N} + C)$ of dimension $4N$ from Theorem 2.8.5 remain irreducible for $C \neq \pm 2$. However, now V^C for such C becomes Y–non-semisimple for integral $k \neq 0$ and Y–semisimple with the two-fold spectrum otherwise.*

Moreover, V^{-2} is irreducible for the half-integral k, as well as the quotient $V_{2N} = \mathcal{P}/(X^N + X^{-N})$ of V^2 for the integral k. Recall that the kernel of the map $V^2 \to V_{2N}$ is isomorphic to the image of V_{2N} under ς_y sending Y to $-Y$.

(ii) If $0 < 2k < N$, then either V^{-2} for $k \in \mathbb{Z}$ or V_{2N} for $k \in 1/2+\mathbb{Z}$ has a unique irreducible nonzero quotient, that is, $V_{2n} = \mathcal{P}/(\varepsilon_{-n})$ of dimension $2n$ for $n = N - 2k$. Its Y-spectrum is simple. The eigenvectors are the images of ε_m for $-n+1 \le m \le n$, which are all well defined. This module coincides with $\mathcal{P}/Rad$ for the radical of the pairing $\{\,,\,\}$ and

$$T(e_{-n}) \;=\; -t^{-1/2}e_{-n} \;=\; Y(e_{-n}). \tag{2.9.3}$$

These statements can be extended to $k = 0$ when $t = 1$ if we take $\varepsilon_{-n} = X^N - X^{-N}$ in place of ε_{-n}.

(iii) Let $k = -1/2 - n$ for integral $0 \le n < (N-1)/2$. Then the module $V_{2N} = \mathcal{P}/(X^N + X^{-N})$ has two non-isomorphic irreducible Y-semisimple quotients, namely, the representations from (2.8.5):

$$V^{\pm}_{2n+1} = \mathcal{P}/(\varepsilon_{n+1} \pm \varepsilon_{-n}),\ \ dim(V^{\pm}_{2n+1}) = |2k| = 2n+1,\ \ V^{-}_{2n+1} = \varsigma_x(V^{+}_{2n+1}).$$

Here the ε_m are well defined when $-2n \le m \le 2n+1$. The binomials $X^{n+1} \pm X^{-n}$ must be taken in place of $\varepsilon_{n+1} \pm \varepsilon_{-n}$ to extend these statements to the boundary case $N = 2n+1$.

The kernel (ε_{-2n-1}) of the map from V_{2N} to the direct sum of V^{+}_{2n+1} and V^{-}_{2n+1} has dimension $2N - |4k|$ and is isomorphic to $V_{2N-4|k|}$ from (ii) under the involution ι sending k to $-k$, $T \mapsto -T$. The vector $e = \varepsilon_{-2n-1}$ satisfies $T(e) \;=\; -t^{-1/2}e,\ Y(e) = t^{-1/2}e$.

(iv) In the last case, that is, $k \in -1-\mathbb{Z}_+$ and $-N/2 \le k < 0$, the module V^{-2} has a unique irreducible nonzero quotient $V_{2N+4|k|}$ (of dimension $2N + 4|k|$). It equals $\mathcal{P}/Rad$ for the radical of the pairing $\{\,,\,\}$, and is isomorphic to $\mathcal{P}/(e)$ for $e = \varepsilon_N - \varepsilon_{-N-2|k|}$, satisfying relations (2.9.3), where the polynomials $\varepsilon_N, \varepsilon_{-N-2|k|}$ are well defined.

This module is also isomorphic to the kernel of the map $V^{-2} \to V_{2N-4|k|}$ from (ii) upon applying the outer involution $\iota\varsigma_y$.

(v) The polynomials ε_m for $\{m = -2|k|, 2|k|+1, -2|k|-1, \ldots, -N+1, N\}$ exist and their images generate the Y-semisimple component of $V_{2N+4|k|}$ (of dimension $2N - 4|k|$). The corresponding Y-weights are

$$\{\lambda = \frac{|k|}{2}, \frac{-|k|-1}{2}, \frac{|k|+1}{2}, \ldots \frac{N-1-|k|}{2}, \frac{|k|-N}{2}\}.$$

The Y-non-semisimple component of $V_{2N+4|k|}$ is the direct sum of $4|k|$ Jordan two-blocks of the total dimension $8|k|$ corresponding to the remaining $4|k|$ weights in the orbit

$$O_\lambda \;=\; \{\lambda = 0,\, -1/2, 1/2, -1, \ldots,\, (N-1)/2,\, -N/2\}.$$

There are two sequences of such non-semisimple λ:

$$\{\frac{-|k|}{2}, \frac{|k|-1}{2}, \ldots, \frac{-1}{2}, \frac{0}{2}\},\ \{\frac{N-k}{2}, \frac{|k|-N-1}{2}, \ldots, \frac{N-1}{2}, \frac{-N}{2}\}.$$

Proof. (i) We use that the X–spectrum is simple in V^C and the intertwiners are always invertible. This readily gives the irreducibility. The Jordan Y–blocks will appear if and only if the Y–spectrum contains 0 (with a reservation about the degenerate case $k = 0$, which must be considered separately).

The space $\widehat{V}(0) = \{v \mid (Y-1)^2 v = 0\}$ is an $\mathcal{H}_Y$–module that is not Y–semisimple. So are all other generalized Y–eigenspaces. This happens precisely for integral $k \neq 0$. This argument has been used quite a few times. We obtain (i).

Concerning the existence of the e–polynomials, the invertibility of the intertwiners is sufficient, but not necessary. The following lemma gives the general construction. In fact, it has already been used before. It will be applied a couple of times, especially in the parts (iv,v). It uses formula (2.9.1).

Let

$$\widehat{V}_0 = \mathbb{Q}_{q,t},\ \widehat{V}_1 = \mathbb{Q}_{q,t}X, \ldots, \widehat{V}_{-m} = B_m \widehat{V}_m,\ \widehat{V}_{m+1} = A_{-m}\widehat{V}_{-m}, \ldots, \tag{2.9.4}$$

where $m > 0$, $A_{-m} = q^{m/2}X\pi$, and B_m denotes the restriction of the intertwiner $t^{1/2}(T + (t^{1/2} - t^{-1/2})/(Y^{-2} - 1))$ to $\widehat{V}_m$, provided that $q^{2\lambda_m} \neq 1$ for $\lambda_m = -m/2 - k/2$. If $q^{2\lambda_m} = 1$ and the denominator of B_m becomes infinity, then we set

$$B_m = t^{1/2}T,\ \ \widehat{V}_{-m} = \widehat{V}_m + T\widehat{V}_m.$$

Lemma 2.9.4. *(i) The space $\widehat{V}_{\pm m}$ is one-dimensional or two-dimensional. In the latter case, it is the Jordan 2–block satisfying $(Y - q^{\pm\lambda_m})^2\widehat{V}_{\pm m} = \{0\}$. If dim$\widehat{V}_{-m} = 1$ then dim$\widehat{V}_{m+1} = 1$ and the generators are*

$$e_{-m} = B_{m-1}\cdots B_1 A_0(1),\ \ e_{m+1} = A_{-m}E_{-m}.$$

If dim$\widehat{V}_{-m} = 2$, then dim$\widehat{V}_{m+1} = 2$ and the e–polynomials e_{-m}, e_{m+1} do not exist, although these spaces contain the e–polynomials of smaller degree.

(ii) Let us assume that either $q^{2\lambda_m} = t$ or $q^{2\lambda_m} = t^{-1}$. Then dim$\widehat{V}_{-m} = 1$ and this space is generated by e_{-m}. If $\widehat{V}_m$ is one-dimensional then, respectively, $(T + t^{-1/2})e_{-m} = 0$ or $(T - t^{1/2})e_{-m} = 0$. If dim$\widehat{V}_m = 2$, then, respectively,

$$(T + t^{-1/2})e_{-m} \ \ \text{or}\ \ (T - t^{1/2})e_{-m}$$

is nonzero and proportional to the (unique) e–polynomial which is contained in the space $\widehat{V}_m$. ❑

If $V = \mathcal{P}/J$ is a proper (neither $\{0\}$ nor $\mathcal{P}$) irreducible $\mathcal{H}$–quotient, then J is an ideal. We may set $J = (e)$ for either

$$(\mathcal{A}):\ e = X^l + \ldots + cX^{-l} \ \ \text{or}\ \ (\mathcal{B}):\ e = X^{l+1} + \ldots + cX^{-l},$$

where $2N > l > 0, c \neq 0$, and, respectively,

$$(\mathcal{A}): \dim V = 2l \quad \text{or} \quad (\mathcal{B}): \dim V = 2l+1.$$

In either case, e is an eigenvector of Y. Indeed, $Y(e) \in J$ has the same type as e with the same degrees. A proper linear combination of e and $Y(e)$ would be lower than e, unless they are proportional. We set $e = e^0 + e^1$, where $e^\alpha(-X) = (-1)^\alpha e^\alpha(X)$. Note that this decomposition is always nontrivial in case $(\mathcal{B})$.

Since Y commutes with the ι_x sending $X \mapsto -X$, then $\{e^0, e^1\}$ are Y–eigenvectors of top X–degrees

$$(\mathcal{A}): \{l, m\} \text{ as } 0 < m < l \text{ for odd } l-m, \quad (\mathcal{B}): \{l+1, l\}.$$

The Y–weights of e^0, e^1 must coincide mod N. They can be readily calculated:

$$(\mathcal{A}): \{(l+k)/2, \pm(m+k)/2\}, \quad (\mathcal{B}): \{-(l+1+k)/2, (l+k)/2\}.$$

Here it suffices to know the leading term of e^α with respect to the ordering $\succ$: $X^{\pm m} \Rightarrow \mp(m+k)/2$ for $m > 0$.

The coincidence mod N immediately gives that only the plus–sign is possible in $(\mathcal{A})$. We arrive at the following relations:

$$(\mathcal{A}): k = -(l+m)/2 \mod N, \quad (\mathcal{B}): k = -1/2 - l \mod N.$$

In either case, $k \in 1/2 + \mathbb{Z}$ if the decomposition $e = e^0 + e^1$ is nontrivial.

We obtain that $(\mathcal{B})$ results in the claims from (iii) from the theorem. Indeed, $0 \le l < 2N$, and l is nothing but $n = k + 1/2$; the representations $V^{\pm}_{2n+1}$ exist and remain irreducible in this case.

We now need to examine $(\mathcal{A})$. We will impose this condition to the end of the proof.

The argument above, which ensured the Y–invariance of e, can be used for T as well. It preserves the type of e (note that it does not hold in case $(\mathcal{B})$). We obtain that $T(e) = \pm t^{\pm 1/2} e$. Therefore $e = e_{-l}$. Indeed, $e - e_{-l}$ must be proportional to e_m for positive $m \le l$. However, such small e_m are never eigenvectors of T, since the chain of intertwiners remains invertible in this range.

We have four subcases of $(\mathcal{A})$:

$$(\alpha)\ T(e) = t^{1/2} e = Y(e), \qquad (\beta)\ T(e) = t^{1/2} e = -Y(e),$$
$$(\gamma)\ T(e) = -t^{-1/2} e = Y(e), \quad (\delta)\ T(e) = -t^{-1/2} e = -Y(e).$$

Concerning (α, β), we obtain that $s(e) = e$, so e is even and the decomposition $e = e^0 + e^1$ is trivial. Then $(l+k)/2 = k/2 \mod N$ for (α) and

$(l+k)/2 = k/2 + N/2 \mod N$ for (β). The former equality results in $l = 0 \mod 2N$, which is impossible. As for (β):

$$\pi(e) = YT^{-1}(e) = -e \Rightarrow e(x+1/2) = -e(x) \Rightarrow e = X^N + X^{-N}.$$

Therefore (β) implies that the quotient $\mathcal{P}/(X^N+X^{-N})$ is an $\mathcal{H}$–module and is irreducible unless it is covered by (iii) of the theorem, which was already considered.

(γ) We obtain $(l+k)/2 = N/2 - k/2 \mod N$ and $l = N - 2k \mod 2N$. Recall that $e = e_{-l}$. If $2k > 0$, we arrive exactly at (ii) of the theorem. In this case, $l = n$. All polynomials e_m are well defined and have nonzero $e_m(-k/2)$ in this case. Otherwise $2k < 0$.

(γ_1) If $2k < 0$, then k cannot be a half-integer. Indeed, $e_{-l}(-k/2) = 0$ in this case (use the evaluation formula), as well as for the generator of the ideal for V^+_{2n+1} in case (b). We see that these two ideals together do not generate the whole $\mathcal{P}$. Since V is irreducible, it has to coincide with V^+_{2n+1}, which is impossible because we assumed that e is in the form (a). Therefore k must be an integer.

(γ_2) We shall consider the integers $-N/2 \le k < 0$.

Lemma 2.9.5. *(a) For integral k such that $-N/2 \le k < 0$, the quotient $V_{2N+4|k|} \stackrel{\mathrm{def}}{=\!=} \mathcal{V}/Rad$ by the radical of the pairing $\{\,,\,\}$ is an irreducible $\mathcal{H}$–module of dimension $2N+4|k|$.*

(b) The polynomials e_m exist and $e_m(-k/2) \neq 0$, i.e., ε_m exist, for the sequences:

$$\begin{aligned} m = &\ \{0, 1, -1, \ldots, -|k|+1, |k|\}, \\ m = &\ \{-2|k|, 2|k|+1, \ldots, -N+1, N\}, \\ m = &\ \{-N, N+1, \ldots, -N-|k|+1, N+|k|\}, \end{aligned}$$

respectively, with $2|k|$, $2(N-2|k|)$, and $2|k|$ elements. They do not exist for the following $2|k|+2|k|$ indices:

$$\begin{aligned} m = &\ \{-|k|, |k|+1, \ldots, -2|k|+1, 2|k|\}, \\ m = &\ \{-N-|k|, N+|k|+1, \ldots, -N-2|k|+1, N+2|k|\}. \end{aligned}$$

(c) The Y–semisimple component of $V_{2N+4|k|}$ of dimension $2N-4|k|$ is linearly generated by e_m for

$$\{m = -2|k|,\ 2|k|+1,\ -2|k|-1,\ \ldots,\ -N+1,\ N\}.$$

The corresponding Y–weights are

$$\{\lambda = \frac{|k|}{2}, \frac{-|k|-1}{2}, \frac{|k|+1}{2}, \ldots, \frac{N-1-|k|}{2}, \frac{|k|-N}{2}\}.$$

(d) The rest of $V_{2N+4|k|}$ is the direct sum of $4|k|$ Jordan two-blocks of the total dimension $8|k|$. There are two series of the corresponding (multiple) weights λ:

$$\{\frac{-|k|}{2}, \frac{|k|-1}{2}, \ldots, \frac{-1}{2}, \frac{0}{2}\},\ \{\frac{N-|k|}{2}, \frac{|k|-N-1}{2}, \ldots, \frac{N-1}{2}, \frac{-N}{2}\}.$$

Proof. We will use the chain of the spaces of generalized eigenvectors

$$\widehat{V}_0 = \mathbb{Q}_{q,t},\ \widehat{V}_1 = \mathbb{Q}_{q,t}X,\ \widehat{V}_{-1},\ \ldots,\ \widehat{V}_m,\ \ldots$$

from (2.9.4) and Lemma 2.9.4. Recall that $m > 0$. The following holds:
(0) the spaces $\widehat{V}_{\pm m}$ are all one-dimensional from 0 to $m = |k|$, i.e., in the sequence $V_0, \ldots, V_{-|k|+1}, V_{|k|}$;
(1) the intertwiner B_m becomes infinity at $m = |k|$, i.e., $B_{|k|} = t^{1/2}T$, and $\dim\widehat{V}_m = 2$ in the range $|k| < m \le 2|k|$;
(2) the intertwiner B_m kills $1 \in \widehat{V}_m$ at $m = 2|k|$, and, after that, $\dim\widehat{V}_m = 1$ for $2|k| < m \le N$;
(3) B_m is proportional to $(T + t^{-1/2})$ at $m = N$, also, $e_{-N} = X^N + X^{-N}$ and $\dim\widehat{V}_m = 1$ as $N < m \le N + |k|$;
(4) the intertwiner B_m becomes infinity again at $m = N + |k|$ and then $\dim\widehat{V}_m = 2$, provided that $N + |k| < m \le N + 2|k|$;
(5) B_m kills e_{-N} at $m = N + 2|k|$ and $B_m(\widehat{V}_m)$ is generated by $e_{-N-2|k|}$, that has the same Y–eigenvalue as e_N does.

Concerning step (5), the polynomials $\epsilon_{-N-2|k|}$ and ϵ_N both exist and the difference $e = \varepsilon_N - \varepsilon_{-N-2|k|}$ belongs to the radical *Rad*, i.e., becomes zero in $V_{2N+4|k|}$.

Note that $(T + t^{1/2})e = 0$, the fact which is important to know to continue the decomposition of $\mathcal{V}$ further. It follows along the same lines.

We see that step (5) is the first one that produces no new elements in $V_{2N+4|k|}$, namely:

$$B_{N+2|k|}(\widehat{V}_{N+2|k|}) \ = \ \mathbb{Q}_{q,t}e_N \ \text{ in } \ V_{2N+4|k|},$$

and we can stop at (5).

The lemma gives that between (2) and (3), the polynomials e_m exist and their images linearly generate the Y–semisimple part of V. It is equivalent to the inequalities $e_m(q^{-\rho_k}) \neq 0$ because these e_m have different Y–eigenvalues.

Apart from (2)–(3), there will be always Jordan two-blocks with respect to Y. Let us check this.

First, we obtain a two-dimensional irreducible representation of $\mathcal{H}_Y = \langle T, Y, \pi \rangle$ in the corresponding $\widehat{V}$–space at step (1). Then we apply the chain of invertible A, B–intertwiners to this space (the weights will go back) and eventually will obtain a two-dimensional $\widehat{V}$–space for the weight coinciding with the initial weight $\lambda = -|k|/2$. Note that the vector $e_0 = 1$ is *not* from the Y–semisimple component of $V_{2N+4|k|}$; it belongs to a Jordan two-block.

Second, the intertwiner from (2) makes the previously obtained $\widehat{V}$–space one-dimensional and make it Y–semisimple (the corresponding eigenvalue is simple in $V_{2N+4|k|}$). It will remain one-dimensional until (3). After step (3), we obtain the Jordan two-blocks. The steps (4)–(5) are parallel to (1)–(2).

The above consideration readily results in the irreducibility of the module $V_{2N+4|k|}$. Indeed, Lemma 2.9.4(ii) gives that if a submodule of $V_{2N+4|k|}$ contains at least one simple Y–eigenvector then it contains the image of 1 and the whole space. Step (5) guarantees that it is always the case, because we can obtain e_N from an *arbitrary* Y–eigenvector. ❑

We arrive at (iv,v).

Let us consider the last subcase.

(δ) Now $l = 2N-2k$. If $2k > 0$, then e_{-l} belongs to the ideal (e_{-n}). Indeed, its image in V_{2n} is zero since there are no eigenvectors of weight $-k/2$ in this module. This contradicts to the irreducibility. Negative k are impossible because $l < 2N$ by assumption. Thus this case is empty.

Concerning the "duality" claim involving ι from (iii), it holds because $e = e_{-2n-1}$, generating the kernel (e) of the map $\mathcal{P} \to V^+_{2n+1} \oplus V^-_{2n-1}$, satisfies the relation (δ): $T(e) = -t^{-1/2}e = -Y(e)$. Note here the reservation about the boundary value $2|k| = N$.

As for the duality (ii)$\leftrightarrow$(iv), the vector $e = e_{-n} = e_{-N-2k}$, generating the kernel of the map $\mathcal{P} \to V_{2N-4k}$, satisfies the relation (γ): $T(e) = -t^{-1/2}e = Y(e)$. The same relation holds for $e = \varepsilon_N - \varepsilon_{-N-2|k|}$ from (iv). Note here the reservation about the boundary value $k = 0$. ❑

2.9.3 Perfect representations

It is important to find out which irreducible modules are $PSL(2,\mathbb{Z})$–invariant. Using the classification, it is clear that the exceptional representations (ii)–(iv) from the theorem are invariant because they can be distinguished by the dimensions. It is also not difficult to describe $PSL(2,\mathbb{Z})$–invariant modules from scratch without using the classification. It is what we are going to do now.

We continue to assume that $q^{1/2}$ is a primitive $2N$–th root of unity and $t = q^k$. It readily results from the description of the Y–spectrum of the polynomial representation that all σ–invariant quotients of $\mathcal{P}$ must be through

$$V' = V^{-C_t} = \mathcal{P}/\left(X^{2N} + X^{-2N} - t^N - t^{-N}\right) = \mathcal{P}/Rad$$

from Theorem 2.8.5. Indeed, the invariance gives that the element $X^{2N} + X^{-2N}$ must act as its σ–image $Y^{2N} + Y^{-2N}$, that is $t^N + t^{-N}$ in $\mathcal{P}$. Recall that $X^{2N} + X^{-2N}$ is central in $\mathcal{H}$, so V' is obviously an $\mathcal{H}$–module.

The module V' is irreducible unless k is integral or half-integral. For generic t, it is not σ–invariant. Instead, the involution

$$X \leftrightarrow Y,\ T \leftrightarrow -T^{-1},\ t^{1/2} \mapsto t^{1/2},\ q^{1/2} \mapsto q^{-1/2}$$

is its inner automorphism. The corresponding map sends the image $1' \in V'$ of 1 to the image $\hat{e}'$ of the polynomial

$$\hat{e} = (X^{2N} + X^{-2N} - t^N - t^{-N})/(X^{-1} - t^{-1/2}),$$

satisfying $X(\hat{e}') = t^{1/2}\hat{e}',\ T(\hat{e}') = -t^{-1/2}\hat{e}'$.

Let us discuss the singular k in detail. The module V' becomes, respectively,

$$V^{-2} = \mathcal{P}/(X^{2N} + X^{-2N} - 2),\ V^2 = \mathcal{P}/(X^{2N} + X^{-2N} + 2),$$

as $k \in \mathbb{Z},\ k \in 1/2 + \mathbb{Z}$, and $t^N = q^{kN} = \pm 1$. Recall that the module V^2 is an extension of $V_{2N} \stackrel{\text{def}}{=\!=} \mathcal{P}/(X^N + X^{-N})$ by its ς_y–image $\varsigma_y(V_{2N})$.

Proposition 2.9.6. *Let $k \in \mathbb{Z}/2, |k| < N/2$. The notation is from Theorem 2.9.3. There are four exact sequences:*

$$0 \to \iota\varsigma_y(V_{2N+4k}) \to V^{-2} \to V_{2N-4k} \to 0 \ \ \textit{for}\ \ k \in \mathbb{Z}_+,$$
$$0 \to \iota(V_{2k}^+ \oplus V_{2k}^-) \to V_{2N} \to V_{2N-4k} \to 0 \ \ \textit{for}\ \ k \in 1/2 + \mathbb{Z}_+.$$

The arrows must be reversed for $k < 0$:

$$0 \to \iota\varsigma_y(V_{2N-4|k|}) \to V^{-2} \to V_{2N+4|k|} \to 0 \ \ \textit{for}\ \ k \in -1 - \mathbb{Z}_+,$$
$$0 \to \iota(V_{2N-4|k|}) \to V_{2N} \to V_{2|k|}^+ \oplus V_{2|k|}^- \to 0 \ \ \textit{for}\ \ k \in -1/2 - \mathbb{Z}_+.$$

In other cases, V^{-2} and V_{2N} are irreducible.

The modules $V_{2|k|}^{\pm}$, $V_{2N-4|k|}$, and $V_{2N+4|k|}$ are irreducible and projective $PSL(2,\mathbb{Z})$–invariant. The modules $V_{2N+4|k|}$ are Y–non-semisimple. The others are semisimple.

Proof. We use Theorem 2.9.3, the structure of the Y–spectrum in the polynomial representation, and that the dimensions of the Jordan Y–blocks are no greater than 2. Concerning the action of $\tau_\pm$ and σ, see Corollary 2.8.6. ❑

We obtain that, up to ι and ς, there are *three* different series of projective $PSL(2,\mathbb{Z})$–invariant irreducible spherical representations at roots of unity, namely,

V_{2N-4k} (integral $N/2 > k > 0$), $V_{2|k|}$ (half-integral $-N/2 < k < 0$), and $V_{2N+4|k|}$ (integral $-N/2 < k < 0$).

If $k = 1$, the subalgebra V_{2N-4}^{sym} of dimension $N-1$ is isomorphic to the usual **Verlinde algebra** of $\widehat{\mathfrak{sl}_2}$ of level N (central charge $= N-2$). The symmetric polynomials $p_m^{(1)}$ are the classical characters of finite dimensional representations of $\mathfrak{sl}_2$. In V_{2N-4}^{sym}, these characters are considered as functions at roots of unity.

Comment. (i) Recently a non-semisimple variant of the Verlinde algebra appeared [FHST] in connection with the fusion procedure for the $(1,p)$ Virasoro algebra, although the connection with the fusion is still not justified. Generally speaking, the fusion procedure for the Virasoro-type algebras and the so-called W–algebras can lead to non-semisimple Verlinde algebras. At least, there are no reasons to expect the existence of a positive hermitian inner product there like the Verlinde pairing for the conformal blocks, because the corresponding physics theories are expected *massless*. Surprisingly, the algebra from [FHST] is defined using the usual (*massive*) Verlinde algebra under certain degeneration. Presumably it coincides with $V_{2N+4|k|}^{sym}$ for $k = -1$ or is very close to it.

(ii) The latter module and its multidimensional generalizations are also expected to be connected with the important problem of describing the complete tensor category of the representations of the Lusztig quantum group at roots of unity. The Verlinde algebra, the symmetric part of $V_{2N-4|k|}$ for $k = 1$, describes the so-called *reduced category*, in a sense, corresponding to the Weyl chamber. The nonsemisimple modules of type $V_{2N+4|k|}$ are expected to appear in the so-called *case of the parallelogram.* ❑

Motivated by the Verlinde algebras [Ver] (see also [KL2]) we will require more structures.

Definition 2.9.7. *We say that a finite dimensional irreducible $\mathcal{H}$–module V is* **perfect** *or a* **nonsymmetric Verlinde algebra** *if the following conditions hold:*

(a) V is **spherical**, *i.e., there exists a surjection $\mathcal{P} \to V$;*

(b) it is $\{\tau_\pm, \epsilon\}$–invariant, which means that there are pullbacks of the $\mathcal{H}$–automorphisms $\tau_\pm$ and ϵ to V satisfying the relations

$$\tau_+\tau_-^{-1}\tau_+ = \tau_-^{-1}\tau_+\tau_-^{-1},\ \epsilon^2 = 1,\ \epsilon\tau_+ = \tau_-\epsilon;$$

*(c) it is also X–***pseudo-unitarity***: there exists a nondegenerate form $\langle \cdot, \cdot \rangle$ on the space V such that the corresponding anti-involution of $\mathcal{H}$ is $\star$ and $\langle e, e\rangle \neq 0$ for any X–eigenvector $e \in V$.*

Comment. Generally, a **quasi-perfect** module is defined as an $\mathcal{H}$–module equipped with a non-degenerate pairing inducing the anti-involution ϕ from (2.5.7). For instance, the polynomial representation is quasi-perfect if the radical of $\{\ ,\ \}$ is zero.

If the pairing is perfect, i.e., establishes an identification of the module and its dual as a vector space, and, moreover, there is a projective action of $PSL(2,\mathbb{Z})$, then the module is called **perfect.**

In this chapter, we consider only finite dimensional *spherical* perfect representations.

So far, the main considered examples of perfect and quasi-perfect modules are spherical, however, non-spherical representations are expected to appear when decomposing the polynomial representation (for special q, t) and in various applications.

In a sense, "perfect representations" in this chapter mean "could not be better." They have all properties of the irreducible modules of the Weyl algebras and are really perfect from the viewpoint of Fourier analysis. ❑

By condition (a), the module V has a structure of a commutative algebra, since it is a quotient of the commutative algebra $\mathcal{P}$. Obviously,

$$V^{sym} \overset{\mathbf{def}}{=\!=} \{v \in V \mid Tv = t^{1/2}v\}$$

is a subalgebra of V, a generalization of the Verlinde algebra.

Due to condition (b), one can replace the X–eigenvectors by the Y–eigenvectors in (c). Thus such an X–semisimple V is Y–semisimple and the other way round.

Also, (c) formally implies that V is X–semisimple. Indeed, if $(X-q^{\xi})v = e$ for e satisfying $(X - q^{\xi})e = 0$, then

$$\langle e, e\rangle = \langle (X - q^{\xi})v, e\rangle = \langle v, (X^{-1} - q^{-\xi})e\rangle = 0.$$

The following theorem directly results from the previous considerations. Actually, we do not need much theory for its justification. We will sketch a straightforward proof.

Theorem 2.9.8. *The perfect representations V are exactly the cases*

$$(a): \ 2k \in \mathbb{Z}_+,\ 0 < k < N/2, \ \textit{and}$$
$$(b): \ k = -1/2 - n,\ n \in \mathbb{Z},\ 0 \le n < N/2$$

from Theorem 2.9.3. Both V are quotients of $\mathcal{P}$ by the radical Rad of the form $\{f, g\} = f(Y^{-1})(g)(-k/2)$. They are isomorphic to Funct($\bowtie'$):

$$(a)\ \bowtie' = \Big\{\frac{k-N+1}{2}, \frac{k-N+2}{2}, \ldots, -\frac{k}{2}, \frac{k+1}{2}, \ldots \frac{N-k}{2}\Big\},$$

$$(b)\ \ \bowtie' = \Big\{\frac{1/2-n}{2}, \frac{3/2-n}{2}, \ldots, -\frac{1}{4}, \frac{1}{4}, \frac{3}{4}, \ldots, \frac{1/2+n}{2}\Big\}.$$

The action of $\mathcal{H}$ is defined here by the formulas from Theorem 2.7.2. The invariant form is (2.8.13):

$$\langle f, g\rangle' \stackrel{\text{def}}{=} \langle f\bar{g}\mu^1\rangle',\ \bar{g}(m_\sharp) = \overline{g(m_\sharp)},\ \langle f\rangle' = \sum_{m_\sharp \in \bowtie'} f(m_\sharp), \tag{2.9.5}$$

$$\mu^1(m_\sharp) = \mu^1((1-m)_\sharp) = q^{-k(m-1)} \prod_{i=1}^{m-1} \frac{1-q^{2k+i}}{1-q^i} \ \textit{as } m > 0.$$

Proof. We firstly check that a perfect V must coincide with $\mathcal{P}/\mathrm{Rad}$. Indeed, using $\langle\,,\,\rangle$ and ϵ on V, we introduce the form $\{\cdot,\cdot\}_\epsilon$ on $\mathcal{P}$, which is the pullback of the form $\langle f, \epsilon g\rangle$. The corresponding anti-involution (in either direction) is $\epsilon\star = \phi$, i.e., the same as for the form $\{\cdot,\cdot\}$. A proper linear combination $\{\cdot,\cdot\}_\dagger$ of $\{\cdot,\cdot\}$ and $\{\cdot,\cdot\}_\epsilon$ satisfies $\{1,1\}_\dagger = 0$. Therefore it vanishes identically and $\{\cdot,\cdot\}$, $\{\cdot,\cdot\}_\epsilon$ are proportional. This gives the desired result.

Here we did not use that the form $\{\cdot,\cdot\}_\epsilon$ is symmetric. It immediately follows from the proportionality. Thus $\langle f, \epsilon g\rangle = \langle \epsilon f, g\rangle$.

Second, conditions (a,b) follow from the existence of the form $(\,,\,)$ from the claim (c) from the definition. One can also use the previous proposition.

Third, using the chain of intertwiners from (2.8.6) and (2.9.2), we obtain that all ε_m for $m_\sharp \in -\bowtie'$ are well defined and their images in V linearly generate an irreducible $\mathcal{H}$–submodule V'. Indeed, the polynomial e_{-n} exists, $e_{-n}(-k/2) = 0$, and therefore it belongs to Rad. Equivalently, one may check that the formulas from Theorem 2.7.2 define an $\mathcal{H}$–action on Funct($\bowtie'$). The orthogonal complement of V' in V with respect to $\langle\,,\,\rangle$ (or $\{\cdot,\cdot\}_\epsilon$) intersects V by zero.

Fourth, the operator X has a simple spectrum on V. The existence of the X–pseudo-unitary pairing guarantees that X is semisimple. Since V is a cyclic module over $\mathbb{C}[X^{\pm 1}]$, each eigenvalue of X has a unique Jordan block. Therefore the spectrum of X in V is simple. Applying ϵ, the same holds for Y. For instance, the image of 1 in V can belong either to V' or to its orthogonal complement constructed above. This is impossible because 1 is a generator of V.

Fifth, we can define τ_+ as the operator of multiplication by the restricted Gaussian $\gamma(m_\sharp) = q^{m_\sharp^2}$ in the realization V =Funct($\bowtie'$). The automorphism ϵ in V is the Fourier transform $\mathbf{E}$: $\widehat{f}(m_\sharp) = \langle \bar{f}\varepsilon_m \mu^1\rangle'$ from (2.8.14). ❑

Note that the action of σ in V_{2n}^{sym} (for $n = N - 2k$ and the integral k) was introduced and calculated for the first time in [Ki1] on the basis of the interpretation of Rogers' polynomials for the integral positive k discovered by Etingof and Kirillov. Verlinde considered only the case $k = 1$, when the classical SL_2–characters are sufficient.

We will finish this section with the following generalization of Theorem 2.3.4. We follow Theorem 2.8.3, where negative half-integers k were considered. See also [C27].

Theorem 2.9.9. *Let* $0 < 2k < N$, $l_\sharp, m_\sharp \in \bowtie'$. *Then*

$$\langle \varepsilon_l \overline{\varepsilon_m} \gamma \mu^1 \rangle' = q^{-\frac{m^2+n^2+2k(|m|+|n|)}{4}} \varepsilon_l(m_\sharp) \langle \gamma \mu^1 \rangle', \tag{2.9.6}$$

$$\langle \gamma \mu^1 \rangle' = \prod_{j=1}^{k} \frac{1}{1-q^j} \sum_{j=0}^{2N-1} q^{j^2/4}, \quad \text{where } k \in \mathbb{Z}, \tag{2.9.7}$$

$$\sum_{j=0}^{2N-1} q^{j^2/4} = (1+\imath)\sqrt{N} \text{ as } q^{1/4} = \exp(\pi\imath/2N),$$

$$\langle \gamma \mu^1 \rangle' = 2q^{1/16} \prod_{j=1}^{k-1/2} \frac{1}{1-q^{1/2-j}} \quad \text{for } k = 1/2 + \mathbb{Z}. \tag{2.9.8}$$

❑

We note an interesting connection of these formulas with Theorem 2.8.4, which describes the **additional series** for generic non-cyclotomic q and $t = -q^{-n/2}$.

The formulas (2.8.20) and (2.8.19) there actually can be used to obtain (2.9.6), and the other way round. Indeed, setting in these formulas $k = N/2 - n/2$ for $n \in \mathbb{N}$, we arrive at the relation $t = -q^{-n/2}$. If N is sufficiently big, then the corresponding q can be treated as "generic."

Then, to establish the connection, we need to rewrite the Gaussian sums using Corollary 2.3.6, which states that

$$\langle \gamma \mu^1 \rangle' = q^{n^2/4} \prod_{j=k+1}^{n} (1-q^j) \qquad \text{if } N = 2n,$$

$$\langle \gamma \mu^1 \rangle' = q^{n^2/4}(1+q^{(2n-1)/4}) \prod_{j=k+1}^{n} (1-q^j) \qquad \text{if } N = 2n+1.$$

Choosing sufficiently big N and providing the desired parity of N, the *additional* representation V_{2n} becomes V_{2N-4k} for such N. This is, essentially, what was observed in [C20](Appendix). ❑

2.10 Little double Hecke algebra

Concerning the above choice $q^{1/4} = \exp(\pi\imath/2N)$, the formulas for the Gaussian sums that do not explicitly contain the imaginary unit $\imath$ hold for any primitive q, since we can apply the Galois automorphisms. However, the "smallest" q is necessary to ensure the positivity of the inner product. See the formulas for $\mu^1(m_\sharp)$. There are some other examples of unitary representations of $\mathcal{H}$ when $q^{1/4}$ is *not* a primitive $4N$-th root of unity; they will be the contents of this section.

All representations considered in this section will be unitary, i.e., with positive invariant $*$–hermitian scalar products.

Recall that the involution $*$ is the complex conjugation trivial on x, where $X = q^x$ and $q^x(m/2) = q^{m/2}$ for $m \in \mathbb{Z}$. Let $q^{1/4}$ be arbitrary (maybe non-primitive) $4N$–th root of unity; $q^{-1/4} = (q^{1/4})^* = (q^{1/4})^{-1}$. We will start with $0 < k < N/2,\ k \in \mathbb{Z}$. The field of constants will be $\mathbb{Q}_q = \mathbb{Q}(q^{1/4})$.

The definition of the functional representation remains the same. See Theorem 2.9.8. For $-N < m \le N$,

$$\begin{aligned} \bowtie' &\stackrel{\text{def}}{=} \{m_\sharp \mid \mu^1(m_\sharp) \neq 0\} \\ &= \Big\{\frac{-N+k+1}{2}, \ldots, -\frac{k}{2}, \frac{k+1}{2}, \ldots, \frac{N-k}{2}\Big\}, \end{aligned} \tag{2.10.1}$$

where the measure is

$$\mu^1(m_\sharp) = \mu^1((1-m)_\sharp) = q^{-k(m-1)} \prod_{j=1}^{m-1} \frac{1-q^{2k+j}}{1-q^j} \quad \text{for} \quad m > 0.$$

The space $\mathcal{F}' = \mathcal{F}'(k) = \mathrm{Funct}(\bowtie', \mathbb{Q}_q)$ is an $\mathcal{H}$–module; the discretization map $\mathcal{P} \ni f \mapsto f(m/2) \in \mathcal{F}'$ is an $\mathcal{H}$–homomorphism. Setting

$$\langle f \rangle' \stackrel{\text{def}}{=} \sum_{m=-N+1}^{N} f(m_\sharp)\mu^1(m_\sharp),$$

the form $\langle f, g\rangle' = \langle fg^* \rangle'$ is $*$–hermitian on $\mathcal{P}$. The quotient of $\mathcal{P}$ by the radical of this form coincides with the whole $\mathcal{F}'$, except in the special case when N is odd and $q^{1/2}$ is an N–th root of unity. In this case, the image of $\mathcal{P}$ upon the discretization mapping is (two times) smaller than $\mathcal{F}'$.

2.10.1 The case of odd N

Let $N = 2n+1$, $q^{1/2} = -\exp(\pi\imath/N)$. When considering the Gaussian, we take $q^{1/4} = \imath \exp(\pi\imath/N)$. Then there are two irreducible components of $\mathcal{F}'$:

$$\begin{aligned} \mathcal{F}' &= (\mathcal{F}')^0 \oplus (\mathcal{F}')^1, \quad \dim(\mathcal{F}')^j = N - 2k, \\ (\mathcal{F}')^j &= \{f \in \mathcal{F}' \mid f(m_\sharp + N/2) = (-1)^j f(m_\sharp),\ m_\sharp < 0\}. \end{aligned} \tag{2.10.2}$$

The component $(\mathcal{F}')^0$ is the image of $\mathcal{P}$.

The hermitian form introduced above is nondegenerate on $\mathcal{F}'$, and, moreover, makes it a $*$–unitary representation. Actually, $\mathcal{F}'$ is unitary for either sign of $q^{1/2}$ because the inner product involves only q.

The Fourier transforms $\mathbf{S}_1, \bar{\mathbf{S}}_1$ defined by the formulas

$$\mathbf{S}_1(f)(m_\sharp) = \langle f \varepsilon_m \rangle' \quad \text{and} \quad \bar{\mathbf{S}}_1(f)(m_\sharp) = \langle f\, \overline{\varepsilon_m} \rangle' \tag{2.10.3}$$

vanish identically on $(\mathcal{F}')^1$. They are zero on the Gaussian when $N = 4d+1$, because it sits in $(\mathcal{F}')^1$ for such N. For $N = 4d+3$, the Gaussian belongs to $(\mathcal{F}')^0$ (see (2.10.2)). Equivalently, as $q^{1/4} = \imath \exp(\pi\imath/2N)$,

$$\sum_{m=-N+1}^{N} q^{m^2/4} = 0 \Leftrightarrow N = 4d+1.$$

To make the above considerations more systematic and cover all classical formulas for the Gaussian sums, we need to diminish the double Hecke algebra and, respectively, the size of its irreducible representations.

2.10.2 Little double *H*

Let $\widetilde{\mathcal{H}}$ be a subalgebra of $\mathcal{H}$ generated by X^2, T, and Y^2. We will assume for a while that q and t are arbitrary nonzero. Respectively, $\mathbb{Q}_{q,t} \stackrel{\text{def}}{=\!=} \mathbb{Q}(q^{1/4}, t^{1/2})$ is the field of constants. The image of $T_0 \stackrel{\text{def}}{=\!=} Y^2T^{-1}$ in the polynomial representation $\mathcal{P}$ can be readily calculated:

$$\widehat{T_0} = t^{1/2}s_0 + (q^{1-2x} - 1)^{-1}(t^{1/2} - t^{-1/2})(s_0 - 1), \ s_0 = s\varpi^2, \tag{2.10.4}$$

where $s(f(x)) = f(-x)$, $\varpi(f(x)) = f(x+1/2)$. This gives that $\widetilde{\mathcal{H}}$ satisfies the PBW theorem, which of course can be checked directly using the abstract relations between the generators.

All the symmetries of $\mathcal{H}$ hold for $\widetilde{\mathcal{H}}$. Moreover, we may restrict its action on $\mathcal{P}$ to the even part $\mathcal{P}^0 \stackrel{\text{def}}{=\!=} \mathbb{Q}_{q,t}[X^2, X^{-2}]$, which is an irreducible $\widetilde{\mathcal{H}}$-module for generic q, t. The odd part $\mathcal{P}^1 \stackrel{\text{def}}{=\!=} X\mathcal{P}^0$ is a tilde-submodule as well, irreducible for generic q, t. Respectively, the spherical polynomials ε_m for even and odd m are even and odd and linearly generate $\mathcal{P}^0$ and $\mathcal{P}^1$.

Similarly, we may diminish $\mathcal{F}$ by introducing the following tilde-module:

$$\widetilde{\mathcal{F}} \stackrel{\text{def}}{=\!=} \mathrm{Funct}_{finite}((2\mathbb{Z})_\sharp, \mathbb{Q}_{q,t}).$$

Correspondingly, the summation in the definition of $\langle\ \rangle_1$ will now be over the set $(2\mathbb{Z})_\sharp$. We write $\langle\ \rangle_{11}$.

Formulas (2.7.20) and (2.9.6) hold for $\langle\ \rangle_{11}$ with arbitrary l, m. The identity (2.3.3) now reads

$$\begin{aligned}\langle q^{x^2}\rangle_{11} &= q^{k^2/4}\sum_{j=0}^{\infty}\frac{1-q^{2j+k}}{1-q^k}\prod_{l=1}^{2j}\frac{1-q^{l+2k-1}}{1-q^l}q^{j^2-jk}\\ &= q^{k^2/4}\prod_{j=1}^{\infty}\frac{1-q^{j+k}}{1-q^j}\sum_{j=-\infty}^{\infty}q^{j^2-jk}.\end{aligned} \tag{2.10.5}$$

Let us take $q = \exp(2\pi\imath/N)$, where $0 < k \le n = [N/2]$, and pick $q^{1/2} = \pm\exp(\pi\imath/N)$. The sign of $q^{1/2}$ can be arbitrary; all formulas hold for either choice.

We obtain the identity

$$\sum_{j=0}^{n-k} \frac{1-q^{2j+k}}{1-q^k} \prod_{l=1}^{2j} \frac{1-q^{l+2k-1}}{1-q^l} q^{j^2-jk}$$
$$= \prod_{j=1}^{k} \frac{1}{(1-q^j)} \sum_{j=0}^{N-1} q^{j^2-jk}, \tag{2.10.6}$$

which is nothing but (2.3.7).

Note that this identity may be $0=0$. It happens exactly when $N=2n$ and $n-k$ is odd.

Setting $k=n$, we come to the formulas

$$\sum_{j=0}^{N-1} q^{j^2} = q^{d^2}\prod_{j=1}^{n}(1-q^j) \ \text{ for } \ N=2n+1,\ d=n/2 \bmod N,$$
$$\sum_{j=0}^{n-1} q^{j^2}(-1)^j = \prod_{j=1}^{n-1}(1-q^j) \ \text{ for } \ N=2n. \tag{2.10.7}$$

The sum in the last formula gives the Legendre symbol.

Using that $\sum_{j=0}^{N-1} q^{(j-k/2)^2}$ does not depend on k, we may simplify (2.10.6) under the assumption that $n-k$ is even for $N=2n$:

$$\sum_{j=0}^{n-k} \frac{1-q^{2j+k}}{1-q^k} \prod_{l=1}^{2j} \frac{1-q^{l+2k-1}}{1-q^l} q^{j^2-jk} \tag{2.10.8}$$
$$= q^{(d-h)(d+h)} \prod_{j=k+1}^{n} (1-q^j) \ \text{ for } \ d\pm h = (n\pm k)/2 \bmod N.$$

Let us discuss the functional representation and, as a by-product, clarify why the case of even N and odd $n-k$ is exceptional. We assume that $k < N/2$ and define the set $\bowtie''$:

$$= \Big\{\frac{-N+k}{2}+1, \ldots, -\frac{k}{2}, \frac{k}{2}+1, \ldots, \frac{N-k}{2}\Big\} \ \text{ if } \ N=2n, \tag{2.10.9}$$
$$= \Big\{\frac{-N+k+1}{2}, \ldots, -\frac{k}{2}, \frac{k}{2}+1, \ldots, \frac{N-k-1}{2}\Big\} \ \text{ if } \ N=2n+1.$$

The dimension of the $\widetilde{\mathcal{H}}$–module $\mathcal{F}'' \stackrel{\text{def}}{=\!=} \text{Funct}(\bowtie'', \mathbb{Q}_q)$ is $N-2k$. It is irreducible for odd N and has two irreducible $\widetilde{\mathcal{H}}$–components otherwise:

$$\mathcal{F}'' = (\mathcal{F}'')^0 \oplus (\mathcal{F}'')^1,\ \ \dim(\mathcal{F}'')^j = n-k \ \text{ for } \ N=2n:$$
$$(\mathcal{F}'')^j = \text{Image}(\mathcal{P}^j) = \{f \in \mathcal{F}'' \mid f(m_\sharp + n) = (-1)^j f(m_\sharp)\}, \tag{2.10.10}$$

where we take only negative $m_\sharp$.

The Gaussian $q^{m_\sharp^2}$ belongs to $(\mathcal{F}'')^0$ precisely when $n-k$ is even. Otherwise it sits in $(\mathcal{F}'')^1$. This explains the degeneration of (2.10.6) observed above. Note that $\mathcal{F}''$ for odd N is nothing but $(\mathcal{F}')^0$ with $q^{1/2} = -\exp(\pi\imath/N)$. It gives that $(\mathcal{F}')^0$ from (2.10.2) remains irreducible upon restriction to $\widetilde{\mathcal{H}}$.

Let us adjust (2.9.6) from Theorem 2.9.9 to the case under consideration. We will need

$$\langle f\rangle'' \stackrel{\mathbf{def}}{=\!=} \sum f(z)\mu^1(z),\ z \in \bowtie'', \text{ and}$$

$$C_k'' \stackrel{\mathbf{def}}{=\!=} q^{d^2} \prod_{j=k+1}^{n} (1-q^j), \text{ where} \tag{2.10.11}$$

$$d = n/2 \text{ as } N = 2n \text{ and } N = 4d \pm 1 \text{ for } N = 2n+1. \tag{2.10.12}$$

Now we involve the spherical polynomials. The images of ε_m in $\mathcal{F}''$ are linearly independent for $m \le N-2k$ and generate the whole space. Picking $1 \le l, m \le N-2k$,

$$\langle \varepsilon_l\, \overline{\varepsilon_m}\, \gamma\rangle'' = C\, q^{-\frac{l^2+m^2+2k(|l|+|m|)}{4}}\, c_l(m_\sharp)\, C_k'', \tag{2.10.13}$$

$$C = \imath^{N(l^2+m^2+2k(|l|+|m|))} \text{ if } N = 2n+1, \tag{2.10.14}$$

$$C = 0 \text{ if } k+l+m+n \text{ is odd for } N = 2n,$$

$$C = 1 \text{ otherwise.}$$

These formulas give a complete description of the Fourier transforms $\mathbf{S}_1$ and $\bar{\mathbf{S}}_1$. See (2.10.3).

2.10.3 Half-integral k

Generally speaking, the identities we discuss here are beyond the classical theory because they depend on k and moreover involve the Macdonald polynomials. However, when $k = 0, 1$ and for other integers $k > 1$ there are strong links to the classical formulas. The transformation $k \mapsto k+1$ can be performed using the shift operator.

The case of the half-integral k does not seem to have direct counterparts in the classical theory of Gaussian sums. The shift operator acts within the set of such k. Thus no immediate connections with the integral k can be expected.

The formulas for the half-integral k are not too complicated to prove directly and sometimes even simpler than the "integral" ones. For instance, the Gaussian sum for $k = 1/2$ is of the collapsing type, in contrast to the classical sums for $k = 0, 1$, that give some nontrivial identities for the theta functions and eta functions.

Note that the value $k = 1/2$ formally corresponds to the symmetric space $SL_2(\mathbb{R})/SO_2$. Thus, in the limit $q \to 1$, the Fourier transform under consideration for the half-integral k is directly connected with the "orthogonal" Harish-Chandra transform. Let us note that there is a classical connection between the orthogonal and symplectic cases, the Howe duality. It could lead to a certain relation between the cases $k = 1/2$ and $k = 2$, classically and in the q-case. Due to the shift operators, it can lead to a connection between the integral and the half-integral cases too. This possibility has not been examined so far.

The half-integers k can be taken positive or negative. At roots of unity, generally speaking, the positive and negative values cannot be distinguished. Moreover, in the case of odd N, half-integers are congruent to integers modulo N. A combination of these two remarks gives that the negative half-integral k are directly connected with the integral positive k for odd N, as we will later use to deform the little Verlinde algebras

Let $k \in 1/2 + \mathbb{Z}$. We follow the procedure which was already used above and begin with the formulas for the Gaussian sums with generic q. By generic, we mean that $q^a t^b$ do not represent nontrivial roots of unity for $a, b \in \mathbb{Z}$. Then we will "specialize" the formulas as q are roots of unity.

The general identities we need are

$$\sum_{j=0}^{\infty} q^{j^2-kj} \frac{1-q^{2j+k}}{1-q^k} \prod_{l=1}^{2j} \frac{1-q^{l+2k-1}}{1-q^l}$$
$$= \prod_{j=1}^{\infty} \frac{(1+q^{-k-1+2j})}{(1+q^{k+2j})} \prod_{j=1}^{\infty} \frac{(1-q^{2k+2j})}{(1-q^{-1+2j})}. \tag{2.10.15}$$

We follow the same procedure and begin with the formulas for the Gaussian sums with generic q, assuming that the $q^a t^b$ do not represent nontrivial roots of unity for $a, b \in \mathbb{Z}$.

Let us denote the double product from the last line by $\Pi\Pi$. It can be simplified. It equals, respectively:

$$\Pi\Pi = \prod_{j=0}^{s} \frac{1+q^{k-2j}}{1-q^{2k-2j}} \quad \text{for} \quad k = 1/2 + s, \ s \in \mathbb{Z}_+, \tag{2.10.16}$$

$$\Pi\Pi = \prod_{j=1}^{s} \frac{1-q^{2k+2j}}{1+q^{k+2j}} \quad \text{for} \quad k = -1/2 - s, \ s \in \mathbb{Z}_+. \tag{2.10.17}$$

Now let us consider $q^{1/2} = \exp(\pi\imath/N)$, provided that $0 < 2k < N$. The analysis of $\mathcal{F}''$ in this case is similar to that for the integral k. However, the roles of even and odd N are inverse. We set $\bowtie'' \stackrel{\text{def}}{=\!=}$

$$\Big\{\frac{-N+k}{2}+1,\ldots,-\frac{k}{2},\ \frac{k}{2}+1,\ldots,\frac{N-k}{2}\Big\} \text{ if } N=2n+1, \tag{2.10.18}$$
$$\Big\{\frac{-N+k+1}{2},\ldots,-\frac{k}{2},\ \frac{k}{2}+1,\ldots,\frac{N-k-1}{2}\Big\} \text{ if } N=2n.$$

The dimension of the $\widetilde{\mathcal{H}}$–module $\mathcal{F}''=\text{Funct}(\bowtie'',\mathbb{Q}_q)$ remains $N-2k$ as in the integral case. It is irreducible for even N and has the following $\widetilde{\mathcal{H}}$–decomposition:

$$\mathcal{F}'' \ = \ (\mathcal{F}'')^0\oplus(\mathcal{F}'')^1,\ \ \dim(\mathcal{F}'')^j=n-k \ \text{ for } \ N=2n+1: \tag{2.10.19}$$
$$(\mathcal{F}'')^j=\text{Image}(\mathcal{P}^j)=\{f\in\mathcal{F}'' \ \mid \ f(m_\sharp+N/2)=(-1)^j f(m_\sharp)\},$$

where $m_\sharp<0$ are taken.

The corresponding reduction of (2.10.15) reads

$$\sum_{j=0}^{[N/2-k]} q^{j^2-kj}\frac{1-q^{2j+k}}{1-q^k}\prod_{l=1}^{2j}\frac{1-q^{l+2k-1}}{1-q^l}$$
$$= \ (1+\imath^{2k-N})\,C''_k,\ \ C''_k \stackrel{\text{def}}{=\!=} \prod_{j=0}^{k-1/2}\frac{1+q^{k-2j}}{1-q^{2k-2j}}. \tag{2.10.20}$$

The main formula then becomes

$$\langle\varepsilon_l\,\overline{\varepsilon_m}\,\gamma\rangle'' \stackrel{\text{def}}{=\!=} \sum_{z\in\bowtie''}\mu^1(z)\varepsilon_l(z)\,\overline{\varepsilon_m}(z)\gamma(z) \tag{2.10.21}$$
$$=(1+\imath^{2k+|l|+|m|-N})\,C''_k\,q^{-\frac{l^2+m^2+2k(|l|+|m|)}{4}}\,\varepsilon_l(m_\sharp).$$

2.10.4 The negative case

Recall the corresponding formulas for $\mathcal{H}$. Taking

$$-N/2\le\bar{k}\stackrel{\text{def}}{=\!=}-1/2-s<0 \ \text{ for } \ s\in\mathbb{Z}_+ \ \text{ and } \ 0\le s<N/2,$$

we set:

$$\bar{\bowtie}' \stackrel{\text{def}}{=\!=}\Big\{\frac{\bar{k}+1}{2},\frac{\bar{k}+2}{2},\ldots,-\frac{\bar{k}+1}{2},-\frac{\bar{k}}{2}\Big\} \tag{2.10.22}$$
$$=\Big\{\frac{\bar{k}+1}{2},\frac{\bar{k}+2}{2},\ \ldots\ ,-\frac{1}{4},\frac{1}{4}=\frac{\bar{k}+s+1}{2},\ldots,-\frac{\bar{k}}{2}=\frac{\bar{k}+2s+1}{2}\Big\}$$
$$=\Big\{\frac{1/2-s}{2},\frac{3/2-s}{2},\ldots,-\frac{1}{4},\frac{1}{4},\frac{3}{4},\ldots,\frac{1/2+s}{2}\Big\}.$$

Then $\bar{\mathcal{F}}' = \text{Funct}(\,\bar{\bowtie}', \mathbb{Q}_q)$ is an irreducible $\mathcal{H}(\bar{k})$–module of dimension $-2\bar{k} = 2s+1$. We put $\mathcal{H}(\bar{k})$ instead of $\mathcal{H}$ to show explicitly the dependence on $\bar{k}$.

Let $q = \exp(2\pi\imath/N)$. To more exact, we pick $q^{1/2} = -\exp(\pi\imath/N)$. The minus–sign in the case under consideration makes the inner product $\langle f, g\rangle'$ positive. Recall that it is the summation of $fg^*\mu^1$ over the set $\bar{\bowtie}'$. All formulas below will hold for either sign, as well as the analysis of the irreducibility. Only the positivity will be missing for the plus–sign.

The module $\bar{\mathcal{F}}'$ is generated by the eigenfunctions of Y, which are the images of ε_m for $m_\sharp \in \bar{\bowtie}'$, and

$$\langle \varepsilon_l \,\overline{\varepsilon_m}\, \gamma\rangle' \;=\; q^{-\frac{l^2+m^2+2\bar{k}(|l|+|m|)}{4}}\,\varepsilon_l(m_\sharp)\, C'_{\bar{k}}, \tag{2.10.23}$$

$$C'_{\bar{k}} \;=\; \sum_{j=s+1}^{2s+1} q^{\frac{(j-\bar{k})^2}{4}}\frac{1-q^{j+\bar{k}}}{1-q^{\bar{k}}}\prod_{l=1}^{j}\frac{1-q^{l+2\bar{k}-1}}{1-q^l} = q^{1/16}\prod_{j=1}^{s}(1-q^{1/2-j}).$$

The case of $\widetilde{\mathcal{H}}$. As above, $\bar{k} = -s - 1/2$ and $q^{1/2} = -\exp(\pi\imath/N)$. The latter ensures the positivity of the hermitian form.

The reduction of (2.10.15) to $\widetilde{\mathcal{H}}$ is straightforward:

$$\sum_{j=0}^{s} q^{j^2-\bar{k}j}\frac{1-q^{2j+\bar{k}}}{1-q^{\bar{k}}}\prod_{l=1}^{2j}\frac{1-q^{l+2\bar{k}-1}}{1-q^l}$$
$$=\prod_{j=1}^{s}\frac{1-q^{2\bar{k}+2j}}{1+q^{\bar{k}+2j}}. \tag{2.10.24}$$

Actually, this formula coincides with the formula for $C'_{\bar{k}}\,q^{-\bar{k}^2/4}$ resulting from (2.10.23). We will skip the identification of the sums. Concerning the coincidence of the products (in the right-hand sides), it is an elementary algebraic exercise:

$$q^{1/16-\bar{k}^2/4}\prod_{j=1}^{s}(1-q^{1/2-j}) = \prod_{j=1}^{s}\frac{1-q^{2\bar{k}+2j}}{1+q^{\bar{k}+2j}}. \tag{2.10.25}$$

Here q can be absolutely arbitrary. This formula is useful in the theory of η–like identities (in the limit $s \to \infty$).

The coincidence has a simple explanation. Indeed, the $\mathcal{H}(\bar{k})$–module $\bar{\mathcal{F}}'$ introduced above for $\bar{k}$ remains irreducible upon the restriction to $\widetilde{\mathcal{H}}(\bar{k})$. The $\bowtie$–set constructed for $\widetilde{\mathcal{H}}(\bar{k})$ and responsible for the summation in formula (2.10.24) is

$$\bar{\bowtie}'' \overset{\text{def}}{=\!=} \Big\{-\frac{\bar{k}}{2} - s, \ldots, -\frac{\bar{k}}{2}, \frac{\bar{k}}{2}+1, \ldots, \frac{\bar{k}}{2}+s\Big\}. \tag{2.10.26}$$

Here the points from the second half ($\bar{k}/2+1$, etc.) sit between consecutive pairs of points from the first half. Ordering them naturally, we obtain

$$\Big\{-\frac{\bar{k}}{2}-s=\frac{1/2-s}{2}, \frac{\bar{k}}{2}+1=\frac{3/2-s}{2}, \text{ and so on }\Big\},$$

which is nothing but the third line for $\bowtie'$ from (2.10.22).

The main formula is equivalent to (2.10.23) and reads as follows:

$$\begin{aligned}&\sum_{j=1}^{s}(q^{-(2j-1)\bar{k}}\prod_{l=1}^{2j-1}\frac{1-q^{l+2\bar{k}}}{1-q^{l}})\,q^{j^2+\bar{k}j}\,\varepsilon_l(\frac{2j+\bar{k}}{2})\,\overline{\varepsilon_m}(\frac{2j+\bar{k}}{2})\\ &+\sum_{j=0}^{s}(q^{-2j\bar{k}}\prod_{l=1}^{2j}\frac{1-q^{l+2\bar{k}}}{1-q^{l}})\,q^{j^2+\bar{k}j}\,\varepsilon_l(-\frac{2j+\bar{k}}{2})\,\overline{\varepsilon_m}(-\frac{2j+\bar{k}}{2})\\ &=q^{-\frac{l^2+m^2+2\bar{k}(|l|+|m|)}{4}}\,\varepsilon_l(m_\sharp)\prod_{j=1}^{s}\frac{1-q^{2\bar{k}+2j}}{1+q^{\bar{k}+2j}}.\end{aligned} \tag{2.10.27}$$

Here we take $l_\sharp, m_\sharp \in \bowtie''$.

2.10.5 Deforming Verlinde algebras

We will conclude this section with the following observation. When $N = 2n+1$, the case $\bar{k} = -1/2-s$ is equivalent to the case of the integral $\hat{k} \stackrel{\mathbf{def}}{=} n-s$. Here we may replace $-1/2$ by n modulo N because $q^{1/2} = -\exp(\pi\imath/N)$ is an N-th root of unity. To be more exact, $\widetilde{\mathcal{H}}(\hat{k})$ coincides with $\widetilde{\mathcal{H}}(\bar{k})$ for such $q^{1/2}$, since the defining relations are the same. The functional representations $\mathcal{F}''(\hat{k})$ and $\mathcal{F}''(\bar{k})$ (see (2.10.9)) are isomorphic too. The dimensions are obviously the same: $\dim\mathcal{F}''(\bar{k}) = -2\bar{k} = 2s+1 = N-2\hat{k} = \dim\mathcal{F}''(\hat{k})$.

So we hit the main sector $0 < \hat{k} < N/2$, $\hat{k} \in \mathbb{Z}$. Recall that the irreducibility of $\mathcal{F}''(\hat{k})$, the positivity of the inner product, and all formulas were independent of the choice of the sign of $q^{1/2}$ for such k.

Comparing (2.10.24) and (2.10.8), we arrive at the identity

$$\prod_{j=1}^{s}\frac{1-q^{1-2j}}{1+q^{n-s+2j}} = q^{-s(s+1)/4}\prod_{j=0}^{s-1}(1-q^{n-j}), \tag{2.10.28}$$

which readily follows from (2.10.25).

It is important that all claims and formulas about $\mathcal{F}''(\bar{k})$ hold for any q. For instance, its dimension is always $2s+1 = -2\bar{k}$, the formula (2.10.27) works, and so on. The particular choice of q does not matter. The constraint $\bar{k} = -1/2-s$ is sufficient. For $|q| \neq 1$, the conjugation $(q^{1/4})^* = q^{-1/4}$ must be understood formally.

We conclude that $\mathcal{F}''(-\frac{1}{2}-s)$ *for generic* q *is a flat* q*–deformation of* $\mathcal{F}''(n-s)$ *as* $q^{1/2}=-\exp(\pi\imath/N)$ *for* $N=2n+1$ *and* $0\le s\le n$.

If one wishes to make the μ–form hermitian and positive, it is necessary to be more specific. The conditions $q^{1/2}=-\exp(\pi\imath\omega)$ and $0<\omega s<1/2$ will do it. It is of course not necessary to assume that q is a root of unity. Note that

$$\omega=\frac{1}{N}\ \text{ and }\ \omega s\le\frac{n}{2n+1}<\frac{1}{2}\ \text{ for }\ q=-e^{\pi\imath/N}.$$

There is an application. Considering the T–plus component $\mathcal{F}''(\bar{k})^+$ as a $PSL(2,\mathbb{Z})$–module, one can calculate its limit as $q\to 1$. Upon proper renormalization, it will become nothing but the restriction of the irreducible representation of $SL_2(\mathbb{C})$ of dimension s to its subgroup $SL_2(\mathbb{Z})$. This readily gives the $PSL(2,\mathbb{Z})$–irreducibility of $\mathcal{F}''(\bar{k})^+$ for generic q. This fact is universal: if the dimension of the T–plus component is fixed, then there can be only finitely many N,q such that the action of $PSL(2,\mathbb{Z})$ in this component is reducible.

Note that one can use $\mathcal{H}$ instead of $\widetilde{\mathcal{H}}$ when discussing the correspondence $\bar{k}\leftrightarrow\hat{k}$ for odd N and $q^{1/2}=-\exp(\pi\imath/N)$. Indeed, the module $(\mathcal{F}')^0(\hat{k})$ is isomorphic to $\mathcal{F}'(\bar{k})$. We remind the reader that both modules remain irreducible and become, respectively, $\mathcal{F}''(\hat{k})$ and $\mathcal{F}''(\bar{k})$ upon the restriction to $\widetilde{\mathcal{H}}$. The identification $\mathcal{F}'(\hat{k})=\mathcal{F}'(\bar{k})$ becomes $\mathcal{F}''(\hat{k})=\mathcal{F}''(\bar{k})$ over $\widetilde{\mathcal{H}}$. Thus the usage of $\mathcal{H}$ here is possible; however, it does not add anything new.

We see that the module $\mathcal{F}''(k)$ for integral $0<k<N/2$ allows a q–deformation when $N=2n+1$, together with all related structures. We can apply it to the **little Verlinde algebra**: the T–plus component $(\mathcal{F}''(1))^+$ of $\mathcal{F}''(k=1)$. Generally speaking, little Verlinde algebras are introduced for the lattice Q of the radical weights. The standard Verlinde algebra is defined for all weights, i.e., for the lattice P.

We obtain that $(\mathcal{F}''(1/2-n))^+$ for generic q is a deformation of $(\mathcal{F}''(1))^+$. The corresponding q–deformation of the "big" Verlinde algebra (for P), which is $(\mathcal{F}'(1))^+$ in our notation, is not known.

Comment. It is worth mentioning that the structural constants of the Verlinde multiplication are integral and positive in the basis formed by the images of the SL_2–characters, i.e., by the discretizations of $p_m^{(1)}$. This holds because multiplication can be interpreted as the restricted tensor (or fusion) product.

This is the only property of the Verlinde algebras that will be lost upon the deformation above. At least we don't know how to reformulate the integrality/positivity for generic q. Everything else will survive. ❑

Let us calculate the first nontrivial deformed little Verlinde algebra, which is $V=(\mathcal{F}''(-3/2))^+$. It is the space of $\mathbb{Q}(q^{1/2})$–valued functions on the set $\{1/4,3/4\}$. Indeed, $s=1$, $\bar{k}=-3/2$, and the set $\bowtie''$ is $\{-1/4,3/4,1/4\}$. Since we consider even functions only, the points $1/4,3/4$ are sufficient.

The "restricted characters" are the images p_0', p_2' of the Rogers polynomials

$$p_0 = 1,\ p_2(x) = q^{2x} + q^{-2x} + 1 + (q + q^{1/2} + 1 + q^{-1/2} + q^{-1}).$$

The restricted multiplication is pointwise. Since $p_0' = 1$, there is only one nontrivial multiplication formula:

$$(p_2')^2 = q^{-3/2}\,(1+q^{1/2})^2\,(1+q)^2\,\big(p_2' - q^{-1}\,(1+q^{1/2})\,(1+q^{3/2})\,p_0'\big).$$

The inner product of functions $f, g \in V$ is

$$\langle f, g\rangle = fg^*(3/4) + (1 - q^{1/2} - q^{-1/2})\, fg^*(1/4).$$

The normalized Gaussian is given by $\gamma(3/4) = 1,\ \gamma(1/4) = q^{1+\tilde{k}} = q^{-1/2}$.

The main formula (the summation with the Gaussian) reads

$$\begin{aligned} &p_l p_m(3/4) + q^{-1/2}(1 - q^{1/2} - q^{-1/2})\, p_l p_m(1/4) = \\ &(1-q^{-1})(1+q^{1/2})^{-1}\, q^{\frac{3(l+m)-l^2-m^2}{4}}\, p_l(m/2 - 3/4)\, p_m(3/4). \qquad (2.10.29) \end{aligned}$$

Here q is arbitrary. If we take $q^{1/2} = -\exp(\pi\imath\omega)$ and $-2/3 < \omega < 2/3$, then the inner product is positive. Note that this inequality is weaker than $0 < \omega s < 1/2$ imposed above.

The space V is not an $\widetilde{\mathcal{H}}$-module, but the elements from $\widetilde{\mathcal{H}}$ that commute with T act there and make it irreducible. Actually, the action of $X+X^{-1}$ and $Y + Y^{-1}$ already results in its irreducibility (it is a general fact).

The reduction to the corresponding Verlinde algebra V' is as follows:

$$N = 5,\ n = 2,\ q^{1/2} = -\exp(\pi\imath/5),\ 1 - q^{1/2} - q^{-1/2} = (q + q^{-1})^{-2}.$$

The latter number appears in the inner product. The polynomial $p_2(x)$ becomes $q^{2x} + q^{-2x} + 1$ with the restricted square $(p_2')^2 = p_2' + p_0'$.

Let me mention that some formulas in the theory of Verlinde algebras can be deformed using filtrations in the spaces of the coinvariants. See [FKLMM].

2.11 DAHA and p–adic theory

This section is devoted to double Hecke algebras in the general setting and their connection with the classical p–adic spherical transform. In contrast to previous sections, we state the results without proofs. The paper [C28] contains a complete theory.

2.11.1 Affine Weyl group

Let $R \subset \mathbb{R}^n$ be a simple reduced root system, $R_+ \subset R$ the set of positive roots, $\{\alpha_1, \ldots, \alpha_n\} \subset R_+$ the corresponding set of simple roots, and $\vartheta \in R_+$ the **maximal coroot** that is also the longest positive root in $R^\vee$. We normalize the scalar product on $\mathbb{R}^n$ by the condition $(\vartheta, \vartheta) = 2$, so ϑ belongs to R as well and it is the longest *short* root there. For any $\alpha \in R$, its dual coroot is $\alpha^\vee = \frac{2\alpha}{(\alpha,\alpha)} = \alpha/\nu_\alpha$. So $\nu_\alpha = (\alpha, \alpha)/2 = 1, 2, 3$. We set $\rho = (1/2)\sum_\alpha \alpha$, $\rho^\vee = (1/2)\sum_\alpha \alpha^\vee$.

The affine roots are

$$\tilde{R} = \{\tilde{\alpha} = [\alpha, \nu_\alpha j]\},\ j \in \mathbb{Z}.$$

Note the appearance of ν_α here. It is because of our nonstandard choice of ϑ. We identify nonaffine roots α with $[\alpha, 0]$ and set $\alpha_0 = [-\vartheta, 1]$. For $\tilde{\alpha} \in \tilde{R}$, $s_{\tilde{\alpha}} \in \mathrm{Aut}(\mathbb{R}^{n+1})$ is the reflection

$$s_{\tilde{\alpha}}([x, \zeta]) = [x, \zeta] - 2\frac{(x, \alpha)}{(\alpha, \alpha)}\tilde{\alpha}, \text{ and}$$
$$W = \langle s_\alpha \mid \alpha \in R\rangle,\ \widetilde{W} = \langle s_{\tilde{\alpha}} \mid \tilde{\alpha} \in \tilde{R}\rangle. \tag{2.11.1}$$

We set $s_i = s_{\alpha_i}$. It is well known that $\widetilde{W}$ is a Coxeter group with the generators $\{s_i\}$. Let $\omega_i \subset \mathbb{R}^n$ be the fundamental weights: $(\omega_i, \alpha_j^\vee) = \delta_{ij}$ for $1 \le i, j \le n$, $P = \oplus_{i=1}^n \mathbb{Z}\omega_i$ the weight lattice, and $P_+ \stackrel{\text{def}}{=} \oplus_{i=1}^n \mathbb{Z}_+\omega_i$ the cone of dominant weights. We call $\widehat{W} \stackrel{\text{def}}{=} W \ltimes P$ the **extended affine Weyl group**:

$$wb([x, \zeta]) = [w(x), \zeta - (b, x)] \text{ for } b \in P, x \in \mathbb{R}^n.$$

The **length function** $l : \widehat{W} \to \mathbb{Z}_+$ is given by the formula

$$l(wb) = \sum_{\substack{\alpha \in R_+ \\ w(\alpha^\vee) \in R_+}} |(b, \alpha^\vee)| + \sum_{\substack{\alpha \in R_+ \\ w(\alpha) \in -R_+}} |(b, \alpha^\vee) + 1|, \tag{2.11.2}$$

where $w \in W$, $b \in P$. Let $Q = \oplus_{i=1}^n \mathbb{Z}\alpha_i$ be the root lattice. Then $\widetilde{W} = W \ltimes Q \subset \widehat{W}$ is a normal subgroup and $\widehat{W}/\widetilde{W} = P/Q$. Moreover, $\widehat{W}$ is the semidirect product $\Pi \ltimes \widetilde{W}$, where

$$\Pi = \{\pi \in \widehat{W} \mid l(\pi) = 0\} = \{\pi \in \widehat{W} \mid \pi : \{\alpha_i\} \mapsto \{\alpha_i\}\},\ 0 \le i \le n.$$

It is isomorphic to P/Q and acts naturally on the affine Dynkin diagram for $R^\vee$ with the reversed arrows. It is not the standard Dynkin diagram for R because of our choice of ϑ. We set $\pi(i) = j$ as $\pi(\alpha_i) = \alpha_j$.

2.11.2 Affine Hecke algebra

We denote the affine Hecke algebra by $\mathcal{H}$. Its generators are T_i for $i = 0, \ldots, n$ and $\pi \in \Pi$; the relations are

$$\underbrace{T_iT_jT_i\ldots}_{\text{order of } s_is_j} = \underbrace{T_jT_iT_j\ldots}_{\text{order of } s_is_j}, \quad \pi T_i\pi^{-1} = T_{\pi(i)}, \tag{2.11.3}$$
$$(T_i - t^{1/2})(T_i + t^{1/2}) = 0 \text{ for } 0 \le i \le n, \ \pi \in \Pi.$$

If $\hat{w} = \pi s_{i_l} \ldots s_{i_1} \in \widehat{W}$ is a reduced expression, i.e., $l(\pi^{-1}w) = l$, we set $T_{\hat{w}} = \pi T_{i_l} \ldots T_{i_1}$. The elements $T_{\hat{w}}$ are well defined and form a basis of $\mathcal{H}$.

Let G be the adjoint split p–adic simple group corresponding to R. Then $\mathcal{H}$ is the convolution algebra of compactly supported functions on G that are left-right invariant with respect to the Iwahori subgroup B, due to Iwahori and Matsumoto. Namely, the T_i are the characteristic functions of the double cosets Bs_iB, where we use a natural embedding $W \to G$. Generally, it is not a homomorphism.

To be more exact, the p–adic quadratic equations are represented in the form $(T_i - 1)(T_i + t_i) = 0$ for the standard normalization of the Haar measure. Here the t_i may depend on the length of α_i and are given in terms of the cardinality of the residue field. We will stick to our normalization of the T and assume that the *parameters t coincide* to simplify the formulas of this section. We will also use

$$\delta_{\hat{w}} \stackrel{\mathbf{def}}{=\!=} t^{-l(\hat{w})/2} T_{\hat{w}},$$

which satisfy the quadratic equation with 1.

Let Δ be the left regular representation of $\mathcal{H}$. In the basis $\{\delta_{\hat{w}}\}$, the representation Δ is given by

$$T_i\delta_{\hat{w}} = \begin{cases} t^{1/2}\delta_{s_i\hat{w}} & \text{if } l(s_i\hat{w}) = l(\hat{w}) + 1, \\ t^{-1/2}\delta_{s_i\hat{w}} + (t^{1/2} - t^{-1/2})\delta_{\hat{w}} & \text{if } l(s_i\hat{w}) = l(\hat{w}) - 1 \end{cases} \tag{2.11.4}$$

and the obvious relations $\pi\delta_{\hat{w}} = \delta_{\pi\hat{w}}$, where $\pi \in \Pi$ and $\hat{w} \in \widehat{W}$.

The **spherical representation** appears as follows. Let

$$\delta^+ \stackrel{\mathbf{def}}{=\!=} (\sum_{w\in W} t^{l(w)})^{-1} \sum_{w\in W} t^{l(w)}\delta_w \in \mathcal{H}.$$

One readily checks that $T_i\delta^+ = t^{1/2}\delta^+$ for $i = 1, \ldots, n$, and $(\delta^+)^2 = \delta^+$. We call δ^+ the t–**symmetrizer**. Then

$$\Delta^+ \stackrel{\mathbf{def}}{=\!=} \Delta\delta^+ = \oplus_{b\in P}\mathbb{C}\delta^+_b, \ \delta^+_{\hat{w}} = \delta_{\hat{w}}\delta^+,$$

is an $\mathcal{H}$–submodule of Δ.

It is nothing but $\mathrm{Ind}_H^{\mathcal{H}}(\mathbb{C}_{t^{1/2}})$, where $H \subset \mathcal{H}$ is the subalgebra generated by $T_i, i = 1, \ldots, n$ and $\mathbb{C}_{t^{1/2}}$ is the one-dimensional representation of H defined by $T_i \mapsto t^{1/2}$.

Due to Bernstein, Zelevinsky, and Lusztig (see, e.g., [Lus1]), we set $Y_a = T_a$ for $a \in P_+$ and extend it to the whole P using $Y_{b-a} = Y_b Y_a^{-1}$ for dominant a, b. These elements are well defined and pairwise commutative. They form the subalgebra $\mathcal{Y} \cong \mathbb{C}[P]$ inside $\mathcal{H}$.

Using Y, one can omit T_0. Namely, the algebra $\mathcal{H}$ is generated by $\{T_i, i > 0, Y_b\}$ with the following relations:

$$T_i^{-1} Y_b T_i^{-1} = Y_b Y_{\alpha_i}^{-1} \text{ if } (b, \alpha_i^\vee) = 1, \qquad (2.11.5)$$
$$T_i Y_b = Y_b T_i \text{ if } (b, \alpha_i^\vee) = 0, \ 1 \le i \le n.$$

The PBW theorem for $\mathcal{H}$ gives that the spherical representation can be canonically identified with $\mathcal{X}$, as δ^+ goes to 1. The problem is to calculate this isomorphism explicitly. It will be denoted by Φ. We come to the definition of Matsumoto's spherical functions:

$$\phi_a(\Lambda) = \Phi(\delta_a^+)(Y \mapsto \Lambda^{-1}), \ a \in P.$$

Here, by $Y \mapsto \Lambda^{-1}$, we mean that Λ_b^{-1} substitutes for Y_b. See [Ma]. In the symmetric (W-invariant) case, the spherical functions are due to Macdonald.

By construction,

$$\phi_a = t^{-l(a)} \Lambda_a^{-1} = t^{-(\rho^\vee, a)} \Lambda_a^{-1} \text{ for } a \in P_+, \ \rho^\vee = (1/2) \sum_{\alpha > 0} \alpha^\vee.$$

The formula $l(a) = (\rho^\vee, a)$ readily results from (2.11.2). So the actual problem is to calculate ϕ_a for non-dominant a.

Example. Consider a root system of type A_1. In this case, $P = \mathbb{Z}\omega$, where $\omega \in \mathbb{Z}$ is the fundamental weight, $Q = 2\mathbb{Z}$, and $\omega = \pi s$ for $s = s_1$. We identify Δ^+ and $\mathcal{Y}$, so $\delta^+ = 1$. Letting $Y = Y_\omega, T = T_1$, we get $Y_m \stackrel{\mathbf{def}}{=\!=} Y_{m\omega} = Y^m$ and

$$\phi_m \stackrel{\mathbf{def}}{=\!=} \phi_{m\omega} = t^{-m/2} \Lambda^{-m}, \text{ for } m \ge 0.$$

Note that $TY^{-1}T = Y$ and $\pi = YT^{-1}$.

Let us check that

$$\Lambda \phi_{-m} = t^{1/2} \phi_{-m-1} - (t^{1/2} - t^{-1/2}) \phi_{m+1}, \ m > 0. \qquad (2.11.6)$$

Indeed, $\phi_{-m} = t^{-m/2} (T\pi)^m (1) \mid_{Y \mapsto \Lambda^{-1}}$ and

$$\begin{aligned} Y^{-1} \phi_{-m} \ &= \ t^{-m/2} (T^{-1}\pi)(T\pi)^m(1) \qquad (2.11.7) \\ =& t^{-m/2}(T - (t^{1/2} - t^{-1/2}))\pi(T\pi)^m(1) \\ =& t^{-m/2}(T\pi)^{m+1}(1) - -t^{-m/2}(t^{1/2} - t^{-1/2})(\pi T)^m \pi(1) \\ =& t^{1/2}\phi_{-m-1}(Y^{-1}) - t^{-m/2}(t^{1/2} - t^{-1/2})(\pi T)^m(t^{-1/2}\pi T)(1) \\ =& t^{1/2}\phi_{-m-1}(Y^{-1}) - (t^{1/2} - t^{-1/2})\phi_{m+1}(Y^{-1}). \end{aligned}$$

2.11.3 Deforming p–adic formulas

The following chain of theorems represents a new vintage of the classical theory. We are not going to prove them here. Actually, all claims that are beyond the classical theory of affine Hecke algebras can be checked by direct and not very difficult calculations, with a reservation about Theorems 2.11.5 and 2.11.6.

Theorem 2.11.1. *Let $\xi \in \mathbb{C}^n$ be a fixed vector and let $q \in \mathbb{C}^*$ be a fixed scalar. We represent $\hat{w} = bw$, where $w \in W, b \in P$. In $\Delta_q^\xi \stackrel{\mathbf{def}}{=\!=} \oplus_{\hat{w}\in\widehat{W}}\mathbb{C}\delta_{\hat{w}}^\xi$, the formulas $\pi\delta_{\hat{w}}^\xi = \delta_{\pi\hat{w}}^\xi$ and*

$$T_i\delta_{\hat{w}}^\xi = \begin{cases} \frac{t^{1/2}q^{(\alpha_i,w(\xi)+b)}-t^{-1/2}}{q^{(\alpha_i,w(\xi)+b)}-1}\delta_{s_i\hat{w}}^\xi - \frac{t^{1/2}-t^{-1/2}}{q^{(\alpha_i,w(\xi)+b)}-1}\delta_{\hat{w}}^\xi & \text{if } i>0, \\ \frac{t^{1/2}q^{1-(\vartheta,w(\xi)+b)}-t^{-1/2}}{q^{1-(\vartheta,w(\xi)+b)}-1}\delta_{s_0\hat{w}}^\xi - \frac{t^{1/2}-t^{-1/2}}{q^{1-(\vartheta,w(\xi)+b)}-1}\delta_{\hat{w}}^\xi & \text{if } i=0 \end{cases}$$

define a representation of the algebra $\mathcal{H}$, provided that all denominators are nonzero, i.e., $q^{(\alpha,b+\xi)} \neq 1$ for all $\alpha \in R$, $b \in P$.

The regular representation Δ with the basis $\delta_{\hat{w}}$ is the limit of representation Δ_q^ξ as $q \to \infty$, provided that ξ lies in the fundamental alcove:

$$(\xi,\alpha_i) > 0 \text{ for } i=1,\ldots,n, \ (\xi,\vartheta) < 1. \tag{2.11.8}$$

We see that the representation Δ_q^ξ is a flat deformation of Δ for such ξ. Moreover, $\Delta_q^\xi \cong \Delta$. This will readily follow from the next theorem.

Note that taking ξ in other alcoves, we get other limits of Δ_q^ξ as $q \to \infty$. They are isomorphic to the same regular representation; however, the formulas do depend on the particular alcove. We see that the regular representation has rather many remarkable systems of basic vectors. They are not quite new in the theory of affine Hecke algebras, but such systems were not studied systematically.

Theorem 2.11.2. *(i) We set $X_b(\delta_{\hat{w}}^\xi) = q^{(\alpha_i,w(\xi)+b)}\delta_{\hat{w}}^\xi$ for $\hat{w} = bw$, where we use the notation $X_{[b,j]} = q^jX_b$. These operators have a simple spectrum in Δ_q^ξ under the conditions of the theorem and satisfy the relations dual to (2.11.5) with $i=0$ added:*

$$\begin{aligned} &T_iX_bT_i = X_bX_{\alpha_i}^{-1} \text{ if } (b,\alpha_i^\vee)=1, \\ &T_iX_b = X_bT_i \text{ if } (b,\alpha_i^\vee)=0, \ 0\le i\le n, \\ &\text{and moreover, } \pi X_b\pi^{-1} = X_{\pi(b)} \ \text{ as } \pi\in\Pi. \end{aligned} \tag{2.11.9}$$

(ii) The **double affine Hecke algebra** *$\mathcal{HH}$ is defined by imposing relations (2.11.3) and (2.11.9). Then the representation Δ_q^ξ is nothing but the induced representation*

$$\mathrm{Ind}_{\mathcal{X}}^{\mathcal{HH}}(\mathbb{C}\delta_{id}^\xi), \ \mathcal{X} \stackrel{\mathbf{def}}{=\!=} \mathbb{C}[X_b],$$

which is isomorphic to Δ as an $\mathcal{H}$–module.

Comment. The operators X_b are in a way the coordinates of the Bruhat-Tits buildings corresponding to the p–adic group G. In the classical theory, we use only their combinatorial variants, namely, the distances between vertices, which are integers. The X–operators clarify dramatically the theory of the p–adic spherical Fourier transform, because, as we will see, they are the "missing" Fourier–images of the Y–operators. Obviously, the X_b do not survive in the limit $q \to \infty$; however, they do not collapse completely. Unfortunately, the Gaussian, which is $q^{x^2/2}$ as $X_b = q^{(x,b)}$, does. ❑

This theorem is expected to be connected with [HO2] and via this paper with [KL1]. Here we will stick to the spherical representations. The following theorem establishes a connection with Δ^+.

Theorem 2.11.3. *We set* $q^\xi = t^{-\rho}$, *i.e.,* $q^{(\xi,b)} \mapsto t^{-(\rho,b)}$ *for all* $b \in P$. *The corresponding representation will be denoted by* $\widetilde{\Delta}$. *It is well defined for generic* q,t. *For any* $b \in P$, *let* π_b *be the minimal length representative in the set* $\{bW\}$. *It equals* bu_b^{-1} *for the length-minimum element* $u_b \in W$ *such that* $u_b(b) \in -P_+$. *Setting* $\delta_b^\sharp \stackrel{\mathbf{def}}{=\!=} \delta_{\pi_b}$ *in* $\widetilde{\Delta}$, *the space* $\Delta^\sharp \stackrel{\mathbf{def}}{=\!=} \oplus_{b\in P}\mathbb{C}\delta_b^\sharp$ *is an* $\mathcal{H}\!\mathcal{H}$–*submodule of* $\widetilde{\Delta}$. *It is isomorphic to* Δ^+ *as an* $\mathcal{H}$–*module.*

The representation $\Delta^\sharp$ is described by the same formulas from Theorem 2.11.1, which vanish *automatically* on $s_i\pi_b$ not in the form π_c, thanks to the special choice of q^ξ. It results directly from the following:

$$s_i\pi_b = \pi_c \Leftrightarrow (\alpha_i, b+d) \neq 0, \text{ where } (\alpha_i, d) \stackrel{\mathbf{def}}{=\!=} \delta_{i0}.$$

Here $c = b - ((\alpha_i, b+d)\alpha_i^\vee$ for $\alpha_0^\vee \stackrel{\mathbf{def}}{=\!=} -\vartheta$.

We define the action $(\!(\ )\!)$ of $\widehat{W}$ on $\mathbb{R}^n$ by the formulas $wa(\!(x)\!) = w(a+x)$. The above c is $s_i(\!(b)\!)$. This action is constantly used in the theory of Kac–Moody algebras. It is very convenient when dealing with $\Delta^\sharp$. Note that

$$\pi_b(\!(c)\!) = bu_b^{-1}(\!(c)\!) = u_b^{-1}(c) + b \text{ for } b, c \in P.$$

Let us calculate the formulas from Theorem 2.11.1 upon $q^\xi \mapsto t^{-\rho}$ as $q \to \infty$. This substitution changes the consideration, but not too much:

$$T_i\delta_b^\sharp = \begin{cases} t^{1/2}\delta^\sharp_{s_i(\!(b)\!)} & \text{if } (\alpha_i, b+d) > 0, \\ t^{-1/2}\delta^\sharp_{s_i(\!(b)\!)} + (t^{1/2} - t^{-1/2})\delta_b^\sharp & \text{as } (\alpha_i, b+d) < 0. \end{cases} \tag{2.11.10}$$

Otherwise it is zero. The formulas $\pi\delta_b^\sharp = \delta^\sharp_{\pi(\!(b)\!)}$ hold for arbitrary $\pi \in \Pi, b \in P$.

Since this calculation is different from that for generic ξ, it is not surprising that (2.11.10) does not coincide with (2.11.4) restricted to $\hat{w} = b$ and multiplied on the right by the t–symmetrizer δ^+. The representations $\lim_{q\to\infty} \Delta^\sharp$ and Δ^+ are equivalent, but the T–formulas with respect to the limit of the basis $\{\delta^\sharp\}$ are different from those in terms of the classical basis $\{\delta_b^+ = \delta_b\delta^+\}$.

2.11.4 Fourier transform

In the first place, Macdonald's nonsymmetric polynomials generalize the Matsumoto spherical functions. We use $\pi_b = bu_b^{-1} = \text{Min-length}\,\{bw,\, w \in W\}$.

Theorem 2.11.4. *(i) Let $\mathcal{P}$ be the representation of the double affine Hecke algebra $\mathcal{H}$ in the space of Laurent polynomials $\mathcal{P} = \mathbb{C}[X_b]$:*

$$T_i \;=\; t_i^{1/2} s_i \;+\; (t_i^{1/2} - t_i^{-1/2})(X_{\alpha_i} - 1)^{-1}(s_i - 1),\; 0 \le i \le n, \qquad (2.11.11)$$
$$X_b(X_c) = X_{b+c},\; \pi(X_b) = X_{\pi(b)},\; \pi \in \Pi,\; \textit{where } X_{[b,j]} = q^j X_b.$$

(ii) For generic q, t, the polynomials ε_b are uniquely defined from the relations:

$$Y_a(\varepsilon_b) = t^{(u(\rho),a)} q^{-(b,a)} \varepsilon_b, \quad \textit{where } \pi_b = bu \textit{ for } u \in W \qquad (2.11.12)$$
$$\varepsilon_b(t^{-\rho}) = 1,\; \textit{where } X_b(t^{-\rho}) = t^{-(b,\rho)}.$$

(iii) Setting $X_b^* = X_{-b}$, $q^* = q^{-1}$, and $t^* = t^{-1}$, the limit of $\varepsilon_b^*(X \mapsto \Lambda)$ as $q \to \infty$ coincides with $\phi_b(\Lambda)$ for $b \in P$.

Comment. Note that the Δ^ξ–formulas from Theorem 2.11.1 are actually the evaluations of (2.11.11) at q^ξ. To be more exact, there is an $\mathcal{H}$–homomorphism from $\mathcal{P}$ to the $\mathcal{H}$–module of functions on Δ^ξ. For instance, Theorem 2.11.1 can be deduced from Theorem 2.11.4. The formulas for the polynomial representation of the double affine Hecke algebra $\mathcal{H}$ are nothing but the **Demazure–Lusztig operators** in the affine setting.

We are going to establish a Fourier-isomorphism $\Delta^\sharp \to \mathcal{P}$, which is a generalization of the Macdonald–Matsumoto inversion formula. We use the constant term functional on Laurent series and polynomials denoted by $\langle\,\rangle$.

The first step is to make both representations unitary using

$$\mu \;=\; \prod_{\tilde\alpha \in \tilde R} \frac{1 - X_{\tilde\alpha}}{1 - tX_{\tilde\alpha}},\; \mu^0 = \mu/\langle\mu\rangle, \qquad (2.11.13)$$
$$\mu^1(\pi_b) = \mu(t^{-\pi_b((\rho))})/\mu(t^{-\rho}),\; \pi_b = bu_b^{-1}.$$

Here we treat μ as a Laurent series to define μ^0. The coefficients of μ^0 are rational functions in terms of q, t. The values $\mu^1(\pi_b)$ are rational functions in terms of $q, t^{1/2}$.

The corresponding pairings are

$$\langle f\,,\, g\rangle_{pol} = \langle f\, T_{w_0} w_0 (g(X^{-1})\, \mu^0\rangle,\; f, g \in \mathcal{P},$$
$$\langle \sum f_b \delta_b^\sharp, \sum g_b \delta_b^\sharp\rangle_{Del} = \sum (\mu^1(\pi_b))^{-1} f_b g_b.$$

Here w_0 is the longest element in W. Note that the element $T_{w_0}^2$ is central in the nonaffine Hecke algebra $\mathbf{H}$ generated by $\{T_i, i > 0\}$. Both pairings are well defined and symmetric. Let us give the formulas for the corresponding anti-involutions:

$$T_i \mapsto T_i,\ X_b \mapsto X_b,\ T_0 \mapsto T_0,\ Y_b \mapsto T_{w_0} Y^{-1}_{w_0(b)} T^{-1}_{w_0} \quad \text{in } \Delta^+,$$
$$T_i \mapsto T_i,\ Y_b \mapsto Y_b,\ T_0 \mapsto T^{-1}_{s_\vartheta} Y_\vartheta,\ X_b \mapsto T^{-1}_{w_0} X^{-1}_{w_0(b)} T_{w_0} \text{ in } \mathcal{P},$$

where $1 \le i \le n$, $b \in P$.

Theorem 2.11.5. *(i) Given $f = \sum_b f_b \delta^\sharp_b \in \Delta^\sharp$, we set $\widehat{f} = \sum_b f_b \varepsilon^*_c \in \mathcal{P}$, where $X^*_b = X^{-1}_b$, $q^* = q^{-1}, t^* = t^{-1}$. The* **inversion** *of this transform is as follows:*

$$f_b \ = \ t^{l(w_0)/2}(\mu^1(\pi_b))^{-1}\langle\, \widehat{f},\, \varepsilon^* \rangle_{pol}. \tag{2.11.14}$$

(ii) The **Plancherel formula** *reads*

$$\langle f\,,\, g\rangle_{Del} \ = \ t^{l(w_0)/2}\langle \widehat{f}\,,\, \widehat{g}\rangle_{pol}. \tag{2.11.15}$$

Both pairings are positive definite over $\mathbb{R}$ if $t = q^k$, $q > 0$, and $k > -1/h$ for the Coxeter number $h = (\rho, \vartheta) + 1$.

(iii) The transform $f = \sum_b f_b \delta^\sharp_b \mapsto \widetilde{f} = \sum_b f^*_b \delta^\sharp_b$ is an involution: $\widetilde{(\widetilde{f})} = f$. To apply it for the second time, we need to replace ε^*_b by the corresponding δ-function, which is $\sum_c \varepsilon^*_b(\pi_c)\mu^1(\pi_c)\delta^\sharp_c$.

Recall that ε^*_b becomes the Matsumoto spherical function ϕ_b in the limit $q \to \infty$ upon the substitution $X \mapsto \Lambda$. It is easy to calculate the limits of μ^0 and $\mu^1(\pi_b)$. We come to a variant of the Macdonald–Matsumoto formula. Claim (iii) has no counterpart in the p–adic theory. Technically, it is because the conjugation $*$ sends $q \mapsto q^{-1}$ and is not compatible with the limit $q \to \infty$. It is equivalent to the non–p–adic self-duality $\varepsilon_b(\pi_c) = \varepsilon_c(\pi_b)$ of the nonsymmetric Macdonald polynomials. The following theorem also has no p–adic counterpart because the Gaussian is missing.

Theorem 2.11.6. *We set $\gamma(\pi_b) = q^{(\,\pi_b((k\rho))\,,\,\pi_b((k\rho))\,)/2}$, where $t = q^k$, use $\nu_\alpha = (\alpha, \alpha)/2$, and $\rho = (1/2)\sum_{\alpha>0}\alpha$. For arbitrary $b, c \in P$,*

$$\langle \varepsilon^*_b\,, \varepsilon^*_c\, \gamma\rangle_{Del} \ = \ \gamma(\pi_0)^2\gamma(\pi_b)^{-1}\gamma(\pi_c)^{-1}\varepsilon^*_c(\pi_b)\langle 1\,,\, \gamma\rangle_{Del}, \tag{2.11.16}$$

$$\langle 1\,,\, \gamma\rangle_{Del} = \Big(\sum_{a\in P}\gamma(\pi_a)\Big) \prod_{\alpha\in R_+}\prod_{j=1}^{\infty}\Big(\frac{1 - t^{(\rho,\alpha)}q^{j\nu_\alpha}}{1 - t^{(\rho,\alpha)-1}q^{j\nu_\alpha}}\Big). \tag{2.11.17}$$

2.11.5 One-dimensional case

Let us use the Pieri formula (2.6.2) from Corollary 2.6.2 for X^{-1} and $m \ge 0$ upon the conjugation $*$:

$$X\varepsilon^*_{-m} = \frac{t^{-1/2-1}q^{-m-1} - t^{1/2}}{t^{-1}q^{-m-1} - 1}\varepsilon^*_{-m-1} - \frac{t^{-1/2} - t^{1/2}}{t^{-1}q^{-m-1} - 1}\varepsilon^*_{1+m}. \tag{2.11.18}$$

Under the limit $q \to \infty$, we obtain exactly (2.11.6) for $\phi_{-m}(\Lambda) \mapsto \varepsilon^*_{-m}(X)$:

$$X\varepsilon^*_{-m} = t^{1/2}\varepsilon^*_{-m-1} - (t^{1/2} - t^{-1/2})\varepsilon^*_{m+1}. \tag{2.11.19}$$

The generalization is Theorem 2.11.4(iii).

Also, $\lim_{q\to\infty} \mu^0_k(x) = \frac{1-X}{1-tX}$. So the limit of the pairing $\langle\, , \,\rangle_{pol}$ reads as

$$\langle f, g\rangle^{\infty}_{pol} = \langle f\, T(g)\frac{1-X}{1-tX}\rangle,\ \ f, g \in \mathcal{P}.$$

Substituting m for $m\omega$ in the indices: $\Delta^\sharp = \oplus_m \mathbb{C}\delta^\sharp_m$. In the limit $q \to \infty$, the operators T and π act here as follows:

$$T\delta^\sharp_m = t^{1/2}\delta^\sharp_{-m},\ T\delta^\sharp_{-m} = t^{-1/2}\delta^\sharp_m + (t^{1/2} - t^{-1/2})\delta^\sharp_{-m},$$

and $\pi\delta^\sharp_m = \delta^\sharp_{1-m}$ for $m \geq 0$.

Concerning $\langle\cdot,\cdot\rangle_{del}$, we use (2.7.9) for $m > 0$:

$$\mu^1(m_\sharp) = \mu^1((1-m)_\sharp) = t^{-(m-1)}\prod_{j=1}^{m-1}\frac{1-t^2q^j}{1-q^j} \tag{2.11.20}$$

$$= \prod_{j=1}^{m-1}\frac{tq^{j/2} - t^{-1}q^{-j/2}}{q^{j/2} - q^{-j/2}} \ \to\ t^{m-1} \text{ as } q \to \infty.$$

Here $\mu^1(m_\sharp) = \mu^1(\pi_{m\omega})$. Therefore the limit of this scalar product is simply

$$\langle\delta^\sharp_m, \delta^\sharp_n\rangle^{\infty}_{del} = \delta_{mn}\begin{cases} t^{1-m} & \text{if } m > 0, \\ t^m & \text{if } m \leq 0. \end{cases}$$

Comment. The latter inner product is different from the classical one, which is calculated as follows. We define the pairing

$$(T_{\hat w}, T_{\hat u}) = \text{ Constant Term } (T_{\hat w}T_{\hat u^{-1}})$$

on the affine Hecke algebra $\langle Y^{\pm 1}, T\rangle$, where the constant term is with respect to the decomposition via $T_{\hat w}$. It is simply $\delta_{\hat u,\hat w}$. Then we switch to $\delta_{\hat w} = t^{-l(\hat w)/2}T_{\hat w}$ and finally calculate (δ^+_m, δ^+_n), which is $\delta_{mn}t^{-|m|}/(1+t)$. ❑

The Fourier transform $f \mapsto \widehat{f}$ from Theorem 2.11.5 is compatible with the limit $q \to \infty$. The inversion formula (2.11.14) and the Plancherel formula (2.11.15) survive as well. We obtain a minor reformulation of the Matsumoto formulas. Upon symmetrization, T disappears from the inversion formula and we come to the Macdonald inversion.

What is completely missing in the limit is (2.11.16). One of the main applications of the double Hecke algebra is adding the Gaussian to the classical p–adic theory. Technically, one does not need $\mathcal{H}$ to do this. The ξ–deformation of the Iwahori-Matsumoto formulas (Theorem 2.11.1) is the main tool. Its justification is elementary. It is surprising that it had not been discovered well before the double Hecke algebras were introduced.

2.12 Degenerate DAHA

The p–adic origin of the double affine Hecke algebra (DAHA) is the most natural to consider, however, the connections with real harmonic analysis, radial parts and Dunkl operators are equally important. They played a key role in the beginning of the theory of DAHA; the *exact* link to the p–adic theory is relatively recent. We now discuss the trigonometric and rational **degenerate DAHAs**, which govern the applications in real harmonic analysis.

This section is a continuation of Section 1.6 from Chapter 1, where we interpreted the Harish-Chandra transform as a map from the trigonometric-differential polynomial representation of the degenerate DAHA (in Laurent polynomials) to the difference-rational polynomial representation.

Affine roots. Continuing from the previous section, let $R = \{\alpha\} \subset \mathbb{R}^n$ be a root system of type $A, B, ..., F, G$ with respect to a euclidean form (z, z') on $\mathbb{R}^n \ni z, z'$, W the Weyl group generated by the reflections s_α, R_+ be the set of positive roots corresponding to (fixed) simple roots roots $\alpha_1, ..., \alpha_n$, Γ the Dynkin diagram with $\{\alpha_i, 1 \le i \le n\}$ as the vertices, $R^\vee = \{\alpha^\vee = 2\alpha/(\alpha,\alpha)\}$ the dual root system, and

$$Q = \oplus_{i=1}^n \mathbb{Z}\alpha_i \subset P = \oplus_{i=1}^n \mathbb{Z}\omega_i,$$

where $\{\omega_i\}$ are the fundamental weights defined by $(\omega_i, \alpha_j^\vee) = \delta_{ij}$ for the simple coroots $\alpha_i^\vee$.

Recall that the form is normalized by the condition $(\alpha,\alpha) = 2$ for the *short* roots. This normalization coincides with that from the tables in [Bo] for A, C, D, E, G. Hence $\nu_\alpha = (\alpha,\alpha)/2$ can be 1, 2, or 3. We write ν_{lng} for long roots ($\nu_{\text{sht}} = 1$) and also set $nu_i = \nu_{\alpha_i}$. Let $\vartheta \in R^\vee$ be the maximal positive *coroot* (it is maximal short in R) and $\rho = (1/2)\sum_{\alpha\in R_+} \alpha \ = \sum_i \omega_i$.

The vectors $\tilde{\alpha} = [\alpha, \nu_\alpha j] \in \mathbb{R}^n \times \mathbb{R} \subset \mathbb{R}^{n+1}$ for $\alpha \in R, j \in \mathbb{Z}$ form the affine root system $\tilde{R} \supset R$ ($z \in \mathbb{R}^n$ are identified with $[z, 0]$). We add $\alpha_0 \stackrel{\text{def}}{=\!=} [-\vartheta, 1]$ to the simple roots. The set $\tilde{R}$ of positive roots is $R_+ \cup \{[\alpha, \nu_\alpha j],\ \alpha \in R,\ j > 0\}$. Let $\tilde{\alpha}^\vee = \tilde{\alpha}/\nu_\alpha$, so $\alpha_0^\vee = \alpha_0$.

The Dynkin diagram Γ of R is completed by α_0 (by $-\vartheta$ to be more exact). The notation is $\tilde{\Gamma}$. It is the completed (affine) Dynkin diagram for $R^\vee$ from [Bo] with the arrows reversed.

The set of the indices of the images of α_0 by all the automorphisms of $\tilde{\Gamma}$ will be denoted by O ($O = \{0\}$ for E_8, F_4, G_2). Let $O' = r \in O, r \neq 0$. The elements ω_r for $r \in O'$ are the so-called minuscule weights: $(\omega_r, \alpha^\vee) \le 1$ for $\alpha \in R_+$.

The affine Weyl group $\widetilde{W}$ is generated by the affine reflections $s_{\tilde{\alpha}}$. This group is the semidirect product $W \ltimes Q$ of its subgroups W and the lattice Q.

The extended Weyl group $\widehat{W}$ is generated by W and P. It is isomorphic to $W \ltimes P$ and, also, isomorphic to $\Pi \ltimes \widetilde{W}$ for the group Π formed by the elements of $\widehat{W}$ leaving $\tilde{\Gamma}$ invariant.

The latter group is isomorphic to P/Q by the natural projection $\{\omega_r \mapsto \pi_r,\ r \in O\}$, where $\omega_r = \pi_r u_r, u_r \in W,\ \pi_0 = \mathrm{id}$. The elements $\{u_r\}$ preserve the set $\{-\vartheta, \alpha_i, i > 0\}$.

Setting $\widehat{w} = \pi_r \widetilde{w} \in \widehat{W},\ \pi_r \in \Pi,\ \widetilde{w} \in \widetilde{W}$, the length $l(\widehat{w})$ is by definition the length of the reduced decomposition $\widetilde{w} = s_{i_l} \ldots s_{i_2} s_{i_1}$ in terms of the simple reflections $s_i, 0 \le i \le n$.

2.12.1 Definition of DAHA

By m, we denote the least natural number such that $(P, P) = (1/m)\mathbb{Z}$. Thus

$$m = 2 \text{ for } D_{2k},\ m = 1 \text{ for } B_{2k} \text{ and } C_k,$$

otherwise $m = |\Pi|$.

The double affine Hecke algebra depends on the parameters $q, t_\nu,\ \nu \in \{\nu_\alpha\}$. The definition ring is $\mathbb{Q}_{q,t} \stackrel{\mathbf{def}}{=} \mathbb{Q}[q^{\pm 1/m}, t^{\pm 1/2}]$, formed by the polynomials in terms of $q^{\pm 1/m}$ and $\{t_\nu^{\pm 1/2}\}$. We set

$$\begin{aligned} &t_{\tilde{\alpha}} = t_\alpha = t_{\nu_\alpha},\ t_i = t_{\alpha_i},\ q_{\tilde{\alpha}} = q^{\nu_\alpha},\ q_i = q^{\nu_{\alpha_i}}, \\ &\text{where } \tilde{\alpha} = [\alpha, \nu_\alpha j] \in \tilde{R},\ 0 \le i \le n. \end{aligned} \tag{2.12.1}$$

It will be convenient in many formulas to switch to the parameters $\{k_\nu\}$ instead of $\{t_\nu\}$, setting

$$t_\alpha = t_\nu = q_\alpha^{k_\nu} \text{ for } \nu = \nu_\alpha, \text{ and } \rho_k = (1/2)\sum_{\alpha>0} k_\alpha \alpha.$$

The notation $k_i = k_{\alpha_i}$ also will be used.

For pairwise commutative $X_1, \ldots, X_n$,

$$\begin{aligned} X_{\tilde{b}} &= \prod_{i=1}^n X_i^{l_i} q^j \text{ if } \tilde{b} = [b, j],\ \widehat{w}(X_{\tilde{b}}) = X_{\widehat{w}(\tilde{b})}. \\ &\text{where } b = \sum_{i=1}^n l_i \omega_i \in P,\ j \in \frac{1}{m}\mathbb{Z},\ \widehat{w} \in \widehat{W}. \end{aligned} \tag{2.12.2}$$

We set $(\tilde{b}, \tilde{c}) = (b, c)$ ignoring the affine extensions in the inner product unless $(\tilde{b}, \tilde{c} + d) = (b, c) + j$ is considered.

Later $Y_{\tilde{b}} = Y_b q^{-j}$ will be needed. Note the negative sign of j.

We will also use that π_r^{-1} is π_{r^*} and u_r^{-1} is u_{r^*} for $r^* \in O$, $u_r = \pi_r^{-1}\omega_r$. The reflection * is induced by the standard **nonaffine involution** of the Dynkin diagram Γ.

Definition 2.12.1. *The double affine Hecke algebra* $\mathcal{H}$ *is generated over* $\mathbb{Q}_{q,t}$ *by the elements* $\{T_i,\ 0 \le i \le n\}$, *pairwise commutative* $\{X_b,\ b \in P\}$ *satisfying (2.12.2), and the group* Π, *where the following relations are imposed:*

(o) $(T_i - t_i^{1/2})(T_i + t_i^{-1/2}) \ = \ 0,\ 0 \ \le \ i \ \le \ n;$
(i) $T_iT_jT_i... \ = \ T_jT_iT_j...,$ *m_{ij} factors on each side;*
(ii) $\pi_r T_i \pi_r^{-1} \ = \ T_j$ *if* $\pi_r(\alpha_i) = \alpha_j;$
(iii) $T_iX_bT_i \ = \ X_bX_{\alpha_i}^{-1}$ *if* $(b, \alpha_i^\vee) = 1,\ 0 \le i \le n;$
(iv) $T_iX_b \ = \ X_bT_i$ *if* $(b, \alpha_i^\vee) = 0$ *for* $0 \le i \le n;$
(v) $\pi_r X_b \pi_r^{-1} \ = \ X_{\pi_r(b)} \ = \ X_{u_r^{-1}(b)} q^{(\omega_{r^*}, b)},\ r \in O'.$ ❑

Given $\widetilde{w} \in \widetilde{W}, r \in O,$ the product

$$T_{\pi_r \widetilde{w}} \stackrel{\mathbf{def}}{=\!=} \pi_r \prod_{k=1}^{l} T_{i_k}, \quad \text{where} \quad \widetilde{w} = \prod_{k=1}^{l} s_{i_k}, l = l(\widetilde{w}), \tag{2.12.3}$$

does not depend on the choice of the reduced decomposition (because $\{T\}$ satisfy the same "braid" relations as $\{s\}$ do). Moreover,

$$T_{\hat{v}} T_{\widehat{w}} \ = \ T_{\hat{v}\widehat{w}} \ \text{ whenever } l(\hat{v}\widehat{w}) = l(\hat{v}) + l(\widehat{w}) \ \text{ for } \ \hat{v}, \widehat{w} \in \widehat{W}. \tag{2.12.4}$$

In particular, we arrive at the pairwise commutative elements from the previous section:

$$Y_b \ = \ \prod_{i=1}^{n} Y_i^{l_i} \ \text{ if } \ b = \sum_{i=1}^{n} l_i \omega_i \in P, \ \text{ where } \ Y_i \stackrel{\mathbf{def}}{=\!=} T_{\omega_i}. \tag{2.12.5}$$

They satisfy the relations

$$\begin{aligned} T_i^{-1} Y_b T_i^{-1} \ &= \ Y_b Y_{\alpha_i}^{-1} \ \text{ if } \ (b, \alpha_i^\vee) = 1, \\ T_i Y_b \ &= \ Y_b T_i \ \text{ if } \ (b, \alpha_i^\vee) = 0, \ 1 \le i \le n. \end{aligned} \tag{2.12.6}$$

For arbitrary nonzero q, t, any element $H \in \mathcal{H}$ has a unique decomposition in the form

$$H = \sum_{w \in W} g_w\, f_w\, T_w, \ g_w \in \mathbb{Q}_{q,t}[X], \ f_w \in \mathbb{Q}_{q,t}[Y], \tag{2.12.7}$$

and five more analogous decompositions corresponding to the other orderings of $\{T, X, Y\}$. It makes the polynomial representation (to be defined next) the $\mathcal{H}$-module induced from the one-dimensional representation $T_i \mapsto t_i^{1/2}$, $Y_i \mapsto Y_i^{1/2}$ of the affine Hecke subalgebra $\mathcal{H}_Y = \langle T, Y \rangle$.

Automorphisms. The following maps can be uniquely extended to automorphisms of $\mathcal{H}$ (see [C20],[C28]):

$$\varepsilon : \ X_i \mapsto Y_i, \ Y_i \mapsto X_i, \ T_i \mapsto T_i^{-1}\,(i \ge 1), \ t_\nu \mapsto t_\nu^{-1}, \ q \mapsto q^{-1}, \tag{2.12.8}$$

$$\tau_+ : X_b \mapsto X_b, \ Y_r \mapsto X_r Y_r q^{-\frac{(\omega_r, \omega_r)}{2}}, \ T_i \mapsto T_i\,(i \ge 1), \ \ t_\nu \mapsto t_\nu, \ q \mapsto q,$$

$$\tau_+ : \ Y_\vartheta \mapsto q^{-1} X_\vartheta T_0^{-1} T_{s_\vartheta}, \ T_0 \mapsto q^{-1} X_\vartheta T_0^{-1}, \ \text{ and} \tag{2.12.9}$$

$$\tau_- \stackrel{\mathbf{def}}{=\!=} \varepsilon \tau_+ \varepsilon, \ \text{ and } \ \sigma \stackrel{\mathbf{def}}{=\!=} \tau_+ \tau_-^{-1} \tau_+ = \tau_-^{-1} \tau_+ \tau_-^{-1} = \varepsilon \sigma^{-1} \varepsilon, \tag{2.12.10}$$

where $r \in O'$. In the definition of $\tau_\pm$ and σ, we need to add $q^{\pm 1/(2m)}$ to $\mathbb{Q}_{q,t}$.

Here the quadratic relation (o) from Definition 2.12.1 may be omitted. Only the group relations matter. Thus these automorphisms act in the corresponding braid groups extended by fractional powers of q treated as central elements.

The elements $\tau_\pm$ generate the projective $PSL(2,\mathbb{Z})$, which is isomorphic to the braid group B_3 due to Steinberg. Adding ε, we obtain the projective $PGL(2,\mathbb{Z})$.

These and the statements below are from [C16]. See also Chapter 3. Note that $\mathcal{H}$, its degenerations below, and the corresponding polynomial representations are actually defined over $\mathbb{Z}$ extended by the parameters of DAHA. We prefer to stick to the field $\mathbb{Q}$, because the Lusztig isomorphisms will require $\mathbb{Q}$.

2.12.2 Polynomials, intertwiners

The Demazure–Lusztig operators are defined as follows:

$$T_i \;=\; t_i^{1/2}s_i \;+\; (t_i^{1/2}-t_i^{-1/2})(X_{\alpha_i}-1)^{-1}(s_i-1),\ 0\le i\le n, \qquad (2.12.11)$$

and obviously preserve $\mathbb{Q}[q,t^{\pm 1/2}][X]$. We note that only the formula for T_0 involves q:

$$\begin{aligned} T_0 &= t_0^{1/2}s_0 \;+\; (t_0^{1/2}-t_0^{-1/2})(qX_\vartheta^{-1}-1)^{-1}(s_0-1),\\ \text{where }\ s_0(X_b) &= X_bX_\vartheta^{-(b,\vartheta)}q^{(b,\vartheta)},\ \alpha_0=[-\vartheta,1]. \end{aligned} \qquad (2.12.12)$$

The map sending T_j to the formula in (2.12.11), $X_b \mapsto X_b$ (see (2.12.2)), and $\pi_r \mapsto \pi_r$ induces a $\mathbb{Q}_{q,t}$–linear homomorphism from $\mathcal{H}$ to the algebra of linear endomorphisms of $\mathbb{Q}_{q,t}[X]$. This $\mathcal{H}$–module, which will be called the **polynomial representation**, is faithful and remains faithful when q,t take any nonzero complex values, assuming that q is not a root of unity.

The images of the Y_b are called the **difference Dunkl operators**. To be more exact, they must be called difference-trigonometric Dunkl operators, because there are also difference-rational Dunkl operators.

Intertwining operators. The Y–intertwiners (see [C23]) are introduced as follows:

$$\begin{aligned} \Psi_i &= T_i + (t_i^{1/2}-t_i^{-1/2})(Y_{\alpha_i}^{-1}-1)^{-1}\ \text{ for }\ 1\le i\le n,\\ \Psi_0 &= X_\vartheta T_{s_\vartheta} - (t_0^{1/2}-t_0^{-1/2})(Y_{\alpha_0}-1)^{-1}\\ &= Y_0T_0X_0 + (t_0^{1/2}-t_0^{-1/2})(Y_{\alpha_0}^{-1}-1)^{-1},\ \text{ for }\ Y_0=Y_{\alpha_0}=q^{-1}Y_\vartheta^{-1},\\ F_i &= \Psi_i(\psi_i)^{-1},\ \psi_i = t_i^{1/2}+(t_i^{1/2}-t_i^{-1/2})(Y_{\alpha_i}^{-1}-1)^{-1}. \end{aligned} \qquad (2.12.13)$$

These formulas are the ε–images of the formulas for the X–intertwiners, which are a straightforward generalization of those in the affine Hecke theory.

The intertwiners belong to $\mathcal{HH}$ extended by the rational functions in terms of $\{Y\}$. The F are called the **normalized intertwiners**. The elements

$$F_i,\ P_r \stackrel{\mathbf{def}}{=\!=} X_r T_{u_r^{-1}},\ 0 \le i \le n,\ r \in O',$$

satisfy the same relations as $\{s_i, \pi_r\}$ do, so the map

$$\widehat{w} \mapsto F_{\widehat{w}} \;=\; P_r F_{i_l} \cdots F_{i_1}, \ \text{ where } \widehat{w} = \pi_r s_{i_l} \cdots s_{i_1} \in \widehat{W}, \tag{2.12.14}$$

is a well defined homomorphism from $\widehat{W}$.

The intertwining property is

$$F_{\widehat{w}} Y_b F_{\widehat{w}}^{-1} = Y_{\widehat{w}(b)} \ \text{ where } \ Y_{[b,j]} \stackrel{\mathbf{def}}{=\!=} Y_b q^{-j}.$$

The P_1 in the case of GL is due to Knop and Sahi.

As to Ψ_i, they satisfy the homogeneous Coxeter relations and those with Π_r. So we may set $\Psi_{\widehat{w}} = P_r \Psi_{i_l} \cdots \Psi_{i_1}$ for the reduced decompositions. They intertwine Y as well.

The formulas for Ψ_i when $1 \le i \le n$ are well known in the theory of affine Hecke algebras. The affine intertwiners, those for $\widehat{w} \in \widehat{W}$, are the raising operators for the Macdonald nonsymmetric polynomials, serve the Harish-Chandra spherical transform and Opdam's nonsymmetric transform, and are a key tool in the theory of semisimple representations of DAHAs.

2.12.3 Trigonometric degeneration

We set $\mathbb{Q}_k \stackrel{\mathbf{def}}{=\!=} \mathbb{Q}[k_\alpha]$. If the integral coefficients are needed, we take $\mathbb{Z}_k \stackrel{\mathbf{def}}{=\!=} \mathbb{Z}[k_\alpha, 1/m]$ as the definition ring.

The **degenerate (graded) double affine Hecke algebra** $\mathcal{HH}'$ is the span of the group algebra $\mathbb{Q}_k\widehat{W}$ and the pairwise commutative

$$y_{\tilde{b}} \stackrel{\mathbf{def}}{=\!=} \sum_{i=1}^{n} (b, \alpha_i^\vee) y_i + u \ \text{ for } \ \tilde{b} = [b, u] \in P \times \mathbb{Z},$$

satisfying the following relations:

$$\begin{aligned} s_j y_b - y_{s_j(b)} s_j \;&=\; -k_j (b, \alpha_j),\ (b, \alpha_0) \stackrel{\mathbf{def}}{=\!=} -(b, \vartheta), \\ \pi_r y_{\tilde{b}} \;&=\; y_{\pi_r(\tilde{b})} \pi_r \ \text{ for } \ 0 \le j \le n,\ r \in O. \end{aligned} \tag{2.12.15}$$

Comment. Without s_0 and π_r, we arrive at the defining relations of the graded affine Hecke algebra from [Lus1]. The algebra $\mathcal{HH}'$ has two natural polynomial representations via the differential-trigonometric and difference-rational Dunkl operators. There is also a third one, the representation in terms of infinite differential-trigonometric Dunkl operators, which generates (at trivial center charge) differential-elliptic W–invariant operators generalizing those due to Olshanetsky–Perelomov. See, e.g., [C23]. We will need in this section only the (most known) differential-trigonometric polynomial representations. ❑

Let us establish a connection with the general DAHA. We set

$$q = \exp(\mathfrak{v}),\ t_j = q_i^{k_i} = q^{\nu_{\alpha_i} k_i},\ Y_b = \exp(-\mathfrak{v} y_b),\ \mathfrak{v} \in \mathbb{C}.$$

Using ε from (2.12.8), the algebra $\mathcal{H}$ is generated by Y_b, T_i for $1 \le i \le n$, and

$$\varepsilon(T_0) = X_\vartheta T_{s_\vartheta},\ \varepsilon(\pi_r) = X_r T_{u_r^{-1}},\ r \in O'.$$

It is straightforward to see that the relations (2.12.15) for the y_b, $s_i (i > 0)$, s_0, and the π_r are the leading coefficients of the $\mathfrak{v}$–expansions of the general relations for this system of generators. Thus $\mathcal{H}'$ is $\mathcal{H}$ in the limit $\mathfrak{v} \to 0$.

When calculating the limits of the Y_b in the polynomial representation, the "trigonometric" derivatives of $\mathbb{Q}[X]$ appear:

$$\partial_a(X_b) \;=\; (a, b)X_b,\ a, b \in P,\ \ w(\partial_b) = \partial_{w(b)},\ w \in W.$$

The limits of the formulas for Y_b acting in the polynomial representation are the **trigonometric Dunkl operators**

$$\mathcal{D}_b \stackrel{\mathbf{def}}{=\!=} \partial_b + \sum_{\alpha \in R_+} \frac{k_\alpha(b, \alpha)}{(1 - X_\alpha^{-1})}(1 - s_\alpha) - (\rho_k, b). \tag{2.12.16}$$

They act on the Laurent polynomials $f \in \mathbb{Q}_k[X]$, are pairwise commutative, and $y_{[b,u]} = \mathcal{D}_b + u$ satisfy (2.12.15) for the following action of the group $\widehat{W}$:

$$w^x(f) = w(f)\ \text{ for }\ w \in W,\ b^x(f) = X_b f\ \text{ for }\ b \in P.$$

For instance, $s_0^x(f) = X_\vartheta s_\vartheta(f)$, and $\pi_r^x(f) = X_r u_r^{-1}(f)$.

Degenerating $\{\Psi\}$, one obtains the intertwiners of $\mathcal{H}'$:

$$\Psi_i' = s_i + \frac{\nu_i k_i}{y_{\alpha_i}},\ 0 \le i \le n,\ \Big(\Psi_0' = X_\vartheta s_\vartheta + \frac{k_0}{1 - y_\vartheta} \text{ in } \mathbb{Q}_k[X]\Big),$$
$$P_r' = \pi_r,\ \big(P_r' = X_r u_r^{-1} \text{ in } \mathbb{Q}_k[X]\big),\ r \in O'. \tag{2.12.17}$$

The operator P_1' in the case of GL (it is of infinite order) plays the key role in [KnS].

Recall that the general normalized intertwiners are

$$F_i = \Psi_i \psi_i^{-1},\ \psi_i = t^{1/2} + (t_i^{1/2} - t^{1/2})(Y_{\alpha_i}^{-1} - 1)^{-1}.$$

Their limits are

$$F_i' = \Psi_i'(\psi_i')^{-1},\ \psi_i' = 1 + \frac{\nu_i k_i}{y_{\alpha_i}}.$$

They satisfy the unitarity condition $(F_i')^2 = 1$, and the products $F_{\widehat{w}}'$ can be defined for any decompositions of $\widehat{w}$. One then has:

$$F_{\widehat{w}}'\, y_b\, (F_{\widehat{w}}')^{-1} \;=\; y_{\widehat{w}(b)}.$$

Equating

$$F_i = F_i' \text{ for } 0 \le i \le n,\ P_r = P_r' \text{ for } r \in O,$$

we come to the formulas for $T_i\,(0 \le i \le n),\ X_r\,(r \in O')$ in terms of s_i, y_b, and $Y_b = \exp(-\mathfrak{v} y_b)$.

These formulas determine the **Lusztig homomorphism** $æ'$ from $\mathcal{H}$ to the completion $\mathbb{Z}_{k,q,t}\mathcal{H}'[[\mathfrak{v} y_b]]$ for $\mathbb{Z}_{k,q,t} \stackrel{\mathbf{def}}{=\!=} \mathbb{Z}_k\mathbb{Z}_{q,t}$. See, e.g., [C23].

For instance, $X_r \in \mathcal{H}$ becomes $\pi_r T_{u_r^{-1}}^{-1}$ in $\mathcal{H}'$, where the T–factor has to be further expressed in terms of s, y. In the degenerate polynomial representation, $æ'(X_r)$ acts as $X_r(æ'(T_{u_r^{-1}})u_r)^{-1}$, not as straightforward multiplication by X_r. However, these two actions coincide in the limit $\mathfrak{v} \to 0$, since T_w become w.

Upon the $\mathfrak{v}$–completion, we obtain an isomorphism

$$æ' : \mathbb{Q}_k[[\mathfrak{v}]] \otimes \mathcal{H} \to \mathbb{Q}_k[[\mathfrak{v}]] \otimes \mathcal{H}'.$$

We will use the notation $(d, [\alpha, j]) = j$. For instance, $(b{+}d, \alpha_0) = 1{-}(b, \vartheta)$.

Treating $\mathfrak{v}$ as a nonzero number, an arbitrary $\mathcal{H}'$–module V' that is a union of finite dimensional Y–modules has a natural structure of an $\mathcal{H}$–module provided that we have

$$\begin{aligned} &q^{(\alpha_i,\xi+d)} = t_i \Rightarrow (\alpha_i, \xi + d) = \nu_i k_i, \\ &q^{(\alpha_i,\xi+d)} = 1 \Rightarrow (\alpha_i, \xi + d) = 0, \quad \text{where} \\ &0 \le i \le n,\ y_b(v') = (b, \xi)v' \ \text{ for } \ \xi \in \mathbb{C}^n,\ 0 \ne v' \in V'. \end{aligned} \tag{2.12.18}$$

For the modules of this type, the map $æ'$ is over the ring $\mathbb{Q}_{k,q,t}$ extended by $(\alpha, \xi + d),\ q^{(\alpha,\xi+d)}$ for $\alpha \in R$, and y–eigenvalues ξ. Moreover, we need to localize by $(1 - q^{(\alpha,\xi+d)}) \ne 0$ and by $(\alpha, \xi + d) \ne 0$. Upon such extension and localization, $æ'$ is defined over $\mathbb{Z}_{k,q,t}$ if the module is y–semisimple. If there are nontrivial Jordan blocks, then the formulas will contain factorials in the denominators.

For instance, let $\mathcal{I}'[\xi]$ be the $\mathcal{H}'$–module induced from the one-dimensional y–module $y_b(v) = (b, \xi)v$. Assuming that q is not a root of unity, the mapping $æ'$ supplies it with a structure of $\mathcal{H}$–module if

$$q^{(\alpha,\xi)+\nu_\alpha j} = t_\alpha \text{ implies } (\alpha, \xi) + \nu_\alpha j = \nu_\alpha k_\alpha$$

for every $\alpha \in R, j \in \mathbb{Z}$, and the corresponding implications hold for t replaced by 1. This means that

$$(\alpha, \xi) - \nu_\alpha k_\alpha \notin \nu_\alpha \mathbb{Z} + \frac{2\pi\imath}{\mathfrak{v}}(\mathbb{Z} \setminus \{0\}) \not\ni (\alpha, \xi) \text{ for all } \alpha \in R. \quad (2.12.19)$$

Generalizing, we obtain that æ$'$ is well defined for any $\mathcal{H}'$–module generated by its y–eigenvectors with the y–eigenvalues ξ satisfying this condition, assuming that $\mathfrak{v} \notin \pi\imath\mathbb{Q}$.

Comment. Actually, there are at least four different variants of æ$'$ because the normalization factors ψ, ψ' may be associated with different one-dimensional characters of the affine Hecke algebra $\langle T, Y\rangle$ and its degeneration. It is also possible to multiply the normalized intertwiners by the characters of $\widehat{W}$ before equating. Note that if we divide the intertwiners Ψ and/or Ψ' by ψ, ψ' on the left in the definition of F, F', it corresponds to switching from $T_i \mapsto t_i$ to the character $T_i \mapsto -t_i^{-1/2}$ together with the multiplication by the sign-character of $\widetilde{W}$. ❑

2.12.4 Rational degeneration

The limit to the **rational Dunkl operators** is as follows. We set $X_b = e^{\mathfrak{w} x_b}$, $d_b(x_c) = (b, c)$, so the above derivatives ∂_b become $\partial_b = (1/\mathfrak{w})d_b$. In the limit $\mathfrak{w} \to 0$, $\mathfrak{w}\mathcal{D}_b$ tends to the operators

$$D_b \stackrel{\text{def}}{=\!=} d_b + \sum_{\alpha \in R_+} \frac{k_\alpha (b, \alpha)}{x_\alpha}(1 - s_\alpha), \quad (2.12.20)$$

which were introduced by Dunkl and, as a matter of fact, gave birth to the DAHA direction, although KZ and the Macdonald theory must be mentioned too as origins of DAHAs.

These operators are pairwise commutative and satisfy the cross-relations

$$D_b x_c - x_c D_b = (b, c) + \sum_{\alpha > 0} k_\alpha (b, \alpha)(c, \alpha^\vee) s_\alpha, \quad \text{for } b, c \in P. \quad (2.12.21)$$

These relations, the commutativity of D, the commutativity of x, and the W–equivariance

$$w\, x_b\, w^{-1} = x_{w(b)}, \ w\, D_b\, w^{-1} = D_b \quad \text{for } b \in P_+, \ w \in W,$$

are the defining relations of the **rational DAHA** $\mathcal{H}''$. The references are [CM] (the case of A_1) and [EG]; however, the key part of the definition is the commutativity of D_b due to Dunkl [Du1]. The Dunkl operators and the operators of multiplication by the x_b form the **polynomial representation** of $\mathcal{H}''$, which is faithful. It readily justifies the PBW theorem for $\mathcal{H}''$.

Note that in contrast to the q,t–setting, the definition of the rational DAHA can be extended to finite groups generated by complex reflections (Dunkl, Opdam and Malle). There is also a generalization due to Etingof–Ginzburg from [EG] (the symplectic reflection algebras).

Comment. Following [CO], there is a one-step limiting procedure from $\mathcal{H}$ to $\mathcal{H}''$. We set

$$Y_b = \exp(-\sqrt{\mathfrak{u}}D_b),\ X_b = e^{\sqrt{\mathfrak{u}}x_b},$$

assuming that $q = e^{\mathfrak{u}}$ and $\mathfrak{u} \to 0$. We come directly to the relations of the rational DAHA and the formulas for D_b. The advantage of this direct construction is that the automorphisms $\tau_\pm$ survive in the limit. Indeed, τ_+ in $\mathcal{H}$ can be interpreted as the formal conjugation by the q–Gaussian $q^{x^2/2}$, where $x^2 = \sum_i x_{\omega_i} x_{\alpha_i^\vee}$. In the limit, it becomes the conjugation by $e^{x^2/2}$, preserving $w \in W$, x_b, and taking D_b to $D_b - x_b$. Respectively, τ_- preserves w and D_b, and sends $x_b \mapsto x_b - D_b$. These automorphisms do not exist in the $\mathcal{H}'$. ❑

The **abstract Lusztig-type map** from $\mathcal{H}'$ to $\mathcal{H}''$ is as follows. Let $w \mapsto w$ and $X_b = e^{\mathfrak{w}x_b}$. We expand X_α in terms of x_α in the formulas for the trigonometric Dunkl operators $\mathcal{D}_b$:

$$\mathcal{D}_b = \frac{1}{\mathfrak{w}}D_b - (\rho_k, b) + \sum_{\alpha\in R_+} k_\alpha(b,\alpha)\sum_{m}^{\infty}\frac{B_m}{m!}(-\mathfrak{w}x_\alpha)^m\,(1-s_\alpha) \qquad (2.12.22)$$

for the Bernoulli numbers B_m. Then we use these formulas as abstract expressions for y_b in terms of the generators of $\mathcal{H}''$: $y_b = \frac{1}{\mathfrak{w}}D_b + \ldots$.

One obtains an isomorphism $\text{æ}'' : \mathbb{Q}[[\mathfrak{w}]]\otimes\mathcal{H}' \to \mathbb{Q}[[\mathfrak{w}]]\otimes\mathcal{H}''$, which maps $\mathcal{H}'$ to the extension of $\mathcal{H}''$ by the formal series in terms of $\mathfrak{w}x_b$. An arbitrary representation V'' of $\mathcal{H}''$ that is a union of finite dimensional $\mathbb{Q}_k[x]$–modules becomes an $\mathcal{H}'$–module provided that

$$\mathfrak{w}\zeta_\alpha \notin 2\pi\imath(\mathbb{Z}\setminus\{0\}) \ \text{ for } \ x_b(v) = (\zeta, b)v, 0 \neq v \in V''. \qquad (2.12.23)$$

Similar to (2.12.19), this constraint simply restricts choosing $\mathfrak{w} \neq 0$. The formulas for y_b become locally finite in any representations of $\mathcal{H}''$, where x_b act locally nilpotent, for instance, in finite dimensional $\mathbf{H}''$–modules. In this case, there are no restrictions for $\mathfrak{w}$.

Comment. Note that the "identification" of $\mathcal{H}'$ and $\mathcal{H}''$ has a common source with the method used in the so-called localization due to Opdam and Rouquier (see [GGOR], [VV2]). They separate the differentials d_b from the formula for the Dunkl operators to define a KZ-type connection with values in double Hecke modules. We equate these differentials in the rational and trigonometric formulas for the Dunkl operators to connect $\mathcal{H}'$ and $\mathcal{H}''$.

❑

Finally, the composition

$$\text{æ} \stackrel{\text{def}}{=} \text{æ}'' \circ \text{æ}' : \mathcal{H}[[\mathfrak{v}, \mathfrak{w}]] \to \mathcal{H}''[[\mathfrak{v}, \mathfrak{w}]]$$

is an isomorphism. Without the completion, it makes an arbitrary finite dimensional $\mathcal{H}''$–module V'' a module over $\mathcal{H}$ as $q = e^{\mathfrak{v}}, t_\alpha = q^{k_\alpha}$ for sufficiently general (complex) nonzero numbers $\mathfrak{v}, \mathfrak{w}$. This isomorphism was discussed in [BEG] (Proposition 7.1).

The finite dimensional representations are the most natural here because, on the one hand, æ$''$ lifts the modules that are unions of finite dimensional x–modules to those for X, on the other hand, æ$'$ maps the $\mathcal{H}'$–modules that are unions of finite dimensional y–modules to those for Y. So one must impose these conditions for both x and y.

We obtain a functor from the category of finite dimensional representations of $\mathcal{H}''$ to that for $\mathcal{H}$. Indeed, it is known now that there are finitely many irreducible objects in the former category. Therefore we can find $\mathfrak{v}, \mathfrak{w}$ serving *all* irreducible representations and their extensions. This functor is *faithful* provided that q is generic in the following sense:

$$q^a t^b = 1 \Rightarrow a + kb = 0 \quad \text{for} \quad a, b \in \mathbb{Q}.$$

This condition ensures the *equivalence of the categories* of finite dimensional representations for $\mathcal{H}$ and $\mathcal{H}''$.

Using æ for infinite dimensional representations is an interesting problem. It makes the theory analytic. For instance, the triple composition æ$'' \circ \mathcal{G} \circ$ æ$'$ for the inverse Opdam transform $\mathcal{G}$ (see [O3] and formula (6.1) from [C24]) embeds $\mathcal{H}$ in $\mathcal{H}''$ and identifies the $\mathcal{H}''$–module $\mathbb{C}_c^\infty(\mathbb{R}^n)$ with the $\mathcal{H}$–module of PW-functions under the condition $\Re k > -1/h$. See [O3, C24] for more detail.

The degenerations above play the role of the Lie algebras in the theory of DAHA. The Lusztig isomorphisms are certain counterparts of the exp-log maps. It is especially true for the composition map from $\mathcal{H}$ to the rational degeneration, because the latter has the projective action of the $PSL(2, \mathbb{Z})$ and some other features that make it close to the general $\mathcal{H}$. The algebra $\mathcal{H}'$ is not projective $PSL(2, \mathbb{Z})$–invariant.

Note that there are some special "rational" symmetries and tools that have no q, t–counterparts. On the other hand, the semisimple representations have no immediate analogs in the rational theory and the perfect representations are simpler to deal with in the q, t–case. Thus it really resembles the relation between Lie groups and Lie algebras.

2.12.5 Diagonal coinvariants

It was conjectured by Haiman [Ha1] that the space of **diagonal coinvariants** for a root system R of rank n has a "natural" quotient of dimension $(1 + h)^n$

for the Coxeter number h. This space is the quotient $\mathbb{C}[x,y]/(\mathbb{C}[x,y]\mathbb{C}[x,y]_o^W)$ for the algebra of polynomials $\mathbb{C}[x,y]$ with the diagonal action of the Weyl group on $x \in \mathbb{C}^n \ni y$ and the ideal $\mathbb{C}[x,y]_o^W \subset \mathbb{C}[x,y]^W$ of the W–invariant polynomials without the constant term.

In [Go], such a quotient was constructed. It coincides with the whole space of the diagonal coinvariants in the A_n–case due to Haiman, but this does not hold for other root systems.

Let $k_{\text{sht}} = -(1+1/h) = k_{\text{lng}}$, h be the Coxeter number $1+(\rho,\vartheta)$. The polynomial representation $\mathbb{Q}[x]$ of $\mathcal{H}''$ has a unique nonzero irreducible quotient-module. It is of dimension $(1+h)^n$, which was checked in [BEG] and [Go], and also follows from [C28] via æ. See Chapter 3.

The application of this representation to the coinvariants of the ring of commutative polynomials $\mathbb{Q}[x,y]$ with the diagonal action of W is as follows.

The polynomial representation $\mathbb{Q}[x]$ is naturally a quotient of the *linear space* $\mathbb{Q}[x,y]$ considered as an induced $\mathcal{H}''$–module from the one-dimensional W–module $w \mapsto 1$. So is V''. The subalgebra $(\mathcal{H}'')^W$ of the W–invariant elements from $\mathcal{H}''$ preserves the image of $\mathbb{Q}\delta$ in V'' for $\delta \stackrel{\text{def}}{=\!=} \prod_{\alpha>0} x_\alpha$.

Let $I_o \subset (\mathbf{H}'')^W$ be the ideal of the elements vanishing at the image of δ in V''. Gordon proves that V'' *coincides with the quotient* $\tilde{V}''$ *of* $\mathcal{H}''(\delta)$ *by the* $\mathcal{H}''$*-submodule* $\mathcal{H}''I_o(\delta)$. It is sufficient to check that $\tilde{V}''$ is irreducible.

The graded space $\mathrm{gr}(V'')$ of V'' with respect to the total x,y–degree of the polynomials is isomorphic as a linear space to the quotient of $\mathbb{Q}[x,y]\delta$ by the graded image of $\mathcal{H}''I_o(\delta)$. The latter contains $\mathbb{Q}[x,y]_o^W\delta$ for the ideal $\mathbb{Q}[x,y]_o^W \subset \mathbb{Q}[x,y]^W$ of the W–invariant polynomials without the constant term. Therefore V'' becomes a certain quotient of $\mathbb{Q}[x,y]/(\mathbb{Q}[x,y]\mathbb{Q}[x,y]_o^W)$. See [Go],[Ha1] about the connection with the Haiman theorem in the A_n–case and related questions for other root systems.

The irreducibility of the $\tilde{V}''$ above is the key fact. The proof from [Go] requires considering a KZ-type local systems and the technique from [GGOR]. We will demonstrate in the next chapter that the irreducibility can be readily proved in the q,t–case by using the passage to the roots of unity and therefore gives an entirely algebraic and simple proof of Gordon's theorem via the æ–isomorphism.

Chapter 3

General theory

3.0 Progenitors

Following [C23], [C28], and the previous chapter, we will study DAHA, its representations, and the q–Fourier transform. The basic representations of this chapter are those in Laurent polynomials, Laurent polynomials multiplied by the Gaussian, and in various spaces of delta functions. We also consider induced, semisimple, spherical, and finite dimensional representations in this chapter. The q–Fourier transform generalizes the Harish-Chandra spherical transform, its p–adic counterpart due to Macdonald–Matsumoto, and the Fourier transforms for the Heisenberg and Weyl algebras. The applications of this transform to Verlinde algebras, Gauss–Selberg integrals and sums, Dedekind–Macdonald-type η–type identities, and the diagonal coinvariants will be discussed.

3.0.1 Fourier theory

There are two major directions in the q–Fourier analysis, compact and non-compact, generalizing the corresponding parts of the harmonic analysis on symmetric spaces. The *compact direction* is based on the imaginary integration and its variant, the constant term functional. The representation of DAHA in Laurent polynonials, called the *polynomial representation*, and in Laurent polynomials multiplied by the Gaussian, a counetrpart of the Schwartz space, are the key here. The theory of Fourier transforms of these spaces includes the Macdonald conjectures (now the theorems), the Mehta–Macdonald formula, and has other applications of combinatorial nature.

Compact case. Concerning the Macdonald conjecture and the Mehta–Macdonald formula, we refer to [C16, C19, C20, C21, M8]. They will not be proven in the chapter, as well as their Jackson counterparts. They were justified using the shift operator. This proof does not add too much to the theory of DAHA, however, the formulas themselves are definitely useful; they help to control the reducibility of the polynomial representation and the singularities of the Fourier transforms.

As a matter of fact, the evaluation and duality formulas follow *directly* from the existence of the *duality involution*, an outer automorphism of DAHA. The latter readily gives the duality formula; the evaluation formula requires the Pieri rules, based on the duality involution too. It is very much similar to the case of A_1 considered in Chapters 1 and 2 in detail. The norm formula can be readily deduced from the evaluation formula using the action of the intertwining operators on the nonsymmetric Macdonald polynomials; we discussed this approach in the chapter.

By the way, it gives the final answer to an old question in the theory of Macdonald polynomials. It was always surprising that the formulas for the evalutaions and norms are given in terms of the Macdonald measure from the definition of the Macdonald polynomials. Now we understand that this reflects the self duality of the Fourier transform in the space of nonsymmetric polynomials.

Thus, we arrive at a new proof of the Macdonald conjectures, reasonably simple and entirely conceptual, that does not require the shift operator and goes uniformly for *arbitrary* k. The k is a DAHA counterpart of the root multiplicity in the theory of symmetric spaces. The shift operator technique is induction with respect to *integral* positive k followed by the analytic continuation to arbitray k.

Noncompact case. The *noncompact direction* is, generally, the theory of integration over $\mathbb{R}^n$. This integration is not considered in this chapter; the theory is not finished at the moment. Instead, we deal with its discrete variant, the *Jackson integration.* The Jackson–Fourier transform is the *inverse* of the "compact" one mentioned above. The Jackson counterparts of the Macdonald norm formula and Mehta–Macdonald integral are needed for the Plancherel formulas. They still remain unknown for the real integration for general root systems (the A_1–case is considered in Section 2.3.5).

Generally, there are *eight* different variants of the DAHA harmonic analysis (without taking into consideration the degenerations of DAHA and different functional spaces):

We have one compact and two noncompact theories (the real and Jackson integrations). Then, the compact and noncompact directions exist in two different variants:

(a) standard, as $0 < q < 1$ or $q > 1$, and (b) unimodular for $|q| = 1$. This gives *six* analytic theories.

The *seventh* theory is at the roots of unity, where the compact and noncompact directions coincide and all irreducible representations are finite dimensional.

Concening the *eighth* theory, there is a challenging parallelism between our cyclotomic Gaussian sums and the so-called modular Gauss–Selberg sums (see, for instance, [Ev]), although almost nothing has been done so far to establish an exact relation. In the modular case, Selberg-type kernels are calculated in finite fields and are embedded "into" the roots of unity right before the summation. Our sums are defined entirely in cyclotomic fields. Presumably they generalize the modular sums. If this is true, DAHAs could be used for the p–adic integration, the eighth theory, directly realted to the Gaussian sums. Connecting the p–adic Gamma with the q–Gamma function seems an important step in this direction.

The analytic aspects are discussed here to put the q–Fourier transform into perspective, however, this chapter is mainly devoted to the algebraic aspects. The basic functional spaces considered in this chapter of of algebraic nature.

We note that, generally, harmonic analysis is about the decomposition of the L^2–space and other standard functional spaces in terms of irreducible representations. It is somewhat different in DAHA theory; an irreducible representation of DAHA is, in a sense, already a "theory." The standard functional spaces become irreducible DAHA–modules, unless q, k, or the corresponding weights are special. Calculating Fourier transforms of arbitrary irreducible representations seems one of the main objectives of the DAHA–Fourier analysis. The decomposition of the whole DAHA, treated as a regular representation, has not been discussed so far.

Similar to the theory of Heisenberg and Weyl algebras, the Fourier transform is associated with a certain outer automorphism of the double Hecke algebra. This fix it uniquely up to proportionality in irreducible representations. Thus, the problem of harmonic analysis is to calculate it explicitly, i.e., to make the Fourier transform an inner homomorphism or automorphism, in a proper *analytic* setting.

Representation theory. The semisimple and unitary representations are of major importance. Their description will be reduced below to a certain combinatorial problem, which can be managed in several cases, including the case of GL_n (apart from the roots of unity). The classification depends on the structural constants $q, t = q^k$ of the double Hecke algebras. For generic q, k, the DAHA modules induced up from the irreducible representations of the affine Hecke algebra are irreducible and we can use the classification from [KL1] (see [C23]). The special, *singular*, k lead to a rich new theory. It is important itself and is a natural step toward the theory at roots of unity (as $q^N = 1$).

This of course does not mean that the "generic case" is not interesting. Even the simplest examples of induced representations give a lot, for instance, the polynomial representation with generic q, k. However, the theory for singular k, when the classification of the irreducible representations is beyond [KL1], seems the most challenging now. One certainly needs a rich "decomposition theory" with respect to the affine Hecke subalgebra. This exactly occurs at singular k.

In the case of DAHAs of type GL_n for singular k and generic q, the semisimple irreducible representations can be described in terms of *periodic skew Young diagrams* using the technique of intertwiners. It is similar to the classification due to Bernstein-Zelevinsky in the affine GL_n–case and papers [C5, C4, Na1]. The non-semisimple case will be touched upon only a little. The technique of this chapter is mainly "semisimple."

Recently Vasserot extended the "geometric approach" from [KL1] (see also [CG]) to the double affine case. His results are for arbitrary root systems. He gives a general geometric classification of irreducible representations including an important interpretation of the case of singular k. Unfortunately, the corresponding algebraic varieties are complicated, and the geometric (K–theoretic, to be more exact) interpretation of the Fourier transform remains unknown.

Arbitrary *reduced* root systems are considered in this chapter. The papers [No2, Di, Sa, St], the book [M8], and [Ra], are devoted to the so-called $C^\vee C$ or Koornwinder case. The corresponding DAHA was defined and studied first by Noumi and Sahi. The prior results are due to Koornwinder. Let me also mention [C12], where the $C^\vee C$ and the corresponding quantum KZ equation was introduced in terms of the pictures of lines in the plane with reflections in two lines.

The root system $C^\vee C$ unifies (and generalizes) the classical root systems and has all the important features of the case of reduced root systems considered in this chapter, including the self-duality and the Fourier-invariance of the Gaussians. It has a record number of defining parameters (five) and important applications.

One-dimensional case. Modern methods are good enough to transfer smoothly the one-dimensional theory from Chapter 2 to arbitrary root systems. Note that DAHAs give quite a few new results and constructions in the rank one case, including new identities and new proofs of the classical formulas. The rank one theory is very rich, expecially in the $C^\vee C$ case or if the roots of unity are considered. The most interesting new one-dimensional development is the theory of q–Fourier transform.

It seems that this transform was not studied before [C19], although its kernel, the basic hypergeometric function, is a classical object. As far as I know, the key point of the new theory, the Fourier-invariance of the Gaussian, remained undiscovered in the works (classical and new) devoted to the basic hypergeometric function, although this fact is not far from some known identities. The symmetry of the basic hypergeometric function responsible for the self-duality of the difference transform was known in the one-dimensional case. The multidimensional Fourier theory is entirely based on DAHAs.

The DAHA theory is essentially algebraic now. We do not discuss the analytic aspects in this chapter. There are some results in this direction in [C21] (the construction of the general spherical functions), [C26] (analytic continuations in terms of k with applications to q–counterparts of Riemann's zeta function), and [KS1, KS2]. The latter two papers are devoted to explicit analytic properties of the one-dimensional Fourier transforms in terms of the basic hypergeometric function, directly connected with our ones.

Nonsymmetric polynomials. The key new development in the theory of orthogonal polynomials, related combinatorics, and related harmonic analysis is the definition of the nonsymmetric Opdam–Macdonald polynomials. The main references are [O3, M5, C20]. Opdam mentions in [O3] that this definition (in the differential setup) was given in Heckman's unpublished lectures.

These polynomials are expected to be, generally speaking, beyond quantum groups and Kac–Moody algebras because of the following metamathematical reason. Major special functions in the Lie and Kac–Moody theory, including the classical and Kac–Moody characters, spherical functions, and conformal blocks, are W–invariant. One can expect these functions to be W–symmetrizations of *fundamental* nonsymmetric functions, but it does not happen in the traditional theory, with a reservation about the Demazure character formula. One needs double Hecke algebras; the Macdonald symmetric polynomials are symmetrizations of their nonsymmetric counterparts, that are of fundamental importance in the DAHA theory.

However, our considerations are not quite new in representation theory. We directly generalize the harmonic analysis on Heisenberg and Weyl algebras and borrow a lot from the affine Hecke algebra technique. The limits of the nonsymmetric Macdonald polynomials as $q \to \infty$ are well known. They are the spherical functions due to Matsumoto. See [Ma] and [O4] about recent developments. The Eisenstein integrals and series in the theory of affine Hecke algebras are also related to the nonsymmetric polynomials.

Number theory does require nonsymmetric functions. For instance, we need all one-dimensional theta functions, not just even ones. However, say, applications of the Kac character formulas to the Dedekind-type identities generally lead to symmetric (even) functions. The nonsymmetric polynomials and their various counterparts can make the connection between representation theory and number theory stronger.

The intertwining operators of the double affine Hecke algebras are the basis of the new technique, at least as far as the semisimple representations are concerned. They were applied in [Kn], [KnS] and further papers to the combinatorial conjectures about the coefficients of Macdonald's polynomials in the GL–case. In [O3] the nonaffine intertwiners were used to define a nonsymmetric counterpart of the Harish-Chandra theory. The affine intertwines made it possible to calculate the Harish-Chandra transform of the coordinates treated as multiplication operators [C24], which was an old open problem in the Harish-Chandra theory.

3.0.2 Perfect representations

A convincing demonstration of the new technique is the theory of *perfect* representations (nonsymmetric Verlinde algebras) with applications to the Verlinde algebras, Gaussian sums and so-called diagonal coinvariants.

Double Hecke algebras generalize Heisenberg and Weyl algebras, so it is not surprising that they are helpful to study Fourier transforms. Perfect representations representations have the simplest possible Fourier theory. Generally, quasi-perfect representations are assumed to posses a non-degenerate pairing which induces the self-duality involution. If the pairing is perfect and there is a projective action of $PSL(2,\mathbb{Z})$, then the modules are called perfect.

In this chapter we consider only semisimple spherical perfect modules. In more detail, perfect modules of this chapter are

(a) unitary (or pseudo-unitary), (b) Fourier-invariant, (c) spherical.

The latter property makes them commutative algebras with a unit. We also suppose that they are

(d) semisimple, and (e) finite dimensional.

We note that the inner product in (a) may be non-positive. For the rational DAHA, it is *always* indefinite and the representations are not of semisimple type. There are examples when (general) perfect representations are not semisimple in the q,t–case as well. Nevertheless, the semisimplicity is an important part of the theory of perfect representations and Verlinde algebras.

The perfect irreducible representations, i.e., those satisfying all five conditions (a), (b), (c), (d), (e), exist either when q is a root of unity or when q is generic but k is singular. The corresponding DAHA has, essentially, only one perfect representation up to isomorphisms and the choice of the character of the affine Hecke algebra; the latter is needed in the definition of spherical representations. This generalizes the well-known uniqueness of the irreducible representation of the Weyl algebra at the roots of unity.

As q is a root of unity, perfect representations lead to a new class of cyclotomic Gauss–Selberg sums. These sums are defined as eigenvalues of the Gaussian with respect to the q–Fourier transform. Interesting *non-cyclotomic* sums appear for generic q and singular k. They are directly related to Dedekind–Macdonald η–type identities. In the rational setting, they describe the degenerate multidimensional Bessel functions.

The subspaces of symmetric (W–invariant) elements of perfect representations are commutative algebras generalizing the Verlinde algebras. The usual Verlinde algebra appera as $k = 1$; it is formed by integrable representations of the Kac–Moody algebras of fixed level with the fusion procedure as multiplication. It can be identified with the *restricted category* of representations of quantum groups at the roots of unity due to Kazhdan–Lusztig [KL2] and Finkelberg. The third approach to the Verlinde algebras is via factors and subfactors.

The perfect representations "double-generalize" the Verlinde algebras and, more importantly, simplify the theory.

First, the Macdonald symmetric polynomials replace the classical compact characters [Ki1, C19]. The parameter k, which is 1 in the Verlinde case [Ver], becomes an arbitrary positive integer (satisfying certain inequalities) or even fractional, namely, from $(1/h)\mathbb{Z}$ for the Coxeter number h subject to some inequalities.

Second, the nonsymmetric polynomials replace the symmetric ones [C20]. It is much more comfortable to deal with the Fourier transform when all functions are available, not only the symmetric ones. It is exactly what perfect representations provide. See Section 3.10 of this chapter.

The final construction is distant from the original Verlinde algebra (and from the Lie and Kac–Moody theories). The structural constants of multiplication are not positive integers any longer. However, all other important features, including the projective action of $PGL(2,\mathbb{Z})$ and the positivity of the inner product, are saved.

Perfect representations for generic q and $k=-1-1/h$ solve the Haiman conjecture about the canonical quotients of the so-called space of the diagonal coinvariants for arbitrary root systems; see [Ha1, Go, C29]. This space is of fundamental importance for the combinatorics of two sets of variables started in [GH].

Macdonald's η–identities. The approach via double Hecke algebras provides interesting q–deformations of the perfect representations at roots of unity and the corresponding cyclotomic Gauss–Selberg sums. We replace the roots of unity q by *any* q such that $|q|=1$. The Gaussian sums remain finite, of the same length. We also deform other structures, for instance, the Verlinde algebra formed by W–invariants, the hermitian form, and the projective $PSL_2(\mathbb{Z})$–action.

The deformation goes as follows. We start with a perfect representation of the *little* double affine Hecke algebras where q is generic and k is *singular*, special rational from $(1/h)\mathbb{Z}$ (see above). They remain irreducible as $q^N=1$ under certain conditions. Provided that $(N,h)=1$, one can pick k such that $k=1 \bmod N$ if we want to deform the standard (little) Verlinde algebra. It gives the desired deformation. Considering little DAHA, for the root lattice instead of the weight lattice, is necessary here.

The representation theory of double Hecke algebras at singular k appeared to be connected with the Macdonald η–identities [M1], that are product formulas for the Gauss-type sums over the root lattices shifted by $(1/h)\rho$. Here

ρ is the half-sum of all positive roots and h is the Coxeter number.

The Dedekind–Macdonald identities are *directly* related to the one-dimensional representation of the double Hecke algebra. Replacing $(1/h)$ by $(m/h)\rho$ for integral $m > 0$ relatively prime to h, we come to new η–type identities associated with general finite dimensional perfect representations.

Here the Fourier-invariance of the perfect representations is important, and also that they are unitary with respect to a hermitian form (not always definite). By unitary, we mean that the generators X, T, Y of the double Hecke algebra are unitary matrices in a given representation, provided that the conjugation sends q to q^{-1}. It is a the standard complex conjugation if $|q| = 1$. The form is positive definite as $|q| = 1$ and the angle of q is sufficiently small. It generalizes the Verlinde pairing, which plays an important role in conformal field theory.

The Verlinde inner product is actually a straightforward variant of the classical paring which makes the characters of compact Lie groups pairwise orthogonal. It uses the same square of the discriminant and summation over a certain finite set ("alcove") versus the integration in the classical compact theory. The inner products in q, t–theory are not that simple; they are given in terms of the "truncated" theta–functions.

3.0.3 Affine Hecke algebras

There is another important inner product that does not involve the conjugation of q and serves the case of real q, i.e., the real harmonic analysis. The operators Y are not normal with respect to this inner product, which diminishes the role of the Y–semisimplicity in the theory.

Actually, we face the same problem as in the theory of representations of affine Hecke algebras. The Y–operators due to Bernstein and Zelevinsky are, generally, not normal with respect to the natural pairing. It is one of the reasons why advanced methods like K–theory [KL1] are needed. It seems better in the double affine case. Many important representations are either Y–semisimple or X–semisimple, although there are some that are neither.

Both variants of the Fourier transform are considered in this chapter, with and without the conjugation $q \mapsto q^{-1}$, $t \mapsto t^{-1}$. The Fourier theory without the conjugation is actually close to that from [O3] and to the p–adic theory due to Macdonald and Matsumoto. In a sense, we replace w_0 in Opdam's paper by T_{w_0}.

In papers [O4, O5], Opdam developed the Matsumoto theory of "nonsymmetric" spherical functions towards the theory of nonsymmetric polynomials. The operator T_{w_0} appears there in the inverse transform in a way similar to that here. Compare [CO] and [C28] with [O4](Proposition 1.12). The p–adic theory and Opdam's construction are the limit of our theory when $q \to \infty$, with t being p or its proper power (in the rigorous p–adic setup). See Section 2.11 of the previous chapter.

It is important to mention here that the double affine Hecke algebra is a quotient of the group algebra of a central extension of the orbifold fundamental group of the elliptic configuration space; see [C13]. The central extension is by certain fractional powers of the parameter q treated as a central element.

The Fourier transform finds a topological interpretation as the transposition of the periods of the elliptic curve. This clarifies the true potential of DAHAs in representation theory: they provide an improvement of the classical harmonic analysis. The self-duality of the Fourier transform does not survive in the p–adic limit, as well as in the Harish-Chandra theory. It is not surprising because the topological interpretation of the affine Hecke algebra is in terms of $\mathbb{C}^*$, that has only one "period".

In the limit $q \to 1$, $t = q^k$, we come to the construction from [O3] and, then, via the symmetrization, to the k–extension of the Harish-Chandra theory. The nonsymmetric generalization of the Harish-Chandra transform in the space of compactly supported C^∞–functions is analyzed in [O3] in detail. The representation of the degenerate double affine Hecke algebra in terms of the Dunkl operators plays the key role. The representation theory of this algebra is actually similar to that in the q, t–case, though the analytic properties of the corresponding Fourier transforms are very different. At the moment, we don't understand how it happens and why the q, t–case is closer analytically to the p–adic theory than to the Harish-Chandra–Opdam theory.

3.0.4 Gauss–Selberg integrals and sums

Selberg's integral was one of the major starting points for the theory of symmetric Macdonald polynomials. These polynomials appeared for the first time in Kadell's (unpublished) work devoted to generalizations of this integral. The *Mehta–Macdonald integrals* is actually a particular case of Selberg's integral for the classical root systems.

Mehta suggested a formula for the integral of $\prod_{1\le i<j\le n}(x_i - x_j)^{2k}$ with respect to the Gaussian measure (see [Meh]). The formula was readily deduced from Selberg's integral by Bombieri. Macdonald extended Mehta's integral in [M2] to arbitrary root systems and verified his conjectures for the classical systems using Selberg's integral too.

Later Opdam found a uniform proof for all root systems using the shift operators in [O1]. Note that neither alternative proof is known for the special root systems. In [C21], the q was added to this theory.

The (classical) Mehta–Macdonald integral is the normalization constant for the generalized Hankel transform introduced by Dunkl [Du1]. The generalized Bessel functions [O2] multiplied by the Gaussian are eigenfunctions of this transform. The eigenvalues are given in terms of this constant [Du1, Je]. This theory can be extended to the q–case.

The Hankel transform is a "rational" degeneration of the Harish-Chandra spherical transform. The symmetric space G/K is replaced by its tangent space $T_e(G/K)$ with the adjoint action of G; see [He3]. Here G is a semisimple Lie group and K is its maximal compact subgroup. Only particular k, given in terms of the root multiplicities, may appear in the Harish-Chandra theory. For instance, $k = 1$ is the so-called group case. The number k is arbitrary real in the quantum many-body problem and in [HO1]. Additional parameters k appear for the non-simply-laced root systems, up to five independent k in the so-called $C^\vee C$–case.

The theory of DAHA unifies the Selberg–Mehta–Macdonald integrals with the classical Gaussian sums at the roots of unity. The latter can be obtained as limits of the q*–analogs* of the former.

Roots of unity always played a fundamental role in mathematics. In the modern theory of Hecke algebras and quantum groups, the roots of unity are the focus of many projects. It does not mean that the case of generic q is not interesting, but quite a few q–developments aim at roots of unity.

The reduction to the Gaussian sums consists of four steps:
(a) the q–deformations of the Mehta–Macdonald integrals,
(b) the transfer from the q–integrals to the Jackson sums,
(c) calculating the limits of the latter at roots of unity,
(d) establishing connections with the classical Gaussian sums.

Let us follow these steps in the one-dimensional case. The aim is to go from the celebrated Euler integral

$$\int_{-\infty}^{\infty} e^{-x^2} x^{2k} dx \;=\; \Gamma(k+1/2),\ \Re k > -1/2, \tag{3.0.1}$$

to the almost equally famous Gauss formula

$$\sum_{m=0}^{2N-1} e^{\frac{\pi m^2}{2N} i} \;=\; (1+i)\sqrt{N},\ \ N \in \mathbf{N}. \tag{3.0.2}$$

Connecting these formulas is of obvious importance, but the real objective will be to incorporate k into the last formula.

Obtaining (3.0.2) from (3.0.1) is a natural problem. However, it was not solved and, it seems, never considered in the classical works on Gaussian sums.

3.0.5 From generic q to roots of unity

Obviously a trigonometric counterpart of (3.0.1) is needed. A natural candidate is the measure-function in the Harish-Chandra theory, where $\sinh(x)^{2k}$

substitutes for x^{2k}. However, such a choice creates problems. Formula (3.0.1) has no straight sinh–counterpart (at least for generic k). We need the q–theory.

The compact case (a). It was demonstrated in [C19, C20, C21] that the self-duality and other important features of the classical Fourier transform hold for the kernel

$$\delta_k(x;q) \stackrel{\mathrm{def}}{=\!=} \prod_{j=0}^{\infty} \frac{(1-q^{j+2x})(1-q^{j-2x})}{(1-q^{j+k+2x})(1-q^{j+k-2x})}, \quad 0<q<1, \ k\in \mathbf{C}. \tag{3.0.3}$$

Actually, the self-duality of the corresponding transform can be expected a priori because the Macdonald truncated theta function δ is a unification of $\sinh(x)^{2k}$ and the Harish-Chandra measure (A_1) serving the inverse spherical transform.

Setting $q=\exp(-1/a),\ a>0$,

$$(-i)\int_{-\infty i}^{\infty i} q^{-x^2}\delta_k\, dx = 2\sqrt{a\pi}\prod_{j=0}^{\infty}\frac{1-q^{j+k}}{1-q^{j+2k}}, \quad \Re k>0. \tag{3.0.4}$$

The limit of (3.0.4) multiplied by $(a/4)^{k-1/2}$ as $a\to\infty$ is (3.0.1) in the imaginary variant. It is the key step. The others are as follows.

Jackson sums (b). Special q–functions have many interesting properties that have no classical counterparts. For instance, we can replace (some) integrals by sums, the Jackson integrals.

Let $\int_\sharp$ be the integration for the path that begins at $z=\epsilon i+\infty$, moves to the left till ϵi, then down through the origin to $-\epsilon i$, and finally returns down the positive real axis to $-\epsilon i+\infty$ (for small $\epsilon>0$).

Then for $|\Im k|<2\epsilon, \Re k>0$,

$$\frac{1}{2i}\int_\sharp q^{x^2}\delta_k\, dx \ = -\frac{a\pi}{2}\prod_{j=0}^{\infty}\frac{(1-q^{j+k})(1-q^{j-k})}{(1-q^{j+2k})(1-q^{j+1})}\times \langle q^{x^2}\rangle_\sharp,$$

$$\langle q^{x^2}\rangle_\sharp \stackrel{\mathrm{def}}{=\!=} \sum_{j=0}^{\infty} q^{\frac{(k-j)^2}{4}}\frac{1-q^{j+k}}{1-q^k}\prod_{l=1}^{j}\frac{1-q^{l+2k-1}}{1-q^l}$$

$$= \prod_{j=1}^{\infty}\frac{(1-q^{j+k})}{(1-q^j)}\sum_{j=-\infty}^{\infty} q^{\frac{(k-j)^2}{4}} = \tag{3.0.5}$$

$$q^{\frac{k^2}{4}}\prod_{j=1}^{\infty}\frac{(1-q^{j/2})(1-q^{j+k})(1+q^{j/2-1/4+k/2})(1+q^{j/2-1/4-k/2})}{(1-q^j)}. \tag{3.0.6}$$

The sum for $\langle q^{x^2}\rangle_\sharp$ is the Jackson integral for a special choice of the starting point. Its convergence is for all k.

Gauss–Selberg sums (c). When $q = \exp(2\pi i/N)$ and k is a positive integer $\le N/2$ we come to the simplest Gauss–Selberg cyclotomic sum:

$$\sum_{j=0}^{N-2k} q^{\frac{(k-j)^2}{4}} \frac{1-q^{j+k}}{1-q^k} \prod_{l=1}^{j} \frac{1-q^{l+2k-1}}{1-q^l} = \prod_{j=1}^{k}(1-q^j)^{-1} \sum_{m=0}^{2N-1} q^{m^2/4}. \tag{3.0.7}$$

Its modular counterpart is formula (1,2b) from [Ev].

Formula (3.0.7) can be deduced directly from (3.0.5) following the classical limiting procedure from [Ch]. Formulas (3.0.4)–(3.0.6) can be verified by elementary methods too (using the Ψ–summation). They are proved in [C21] for arbitrary root systems, as well as the generalizations of these formulas involving the Macdonald polynomials. The double Hecke algebra technique is needed even in the one-dimensional case when two Rogers' polynomials are added to the integrand. The nonsymmetric formulas and the multidimensional degenerations are based on DAHA; see [C21].

New proof of the Gauss formula (d). Substituting $k = [N/2]$ in the last formula (not $k = 0$, as one may expect, which makes (3.0.7) a trivial identity) we obtain the product formula for $\sum_{m=0}^{2N-1} q^{m^2/4}$, which can be readily calculated as $q = e^{2\pi i/N}$ and quickly results in (3.0.2).

Let us demonstrate it in the case of $N = 2k$ (odd $N = 2k+1$ are quite similar). We obtain

$$\sum_{m=0}^{2N-1} q^{\frac{m^2}{4}} = q^{\frac{k^2}{4}} \Pi \ \text{ for } \ \Pi = \prod_{j=1}^{k}(1-q^j).$$

First, $\Pi\bar{\Pi} = (X^N-1)(X+1)(X-1)^{-1}(1) = 2N$. Second, $\arg(1-e^{i\phi}) = \phi/2 - \pi/2$ when $0 < \phi < 2\pi$, and therefore

$$\arg \Pi = \frac{\pi}{N}\frac{k(k+1)}{2} - \frac{\pi k}{2} = \frac{\pi(1-k)}{4}.$$

Here we can switch to arbitrary primitive $q = \exp(2\pi i l/N)$ as $(l, N) = 1$ if we can control the set of $\arg(q^j)$ for $1 \le j \le k$. This can be used to justify the quadratic reciprocity (see [CO]). We do not discuss this direction here.

Finally,

$$\arg(q^{\frac{k^2}{4}}\Pi) = \pi(1-k)4 + \frac{\pi k}{4} = \frac{\pi}{4}, \ \text{ and } \ q^{\frac{k^2}{4}}\Pi = \sqrt{N}(1+i).$$

3.0.6 Structure of the chapter

The first three sections mainly contain the basic theory, including the scalar products with and without the conjugation $q \mapsto q^{-1}$, and the case of roots of unity.

Sections 3.4–3.5 are about the general theory of the Fourier transforms in the compact and noncompact cases. They act on polynomials, polynomials multiplied by the Gaussian, and in spaces of delta functions (called the functional representations). The application to the η–identities concludes Section 3.5.

In Section 3.6, we find necessary and sufficient conditions for a representation with the cyclic vector to be semisimple and pseudo-unitary. The consideration of the case of GL_n is the subject of the next Section 3.7. The main theorem of Section 3.6 is the key tool for the next Sections 3.8–3.10, which are devoted to spherical, cospherical, self-dual, and perfect representations. The application to the diagonal coinvariants and the universal Dunkl operators is discussed in the last Section 3.11.

The exposition is essentially self-contained. The previous chapters and the one-dimensional papers [CM, C27, CO] combined with the second part of [C25] could be a reasonable introduction. The papers [C23] and [C21] contain additional results on the nonsymmetric polynomials and q–spherical functions. We also recommend the book [M8].

Acknowledgments. *The author thanks V. Kac and I. Macdonald for useful comments on the material of this section He also thanks P. Etingof and E. Vasserot for discussing the case of GL_n. I acknowledge my indebtedness to E. Opdam for reading the chapter and making important suggestions that especially influenced Section 3.8. I am also thankful to G. Heckman for discussing Section 3.9.*

A significant part of the chapter was prepared during my visit to Harvard University (spring 2001). I am grateful to D. Kazhdan and P. Etingof for inviting me.

3.1 Affine Weyl groups

Let $R = \{\alpha\} \subset \mathbb{R}^n$ be a root system of type $A, B, ..., F, G$ with respect to a euclidean form (z, z') on $\mathbb{R}^n \ni z, z'$, W the **Weyl group** generated by the reflections s_α, R_+ the set of positive roots (also, $R_- = -R_+$) corresponding to the (fixed) simple roots $\alpha_1, ..., \alpha_n$ and Γ be the Dynkin diagram with $\{\alpha_i, 1 \leq i \leq n\}$ as the vertices.

We will also use the dual roots (coroots) and the dual root system:

$$R^\vee = \{\alpha^\vee = 2\alpha/(\alpha, \alpha)\}.$$

The root lattice and the weight lattice are

$$Q = \oplus_{i=1}^n \mathbb{Z}\alpha_i \subset P = \oplus_{i=1}^n \mathbb{Z}\omega_i,$$

where $\{\omega_i\}$ are fundamental weights: $(\omega_i, \alpha_j^\vee) = \delta_{ij}$ for the simple coroots $\alpha_i^\vee$.

Replacing $\mathbb{Z}$ by $\mathbb{Z}_\pm = \{m \in \mathbb{Z}, \pm m \geq 0\}$ we obtain $Q_\pm, P_\pm$. Note that $Q \cap P_+ \subset Q_+$, i.e., the coefficients of the fundamental weights in terms of the simple roots are non-negative. Moreover, each ω_j has *all nonzero* coefficients (sometimes rational) when expressed in terms of $\{\alpha_i\}$; see [Bo].

The form will be normalized by the condition $(\alpha, \alpha) = 2$ for the *short* roots. This normalization coincides with that from the tables in [Bo] for A, C, D, E, G. Thus, $\nu_\alpha \stackrel{\mathbf{def}}{=\!=} (\alpha, \alpha)/2$ can be either 1 or in $\{1, 2\}$ or in $\{1, 3\}$.

This special normalization leads to the inclusions $Q \subset Q^\vee$ and $P \subset P^\vee$, where $P^\vee$ is generated by the fundamental coweights $\omega_i^\vee$. We will use the notation $\nu_{\rm lng}$ for the long roots ($\nu_{\rm sht} = 1$).

Setting $\nu_i = \nu_{\alpha_i}$, $\nu_R = \{\nu_\alpha, \alpha \in R\}$, one has

$$\rho_\nu \stackrel{\mathbf{def}}{=\!=} (1/2) \sum_{\nu_\alpha = \nu} \alpha = \sum_{\nu_i = \nu} \omega_i, \text{ where } \alpha \in R_+,\ \nu \in \nu_R. \tag{3.1.1}$$

Note that $(\rho_\nu, \alpha_i^\vee) = 1$ as $\nu_i = \nu$. We will call ρ_ν the **partial** ρ.

3.1.1 Affine roots

Let $\vartheta \in R^\vee$ be the *maximal positive coroot.* All simple coroots appear in its decomposition. Also note that $2 \geq (\vartheta, \alpha^\vee) \geq 0$ for $\alpha > 0$, $(\vartheta, \alpha^\vee) = 2$ only for $\alpha = \vartheta$, and $s_\vartheta(\alpha) < 0$ if $(\vartheta, \alpha) > 0$.

See [Bo] to check that ϑ considered as a root (it belongs to R because of the choice of normalization) is maximal among all short positive roots of R. For the sake of completeness, let us prove another defining property of ϑ (see Proposition 3.1.7 below for a uniform proof).

Proposition 3.1.1. *The least nonzero element in $Q_+^+ = Q_+ \cap P_+ = Q \cap P_+$ with respect to Q_+ is ϑ, i.e., $b - \vartheta \in Q_+$ for all $b \in Q_+^+$.*

Proof. If $a \in Q_+^+$, then all coefficients of the decomposition of a in terms of α_i are nonzero. Indeed, if α_i does not appear in this decomposition, then its neighbors (in the Dynkin diagram) have negative scalar products with a. This readily gives the claim when $\vartheta = \sum_{i=1}^n \alpha_i$, i.e., for the systems A, B. In these cases, ϑ is $\omega_1 + \omega_n$ for A and ω_1 for B. Otherwise, it is ω_2 (C, D, E_6), ω_1 (E_7), ω_8 (E_8), ω_4 (F_4), and ω_1 (G_2), respectively (in the notation from [Bo]). Let us denote the corresponding subscripts of ω $(1, 2, 1, 8, 4, 1)$ by $\tilde{o}$. They are the indices of the simple roots neighboring the root $\alpha_0^\vee$ in the affine (completed) Dynkin diagram for $R^\vee$.

The proposition holds for E_8, F_4, and G_2, because in these cases $P = Q$ and $\omega_j - \omega_{\tilde{o}} \in Q_+$ (use the tables from [Bo]). So we need to examine C, D, and $E_{6,7}$.

As to C, one verifies that $\omega_j - \omega_2 \in (1/2)Q_+$ for $j \geq 2$ and Q_+^+ is generated by ω_{2i} and $\omega_{2i+1} + \omega_{2j+1}$ for $1 \leq i \leq j \leq [n/2]$. Thus the relation $2\omega_1 - \omega_2 \in Q_+$

proves the claim. In the D–case, $a - \omega_2$ for $a \in Q_+^+$ can be apart from $(1/4)Q_+$ only if it is a linear combination of ω_1, ω_{n-1}, and ω_n. However, such a combination "contains" either $\omega_{n-1}+\omega_n$ or $2\omega_1$, which is sufficient to make it greater than ω_2 (with respect to Q_+).

For $R = E_6$, $\omega_j - \omega_2 \in (1/3)Q_+$ when $j \geq 2 = \tilde{o}$ except for $j = 6$. The intersection of Q and $\mathbb{Z}_+\omega_1 + \mathbb{Z}_+\omega_6$ is generated by $3\omega_1$, $3\omega_6$, $\omega_1 + \omega_6$. Either weight is greater than ω_2. In the case of E_7, $\tilde{o} = 1$, $\omega_j - \omega_1 \in (1/2)Q_+$ if $j \leq 6$, and $2\omega_7$ is greater than ω_1. ❑

The vectors $\tilde{\alpha} = [\alpha, \nu_\alpha j] \in \mathbb{R}^n \times \mathbb{R} \subset \mathbb{R}^{n+1}$ for $\alpha \in R, j \in \mathbb{Z}$ form the **affine root system** $\tilde{R} \supset R$ ($z \in \mathbb{R}^n$ are identified with $[z, 0]$). We add $\alpha_0 \stackrel{\text{def}}{=\!=} [-\vartheta, 1]$ to the simple roots for the **maximal short root** ϑ. The corresponding set $\tilde{R}$ of positive roots coincides with $R_+ \cup \{[\alpha, \nu_\alpha j],\ \alpha \in R,\ j > 0\}$.

We complete the Dynkin diagram Γ of R by α_0 (by $-\vartheta$ to be more exact). The notation is $\tilde{\Gamma}$. One can obtain it from the completed Dynkin diagram for $R^\vee$ from [Bo] by reversing the arrows. The number of laces between α_i and α_j in $\tilde{\Gamma}$ is denoted by m_{ij}.

The set of indices of the images of α_0 by all the automorphisms of $\tilde{\Gamma}$ will be denoted by O ($O = \{0\}$ for E_8, F_4, G_2). Let $O' = r \in O, r \neq 0$. The elements ω_r for $r \in O'$ are the so-called minuscule weights: $(\omega_r, \alpha^\vee) \leq 1$ for $\alpha \in R_+$.

Given $\tilde{\alpha} = [\alpha, \nu_\alpha j] \in \tilde{R},\ b \in B$, let

$$s_{\tilde{\alpha}}(\tilde{z}) \ = \ \tilde{z} - (z, \alpha^\vee)\tilde{\alpha}, \quad b'(\tilde{z}) \ = \ [z, \zeta - (z, b)] \tag{3.1.2}$$

for $\tilde{z} = [z, \zeta] \in \mathbb{R}^{n+1}$.

The **affine Weyl group** $\widetilde{W}$ is generated by all $s_{\tilde{\alpha}}$ (we write $\widetilde{W} = \langle s_{\tilde{\alpha}}, \tilde{\alpha} \in \tilde{R}_+ \rangle$). One can take the simple reflections $s_i = s_{\alpha_i}$ $(0 \leq i \leq n)$ as its generators and introduce the corresponding notion of the length. This group is the semidirect product $W \ltimes Q'$ of its subgroups $W = \langle s_\alpha, \alpha \in R_+ \rangle$ and $Q' = \{a', a \in Q\}$, where

$$\alpha' = \ s_\alpha s_{[\alpha, \nu_\alpha]} = \ s_{[-\alpha, \nu_\alpha]} s_\alpha \ \text{ for } \ \alpha \in R. \tag{3.1.3}$$

The **extended Weyl group** $\widehat{W}$ generated by W and P' (instead of Q') is isomorphic to $W \ltimes P'$:

$$(wb')([z, \zeta]) \ = \ [w(z), \zeta - (z, b)] \ \text{ for } \ w \in W, b \in B. \tag{3.1.4}$$

From now on, b and b', P and P' will be identified.

Given $b \in P_+$, let w_0^b be the longest element in the subgroup $W_0^b \subset W$ of the elements preserving b. This subgroup is generated by simple reflections. We set

$$u_b = w_0 w_0^b \in W, \ \pi_b = b(u_b)^{-1} \ \in \ \widehat{W}, \ u_i = u_{\omega_i}, \pi_i = \pi_{\omega_i}, \tag{3.1.5}$$

where w_0 is the longest element in W, $1 \le i \le n$.

The elements $\pi_r \stackrel{\mathbf{def}}{==} \pi_{\omega_r}$ for $r \in O'$ and $\pi_0 \stackrel{\mathbf{def}}{==} \mathrm{id}$ form a group denoted by Π, which is isomorphic to P/Q by the natural projection $\{\omega_r \mapsto \pi_r\}$; these elements leave $\tilde{\Gamma}$ invariant. As to $\{u_r\}$, they preserve the set $\{-\vartheta, \alpha_i, i > 0\}$. The relations $\pi_r(\alpha_0) = \alpha_r = (u_r)^{-1}(-\vartheta)$ distinguish the indices $r \in O'$. Moreover (see, e.g., [C16]):

$$\widehat{W} = \Pi \ltimes \widetilde{W}, \quad \text{where } \pi_r s_i \pi_r^{-1} = s_j \ \text{ if } \pi_r(\alpha_i) = \alpha_j, \ 0 \le j \le n. \tag{3.1.6}$$

3.1.2 Affine length function

Setting $\widehat{w} = \pi_r \widetilde{w} \in \widehat{W}$ for $\pi_r \in \Pi$ and $\widetilde{w} \in \widetilde{W}$, the length $l(\widehat{w})$ is, by definition, the length of the reduced decomposition $\widetilde{w} = s_{i_l} \ldots s_{i_2} s_{i_1}$ in terms of the simple reflections $s_i, 0 \le i \le n$. The number of s_i in this decomposition such that $\nu_i = \nu$ is denoted by $l_\nu(\widehat{w})$.

The **length** also can be defined as the cardinality $|\lambda(\widehat{w})|$ of

$$\lambda(\widehat{w}) \stackrel{\mathbf{def}}{==} \tilde{R}_+ \cap \widehat{w}^{-1}(\tilde{R}_-) = \{\tilde{\alpha} \in \tilde{R}_+, \ \widehat{w}(\tilde{\alpha}) \in \tilde{R}_-\}, \ \widehat{w} \in \widehat{W}.$$

Respectively,

$$\lambda(\widehat{w}) = \cup_\nu \lambda_\nu(\widehat{w}), \ \lambda_\nu(\widehat{w}) \stackrel{\mathbf{def}}{==} \{\tilde{\alpha} \in \lambda(\widehat{w}), \nu(\tilde{\alpha}) = \nu\}. \tag{3.1.7}$$

The coincidence with the previous definition is based on the equivalence of the *length equality*

$$(a) \ \ l_\nu(\widehat{w}\hat{u}) = l_\nu(\widehat{w}) + l_\nu(\hat{u}) \ \text{ for } \ \widehat{w}, \hat{u} \in \widehat{W}$$

and the *cocycle relation*

$$(b) \ \ \lambda_\nu(\widehat{w}\hat{u}) = \lambda_\nu(\hat{u}) \cup \hat{u}^{-1}(\lambda_\nu(\widehat{w})), \tag{3.1.8}$$

which, in its turn, is equivalent to the *positivity condition*

$$(c) \ \ \hat{u}^{-1}(\lambda_\nu(\widehat{w})) \subset \tilde{R}_+.$$

Formula (3.1.8) obviously includes the positivity condition (c). It also implies that

$$\lambda_\nu(\hat{u}) \cap \hat{u}^{-1}(\lambda_\nu(\widehat{w})) \ = \ \hat{u}^{-1}\big(\hat{u}(\lambda_\nu(\hat{u})) \cap \lambda_\nu(\widehat{w})\big) \ = \ \emptyset,$$

thanks to the general formula

$$\lambda_\nu(\widehat{w}^{-1}) = -\widehat{w}(\lambda_\nu(\widehat{w})).$$

Thus it results in the equality $l_\nu(\widehat{w}\hat{u}) = l_\nu(\widehat{w}) + l_\nu(\hat{u})$ and we have the implications $(a) \Leftarrow (b) \Rightarrow (c)$.

The remaining implications $(a) \Rightarrow (b) \Leftarrow (c)$ are based on the following simple general fact:

$$\lambda_\nu(\widehat{w}\hat{u}) \setminus \{\lambda_\nu(\widehat{w}\hat{u}) \cap \lambda_\nu(\hat{u})\} = \hat{u}^{-1}(\lambda_\nu(\widehat{w})) \cap \tilde{R}_+ \ for \ \ any \ \ \hat{u}, \widehat{w}. \tag{3.1.9}$$

For instance, the length equality (a) readily implies (3.1.8). For the sake of completeness, let us deduce (3.1.8) from the positivity condition (c) above. We follow [C16].

It suffices to check that $\lambda_\nu(\widehat{w}\hat{u}) \supset \lambda_\nu(\hat{u})$. If there exists a positive $\tilde{\alpha} \in \lambda_\nu(\hat{u})$ such that $(\widehat{w}\hat{u})(\tilde{\alpha}) \in \tilde{R}_+$, then

$$\widehat{w}(-\hat{u}(\tilde{\alpha})) \in \tilde{R}_- \ \Rightarrow \ -\hat{u}(\tilde{\alpha}) \in \lambda_\nu(\widehat{w}) \ \Rightarrow \ -\tilde{\alpha} \in \hat{u}^{-1}(\lambda_\nu(\widehat{w})).$$

We come to a contradiction with the positivity. Hence $(a) \Leftrightarrow (b) \Leftrightarrow (c)$.

Applying (3.1.8) to the reduced decomposition $\widehat{w} = \pi_r s_{i_l} \cdots s_{i_2} s_{i_1}$,

$$\begin{aligned} \lambda(\widehat{w}) =& \{\tilde{\alpha}^1 = \alpha_{i_1},\ \tilde{\alpha}^2 = s_{i_1}(\alpha_{i_2}),\ \tilde{\alpha}^3 = s_{i_1}s_{i_2}(\alpha_{i_3}), \ldots \\ & \ldots, \tilde{\alpha}^l = \widetilde{w}^{-1} s_{i_l}(\alpha_{i_l})\}. \end{aligned} \tag{3.1.10}$$

The cardinality l of the set $\lambda(\widehat{w})$ equals $l(\widehat{w})$.

This set can be introduced for nonreduced decompositions as well. Let us denote it by $\tilde{\lambda}(\widehat{w})$ to differentiate it from $\lambda(\widehat{w})$. It always contains $\lambda(\widehat{w})$ and, moreover, can be represented in the form

$$\begin{aligned} \tilde{\lambda}(\widehat{w}) &= \lambda(\widehat{w}) \cup \tilde{\lambda}^+(\widehat{w}) \cup -\tilde{\lambda}^+(\widehat{w}), \\ \text{where } \tilde{\lambda}^+(\widehat{w}) &= (\tilde{R}_+ \cap \tilde{\lambda}(\widehat{w})) \setminus \lambda(\widehat{w}). \end{aligned} \tag{3.1.11}$$

The coincidence with $\lambda(\widehat{w})$ is for reduced decompositions only.

Let us prove another standard property of the λ–sets:

$$\lambda_\nu(\widehat{w}) \ = \ \{\tilde{\alpha} > 0,\ l_\nu(\widehat{w}s_{\tilde{\alpha}}) < l_\nu(\widehat{w})\}. \tag{3.1.12}$$

It suffices to consider $\lambda(\widehat{w})$. Thanks to (3.1.10), $\{\tilde{\alpha} > 0,\ l(\widehat{w}s_{\tilde{\alpha}}) < l(\widehat{w})\}$ contains $\lambda(\widehat{w})$. Obviously (3.1.12) holds for $s_{\tilde{\alpha}} = s_{\alpha_i}$ $(0 \le i \le n)$.

An arbitrary $s_{\tilde{\alpha}}$ has a *reduced* decomposition in the form:

$$s_{\tilde{\alpha}} = s_{i_1} s_{i_2} \cdots s_{i_p} s_m s_{i_p} \cdots s_{i_2} s_{i_1},\ i_\bullet, m \ge 0.$$

Indeed, $\alpha_i \in \lambda(\tilde{\alpha})$ if and only if $(\tilde{\alpha}, \alpha_i) > 0$ because $\{\alpha_i\}$ are minimal positive affine roots. Given such α_i, the reflection $s_{\tilde{\alpha}}$ is divisible by s_i on the right (i.e., has a reduced decomposition in the form $\cdots s_i$) and $l(s_{\tilde{\beta}}) \le l(s_{\tilde{\alpha}})$ for $\tilde{\beta} \stackrel{\mathbf{def}}{=\!=} s_i(\tilde{\alpha})$. If $l(s_{\tilde{\beta}}) = l(s_{\tilde{\alpha}})$, then $s_{\tilde{\beta}} = s_i s_{\tilde{\alpha}} s_i = s_{\tilde{\beta}}^{-1}$ is divisible by s_i on the left or on the right, which contradicts the inequality $(\tilde{\beta}, \alpha_i) = -(\tilde{\alpha}, \alpha_i) < 0$. Therefore $l(s_{\tilde{\beta}}) < l(s_{\tilde{\alpha}})$ and we can proceed by induction. We also have obtained that the desired decomposition can be started with an arbitrary $\alpha_{i_1} = \alpha_i \in \lambda(\tilde{\alpha})$.

Now, we will prove (3.1.12) by induction with respect to the length $l(s_{\tilde{\alpha}})$ for $\tilde{\alpha}$ such that $l(\widehat{w}s_{\tilde{\alpha}}) < l(\widehat{w})$. As we already noticed, it holds for simple reflections.

Let $\alpha_i \in \lambda(s_{\tilde{\alpha}})$. If $\alpha_i \notin \lambda(\widehat{w})$, then

$$l((\widehat{w}s_i)(s_i s_{\tilde{\alpha}} s_i)) < l(\widehat{w}) + 1 = l(\widehat{w}s_i),$$

and the induction statement gives that

$$\tilde{\beta} \stackrel{\mathbf{def}}{=\!=} s_i(\tilde{\alpha}) \in \lambda(\widehat{w}s_i), \;\; s_i(\tilde{\alpha}) \in s_i(\lambda(\widehat{w})), \quad \text{and} \;\; \tilde{\alpha} \in \lambda(\widehat{w}).$$

Otherwise $\alpha_i \in \lambda(\widehat{w})$ and therefore $\widehat{w}(\alpha_i) < 0$. Since $\alpha_i \in \lambda(s_{\tilde{\alpha}})$ and $(\tilde{\alpha}, \alpha_i) > 0$, we obtain:

$$(\widehat{w}s_{\tilde{\alpha}})(\alpha_i) = \widehat{w}(\alpha_i - 2\frac{(\tilde{\alpha}, \alpha_i)}{(\tilde{\alpha}, \tilde{\alpha})}\tilde{\alpha}) < 0 \;\text{ for }\; \widehat{w}(\tilde{\alpha}) > 0.$$

The opposite case $\widehat{w}(\tilde{\alpha}) < 0$ means that $\tilde{\alpha} \in \lambda(\widehat{w})$, which gives the desired fact.

Thus it suffices to consider the case when $\alpha_i \in \lambda(\widehat{w}s_{\tilde{\alpha}})$ for the above $\alpha_i \in \lambda(\tilde{\alpha})$, which gives:

$$l((\widehat{w}s_i)s_{\tilde{\beta}}) = l(\widehat{w}s_{\tilde{\alpha}}s_i) \le l(\widehat{w}s_{\tilde{\alpha}}) - 1 < l(\widehat{w}) - 1.$$

By induction, $\tilde{\beta} \in \lambda(\widehat{w}s_i)$ because now $\alpha_i \in \lambda(\widehat{w}s_i)$ (the only case left unchecked) and therefore $l(\widehat{w}s_i) = l(\widehat{w}) - 1$. Hence, $\tilde{\beta} \in s_i(\lambda(\widehat{w}))$ and, finally, $\tilde{\alpha} \in \lambda(\widehat{w})$.

3.1.3 Reduction modulo *W*

The following proposition is from [C20]. It generalizes the construction of the elements π_b for $b \in P_+$.

Proposition 3.1.2. *Given $b \in P$, there exists a unique decomposition $b = \pi_b u_b$, $u_b \in W$ satisfying one of the following equivalent conditions:*

(i) $l(\pi_b) + l(u_b) = l(b)$ *and* $l(u_b)$ *is the greatest possible,*

(ii) $\lambda(\pi_b) \cap R = \emptyset$.

The latter condition implies that $l(\pi_b) + l(w) = l(\pi_b w)$ *for any* $w \in W$. *In addition, the relation* $u_b(b) \stackrel{\mathbf{def}}{=\!=} b_- \in P_- = -P_+$, *holds, which, in its turn, determines* u_b *uniquely if one of the following equivalent conditions is imposed:*

(iii) $l(u_b)$ *is the smallest possible,*

(iv) if $\alpha \in \lambda(u_b)$, *then* $(\alpha, b) \neq 0$. ❑

Since $\pi_b = bu_b^{-1} = u_b^{-1}b_-$, the set $\pi_P = \{\pi_b, b \in P\}$ can be described in terms of P_-:

$$\pi_P = \{u^{-1}b_- \text{ for } b_- \in P_-,\ u \in W \\ \text{such that } \alpha \in \lambda(u^{-1}) \Rightarrow (\alpha, b_-) \neq 0\}. \tag{3.1.13}$$

Indeed,

$$\alpha \in \lambda(u_b^{-1}) = -u_b(\lambda(u_b)) \ \Rightarrow\ (u_b^{-1}(\alpha), b) = (\alpha, u_b(b)) = (\alpha, b_-) \neq 0.$$

Using the longest element w_0' in the centralizer W_0' of b_-, such u constitute the set

$$\{u, \mid l(u^{-1}w_0') \ =\ l(w_0') + l(u^{-1})\}.$$

Their number is $|W|/|W_0'|$.

For $\tilde\alpha = [\alpha, \nu_\alpha j] \in \tilde R_+$, one has:

$$\lambda(b) = \{\tilde\alpha,\ (b, \alpha^\vee) > j \geq 0 \text{ if } \alpha \in R_+, \\ (b, \alpha^\vee) \geq j > 0 \text{ if } \alpha \in R_-\}, \tag{3.1.14}$$

$$\lambda(\pi_b) = \{\tilde\alpha,\ \alpha \in R_-,\ (b_-, \alpha^\vee) > j > 0 \text{ if } u_b^{-1}(\alpha) \in R_+, \\ (b_-, \alpha^\vee) \geq j > 0 \text{ if } u_b^{-1}(\alpha) \in R_-\}, \tag{3.1.15}$$

$$\lambda(\pi_b^{-1}) = \{\tilde\alpha,\ -(b, \alpha^\vee) > j \geq 0\}, \tag{3.1.16}$$

$$\lambda(u_b) = \{\alpha \in R_+,\ (b, \alpha^\vee) > 0\}. \tag{3.1.17}$$

For instance, $l(b) = l(b_-) = -2(\rho^\vee, b_-)$ for $2\rho^\vee = \sum_{\alpha>0} \alpha^\vee$.

Let us also calculate the set $\lambda(s_{\tilde\beta})$ for $\tilde\beta = [-\beta, l\nu_\beta]$, where $\beta \in R_+, l > 0$. One has:

$$(-l\beta)' = s_\beta s_{\tilde\beta},\ \lambda(s_\beta) = s_\beta(R_-) \cap R_+,\ s_{\tilde\beta}(\lambda(s_\beta)) = \{[-\alpha, l(\beta, \alpha)],\ \alpha \in \lambda(s_\beta)\}.$$

Here (β, α) must be greater than zero. Therefore the above decomposition of $-l\beta \in Q$ is *reduced*, thanks to the positivity condition (c) above.

Note that $(\beta, \alpha) > 0$ does not guarantee that $\alpha \in \lambda(s_\beta)$. However, for some elements in W it holds, for instance, for $\alpha = \vartheta$.

Using (3.1.14),

$$\lambda(-l\beta) = \{\tilde\alpha = [-\alpha, \nu_\alpha j] \mid l(\alpha^\vee, \beta) > j \geq 0 \text{ as } \alpha \in R_-, \\ l(\alpha^\vee, \beta) \geq j > 0 \text{ as } \alpha \in R_+\}.$$

Finally,

$$\lambda(s_{\tilde\beta}) = \{[-\alpha, \nu_\alpha j] \in \tilde R_+ \mid l(\alpha^\vee, \beta) > j\} \cup \\ \{[-\alpha, l(\beta, \alpha)] \mid \alpha > 0 < s_\beta(\alpha),\ (\beta, \alpha) > 0\}. \tag{3.1.18}$$

We will need later the following **affine action** of $\widehat{W}$ on $z \in \mathbb{R}^n$:

$$\begin{aligned}(wb)((z)) &= w(b+z),\ w \in W, b \in P,\\ s_{\tilde\alpha}((z)) &= z - ((z,\alpha)+j)\alpha,\ \tilde\alpha = [\alpha, \nu_\alpha j] \in \tilde R.\end{aligned} \tag{3.1.19}$$

Also,

$$(bw)((z)) = b + w(z),\ (bw)((0)) = b \quad \text{for any} \quad w \in W.$$

The relation to the above action is given in terms of the **affine pairing** $([z,l], z'+d) \stackrel{\mathbf{def}}{=\!=} (z,z')+l$:

$$(\widehat{w}([z,l]), \widehat{w}((z'))+d) = ([z,l], z'+d) \ \text{for}\ \widehat{w} \in \widehat{W}, \tag{3.1.20}$$

where we treat d formally.

Introducing the **affine Weyl chamber**

$$\mathfrak{C}^a = \bigcap_{i=0}^{n} \mathfrak{L}_{\alpha_i},\ \mathfrak{L}_{[\alpha,\nu_\alpha j]} = \{z \in \mathbb{R}^n,\ (z,\alpha)+j > 0\},$$

we come to another interpretation of the λ–sets:

$$\lambda_\nu(\widehat{w}) = \{\tilde\alpha \in \tilde R_+,\ \mathfrak{C}^a \not\subset \widehat{w}((\mathfrak{L}_{\tilde\alpha})),\ \nu_\alpha = \nu\}. \tag{3.1.21}$$

For instance, Π is the group of all elements of $\widehat{W}$ preserving $\mathfrak{C}^a$ with respect to the affine action. Similarly, the elements π_b for $b \in P$ are exactly those sending the negative $-\mathfrak{C}^a$ of $\mathfrak{C}^a$ to the negative $-\mathfrak{C}$ of the nonaffine Weyl chamber $\mathfrak{C} \stackrel{\mathbf{def}}{=\!=} \{z \in \mathbb{R}^n, (z,\alpha_i) > 0 \text{ as } i > 0\}$. More generally, we have the following proposition.

Proposition 3.1.3. *Given two finite sets of positive affine roots* $\{\tilde\beta = [\beta, \nu_\beta i]\}$ *and* $\{\tilde\gamma = [\gamma, \nu_\gamma j]\}$, *let* $L_{\beta,\gamma}$ *be the closure of the union of* $\widehat{w}((-\mathfrak{C}^a))$ *over* $\widehat{w} \in \widehat{W}$ *such that* $\tilde\beta \not\in \lambda(\widehat{w}) \ni \tilde\gamma$. *Then*

$$L_{\beta,\gamma} = \{z \in \mathbb{R}^n,\ (z,\beta)+i \le 0,\ (z,\gamma)+j \ge 0\ \ \textit{for all}\ \tilde\beta, \tilde\gamma\}.$$

The same holds in the nonaffine variant for $\mathfrak{C}$ *in place of* $\mathfrak{C}^a$. ❑

We will also need the following "affine" variant of Proposition 3.1.2. Given $z \in \mathbb{R}^n$, there exists a unique element $\tilde w = u_z a_z$ with $a_z \in Q$ and $u_z \in W$ satisfying the relations

$$\begin{aligned}&z_- \stackrel{\mathbf{def}}{=\!=} \tilde w((z)) \in -\bar{\mathfrak{C}}^a, \\ &(z_-, \vartheta) = -1 \Rightarrow u_z^{-1}(\vartheta) \in R_-, \text{ and}\\ &(\alpha_i, z_-) = 0 \Rightarrow u_z^{-1}(\alpha_i) \in R_+,\ i > 0,\end{aligned} \tag{3.1.22}$$

where $-\bar{\mathfrak{C}}^a$ is the negative of the closure $\bar{\mathfrak{C}}^a$ of $\mathfrak{C}^a$.

The element $b_- = u_b(b)$ is a unique element from P_- which belongs to the orbit $W(b)$. So the equality $c_- = b_-$ means that b, c belong to the same orbit. We will also use $b_+ \stackrel{\mathbf{def}}{=} w_0(b_+)$, a unique element in $W(b) \cap P_+$. In terms of the elemnts π_b,

$$u_b \pi_b = b_-, \ \pi_b u_b = b_+.$$

Note that $l(\pi_b w) = l(\pi_b) + l(w)$ for all $b \in P$, $w \in W$. For instance,

$$\begin{aligned} &l(b_- w) = l(b_-) + l(w), \ l(w b_+) = l(b_+) + l(w), \\ &l(u_b \pi_b w) = l(u_b) + l(\pi_b) + l(w) \ \text{ for } \ b \in P, \ w \in W. \end{aligned} \tag{3.1.23}$$

We will use these relations together with the following proposition when calculating the conjugations of the nonsymmetric Macdonald polynomials.

Proposition 3.1.4. *The set $\lambda(\widehat{w})$ for $\widehat{w} = wb$ consists of $\tilde{\alpha} = [\alpha, \nu_\alpha j]$ with positive α if and only if*

$$(b, \alpha^\vee) \geq -1 \ \text{ and } \ (b, \alpha^\vee) = -1 \Rightarrow \alpha \in \lambda(w) \tag{3.1.24}$$

for all $\alpha \in R_+$, $b \in P$. The elements $\widehat{w} \in W \cdot P_+$ are of this type.

Proof. We use (3.1.11):

$$\begin{aligned} \lambda(wb) &= \lambda(b) \cup ({}^{-b}\lambda(w)) \setminus (\tilde{\lambda}^+ \cup -\tilde{\lambda}^+), \\ \text{where } \ \tilde{\lambda}^+ &= \tilde{R}_- \cap ({}^{-b}\lambda(w)). \end{aligned} \tag{3.1.25}$$

The latter set is

$$\{[\alpha, \nu_\alpha(b, \alpha^\vee)]\} \text{ such that } (b, \alpha^\vee) < 0, \ \alpha \in \lambda(w).$$

Let us calculate $\lambda(wb) \cap \lambda(b)$. We need to remove from $\lambda(b)$ the roots

$$-[\alpha, \nu_\alpha(b, \alpha^\vee)] \ \text{ for } \ \alpha > 0 \text{ such that } (b, \alpha^\vee) < 0, \ \alpha \in \lambda(w).$$

The roots in the form $[-\alpha, \nu_\alpha j]$ belong to $\lambda(b)$ exactly for $0 < j \leq -(b, \alpha^\vee)$. Therefore

$$\begin{aligned} &\lambda(wb) \cap \lambda(b) = \{ \ [\alpha, \nu_\alpha j], \ 0 \leq j < (b, \alpha^\vee) \text{ as } (b, \alpha^\vee) > 0, \\ &[-\alpha, \nu_\alpha j], \ 0 < j < -(b, \alpha^\vee) \text{ as } (b, \alpha^\vee) < 0 \text{ and } \alpha \in \lambda(w), \\ &[-\alpha, \nu_\alpha j], \ 0 < j \leq -(b, \alpha^\vee) \text{ as } (b, \alpha^\vee) < 0 \text{ and } \alpha \notin \lambda(w) \ \} \end{aligned} \tag{3.1.26}$$

for $\alpha \in R_+$.

Now let us assume that the nonaffine components of the roots from $\lambda(wb)$ are all positive. Then $(b, \alpha^\vee) \geq -1$ for all $\alpha > 0$ and $\lambda(w)$ contains all α making -1 in the scalar product with b above. These conditions are also sufficient. ❑

3.1.4 Partial ordering in P

The main application of this section is the ordering needed in the theory of nonsymmetric polynomials; see [O3, M5]. This ordering was also used in [C16] in the process of calculating the coefficients of the Y–operators. The definition is as follows:

$$b \le c, c \ge b \ \text{ for } \ b, c \in P \ \text{ if } \ c - b \in Q_+, \tag{3.1.27}$$
$$b \preceq c, c \succeq b \ \text{ if } \ b_- < c_- \ \text{or} \ \{b_- = c_- \ \text{and} \ b \le c\}. \tag{3.1.28}$$

Recall that $b_- = c_-$ means that b, c belong to the same W–orbit. We write $<, >, \prec, \succ$, respectively, if $b \neq c$.

The following sets

$$\sigma(b) \stackrel{\mathbf{def}}{=\!=} \{c \in P, c \succeq b\}, \ \sigma_*(b) \stackrel{\mathbf{def}}{=\!=} \{c \in P, c \succ b\},$$
$$\sigma_-(b) \stackrel{\mathbf{def}}{=\!=} \sigma(b_-), \ \sigma_+(b) \stackrel{\mathbf{def}}{=\!=} \sigma_*(b_+) = \{c \in P, c_- > b_-\} \tag{3.1.29}$$

are convex. By **convex**, we mean that if $c, d = c + r\alpha \in \sigma$ for $\alpha \in R_+, r \in \mathbb{Z}_+$, then

$$\{c, \ c + \alpha, ..., c + (r-1)\alpha, \ d\} \subset \sigma. \tag{3.1.30}$$

The convexity of the intersections $\sigma(b) \cap W(b), \sigma_*(b) \cap W(b)$ is by construction. For the sake of completeness, let us check the convexity of the sets $\sigma_\pm(b)$.

Both sets are W–invariant. Indeed, $c_- > b_-$ if and only if $b_+ > w(c) > b_-$ for all $w \in W$. The set $\sigma_-(b)$ is the union of σ_+ and the orbit $W(b)$. Here we use that b_+ and b_- are the greatest and the least elements of $W(b)$ with respect to "$>$." This is known (and can be readily checked by the induction with respect to the length; see, e.g., [C16]).

If the end points of (3.1.30) are between b_+ and b_-, then it is true for the orbits of all inner points even if $w \in W$ changes the sign of α (and the order of the end points). Also the elements from $\sigma(b)$ strictly between c and d (i.e., $c + q\alpha, \ 0 < q < r$) belong to $\sigma_+(b)$. This gives the required fact.

The next two propositions are essentially from [C23].

Proposition 3.1.5. *(i) Let $c = \hat{u}((0))$, where $\hat{u}$ is obtained by striking out any number of $\{s_j\}$ from a reduced decomposition of $\pi_b (b \in P)$. Then $c \succ b$. Generally speaking, the converse is not true even if $c \in W(b)$ (I. Macdonald). In other words, the Bruhat order of $\widehat{W}$ when restricted to $\{\pi_b, b \in P\}$ is stronger than $\succ$.*

(ii) Letting $b = s_i((c))$ for $0 \le i \le n$, the element $s_i\pi_c$ can be represented in the form π_b for some $b \in P$ if and only if $(\alpha_j, c + d) \neq 0$. More exactly, the following three conditions are equivalent:

$$\{c \succ b\} \Leftrightarrow \{(\alpha_i^\vee, c + d) > 0\} \Leftrightarrow \{s_i\pi_c = \pi_b, \ l(\pi_b) = l(\pi_c) + 1\}. \tag{3.1.31}$$

❑

The following lemma from [C20] describes the case $(\alpha_i^\vee, c+d) = 0$.

Lemma 3.1.6. *If $(\alpha_i, c+d) = 0$ for an index $0 \le i \le n$, then $u_c(\alpha_i) = \alpha_j$ as $i > 0$ and $u_c(-\vartheta) = \alpha_j$ as $i = 0$ for proper $j > 0$. Equivalently, $s_i\pi_c \notin \pi_P$ provided that $l(s_i\pi_c) = 1 + l(\pi_c)$ if and only if*

$$k + (\alpha_i^\vee, \pi_c((-k\rho)) + d) = 0 \quad for \ \ any \ \ k \in \mathbb{C},\, i \ge 0. \tag{3.1.32}$$

Proof. If $i > 0$, then $\alpha = u_c(\alpha_i) > 0$ and $(\alpha, c_-) = 0$. If $\alpha = \beta + \gamma$ for positive roots β, γ, then $(\beta, c_-) = 0 = (\gamma, c_-)$. Hence $\beta' = u_c^{-1}(\beta) > 0$, $\gamma' = u_c^{-1}(\gamma) > 0$, and $\alpha_i = \beta + \gamma$, which is impossible. Thus α is simple.

Similarly, $(\vartheta, c) = 1$ implies that $\alpha = u_c(-\vartheta) > 0, (\alpha, c_-) = -1$. Let $\alpha = \beta + \gamma$, where $\beta > 0 < \gamma$. Since ϑ and α are short, we can assume that at least one of them is short, but it will follow automatically. Then, transposing $\beta \leftrightarrow \gamma$, if necessary, $(\beta, c_-) = -1$ and $(\gamma, c_-) = 0$. Since $(\beta, c_-) = \nu_\beta(\beta^\vee, c_-)$, we conclude that $\nu_\beta = 1$ and β is short. Hence $\beta' = -u_c^{-1}(\beta) > 0$, $\gamma' = -u_c^{-1}(\gamma) < 0$ and $\vartheta = \beta' + \gamma' < \beta'$, which gives that $\vartheta^\vee < (\beta')^\vee$. The latter is impossible because $\vartheta^\vee$ is the maximal positive root in $R_+^\vee$.

Using $\pi_c = cu_c^{-1}$ and $\pi_c(z) = c + u_c^{-1}(z)$, we obtain that

$$k + (\alpha_i^\vee, \pi_c((-k\rho)) + d) = k - k(u_c(\alpha_i), \rho) + (\alpha_i, c+d)$$

and come to (3.1.32). ❑

Proposition 3.1.7. *(i) Assuming (3.1.31), let $i > 0$. Then $b = s_i(c)$, $b_- = c_-$, and $u_b = u_c s_i$. The set $\lambda(\pi_b)$ is obtained from $\lambda(\pi_c)$ by replacing the strict inequality $(c_-, \alpha^\vee) > j > 0$ for $\alpha = u_c(\alpha_i)$ (see (3.1.15)) with $(c_-, \alpha^\vee) \ge j > 0$. Here $\alpha \in R_-$ and $(c_-, \alpha^\vee) = (c, \alpha_i^\vee) > 0$.*

(ii) In the case $i = 0$, the following holds: $b = \vartheta + s_\vartheta(c)$, the element c is from $\sigma_+(b)$, $b_- = c_- - u_c(\vartheta) \in P_-$, and $u_b = u_c s_\vartheta$. For $\alpha = u_c(-\vartheta) = \alpha^\vee$, the λ-inequality $(c_-, \alpha^\vee) \ge j > 0$ is replaced with the strict inequality $(c_-, \alpha^\vee) + 2 > j > 0$. Here $\alpha \in R_-$ and $(c_-, \alpha^\vee) = -(c, \vartheta) \ge 0$.

(iii) For any $c \in P, r \in O'$, $\pi_r\pi_c = \pi_b$, where $b = \pi_r((c))$. Respectively, $u_b = u_c u_r$, $b = \omega_r + u_r^{-1}(c)$, $b_- = c_- + u_c w_0(\omega_r)$. In particular, the latter weight always belongs to P_-.

Proof. Let us check (i). First, $\pi_b = s_i\pi_c = s_i c u_c^{-1} = s_i(c)(u_c s_i)^{-1}$. The uniqueness of the latter decomposition gives the coincidence $u_b = u_c s_i$. Second, $\lambda(\pi_b)$ is the union of $\lambda(\pi_c)$ and $(u_c c^{-1})(\alpha_i) = [\alpha, (c, \alpha_i)]$ for $\alpha = u_c(\alpha_i) \in R_-$ (see (3.1.8)). Third, the inequality with $\alpha^\vee$ in (3.1.16) is strict for c because $u_c^{-1}(\alpha) = \alpha_i \in R_+$ and becomes non-strict for b, since $u_b^{-1}(\alpha) = -\alpha_i \in R_-$.

For (ii), it is the other way round. Namely, the extra affine root from $\lambda(\pi_b) \setminus \lambda(\pi_c)$ is $\pi_c^{-1}(\alpha_0) = [u_c(-\vartheta), 1 - (c, \vartheta)]$. Therefore $\alpha = u_c(-\vartheta) \in R_-$ and the λ-inequality for π_c is nonstrict. As for π_b, $u_b^{-1}(\alpha) = (u_c s_\vartheta)^{-1}(\alpha) = \vartheta \in R_+$ and the inequality becomes strict. Explicitly, $(b_-, \alpha) = (b, -\vartheta) = (s_\vartheta(c - \vartheta), -\vartheta) = 2 - (c, \vartheta)$. ❑

3.1.5 Arrows in P

We write $c \to b$ or $b \leftarrow c$ in the cases (i) or (ii) from the proposition above and use the left-right arrow $c \leftrightarrow b$ for (iii) or when b and c coincide.

By $c \twoheadrightarrow b$, we mean that b can be obtained from c by a chain of (simple) right arrows. Respectively, $c \leftrightarrow\!\!\!\to b$ indicates that $\leftrightarrow$ can be used in the chain.

Actually, *one* left-right arrow of the type $c \leftrightarrow\!\!\!\to b$ is always sufficient because *one* element π_r is sufficient in a reduced decomposition and it can be placed at the end of the decomposition. If such double arrows are excluded, then the ordering "$\to$" is obviously stronger than the Bruhat order given by the procedure from Proposition 3.1.5(i), which, in its turn, is stronger than "$\succ$."

If $l(\pi_b) = l(\pi_c) + l(\pi_b\pi_c^{-1})$, then the reduced decomposition of $\widehat{w} = \pi_b\pi_c^{-1}$ readily produces a chain of simple arrows from c to b. The number of right arrows is precisely $l(\widehat{w})$ (see (3.1.31)). Only transforms of types (ii) and (iii) will change the W–orbits, adding negative short roots to the corresponding c_- for (ii) and the weights in the form $w(\omega_r)$ $(w \in W)$ for (iii).

For instance, let $c, b, b-c \in P_-$. Then $\pi_c = c$, $\pi_b = b$, $\widehat{w} = b - c$ is of length $l(b) - l(c)$, and we need to decompose $b-c$. If $c = 0$, then $b - \omega_r \in Q$, the reduced decomposition of b begins with π_r $(r \in O')$, and the first new c_- is $w_0(\omega_r)$. When there is no π_r and $b \in Q \cap P_-$, the chain always starts with $-\vartheta$; it gives another proof of Proposition 3.1.1.

Let us now examine the arrows $c \to b$ from the viewpoint of the W–orbits. The following proposition is useful for the classification and description of the perfect representation.

Proposition 3.1.8. *(i) Given $c_- \in P_-$, any element in the form $c_- + u(\omega_r) \in P_-$ can be represented as b_- for proper $b \leftrightarrow c$. Respectively, $c_- + u(\vartheta) \in P_-$ for $u \in W$ such that $u(\vartheta) \in R_-$ can be represented as b_- for proper b such that $W(c_-) \ni c \to b$.*

(ii) In the case of A, D, E, any element b_- such that $b_- \prec c_-$ (both belong to P_-) can be obtained from $c_- \in P_-$ using consecutive arrows $W(c_-) \ni c \to b \in W(b_-)$. This cannot be true for all root systems because only short roots may be added to c_- using such a construction.

Proof. Let $c' = c_- + u(\omega_r) \in P_-$. Given $a = u(\omega_r)$, one can assume that u is the greatest possible, i.e., $u(\beta) \in R_-$ if $\beta \in R_+$ and $(\beta^\vee, \omega_r) = 0$. Explicitly, $u = u_a^{-1} w_0$.

We use the inequalities $(c', \alpha) \le 0$ $(\alpha \in R_+)$ for α such that $(\alpha, c_-) = 0$ (if such α exist). This gives that $(c', \alpha) = (\omega_r, u^{-1}(\alpha))$ is either 0 or negative (actually only -1 may appear). In the latter case, $u^{-1}(\alpha)$ contains α_r with a negative coefficient, and therefore $u^{-1}(\alpha) \in R_-$. If the scalar product is zero, then $u^{-1}(\alpha)$ is negative too because of the maximality of u.

Thus w_0u^{-1} sends such α to R_+ and therefore can be represented as u_c^{-1} for proper $c \in W(c_-)$ (see Proposition 3.1.2 and (3.1.13)). This gives the first part of (i).

If $c' = c_- + u(\vartheta) \in P_-$, we may apply the same argument: $(\vartheta, u^{-1}(\alpha)) < 0$ implies that $u^{-1}(\alpha)$ is negative. So u can be represented as u_cw_0 and $c' = c_- - u_c(\vartheta)$ for proper c. Then we observe that $(c_-, u_c(\vartheta)) = (c, \vartheta) \leq 0$ because $u(\vartheta)$ was assumed to be negative. This proves (i).

It is worth mentioning that the direct statement (that $c' \in P_-$ for u from Proposition 3.1.7) also holds and can be readily checked. One needs to use that the weights $\{\omega_r\}$ are minuscule and $(\vartheta, \alpha^\vee) \geq 2$ $(\alpha \in R)$, with the equality exactly for $\alpha = \vartheta$ $(\alpha \in R)$.

Now let us assume that $c, b \in P_-$ and $0 \neq x = b - c \in Q_-$ in the simply-laced case A, D, E. We pick a connected component I_t of $I = \{i \mid (x, \alpha_i) \leq 0\}$ in the Dynkin diagram Γ and set $x_t = \sum_{i \in I_t} l_i\alpha_i$ for $x = \sum_{i=1}^n l_i\alpha_i$. Then $(x_t, \alpha_i) \leq 0$ for $i \in I_t$, since $(\alpha_i, \alpha_j) < 0$ as $i \in I_t \not\ni j$. Obviously, $x_t \geq x$.

Let ϑ_t be the maximal short root for the root subsystem $R_t \subset R$ generated by $\{\alpha_i, i \in I_t\}$ as simple roots. Then $-\vartheta_t \geq x_t$, thanks to Proposition 3.1.1. Moreover, $(\vartheta_t, \alpha_j) \geq -1$ for $j \notin I_t$. Here we use that the coefficient of α_i in ϑ_c is 1 if α_i is an end point of I_t and at the same time an inner point in Γ. This can be readily checked using the tables of [Bo]. Note that it is not necessary to assume that α_i is inner in Γ if I_t is of type A, D.

Here the value -1 is reached precisely for the neighbors of the end points of I_t in $\Gamma \setminus I_t$ (otherwise the scalar product is zero). Such points α_j do not belong to I because I_t is a connected component of I. Therefore $(c, \alpha_j) < 0$, since $(x, \alpha_j) > 0$ and $(b + x, \alpha_j) \leq 0$. We see that $c > c' = c - \vartheta_t \geq b$ (all three weights are from P_-) and can continue by induction. ❑

3.2 Double Hecke algebras

We keep the notations from the previous section. For the sake of uniformity,

$$([b, l], [b', l']) = (b, b'), \ [\alpha, \nu_\alpha j]^\vee = [\alpha^\vee, j], \ X_0 = X_{\alpha_0} = q_\vartheta X_\vartheta^{-1},$$
$$(d, [\alpha, j]) = j, \text{ for instance, } (b + d, \alpha_0^\vee) = 1 - (b, \vartheta).$$

By m, we denote the least natural number such that $(P, P) = (1/m)\mathbb{Z}$; namely, $m = 2$ for D_{2k}, $m = 1$ for B_{2k}, C_k, otherwise $m = |\Pi|$.

3.2.1 Main definition

The double affine Hecke algebra depends on the parameters $q, \{t_\nu, \nu \in \nu_R\}$. The definition ring will be $\mathbb{Q}_{q,t} \stackrel{\text{def}}{=\!=} \mathbb{Q}[q^{\pm 1/(2m)}, t^{\pm 1/2}]$, formed by polynomials in terms of $q^{\pm 1/(2m)}$ and $\{t_\nu^{\pm 1/2}\}$. If q, t are regarded to be complex numbers (e.g., when q is a root of unity), then the fractional powers of q, t have to be fixed somehow. We set

$$t_{\tilde{\alpha}} = t_\alpha = t_{\nu_\alpha},\ t_i = t_{\alpha_i},\ q_{\tilde{\alpha}} = q^{\nu_\alpha},\ q_i = q^{\nu_{\alpha_i}},$$
$$\text{where } \tilde{\alpha} = [\alpha, \nu_\alpha j] \in \tilde{R},\ 0 \le i \le n,\ \nu_\alpha = \frac{(\alpha,\alpha)}{2}. \tag{3.2.1}$$

To simplify the formulas we will use the k–**function**, $\{k_\nu\}$, together with the $\{t_\nu\}$, setting $\{k_\alpha = k_{\nu_\alpha}\}$,

$$t_\alpha = t_\nu = q_\alpha^{k_\nu} \text{ for } \nu = \nu_\alpha, \text{ and } \rho_k = \sum_\nu k_\nu \rho_\nu = \sum_{\alpha>0} k_\alpha \alpha/2. \tag{3.2.2}$$

We will call ρ_k the **weighted** ρ; see (3.1.1).

For instance, by $q^{(\rho_k,\alpha)}$, we mean $\prod_{\nu\in\nu_R} t_\nu^{(\rho_\nu,\alpha^\vee)}$; here $\alpha \in R$ and therefore this product contains only *integral* powers of t_{sht} and t_{lng}. Note that $(\rho_k, \alpha_i^\vee) = k_i = k_{\alpha_i}$ for $i > 0$.

Let $X_1, \ldots, X_n$ be pairwise commutative and algebraically independent. We set

$$X_{\tilde{b}} = \prod_{i=1}^n X_i^{l_i} q^j \text{ if } \tilde{b} = [b, j], \tag{3.2.3}$$
$$\text{where } b = \sum_{i=1}^n l_i \omega_i \in P,\ j \in \frac{1}{m}\mathbb{Z}.$$

The elements $\widehat{w} \in \widehat{W}$ act in the ring $\mathbb{Q}[q^{\pm 1/m}][X]$ of polynomials in terms of $X_b (b \in P)$ and $q^{\pm 1/m}$ by the formulas:

$$\widehat{w}(X_{\tilde{b}}) = X_{\widehat{w}(\tilde{b})}. \tag{3.2.4}$$

In particular,

$$\pi_r(X_b) = X_{u_r^{-1}(b)} q^{(\omega_{r^*}, b)} \text{ for } \alpha_{r^*} \overset{\mathbf{def}}{=\!=} \pi_r^{-1}(\alpha_0),\ r \in O'. \tag{3.2.5}$$

The involution $r \mapsto r^*$ of the set O' satisfies the relations $u_r u_{r^*} = 1 = \pi_r \pi_{r^*}$. Moreover, $w_0(\omega_r) = -\omega_{r^*}$ for the longest element $w_0 \in W$. Thus $\alpha_r \mapsto \alpha_{r^*}$ is nothing but the automorphism of the nonaffine Dynkin diagram (preserving α_0). This can be readily seen from the tables of [Bo], where only the case of D_n requires some consideration.

Definition 3.2.1. *The* **double affine Hecke algebra** $\mathcal{H}$ *(see [C16]) is generated over the field $\mathbb{Q}_{q,t}$ by the elements $\{T_i,\ 0 \le i \le n\}$, pairwise commutative $\{X_b,\ b \in P\}$ satisfying (3.2.3), and the group Π, where the following relations are imposed:*

(o) $(T_i - t_i^{1/2})(T_i + t_i^{-1/2}) = 0,\ 0 \le i \le n;$

(i) $T_iT_jT_i... = T_jT_iT_j...,$ m_{ij} *factors on each side;*
(ii) $\pi_r T_i \pi_r^{-1} = T_j$ *if* $\pi_r(\alpha_i) = \alpha_j;$
(iii) $T_iX_bT_i = X_bX_{\alpha_i}^{-1}$ *if* $(b, \alpha_i^\vee) = 1,\ 0 \le i \le n;$
(iv) $T_iX_b = X_bT_i$ *if* $(b, \alpha_i^\vee) = 0$ *for* $0 \le i \le n;$
(v) $\pi_r X_b \pi_r^{-1} = X_{\pi_r(b)} = X_{u_r^{-1}(b)} q^{(\omega_{r^*}, b)},\ r \in O'.$ ❑

Given $\widetilde{w} \in \widetilde{W}, r \in O,$ the product

$$T_{\pi_r \widetilde{w}} \stackrel{\mathbf{def}}{=} \pi_r \prod_{k=1}^{l} T_{i_k}, \quad \text{where } \widetilde{w} = \prod_{k=1}^{l} s_{i_k}, l = l(\widetilde{w}), \tag{3.2.6}$$

does not depend on the choice of the reduced decomposition (because the $\{T\}$ satisfy the same "braid" relations as do $\{s\}$). Moreover,

$$T_{\hat{v}} T_{\widehat{w}} = T_{\hat{v}\widehat{w}} \text{ whenever } l(\hat{v}\widehat{w}) = l(\hat{v}) + l(\widehat{w}) \text{ for } \hat{v}, \widehat{w} \in \widehat{W}. \tag{3.2.7}$$

In particular, we arrive at the pairwise commutative elements

$$Y_b = \prod_{i=1}^{n} Y_i^{l_i} \text{ if } b = \sum_{i=1}^{n} l_i \omega_i \in P, \text{ where } Y_i \stackrel{\mathbf{def}}{=} T_{\omega_i}, \tag{3.2.8}$$

satisfying the relations

$$\begin{aligned} T_i^{-1} Y_b T_i^{-1} &= Y_b Y_{a_i}^{-1} \text{ if } (b, \alpha_i^\vee) = 1, \\ T_i Y_b &= Y_b T_i \text{ if } (b, \alpha_i^\vee) = 0,\ 1 \le i \le n. \end{aligned} \tag{3.2.9}$$

3.2.2 Automorphisms

The following maps can be uniquely extended to automorphisms of $\mathcal{H\!H}$ (see [C23, C20]):

$$\varepsilon: X_i \mapsto Y_i,\ Y_i \mapsto X_i,\ T_i \mapsto T_i^{-1},\ t_\nu \mapsto t_\nu^{-1}, q \mapsto q^{-1}, \tag{3.2.10}$$

$$\tau_+: X_b \mapsto X_b,\ Y_r \mapsto X_r Y_r q^{-(\omega_r, \omega_r)/2},\ T_i \mapsto T_i,\ t_\nu \mapsto t_\nu,\ q \mapsto q,$$

$$\tau_+: Y_0 \mapsto Y_0 T_0^2 X_0 = q^{-1} T_{s_\vartheta}^{-1} T_0 X_0,\ Y_0 = Y_{\alpha_0} \stackrel{\mathbf{def}}{=} q^{-1} Y_\vartheta^{-1}, \tag{3.2.11}$$

where $1 \le i \le n,\ r \in O'$. The formulas for the images of $\{Y_r, Y_0\}$ readily give:

$$\begin{aligned} \varepsilon(T_0) &= (Y_0 T_0 X_0)^{-1} = X_\vartheta T_{s_\vartheta},\ \tau_+(T_0) = X_0^{-1} T_0^{-1}, \\ \varepsilon(\pi_r) &= X_r T_{u_r^{-1}},\ \tau_+(\pi_r) = q^{-(\omega_r, \omega_r)/2} X_r \pi_r = q^{(\omega_r, \omega_r)/2} \pi_r X_{r^*}^{-1}, \\ \text{for } \pi_r X_{r^*} \pi_r^{-1} &= q^{(\omega_r, \omega_r)} X_r^{-1},\ X_{r^*} T_{u_r} X_r = T_{u_{r^*}}^{-1}. \end{aligned} \tag{3.2.12}$$

Theorem 2.3 from [C20] states that the mapping

$$\begin{pmatrix} 0 & -1 \\ -1 & 0 \end{pmatrix} \mapsto \varepsilon,\ \begin{pmatrix} 1 & 1 \\ 0 & 1 \end{pmatrix} \mapsto \tau_+ \tag{3.2.13}$$

can be extended to a homomorphism from $GL_2(\mathbb{Z})$ to the group of automorphisms of $\mathcal{H}$ modulo conjugations by the central elements from the group generated by $T_1,\ldots,T_n$. In this statement, the quadratic relation (o) from Definition 3.2.1 may be omitted. Only the group relations matter.

We will also need the following automorphisms of $\mathcal{H}$:

$$\tau_- \stackrel{\mathbf{def}}{=\!=} \varepsilon\tau_+\varepsilon, \ \text{ and } \ \sigma \stackrel{\mathbf{def}}{=\!=} \tau_+\tau_-^{-1}\tau_+ = \tau_-^{-1}\tau_+\tau_-^{-1} = \varepsilon\sigma^{-1}\varepsilon. \qquad (3.2.14)$$

They preserve $T_1,\ldots,T_n,\ t,q$ and are uniquely defined by the following relations:

$$\begin{aligned}
&\tau_- : Y_b \mapsto Y_b,\ X_r \mapsto Y_r X_r q^{\frac{(\omega_r,\omega_r)}{2}},\\
&\tau_- : X_0 \mapsto (T_0X_0Y_0T_0)^{-1} = qT_{s_\vartheta}X_0^{-1}T_0^{-1}, \qquad (3.2.15)\\
&\sigma : X_b \mapsto Y_b^{-1},\ Y_r \mapsto Y_r^{-1}X_rY_rq^{-(\omega_r,\omega_r)},\ Y_0 \mapsto T_{s_\vartheta}^{-1}X_0^{-1}T_{s_\vartheta},
\end{aligned}$$

where $b \in P,\ r \in O'$. Besides,

$$\begin{aligned}
\tau_-(T_0) &= T_0,\ \tau_-(\pi_r) = \pi_r\ (r \in O'),\\
\sigma(T_0) &= T_{s_\vartheta}^{-1}X_\vartheta^{-1},\ \sigma(\pi_r) = Y_r^{-1}X_r\pi_rq^{-(\omega_r,\omega_r)} = T_{u_r}^{-1}X_{r^*}^{-1}. \qquad (3.2.16)
\end{aligned}$$

Thus, τ_- corresponds to $\begin{pmatrix}1&0\\1&1\end{pmatrix}$, σ to $\begin{pmatrix}0&1\\-1&0\end{pmatrix}$. Note that

$$\sigma\tau_+^{-1}\sigma^{-1} = (\tau_+\tau_-^{-1}\tau_+)\tau_+^{-1}(\tau_-\tau_+^{-1}\tau_-) = \tau_-. \qquad (3.2.17)$$

The *braid* relation $\tau_+\tau_-^{-1}\tau_+ = \tau_-^{-1}\tau_+\tau_-^{-1}$ is exactly the definition of the projective $PSL(2,\mathbb{Z})$ due to Steinberg. It formally gives that σ^2 commutes with $\tau_\pm$. Generally speaking, σ^2 is not inner in $\mathcal{H}$. It is the conjugation by $T_{w_0}^{-1}$ if $w_0 = -1$; see below. Always, σ^4 is the conjugation by $T_{w_0}^{-2}$, which is central in the **nonaffine Hecke algebra** $\mathbf{H} = \langle T_i,\ 1 \le i \le n\rangle$.

All automorphisms introduced above are unitary. An automorphism ω of $\mathcal{H}$ is called **unitary** if $\star\omega\star = \omega$ for the main anti-involution $\star$ from [C16]:

$$\begin{aligned}
X_i^\star &= X_i^{-1},\ Y_i^\star = Y_i^{-1},\ T_i^\star = T_i^{-1},\\
t_\nu &\mapsto t_\nu^{-1},\ q \mapsto q^{-1},\ 0 \le i \le n,\ (AB)^\star = B^\star A^\star. \qquad (3.2.18)
\end{aligned}$$

The commutativity with $\star$ is obvious because this anti-involution is the inversion with respect to the multiplicative structure of $\mathcal{H}$.

The following automorphism (involution) of $\mathcal{H}^\flat$ will play an important role in the theory: $\eta \stackrel{\mathbf{def}}{=\!=} \varepsilon\sigma$. It conjugates q,t and is uniquely defined from the relations

$$\begin{aligned}
\eta : T_i \mapsto T_i^{-1},\ X_b \mapsto X_b^{-1},\ \pi_r \mapsto \pi_r,\\
\text{where } 0 \le i \le n,\ b \in P,\ r \in O'. \qquad (3.2.19)
\end{aligned}$$

This involution extends the **Kazhdan–Lusztig involution** on the affine Hecke algebra generated by $\{T_i,\ i \geq 0\}$.

Here we have used (3.2.15) to calculate the images of X, T_i $(i > 0)$, namely, applying (3.2.16),

$$\begin{aligned}\varepsilon\sigma(T_0) &= \varepsilon(T_{s_\vartheta}^{-1}X_\vartheta) = T_{s_\vartheta}Y_\vartheta^{-1} = T_0^{-1}, \quad \varepsilon\sigma(\pi_r)\\ =&\varepsilon(T_{u_r}^{-1}X_{r^*}^{-1}) = T_{u_r^{-1}}Y_{r^*}^{-1} = T_{u_{r^*}}Y_{r^*}^{-1} = \pi_{r^*}^{-1} = \pi_r.\end{aligned}$$

Recall that $\pi_{r^*} = \pi_r^{-1}$ and $u_{r^*} = u_r^{-1}$, where $r \in O$ and $\alpha_{r^*} = \pi_r^{-1}(\alpha_0)$. The map $r \mapsto r^*$ corresponds to the standard automorphism of the nonaffine Dynkin diagram. It is straightforward to check that

$$\eta(Y_r) = T_{w_0}Y_{r^*}^{-1}T_{w_0}^{-1} \quad \text{for} \quad r \in O'.$$

We are going to calculate the η–images of arbitrary Y_b.

We will use the same symbol ς for two different maps. The first on is

$$\begin{aligned} b \mapsto b^\varsigma \ &= \ -w_0(b),\ b \in P, \quad s_i \mapsto s_{\varsigma(i)}, \quad \text{where} \qquad (3.2.20)\\ \varsigma(\alpha_i) =&\alpha_{\varsigma(i)} = -w_0(\alpha_i) \ \text{ for } \ 1 \leq i \leq n,\ \varsigma(0) = 0,\ \varsigma(\alpha_0) = \alpha_0,\end{aligned}$$

which is extended naturally to $\widehat{W}$. The second one is the corresponding automorphism of $\mathcal{H\!H}$:

$$\varsigma: \ X_b \mapsto X_{\varsigma(b)},\ Y_b \mapsto Y_{\varsigma(b)},\ T_i \mapsto T_{\varsigma(i)},\ \pi_r \mapsto \pi_{r^*}. \qquad (3.2.21)$$

The automorphism from (3.2.21) commutes with all previously considered DAHA automorphisms.

Proposition 3.2.2.

$$\begin{aligned}\eta(Y_b) \ &= \ T_{w_0}Y_{w_0(b)}T_{w_0}^{-1},\ \sigma(Y_b) = T_{w_0}^{-1}X_{w_0(b)}T_{w_0}, \qquad (3.2.22)\\ \sigma^2(X_b) &= T_{w_0}^{-1}X_{w_0(b)}^{-1}T_{w_0},\ \sigma^2(Y_b) = T_{w_0}^{-1}Y_{w_0(b)}^{-1}T_{w_0},\\ \sigma^2(T_0) &= T_{w_0}^{-1}T_0T_{w_0},\ T_{w_0}^{-1}Y_\vartheta T_{w_0} = T_{s_\vartheta}^{-1}Y_\vartheta T_{s_\vartheta},\\ T_{\varsigma(i)} &= T_{w_0}^{-1}T_iT_{w_0} \ \textit{ for } \ i > 0,\ T_{w_0}^{-1}T_{s_\vartheta}T_{w_0} = T_{s_\vartheta},\\ \sigma^2(H) &= T_{w_0}^{-1}\varsigma(H)T_{w_0},\ \sigma^4(H) = T_{w_0}^{-2}HT_{w_0}^2 \ \textit{ for } \ H \in \mathcal{H\!H}.\end{aligned}$$

Proof. The formulas for σ easily follow from those for η, so it suffices to check that $\eta(Y_b) = T_{-b}^{-1}$ for $b \in P_+$ coincides with

$$T_{w_0}Y_{w_0(b)}T_{w_0}^{-1} = T_{w_0}Y_{\varsigma(b)}^{-1}T_{w_0}^{-1} = T_{w_0}T_{\varsigma(b)}^{-1}T_{w_0}^{-1},$$

i.e., we need the relation $T_{-b} = T_{w_0}T_{\varsigma(b)}T_{w_0}^{-1}$.

Since $-b \in P_-$, $T_{-b}T_{w_0} = T_{(-b)\cdot w_0}$. Here we apply (3.1.23). Similarly, $\varsigma(b) \in P_+$ and

$$T_{w_0}T_{\varsigma(b)} \ = \ T_{w_0\cdot(\varsigma(b))}.$$

However, $w_0 \cdot (\varsigma(b)) = w_0 \cdot w_0 \cdot (-b) \cdot w_0 = (-b) \cdot w_0$. ❑

3.2.3 Demazure–Lusztig operators

Following [Lus1, KK, C16], the operators

$$\hat{T}_i = t_i^{1/2} s_i + (t_i^{1/2} - t_i^{-1/2})(X_{\alpha_i} - 1)^{-1}(s_i - 1) \ (0 \le i \le n) \qquad (3.2.23)$$

are well defined in the space of Laurent polynomials $\mathbb{Q}[q, t^{\pm 1/2}][X]$. We note that only T_0 involves q:

$$\begin{aligned} \hat{T}_0 &= t_0^{1/2} s_0 + (t_0^{1/2} - t_0^{-1/2})(qX_\vartheta^{-1} - 1)^{-1}(s_0 - 1), \\ \text{where } s_0(X_b) &= X_b X_\vartheta^{-(b,\vartheta)} q^{(b,\vartheta)}, \ \alpha_0 = [-\vartheta, 1]. \end{aligned} \qquad (3.2.24)$$

Theorem 3.2.1. *(i) The map $T_j \mapsto \hat{T}_j$, $X_b \mapsto X_b$ (see (3.2.3)), $\pi_r \mapsto \pi_r$ (see (3.2.5)) induces a $\mathbb{Q}_{q,t}$–linear homomorphism from $\mathcal{H}$ to the algebra of linear endomorphisms of $\mathbb{Q}_{q,t}[X]$. This representation, which will be called the* **polynomial representation,** *is faithful and remains faithful when q, t take any nonzero complex values, assuming that q is not a root of unity.*

(ii) For arbitrary nonzero q, t, any element $H \in \mathcal{H}$ has a unique decomposition in the form

$$H = \sum_{w \in W} g_w T_w f_w, \ g_w \in \mathbb{Q}_{q,t}[X], \ f_w \in \mathbb{Q}_{q,t}[Y], \qquad (3.2.25)$$

and five more analogous decompositions corresponding to the other orders of $\{T, X, Y\}$.

(iii) The image $\hat{H}$ of $H \in \mathcal{H}$ is uniquely determined from the following condition:

$$\hat{H}(f(X)) = g(X) \ \text{for} \ H \in \mathcal{H}, \ \text{if} \ Hf(X) - g(X) \in$$
$$\{\sum_{i=0}^{n} H_i(T_i - t_i^{1/2}) + \sum_{r \in O'} H_r(\pi_r - 1), \ \text{where} \ H_i, H_r \in \mathcal{H}\}. \qquad (3.2.26)$$

The automorphism τ_- preserves the right-hand side of (3.2.26) and therefore acts in the polynomial representation.

Proof. This theorem is essentially from [C16] and [C19]. One only needs to extend (ii) to roots of unity (it was not formulated in [C16] in the complete generality). In the first place, the existence of such a decomposition is true for all q, which follows directly from the defining relations. Secondly, the uniqueness holds for generic q because the polynomial representation is faithful (a similar argument was used in [C16]). Finally, the number of linearly independent expressions in the form (3.2.25) (when the degrees of f, g are bounded) may not become smaller for special q than for generic q (here by special we mean the roots of unity). ❑

Intermediate subalgebras. We will also use the above statements for the subalgebras of $\mathcal{H}$ with P replaced by its subgroups containing Q. Let B be any lattice between Q and P, respectively, $\Pi^\flat$ be the preimage of B/Q in Π, and $\widehat{W}^\flat = \Pi^\flat \cdot \widetilde{W} = B \cdot W$.

The **intermediate subalgebra** $\mathcal{H}^\flat$ is the subalgebra of $\mathcal{H}$ generated by $X_b(b \in B), \pi \in \Pi^\flat$, and $T_i(0 \le i \le n)$. Thus $Y_b \in \mathcal{H}^\flat$ when $b \in B$. Actually, the lattices for X_b and Y_b can be different, but, when discussing the Fourier transform, it is technically convenient to impose the coincidence of these lattices.

Proposition 3.2.3. *The algebra $\mathcal{H}^\flat$ satisfies all claims of Theorem 3.2.1 for B instead of P and the polynomial representations defined as $\mathbb{Q}_{q,t}[X_b, b \in B]$. The automorphisms $\varepsilon, \tau_\pm, \sigma, \eta$ and the anti-involutions $\star, \phi$ preserve this subalgebra.*

Proof. The compatibility of the definition of $\mathcal{H}^\flat$ with the automorphisms $\tau_\pm$ follows directly from the formulas for their action on $X_r, Y_r (r \in O')$ and T_0 (see (3.2.12, 3.2.15)). Since $\mathcal{H}^\flat$ is generated by $X_b, Y_b (b \in B)$ and $\{T_1, ..., T_n\}$, this holds for ε as well. Claim (i) from Theorem 3.2.1 remains true for the polynomials in $X_b(b \in B)$ because the formulas for $\hat{T}_i$ involve X_α only.

To check (ii) we need a more complete version of the relations (iii,iv) from Definition 3.2.1 (cf. [Lus1] and formula (3.2.23) above). Namely, for all $b \in P$,

$$T_i X_b - X_{s_i(b)} T_i \;=\; (t_i^{1/2} - t_i^{-1/2}) \frac{s_i(X_b) - X_b}{X_{\alpha_i} - 1}, \quad 0 \le i \le n. \qquad (3.2.27)$$

This relation and its dual counterpart for Y instead of X ensure the existence of the decompositions (3.2.25). Using the fact that the polynomial representation is faithful (for generic q) we establish the uniqueness (for all $q \neq 0$). Claim (iii) formally results from (ii). ❑

Concerning the notations, we, naturally, set $\widehat{W}^\flat = B \cdot W \subset \widehat{W}$, $\mathbb{Q}_{q,t}[X_b] = \mathbb{Q}_{q,t}[X_b, b \in B]$, and replace m by the least $\tilde{m} \in \mathbb{N}$ such that $\tilde{m}(B, B) \subset \mathbb{Z}$ in the definition of the $\mathbb{Q}_{q,t}$, that is, the ring of constants of $\mathcal{H}^\flat$. The same notation $\mathbb{Q}_{q,t}$ will be used in the $\flat$–case.

3.2.4 Filtrations

The following proposition is essentially from [C20] (Proposition 3.3). In the notation from (3.1.29), we set

$$\begin{aligned} \Sigma(b) &\overset{\text{def}}{=\!=} \oplus_{c \in \sigma(b)} \mathbb{Q}_{q,t} x_c \,,\, \Sigma_*(b) \overset{\text{def}}{=\!=} \oplus_{c \in \sigma_*(b)} \mathbb{Q}_{q,t} x_c \\ \Sigma_-(b) &= \Sigma(b_-),\ \Sigma_+(b) \;=\; \Sigma_*(b_+),\ b \in B. \end{aligned} \qquad (3.2.28)$$

Let us denote the maps $b \mapsto -b$ by $\imath$. Recall that $b \mapsto -w_0(b)$ is denoted by ς. Respectively,

$$\imath : X_b \mapsto X_{-b} = X_b^{-1},\ \varsigma : X_b \mapsto X_b^{\varsigma} = \varsigma(X_b) = X_{-w_0(b)}.$$

Then ς preserves the ordering and sends $\Sigma(b)$ and $\Sigma_*(b)$ to $\Sigma(\varsigma(b))$ and $\Sigma_*(\varsigma(b))$, respectively.

Proposition 3.2.4. *(i) Given $b \in B$, $\tilde\alpha = [\alpha, \nu_\alpha j] \in \tilde R$ with $\alpha > 0$, the operator*

$$\mathcal{R}_{\tilde\alpha} \ = \ t_\alpha^{1/2} + \ (t_\alpha^{1/2} - t_\alpha^{-1/2})(X_{\tilde\alpha}^{-1} - 1)^{-1}(1 - s_{\tilde\alpha}) \tag{3.2.29}$$

preserves $\Sigma(b), \Sigma_*(b), \Sigma_\pm(b)$. Moreover,

$$\begin{aligned} \mathcal{R}_{\tilde\alpha}(X_b)\, mod\, \Sigma_+(b) \ &= \ t_\alpha^{1/2} X_b + \ (t_\alpha^{1/2} - t_\alpha^{-1/2}) s_{\tilde\alpha}(X_b) \ \ \textit{if} \ \ (b,\alpha) < 0, \\ &= \ t_\alpha^{-1/2} X_b \ \ \textit{if} \ \ (b,\alpha) > 0, \\ &= \ t_\alpha^{1/2} X_b \ \ \textit{if} \ \ (b,\alpha) = 0. \end{aligned} \tag{3.2.30}$$

(ii) The operators $\widehat{w}^{-1} T_{\widehat{w}}$ preserve $\Sigma(b), \Sigma_*(b)$, and $\Sigma_\pm(b)$ provided that all affine roots $\tilde\alpha = [\alpha, \nu_\alpha j] \in \lambda(\widehat{w})$ have positive α. The elements $\widehat{w} \in W \cdot B_+$ have this property. For instance, Y_c leave the Σ-sets invariant for all $c \in B$ and

$$\imath \cdot T_{w_0} : \Sigma(b) \to \Sigma(\varsigma(b)),\ \Sigma_*(b) \to \Sigma_*(\varsigma(b)).$$

(iii) Assuming (ii),

$$\begin{aligned} \widehat{w}^{-1} T_{\widehat{w}}(X_b)\, mod\, \Sigma_*(b) \ &= \ \prod_{(\tilde\alpha,b)\le 0} t_\alpha^{1/2} \prod_{(\tilde\alpha,b)>0} t_\alpha^{-1/2} X_b \\ &\textit{multiplied over } \tilde\alpha = [\alpha, \nu_\alpha j] \in \lambda(\widehat{w}). \end{aligned} \tag{3.2.31}$$

In particular,

$$\begin{aligned} Y_c(X_b)\, mod\, \Sigma_*(b) \ &= \ q^{(c, u_b^{-1}(\rho_k) - b)} X_b, \qquad (3.2.32) \\ \imath \cdot T_{w_0}(X_b)\, mod\, \Sigma_*(b) &= \prod_{\nu \in \nu_R} t_\nu^{l_\nu(w_0)/2 - l_\nu(u_b)} X_{\varsigma(b)}\, mod\, \Sigma_*(\varsigma(b)). \end{aligned}$$

❑

The operators $\{T_j, 0 \le j \le n\}$ may not fix the spaces $\Sigma(b), \Sigma_*(b)$. However, they leave $\Sigma_-(b)$ and $\Sigma_+(b)$ invariant. Hence the **affine Hecke algebra** $\mathcal{H}_Y^\flat$ generated by $T_i (0 \le i \le n)$ and the group $\Pi^\flat$ act in the space

$$V(b) \stackrel{\mathbf{def}}{=\!=} \Sigma_-(b)/\Sigma_+(b) \ \cong \ \mathbb{Q}_{q,t}[X_c, c \in W(b)]. \tag{3.2.33}$$

This representation is the (parabolically) induced $\mathcal{H}_Y^\flat$–module generated by the image X_+ of X_{b_+} in $V(b)$ subject to

$$\begin{aligned} T_i(X_+) &= t_i^{1/2}X_+ \text{ if } (b_+,\alpha_i)=0,\ 1\le i\le n,\\ Y_a(X_+) &= q^{(a,u_{b_+}^{-1}(\rho_k)-b_+)}X_+,\ a\in B. \end{aligned} \tag{3.2.34}$$

This is true for arbitrary nonzero q,t.

Note the following explicit formulas:

$$\begin{aligned} s_i(X_c) &= (t_i^{1/2}T_i(X_c)) \text{ if } (c,\alpha_i)>0,\\ &= (t_i^{1/2}T_i(X_c))^{-1} \text{ if } (c,\alpha_i)<0,\\ T_i(X_c) &= t_i^{1/2}X_c \text{ if } (c,\alpha_i)=0,\\ \pi_r(X_c) &= X_{\pi_r(c)},\ c\in W(b),\ r\in O'. \end{aligned} \tag{3.2.35}$$

Here $0\le i\le n,\ (c,\alpha_0)=-(c,\vartheta)$.

3.3 Macdonald polynomials

Continuing from the previous section, we set

$$\mu = \mu^{(k)} = \prod_{\alpha\in R_+}\prod_{i=0}^{\infty}\frac{(1-X_\alpha q_\alpha^i)(1-X_\alpha^{-1}q_\alpha^{i+1})}{(1-X_\alpha t_\alpha q_\alpha^i)(1-X_\alpha^{-1}t_\alpha q_\alpha^{i+1})}. \tag{3.3.1}$$

It is considered as a Laurent series with coefficients in $\mathbb{Q}[t_\nu][[q_\nu]]$ for $\nu\in\nu_R$.

We denote the constant term of $f(X)$ by $\langle f\rangle$. Let $\mu_\circ \stackrel{\mathbf{def}}{=} \mu/\langle\mu\rangle$, where

$$\langle\mu\rangle = \prod_{\alpha\in R_+}\prod_{i=1}^{\infty}\frac{(1-q^{(\rho_k,\alpha)+i})^2}{(1-t_\alpha q^{(\rho_k,\alpha)+i})(1-t_\alpha^{-1}q^{(\rho_k,\alpha)+i})}. \tag{3.3.2}$$

Recall that $q^{(\rho_k,\alpha)} = q_\alpha^{(\rho_k,\alpha^\vee)}$, $t_\alpha = q_\alpha^{k_\alpha}$. This formula is from [C16]. It is nothing but the Macdonald constant term conjecture from [M2].

The coefficients of the Laurent series $\mu_\circ$ are actually from the field of rationals

$$\mathbb{Q}(q,t) \stackrel{\mathbf{def}}{=} \mathbb{Q}(q_\nu,t_\nu),\quad \text{where}\quad \nu\in\nu_R.$$

One also has $\mu_\circ^* = \mu_\circ$ for the involution

$$X_b^* = X_{-b},\ t^* = t^{-1},\ q^* = q^{-1}.$$

This involution is the restriction of the anti-involution $\star$ from (3.2.18) to X–polynomials (and Laurent series). These two properties of $\mu_\circ$ can be directly seen from the difference relations for μ. See the next two propositions.

3.3.1 Definitions

Setting

$$\langle f, g\rangle_\circ \stackrel{\text{def}}{=} \langle \mu_\circ f\, g^*\rangle \;=\; \langle g, f\rangle_\circ^* \text{ for } f, g \in \mathbb{Q}(q,t)[X], \tag{3.3.3}$$

we introduce the **nonsymmetric Macdonald polynomials** $E_b(x) = E_b^{(k)} \in \mathbb{Q}(q,t)[X]$ for $b \in P$ by means of the conditions

$$E_b - X_b \in \oplus_{c\succ b}\mathbb{Q}(q,t)X_c,\ \langle E_b, X_c\rangle_\circ = 0 \text{ for } P \ni c \succ b. \tag{3.3.4}$$

They are well defined because the pairing is nondegenerate and form a basis in $\mathbb{Q}(q,t)[X]$.

This definition is due to Macdonald (for $k_{\text{sht}} = k_{\text{lng}} \in \mathbb{Z}_+$), who extended Opdam's nonsymmetric polynomials introduced in the differential case in [O3] (Opdam mentions Heckman's contribution in [O3]). The general case was considered in [C20].

Another approach is based on the Y–operators. See formulas (3.2.34) and (3.2.32) above, and also Proposition 3.2.4.

In the previous section we denoted Y, T acting in the polynomial representation by $\hat{Y}, \hat{T}$. From now on, we will mainly omit the hat (if it does not lead to confusion).

Proposition 3.3.1. *The polynomial representation is $\star$–unitary:*

$$\langle H(f), g\rangle_\circ \;=\; \langle f, H^\star(g)\rangle_\circ \ \textit{for}\ H \in \mathcal{H},\ f \in \mathbb{Q}_{q,t}[X]. \tag{3.3.5}$$

The polynomials $\{E_b, b \in P\}$ are unique (up to proportionality) eigenfunctions of the operators $\{L_f \stackrel{\text{def}}{=} f(Y_1, \ldots, Y_n), f \in \mathbb{Q}[X]\}$ acting in $\mathbb{Q}_{q,t}[X]$:

$$L_f(E_b) \;=\; f(q^{-b_\sharp})E_b, \ \textit{where}\ b_\sharp \stackrel{\text{def}}{=} b - u_b^{-1}(\rho_k), \tag{3.3.6}$$

$$X_a(q^b) \;=\; q^{(a,b)} \ \textit{for}\ a, b \in P,\ u_b = \pi_b^{-1}b \textit{ is from Section 3.1.} \tag{3.3.7}$$

❑

Thus we have two equivalent definitions of the nonsymmetric polynomials. Both are compatible with the transfer to the intermediate subalgebras $\mathcal{H}^\flat$ and subspaces $\mathbb{Q}_{q,t}[X_b] \stackrel{\text{def}}{=} \mathbb{Q}_{q,t}[X_b, b \in B] \subset \mathbb{Q}_{q,t}[X]$.

Let us check that the coefficients of $\mu_\circ$ and the Macdonald polynomials are rational functions in terms of q_ν, t_ν. We need to make the construction of $\langle\ ,\ \rangle_\circ$ more abstract. A form $(f\,,\,g)$ will be called $*$–**bilinear** if

$$(rf\,,\,g) \;=\; r\,(f\,,\,g) \;=\; (f\,,\,r^*g) \text{ for } r \in \mathbb{Q}_{q,t}. \tag{3.3.8}$$

Proposition 3.3.2. *(i) Forms $\langle f\,,\,g\rangle$ on $\mathbb{Q}_{q,t}[X]$ satisfying (3.3.5) and which are $*$–bilinear are in one-to-one correspondence with $\mathbb{Q}_{q,t}$–linear maps ϖ : $\mathbb{Q}_{q,t}[X] \to \mathbb{Q}_{q,t}$ such that*

$$\varpi(T_i(u)) = t_i^{1/2}\,\varpi(u),\ \ \varpi(Y_a(u)) = q^{(a\,,\,\rho_k)}\varpi(u),\ \ i \geq 0,\ \ a \in P. \tag{3.3.9}$$

Such a form is $*$–hermitian in the sense of (3.3.3) if and only if $\varpi(f^*) = (\varpi(f))^*$. Given ϖ, the corresponding form is $\langle f\,,\,g\rangle = \varpi(fg^*)$.

(ii) Replacing $\mathbb{Q}_{q,t}$ by the field of rationals $\widetilde{\mathbb{Q}}(q,t) = \mathbb{Q}(q^{1/(2m)}, t^{1/2})$ in the definition of $\mathcal{H}$ and in the polynomial representation, there is a unique nonzero ϖ up to proportionality. Namely, it is the restriction to

$$\widetilde{\mathbb{Q}}(q,t)[X] \subset \mathcal{H} \otimes_{\mathbb{Q}} \widetilde{\mathbb{Q}}(q,t)$$

of the linear map $\varpi_{ext} : \mathcal{H} \to \widetilde{\mathbb{Q}}(q,t)$ uniquely determined from the relation $H - \varpi_{ext}(H)\mathcal{I} \subset \mathfrak{J} \stackrel{\mathbf{def}}{=\!=}$

$$\{\sum_{r\in O'}(H_r(\pi_r - 1) + (\pi_r - 1)H_r') + \sum_{i=0}^{n}(H_i(T_i - t_i^{1/2}) + (T_i - t_i^{1/2})H_i')\},$$

where $H_i, H_i', H_r, H_r' \in \mathcal{H}$. Equivalently,

$$f(X) \ = \ \varpi(f) + \sum_{r\in O'}(\pi_r - 1)(h_r) + \sum_{i=0}^{n}(\hat{T}_i - t_i^{1/2})(h_i) \tag{3.3.10}$$

for $h_i,\ h_r \in \widetilde{\mathbb{Q}}(q,t)[X]$.

Proof. In the first place, given a form $\langle f, g\rangle$ satisfying (3.3.5), the corresponding linear form is of course $\varpi(f) = \langle f, 1\rangle$. It obviously satisfies (3.3.10). Its extension to $\mathcal{H}$ is $\varpi_{\text{ext}}(H) = \varpi(\hat{H}(1))$.

Now let ϖ_{unv} be the projection $\mathcal{H} \to \mathcal{H}/\mathfrak{J}$. We set

$$\langle A\,,\,B\rangle_{\text{unv}} = \ \varpi_{\text{unv}}(AB^\star) \ = \ \langle B\,,\,A\rangle^*_{\text{unv}},$$

where the anti-involution $\star$ acts naturally on $\mathcal{H}/\mathfrak{J}$ thanks to the $\star$–invariance of $\mathfrak{J}$.

By construction,

$$\langle HA\,,\,B\rangle_{\text{unv}} \ = \ \varpi_{\text{unv}}(HAB^\star) \ = \ \varpi_{\text{unv}}(AB^\star H) \ = \ \langle A\,,\,H^\star B\rangle_{\text{unv}}$$

for $H = T_i, \pi_r$. Upon the restriction to $\mathbb{Q}_{q,t}[X]$,

$$\langle \hat{H}(f)\,,\,g\rangle \ = \ \langle f\,,\,\widehat{H^\star(g)}\rangle \ \text{ for } \langle f\,,\,g\rangle \stackrel{\mathbf{def}}{=\!=} \varpi_{\text{unv}}(f(X)g(X)^*) \tag{3.3.11}$$

and such H. Here $\hat{H}$ is the action of H in the polynomial representation. Obviously (3.3.11) holds when H is a multiplication by a polynomial. So it is true for all $H \in \mathcal{H}$. Thus the form $\langle f\,,\,g\rangle$ satisfies the same relations as $\langle f\,,\,g\rangle_\circ$. Namely, it is hermitian with respect to * and $\mathcal{H}$–invariant with respect to $\star$. Recall that its values are in the vector space $\mathcal{H}/\mathfrak{J}$ with a natural action of $\star$.

The form $\langle f\,,\,g\rangle$ is universal among such forms. To be more exact, an arbitrary $\mathbb{Q}_{q,t}$–valued linear form on $\mathbb{Q}_{q,t}[X]$ obeying the same relations as ϖ_{unv} is the composition of ϖ_{unv} and a homomorphism $\dot{\omega} : \mathcal{H}/\mathfrak{J} \to \mathbb{Q}_{q,t}$. The latter has to satisfy the $\star$–invariance relations $\omega(H^\star) = \omega(H)^*$ for $H \in \mathcal{H}$ to make the corresponding bi-form $*$–hermitian.

Let us switch from $\mathbb{Q}_{q,t}$ to the field of rationals $\mathbb{Q}(q,t)$. We already know that at least one form $\langle\ ,\ \rangle$ exists for generic q,t with values in a proper completion of $\mathbb{Q}(q,t)$. It is given by (3.3.3) and is unique up to proportionality. Indeed, the polynomial representation is irreducible, the Y–operators are diagonalizable there, and the Y–spectrum is simple. Thus the space $\mathcal{H}/\mathfrak{J}$ is one-dimensional upon proper completion, so it has to be one-dimensional over $\mathbb{Q}(q,t)$ as well. We obtain the rationality of the coefficients of $\mu_\circ$ and therefore the rationality of the coefficients of the nonsymmetric Macdonald polynomials. ❑

Symmetric polynomials. Following Proposition 3.3.1, the **symmetric Macdonald polynomials** $P_b = P_b^{(k)}$ can be introduced as eigenfunctions of the W–invariant operators $L_f = f(Y_1, \ldots, Y_n)$ defined for *symmetric*, i.e., W–invariant, polynomials f:

$$L_f(P_b) = f(q^{-b+\rho_k})P_b,\quad b \in P_-,\ P_b = \sum_{c\in W(b)} X_c \mod \Sigma_+(b). \tag{3.3.12}$$

Here it suffices to take the **monomial symmetric functions**, namely, $f_b = \sum_{c\in W(b)} X_c$ for $b \in P_-$.

Actually, if a polynomial P is an eigenfunction of $\{L_f\}$, then the corresponding "eigenvalue" is represented in the form of $q^{-b+\rho_k}$ for a certain $b \in P_-$, which automatically results in $P_b = \text{Const}\sum_{c\in W(b)} X_c \mod \Sigma_+(b)$. Thus, it is sufficient to say that $\{P\}$ are eigenfunctions of $\{L\}$.

For *minuscule* $b = -\omega_r$ and for $b = -\vartheta$, the restrictions of the operators L_{f_b} to symmetric polynomials are the **Macdonald operators** from [M3, M4, Ru].

The P–polynomials are pairwise orthogonal with respect to $\langle\ ,\ \rangle_\circ$, the inner product that makes the nonsymmetric polynomials $\{E\}$ pairwise orthogonal. However, since they are W–invariant, μ can be replaced by the measure–function due to Macdonald, the symmetric **truncated theta function**:

$$\delta = \delta^{(k)} = \prod_{\alpha\in R_+}\prod_{i=0}^{\infty} \frac{(1 - X_\alpha q_\alpha^i)(1 - X_\alpha^{-1} q_\alpha^i)}{(1 - X_\alpha t_\alpha q_\alpha^i)(1 - X_\alpha^{-1} t_\alpha q_\alpha^i)}. \tag{3.3.13}$$

The corresponding pairing remains $*$–hermitian because $\delta_\circ$ is $*$–invariant.

These polynomials were introduced in [M3, M4]. They were used for the first time in Kadell's unpublished work (classical root systems). In the one-dimensional case, they are due to Rogers (see [AI]).

The connection between E and P is as follows:

$$\begin{aligned} P_{b_-} &= \mathbf{P}_{b_+} E_{b_+},\ b_- \in P_-,\ b_+ = w_0(b_-), \\ \mathbf{P}_{b_+} &\stackrel{\text{def}}{=} \sum_{c \in W(b_+)} \prod_{\nu} t_\nu^{l_\nu(w_c)/2} \hat{T}_{w_c}, \end{aligned} \tag{3.3.14}$$

where $w_c \in W$ is the element of the least length such that $c = w_c(b_+)$. See [O3, M5, C20].

There is another formula [M5, C20]

$$P_{b_-} = \sum_{c \in W(b_+)} \prod_{(\alpha^\vee, c) > 0, \alpha \in R_+} \frac{t_\alpha - X_\alpha(q^{c_\sharp})}{1 - X_\alpha(q^{c_\sharp})} E_c. \tag{3.3.15}$$

The demonstration is based on the technique of intertwiners, that will be considered below. For instance, it gives that P_{b_-} is E_{b_-} modulo lower terms, that is, a part of the definition of P_{b_-}.

3.3.2 Spherical polynomials

We will mainly use the following renormalization of the E–polynomials (see [C20]):

$$\mathcal{E}_b \stackrel{\text{def}}{=} E_b(X)(E_b(q^{-\rho_k}))^{-1}, \quad \text{where}$$

$$E_b(q^{-\rho_k}) = q^{(\rho_k, b_-)} \prod_{[\alpha,j] \in \lambda'(\pi_b)} \Big(\frac{1 - q_\alpha^j t_\alpha X_\alpha(q^{\rho_k})}{1 - q_\alpha^j X_\alpha(q^{\rho_k})} \Big), \tag{3.3.16}$$

$$\lambda'(\pi_b) = \{[\alpha, j] \mid [-\alpha, \nu_\alpha j] \in \lambda(\pi_b)\}. \tag{3.3.17}$$

We call them **spherical polynomials**. Explicitly (see (3.1.15)),

$$\begin{aligned} \lambda'(\pi_b) = \{[\alpha, j] \mid\ & \alpha \in R_+, \\ & -(b_-, \alpha^\vee) > j > 0 \ \text{ if } \ u_b^{-1}(\alpha) \in R_-, \\ & -(b_-, \alpha^\vee) \geq j > 0 \ \text{ if } \ u_b^{-1}(\alpha) \in R_+\}. \end{aligned} \tag{3.3.18}$$

Formula (3.3.16) is the Macdonald **evaluation conjecture** in the nonsymmetric variant.

Note that one has to consider only long α (resp., short) if $k_{\text{sht}} = 0$ (resp., $k_{\text{lng}} = 0$) in the λ'–set. All of the formulas below involving λ or λ' have to be modified correspondingly in such a case.

We have the following **duality formula**:

$$\mathcal{E}_b(q^{c_\sharp}) \;=\; \mathcal{E}_c(q^{b_\sharp}) \;\text{ for }\; b,c\in P,\; b_\sharp = b - u_b^{-1}(\rho_k), \tag{3.3.19}$$

justifying the definition above; see [C20]. The proof is based on the following anti-involution

$$\phi \stackrel{\textbf{def}}{=\!=} \varepsilon\star = \star\varepsilon:\; X_b \mapsto Y_b^{-1},\; T_i \mapsto T_i\;(1\le i\le n). \tag{3.3.20}$$

In contrast to the previous chapter, the latter anti-involution will not be used too much in this chapter, because mainly we study the inner product based on the $\star$–involution, inducing the conjugation $q\mapsto q^{-1}, t\mapsto t^{-1}$.

The following "derivative" $\heartsuit = \sigma\cdot\phi$ of ϕ is a counterpart of $\star$ in the theory without the conjugation, i.e., for the harmonic analysis in the case of real q. It is defined as an anti-involution fixing q,t and their fractional powers, and sending

$$\heartsuit:\; T_i\mapsto T_i,\; \pi_r\mapsto\pi_r,\; Y_b\mapsto Y_b,\; X_b\mapsto T_{w_0}^{-1}X_{\varsigma(b)}T_{w_0} \tag{3.3.21}$$

for $0\le i\le n,\; b\in P$.

Comment. The evaluation formula can be obtained from the duality via the Pieri rules. This derivation in the case of symmetric Macdonald polynomials of type A is due to Koornwinder. The norm formula, including the celebrated Macdonald constant term conjecture, follows from the duality and the evaluation formula by using the technique of intertwiners. See, e.g., [C23]. It is the simplest known approach. ❑

The **symmetric duality and evaluation formulas** from [C19] are

$$P_b(q^{c-\rho_k})P_c(q^{-\rho_k}) \;=\; P_c(q^{b-\rho_k})P_b(q^{-\rho_k}),\; b,c\in P_-, \tag{3.3.22}$$

$$P_b(q^{-\rho_k}) = P_b(q^{+\rho_k}) = X_b(q^{\rho_k})\prod_{\alpha\in R_+}\prod_{j=1}^{-(b,\alpha^\vee)}\Big(\frac{1-q_\alpha^{j-1}t_\alpha X_\alpha(q^{\rho_k})}{1-q_\alpha^{j-1}X_\alpha(q^{\rho_k})}\Big). \tag{3.3.23}$$

Conjugations. We now come to the formulas for the $*$–conjugations of the nonsymmetric polynomials, which are important in the theory of Fourier transforms. The conjugation $\{\,\}^*$ is well defined on the polynomials E_b because their coefficients are rational functions in terms of q,t. Note that, generally, conjugating infinite series in terms of q,t reverses their "direction" and creates problems.

Proposition 3.3.3. *The conjugation $*$ in the polynomial representation is induced by the involution η of $\mathcal{H}^\flat$. The paring $\langle fg\mu_0\rangle$ induces on $\mathcal{H}$ the anti-involution $\star\cdot\eta = \eta\cdot\star$. For $b\in B$,*

$$E_b^* \;=\; \prod_{\nu\in\nu_R} t_\nu^{l_\nu(u_b)-l_\nu(w_0)/2}\,T_{w_0}(E_{\varsigma(b)}),\;\; \textit{where}\;\; \varsigma(b) = -w_0(b). \tag{3.3.24}$$

Proof. The conjugation $*$ sends $\hat{T}_i$ to $\hat{T}_i^{-1}$ for $0 \le i \le n$, i.e., it coincides with the action of $\star$ on T_i. This can be checked by a simple direct calculation. However, $*$ is an *involution*, in contrast to $\star$ which is an *anti-involution*. By definition, the conjugation takes X_b to X_b^{-1} and fixes π_r for $r \in O$. Using (3.2.19), we conclude that $*$ is induced by the involution η of $\mathcal{H}^\flat$.

Applying (3.2.32) to $\varsigma(b)$,

$$T_{w_0}(E_{\varsigma(b)}) \bmod \Sigma_*(b) = \prod_{\nu\in\nu_R} t_\nu^{l_\nu(w_0)/2 - l_\nu(u_{\varsigma(b)})} X_{-b} \bmod \imath(\Sigma_*(b)).$$

Here $\imath$ sends $X_b \mapsto X_{-b}$ and fixes q,t. Obviously $\imath(\Sigma_*(b))$ coincides with $(\Sigma_*(b))^*$ and $l_\nu(u_{\varsigma(b)}) = l_\nu(u_b)$. This gives the coefficient of proportionality in (3.3.24).

Conjugating,

$$\eta(Y_c)(E_b^*) = q^{(c,b_\sharp)} E_b^* = q^{(w_0(c),w_0(b_\sharp))} E_b^* = q^{-(w_0(c),w_0(-b)_\sharp)} E_b^*.$$

The last transformation is possible because the automorphism $\varsigma = -w_0$ is compatible with the representations $b = \pi_b u_b$. Finally, $\eta(Y_c) = T_{w_0} Y_{w_0(c)} T_{w_0}^{-1}$ due to Proposition 3.2.2 (formula (3.2.22)), and $T_{w_0}^{-1}(E_b^*)$ has to be proportional to $E_{w_0(-b)}$. ❑

For the symmetric polynomials, $P_b^* = P_{\varsigma(b)}$, $b \in P_-$, which readily results from the orthogonality property and the second relation from (3.3.12).

The conjugation formula (3.3.24) gets simpler in the spherical normalization. To recalculate it we use the equality

$$q^{(\rho_k, b_-)} = \prod_{\nu\in\nu_R} t_\nu^{-l_\nu(b)/2}$$

and present the evaluation formula (3.3.16) in the following form:

$$E_b(q^{-\rho_k}) = \prod_\nu t_\nu^{-l_\nu(u_b)/2} M_b, \quad \text{where} \tag{3.3.25}$$

$$M_b \stackrel{\text{def}}{=\!=} \prod_{[\alpha,j]\in\lambda'(\pi_b)} \Big(\frac{t_\alpha^{-1/2} - q_\alpha^j t_\alpha^{1/2} X_\alpha(q^{\rho_k})}{1 - q_\alpha^j X_\alpha(q^{\rho_k}).}\Big),$$

The factor M_b is "real," i.e., $*$–invariant. Hence,

$$\begin{aligned}\mathcal{E}_b^* &= M_b^{-1}(E_b \prod_\nu t_\nu^{l_\nu(u_b)/2})^* \\ &= \prod_\nu t_\nu^{l_\nu(u_b)/2 - l_\nu(w_0)/2} T_{w_0}(E_{\varsigma(b)}) \\ &= \prod_\nu t_\nu^{-l_\nu(w_0)/2} T_{w_0}(\mathcal{E}_{\varsigma(b)}).\end{aligned} \tag{3.3.26}$$

We will give later a better proof of this relation without using (3.3.16). It will be based on the intertwining operators and the Fourier transform.

3.3.3 Intertwining operators

The X–**intertwiners** (see, e.g., [C23]) are introduced as follows:

$$
\begin{aligned}
\Phi_i &= T_i + (t_i^{1/2} - t_i^{-1/2})(X_{\alpha_i} - 1)^{-1},\\
S_i &= (\phi_i)^{-1}\Phi_i,\ G_i = \Phi_i(\phi_i)^{-1},\ X_0 = qX_\vartheta^{-1},\\
\phi_i &= t_i^{1/2} + (t_i^{1/2} - t_i^{-1/2})(X_{\alpha_i} - 1)^{-1},
\end{aligned}
\tag{3.3.27}
$$

for $0 \le i \le n$. They belong to $\mathcal{H}$ extended by the field $\mathbb{Q}_{q,t}(X)$ of rational functions in $\{X\}$. The elements S_i and G_i satisfy the same relations as $\{s_i, \pi_r\}$. Hence the map

$$
\widehat{w} \mapsto S_{\widehat{w}} = \pi_r S_{i_l} \cdots S_{i_1}, \ \text{ where } \ \widehat{w} = \pi_r s_{i_l} \cdots s_{i_1} \in \widehat{W}, \tag{3.3.28}
$$

is a well defined homomorphism from $\widehat{W}$. The same holds for $G_{\widehat{w}}$ and $S_{\widehat{w}}$. The latter are called the **normalized intertwiners**.

It suffices to check it for $S_{\widehat{w}}$ and in the polynomial representation, where $S_{\widehat{w}} = \widehat{w}$. We do not need to assume here that the representation is reduced, i.e., $l = l(\widehat{w})$.

As to Φ_i, they satisfy the relations for $\{T_i\}$, i.e., the homogeneous Coxeter relations, and those with π_r. So the decomposition in (3.3.28) has to be reduced to make $\Phi_{\widehat{w}}$ well defined.

A direct proof of the latter is based on the following *intertwining property* of $\{\Phi\}$:

$$
\Phi_{\widehat{w}} X_b = X_{\widehat{w}(b)} \Phi_{\widehat{w}}, \ \widehat{w} \in \widehat{W}, \tag{3.3.29}
$$

which fixes $\Phi_{\widehat{w}}$ uniquely up to left (or right) multiplications by functions of X. Therefore $\Phi^{(1)}_{\widehat{w}} = \Phi^{(2)}_{\widehat{w}} \Lambda(X)$ for a Laurent polynomial $\Lambda(X)$ if we obtain $\Phi^{(1,2)}_{\widehat{w}}$ using the formula $\Phi_{\widehat{w}} = \Phi_{\pi_r}\Phi_{s_{i_l}} \cdots \Phi_{s_{i_1}}$ for two different decompositions of $\widehat{w}$ (maybe nonreduced). The normalization of Φ guarantees that the difference $\Phi_{\widehat{w}} - T_{\widehat{w}}$ is a combination of $T_{\widehat{w}'}$ for $l(\widehat{w}') < l(\widehat{w})$ for any *reduced* decompositions of $\widehat{w}$. Therefore the $\Phi_{\widehat{w}}$ are all well defined (do not depend on the choice of the reduced decomposition) and the map Φ has the desired multiplicative property.

We will also use that Φ_i, ϕ_i are self-adjoint with respect to the anti-involution (3.2.18) and therefore

$$
\Phi^\star_{\widehat{w}} = \Phi_{\widehat{w}^{-1}}, \ S^\star_{\widehat{w}} = G_{\widehat{w}^{-1}}, \ \widehat{w} \in \widehat{W}. \tag{3.3.30}
$$

This gives that G and S are $\star$–unitary up to an X–functional coefficient of proportionality. Explicitly:

$$
(G_{\widehat{w}})^\star G_{\widehat{w}} = \big((S_{\widehat{w}})^\star S_{\widehat{w}}\big)^{-1} = \tag{3.3.31}
$$

$$
\prod_{\alpha,j} \Big(\frac{t_\alpha^{1/2} - q_\alpha^j t_\alpha^{-1/2} X_\alpha}{t_\alpha^{-1/2} - q_\alpha^j t_\alpha^{1/2} X_\alpha} \Big), \ [\alpha, \nu_\alpha j] \in \lambda(\widehat{w}).
$$

To define the Y–**intertwiners** we apply the involution ε to $\Phi_{\widehat{w}}$ and to S, G. Thus $\varepsilon(\Phi_{\widehat{w}})Y_b = Y_{\widehat{w}(b)}\Phi_{\widehat{w}}$, for $\widehat{w} \in \widehat{W}$, where $Y_{[b,j]} \stackrel{\mathbf{def}}{=\!=} q^{-j}Y_b$.

The Y–intertwiners satisfy the same $\star$–relations (3.3.30). The formulas can be easily calculated using (3.2.12). In the case of GL_n, one obtains the intertwiners from [KnS]. For $i > 0$, the Y–intertwiners $\varepsilon(\Phi_{s_i})$ are the X–intertwiners Φ_{s_i} with X_{α_i} replaced by $Y_{\alpha_i}^{-1}$ without touching T and t. Here we use that ε sends nonaffine T_i to T_i^{-1}, $t_\nu^{1/2}$ to $t_\nu^{-1/2}$, and transposes X and Y.

Using (3.2.12), $\varepsilon(T_0) = X_\vartheta T_{s_\vartheta} = (Y_0T_0X_0)^{-1}$ and we obtain the relations

$$\begin{aligned}
\varepsilon(\Phi_i) &= T_i + (t_i^{1/2} - t_i^{-1/2})(Y_{\alpha_i}^{-1} - 1)^{-1} \ \text{ for } i > 0,\\
\varepsilon(\Phi_0) &= X_\vartheta T_{s_\vartheta} - (t_0^{1/2} - t_0^{-1/2})(Y_{\alpha_0} - 1)^{-1}\\
&= Y_0T_0X_0 + (t_0^{1/2} - t_0^{-1/2})(Y_{\alpha_0}^{-1} - 1)^{-1}, \ \text{ for } Y_0 = q^{-1}Y_\vartheta^{-1},\\
\varepsilon(S_i) &= \varepsilon((\phi_i)^{-1}\Phi_i), \ \ \varepsilon(G_i) = \varepsilon(\Phi_i(\phi_i)^{-1}),\\
\varepsilon(\phi_i) &= t_i^{1/2} + (t_i^{1/2} - t_i^{-1/2})(Y_{\alpha_i}^{-1} - 1)^{-1}, \ i \geq 0.
\end{aligned} \tag{3.3.32}$$

When applying the Y–intertwiners to the spherical polynomials (or any eigenfunctions of the Y–operators), it is more convenient to use τ_+ instead of ε. Let us check that for $i \geq 0$ and $r \in O'$,

$$\Psi_i \stackrel{\mathbf{def}}{=\!=} \tau_+(T_i) + \frac{t_i^{1/2} - t_i^{-1/2}}{Y_{\alpha_i}^{-1} - 1}, \ \text{ and } \ P_r \stackrel{\mathbf{def}}{=\!=} \tau_+(\pi_r) \tag{3.3.33}$$

intertwine $\{Y_b\}$. Recall that $\tau_+(T_0) = X_0^{-1}T_0^{-1} = q^{-1}X_\vartheta T_0^{-1}$, $\tau_+(\pi_r) = q^{-(\omega_r,\omega_r)/2}X_r\pi_r$.

Thanks to the relations $\sigma(X_b) = Y_b^{-1}$, $\tau_-(Y_b) = Y_b$:

$$\tau_-\sigma(\Phi_i X_b) = \tau_-\sigma(X_c\Phi_i) \ \Rightarrow \ \tau_-\sigma(\Phi_i)(Y_b^{-1}) = Y_c^{-1}\tau_-\sigma(\Phi_i)$$

for $c = s_i(b)$. Thus $\tau_-\sigma(\Phi_i)$ intertwine $\{Y_b\}$. So do $\tau_-\sigma(\pi_r)$. However,

$$\tau_-\sigma(\Phi_i) = \tau_+(T_i) + (t_i^{1/2} - t_i^{-1/2})(Y_{\alpha_i}^{-1} - 1)^{-1}$$

because $\tau_-\sigma = \tau_+\tau_-^{-1}$ and τ_- preserves T_i. Similarly, $\tau_-\sigma(\pi_r) = \tau_+(\pi_r)$.

The automorphism τ_- acts in $\mathcal{V}$ and commutes with the Y–operators. The following proposition describes its action on the Ψ–intertwiners.

Proposition 3.3.4. *(i) For generic q, t or for arbitrary q, t provided that the polynomial E_b for $b \in P$ is well defined,*

$$\tau_-(E_b) = q^{-\frac{(b_-,b_-)}{2} + (b_-,\rho_k)} E_b \ \ \textit{for} \ \ P_- \ni b_- \in W(b). \tag{3.3.34}$$

(ii) For any q, t, $Y_0 = q^{-1}Y_\vartheta^{-1}$, *and* Ψ_i *from (3.3.33),*

$$\tau_-(T_i) = T_i\,(i>0),\quad \tau_-(\tau_+(T_0)) = \tau_+(T_0)^{-1}Y_0, \tag{3.3.35}$$
$$\tau_-(\Psi_i) = \Psi_i(i>0),\quad \tau_-(\Psi_0) = \Psi_0 Y_0 = Y_0^{-1}\Psi_0,$$
$$\tau_-(\tau_+(\pi_r)) = q^{(\omega_r,\omega_r)/2}Y_r\tau_+(\pi_r) = q^{-(\omega_r,\omega_r)/2}\tau_+(\pi_r)Y_{r^*}^{-1}.$$

Proof. The claims can be checked using directly the definitions. One can also identify τ_- in $\mathcal{V}$ with the operator of multiplication by $\widetilde{\tau_-} = \gamma_y(1)^{-1}\gamma_y$ for the Y*–Gaussian* defined as follows:

$$\gamma_y \stackrel{\mathbf{def}}{=\!=} \sum_{b\in P} q^{(b,b)/2}Y_b,\ \gamma_y(1) = \sum_{b\in P} q^{(b,b)/2}q^{(b,\rho_k)}. \tag{3.3.36}$$

Here we assume that $0<q<1$ and use that $\mathcal{V}$ is a union of finite dimensional spaces preserved by the Y–operators.

The irreducibility of $\mathcal{V}$ for generic q, t is sufficient to conclude that $\widetilde{\tau_-} = \tau_-$, because the conjugation by $\widetilde{\tau_-}$ indices τ_- in $\mathcal{HH}$. This implies their coincidence for generic q (apart from roots of unity) and arbitrary t_ν, and, finally, results in (3.3.35) for *arbitrary* q, t. ❑

Setting

$$\Phi_i^b \;=\; \Phi_i(q^{b_\sharp}) \;=\; T_i + (t_i^{1/2} - t_i^{-1/2})(X_{\alpha_i}(q^{b_\sharp}) - 1)^{-1}, \tag{3.3.37}$$
$$G_i^b \;=\; (\Phi_i\phi_i^{-1})(q^{b_\sharp}) \;=\; \frac{T_i + (t_i^{1/2} - t_i^{-1/2})(X_{\alpha_i}(q^{b_\sharp}) - 1)^{-1}}{t_i^{1/2} + (t_i^{1/2} - t_i^{-1/2})(X_{\alpha_i}(q^{b_\sharp}) - 1)^{-1}}, \tag{3.3.38}$$

we come to the Main theorem 5.1 from [C23].

Proposition 3.3.5. *Given* $c\in P$, $0\le i\le n$ *such that* $(\alpha_i, c+d) > 0$,

$$E_b q^{-(b,b)/2} \;=\; t_i^{1/2}\tau_+(\Phi_i^c)(E_c)q^{-(c,c)/2}\quad for\ \ b = s_i((c)) \tag{3.3.39}$$

and the automorphism τ_+ *from (3.2.11). If* $(\alpha_i, c+d) = 0$, *then*

$$\tau_+(T_i)(E_c) \;=\; t_i^{1/2}E_c,\ 0\le i\le n, \tag{3.3.40}$$

which results in the relations $s_i(E_c) = E_c$ *as* $i>0$. *For* $b = \pi_r((c))$, *where the indices* r *are from* O',

$$q^{-(b,b)/2+(c,c)/2}E_b \;=\; \tau_+(\pi_r)(E_c) \;=\; X_{\omega_r}q^{-(\omega_r,\omega_r)/2}\pi_r(E_c). \tag{3.3.41}$$

❑

3.3.4 Some applications

We will reformulate the proposition using the spherical polynomials. One can also replace P by any lattice $Q \subset B \subset P$. The above considerations are compatible with the reduction $P \to B$. Let

$$\widehat{E}_b \stackrel{\mathbf{def}}{=\!=} \tau_+(\pi_r G_{i_l}^{c_l} \dots G_{i_1}^{c_1})(1), \quad \text{where} \tag{3.3.42}$$
$$c_1 = 0, c_2 = s_{i_1}((c_1)), \dots, c_l = s_{i_l}((c_{l-1})), \quad \text{for} \quad \pi_b = \pi_r s_{i_l} \dots s_{i_1}.$$

These polynomials do not depend on the particular choice of the decomposition of π_b (not necessarily reduced) and are proportional to E_b for all $b \in B$:

$$\begin{aligned} E_b q^{-(b,b)/2} &= \prod_{1 \le p \le l} \left(t_{i_p}^{1/2} \phi_{i_p}(q^{c_p})\right) \widehat{E}_b \\ &= \prod_{[\alpha,j] \in \lambda'(\pi_b)} \left(\frac{1 - q_\alpha^j t_\alpha X_\alpha(q^{\rho_k})}{1 - q_\alpha^j X_\alpha(q^{\rho_k})} \right) \widehat{E}_b. \end{aligned} \tag{3.3.43}$$

We conclude that

$$\mathcal{E}_b = q^{(-\rho_k + b_-, b_-)/2} \widehat{E}_b. \tag{3.3.44}$$

Numerically, it is a straightforward combination of (3.3.43) and (3.3.16). As a matter of fact, it is a simple calculation with the intertwining operators (see [C23] and Corollary 3.3.6 below), so the evaluation formula is not needed here.

As an application we obtain that, given $b \in P$, the spherical polynomial $\mathcal{E}_b$ is well defined for $q, t \in \mathbb{C}^*$ if

$$\prod_{[\alpha,j] \in \lambda'(\pi_b)} \left(1 - q_\alpha^j t_\alpha X_\alpha(q^{\rho_k})\right) \neq 0. \tag{3.3.45}$$

Similarly (see [C23], Corollary 5.3), the polynomial E_b is well defined if

$$\prod_{[\alpha,j] \in \lambda'(\pi_b)} \left(1 - q_\alpha^j X_\alpha(q^{\rho_k})\right) \neq 0.$$

If $b = b_- \in P_-$ and the latter inequality holds for $b_+ = w_0(b_-) \in P_+$, then the corresponding symmetric polynomial P_{b_-} is well defined.

Another application is the following variant of the evaluation formula (3.3.16), which requires only basic properties of the intertwining operators and readily leads to a justification of the evaluation formula. It can be used for arbitrary q, t. We follow Proposition 3.1.7.

Corollary 3.3.6. *Let us assume that E_c is well defined. We will introduce E_b using formula (3.3.39) provided that $l(\pi_b) = 1 + l(\pi_c)$ for $b = s_i((c)), i \geq 0$, and using (3.3.41) for $b = \pi_r((c))$. Then*

$$q^{-(\rho_k, b_-)} E_b(q^{-\rho_k}) = q^{-(\rho_k, c_-)} E_c(q^{-\rho_k}) \frac{1 - q_\alpha^{(\tilde{\alpha}^\vee, c_- + d)} t_\alpha X_\alpha(q^{\rho_k})}{1 - q_\alpha^{(\tilde{\alpha}^\vee, c_- + d)} X_\alpha(q^{\rho_k})}, \tag{3.3.46}$$

$$\tilde{\alpha} = u_c(\alpha_i),\quad \alpha = u_c(\alpha_i) \text{ for } i > 0,\quad \alpha = u_c(-\vartheta) \text{ for } i = 0,$$

$$q^{-(\rho_k, b_-)} E_b(q^{-\rho_k}) = q^{-(\rho_k, c_-)} E_c(q^{-\rho_k}) \quad \text{for} \quad b = \pi_r((c)). \tag{3.3.47}$$

❑

Symmetric polynomials. Recall that the connection of P_{b_-} and $\{E_b\}$ is given in (3.3.14) and (3.3.15). Using the G–intertwiners, the formulas read

$$P_{b_-} = \mathbf{P}_{b_+} E_{b_+} = \mathbf{G}_{b_+} E_{b_-}, \text{ where} \tag{3.3.48}$$
$$\mathbf{P}_{b_+} = \sum_{c \in W(b_+)} \prod_\nu t_\nu^{l_\nu(w_c)/2} T_{w_c},\quad \mathbf{G}_{b_+} = \sum_{c \in W(b_+)} G_{u_c}.$$

These formulas are verified as follows.

One checks that $\mathbf{G} \overset{\mathbf{def}}{=\!=} \sum_{c \in W} G_w$ is divisible by $T_j + t_j^{-1/2}$ on the left for $j > 0$, i.e., $\mathbf{G} \in (T_j + t_j^{-1/2}) \cdot \mathbf{H}$ for the nonaffine Hecke algebra $\mathbf{H}$. Indeed, $\mathbf{G}$ is divisible by $(s_j + 1)$ on the left in the polynomial representation, where it becomes simply the symmetrizer $\sum_{w \in W} w$.

The operator $\mathbf{P}_{b_+}$ for generic b_+ becomes $\mathbf{P} \overset{\mathbf{def}}{=\!=} \sum_{c \in W} \prod_\nu t_\nu^{l_\nu(w)/2} T_w$. The $\mathbf{P}$ is divisible by $T_j + t_j^{-1/2}$ on the left. Moreover, due to the formula

$$\hat{T}_j + t_j^{-1/2} = (s_j + 1) \frac{t_j^{1/2} X_{\alpha_j}^{-1} - t_j^{-1/2}}{X_{\alpha_j}^{-1} - 1},$$

the operator $\mathbf{P}$ is also divisible by $(s_j + 1)$ on the left *in the polynomial representation.* Recall that by $\hat{T}_j$, we denote the action of T_j in $\mathcal{V}$. See, e.g., [C16] (Corollary 4.7). This gives that $\mathbf{P}$ transfers *any* polynomials to symmetric ones.

Thus $\mathbf{P}$ has the same divisibility property as $\mathbf{G}$, and they must coincide up to proportionality as elements in $\mathbf{H}$. Applying $\mathbf{P}$ or $\mathbf{G}$ to an arbitrary E_b with $b \in W(b_+)$, we obtain P_{b_-} up to proportionality. The formula (3.3.48) reflects the normalization and that b can have a nontrivial W–stabilizer.

These formulas can be used to deduce the evaluation formulas and the norm formulas (see Proposition 3.4.1 below and (3.4.3)) in the symmetric case from those for the E–polynomials. Actually, directly proving the symmetric formulas using the shift operator is not difficult either.

We note that there is a similar expression for the **anti-symmetric Macdonald polynomial** $P_{b_-}^-$ in [C20]. This polynomial belongs to the same space linearly generated by E_c $(c \in W(b_+))$ and satisfies the t–anti-symmetric relations $(T_j + t_j^{-1/2})P_{b_-}^- = 0$ for all $j > 0$. The formula for $P_{b_-}^-$ is useful in the theory of the shift operators.

3.4 Polynomial Fourier transforms

We will begin with the **norm formula** for the spherical polynomials. See [C23] (Theorem 5.6) and [M4, C16, M5, C20, M7].

3.4.1 Norm formulas

From here on, B will be any lattice between Q and P, $\{b_i, 1 \le i \le n\}$ a basis of B, and $\mathcal{H}^\flat$ the corresponding intermediate subalgebra of $\mathcal{H}$. We also set $\widehat{W}^\flat = B \cdot W \subset \widehat{W}$, $\mathbb{Q}_{q,t}[X_b] = \mathbb{Q}_{q,t}[X_b, b \in B]$, and replace m by the least $\tilde{m} \in \mathbb{N}$ such that $\tilde{m}(B, B) \subset \mathbb{Z}$ in the definition of the $\mathbb{Q}_{q,t}$.

Proposition 3.4.1. *For $b, c \in B$ and the Kronecker delta δ_{bc},*

$$\begin{aligned}\langle \mathcal{E}_b, \mathcal{E}_c\rangle_\circ \;=\; \langle \widehat{E}_b, \widehat{E}_c\rangle_\circ \;&=\; \delta_{bc}\mu^{-1}(q^{h_\sharp})\mu(q^{-\rho_k}) = \\ \delta_{bc} \prod_{[\alpha,j]\in\lambda'(\pi_b)} &\Big(\frac{t_\alpha^{1/2} - q_\alpha^j t_\alpha^{-1/2} X_\alpha(q^{\rho_k})}{t_\alpha^{-1/2} - q_\alpha^j t_\alpha^{1/2} X_\alpha(q^{\rho_k})}\Big).\end{aligned} \tag{3.4.1}$$

Proof. Using (3.3.42) and that τ_+ is $\star$–unitary,

$$\langle \widehat{E}_b, \widehat{E}_b\rangle_\circ = \langle \tau_+(G_{\pi_b})(1)\,,\, \tau_+(G_{\pi_b})(1)\rangle_\circ = \langle G^\star_{\pi_b} G_{\pi_b}(1)\,,\, 1\rangle_\circ.$$

Hence we can apply (3.3.31), substituting $X_\alpha \mapsto X_\alpha(q^{-\rho_k})$. ❑

Let us use the evaluation formulas to recalculate (3.4.1) in terms of the E–polynomials. One has:

$$\langle E_b, E_c\rangle_\circ \;=\; \delta_{bc} \prod_{[\alpha,j]\in\lambda'(\pi_b)} \Big(\frac{(1 - q_\alpha^j t_\alpha^{-1} X_\alpha(q^{\rho_k}))(t_\alpha^{-1} - q_\alpha^j X_\alpha(q^{\rho_k}))}{(1 - q_\alpha^j X_\alpha(q^{\rho_k}))(1 - q_\alpha^j X_\alpha(q^{\rho_k}))}\Big). \tag{3.4.2}$$

The corresponding formula for the symmetric polynomials, the Macdonald norm conjecture, proved in [C16] in full generality, reads as:

$$\begin{aligned}&\langle P_{b_-}, P_{c_-}\rangle_\circ \\ &= \delta_{b_- c_-} \prod_{\alpha>0} \prod_{j=0}^{-(\alpha^\vee,b)-1} \frac{(1 - q_\alpha^{j+1} t_\alpha^{-1} X_\alpha(q^{\rho_k}))(1 - q_\alpha^j t_\alpha X_\alpha(q^{\rho_k}))}{(1 - q_\alpha^j X_\alpha(q^{\rho_k}))(1 - q_\alpha^{j+1} X_\alpha(q^{\rho_k}))}.\end{aligned} \tag{3.4.3}$$

3.4.2 Discretization

This proposition will be interpreted as a calculation of the Fourier transform acting from the polynomial representation to the functional representation of $\mathcal{H}^\flat$. The functional representation depends on n independent parameters, the exponentials q^{ξ_i} for $\{1 \le i \le n\}$.

We set

$$
\begin{aligned}
X_a(q^\xi) &\overset{\mathbf{def}}{=\!=} \prod_{i=1}^n q^{l_i\xi_i}, \quad \text{where } a = \sum_{i=1}^n l_i b_i \in B,\\
X_a(bw) &= X_a(q^{b+w(\xi)}) = q^{(a,b)} X_{w^{-1}(a)}(q^\xi),\\
\hat{u}(g)(bw) &= g(\hat{u}^{-1}bw), \ b \in B, w \in W. \qquad (3.4.4)
\end{aligned}
$$

Note that $b + w(\xi) = (bw)((\xi))$, using the affine action from (3.1.19).

These formulas determine the **discretization homomorphism** $g(X) \mapsto \delta(g)(\widehat{w})$ (depending on ξ) that maps the space of rational functions $g \in \mathbb{Q}_{q,t}(X)$ or more general functions of X to

$$\mathfrak{F}[\xi] \overset{\mathbf{def}}{=\!=} \mathrm{Funct}(\widehat{W}^\flat, \mathbb{Q}_\xi),$$

formed by $\mathbb{Q}_\xi$–valued functions on $\widehat{W}^\flat$.

From here on, $\mathbb{Q}_\xi \overset{\mathbf{def}}{=\!=} \mathbb{Q}_{q,t}(q^{\xi_1}, \ldots, q^{\xi_n})$. We also omit δ and write g instead of $\delta(g)$ when it is clear that $\mathfrak{F}[\xi]$ is considered.

This homomorphism can be naturally extended to the operator algebra

$$\mathfrak{A} \overset{\mathbf{def}}{=\!=} \oplus_{\hat{u}\in\widehat{W}^\flat} \mathbb{Q}_{q,t}(X)\hat{u}.$$

For instance, the discretizations $\delta(\hat{H})$ of the operators $\hat{H}$ for $H \in \mathcal{H}^\flat$ are well defined and we obtain the action of $\mathcal{H}^\flat$ in $\mathfrak{F}_\xi$. We will call the latter the **functional representation**. It contains the image of the polynomial representation as an $\mathcal{H}^\flat$–submodule.

Explicitly, $\delta(X_a)$ is (multiplication by) the discretization of X_a, $\delta(\pi_r) = \pi_r$, for $\pi_r \in \Pi^\flat$, and

$$
\begin{aligned}
\delta(T_i(g))(\widehat{w})) &= \frac{t_i^{1/2} X_{\alpha_i}(q^{w(\xi)})q^{(\alpha_i,b)} - t_i^{-1/2}}{X_{\alpha_i}(q^{w(\xi)})q^{(\alpha_i,b)} - 1}\, g(s_i\widehat{w})\\
&-\frac{t_i^{1/2} - t_i^{-1/2}}{X_{\alpha_i}(q^{w(\xi)})q^{(\alpha_i,b)} - 1}\, g(\widehat{w}) \ \text{ for } \ 0 \le i \le n, \ \widehat{w} = bw. \qquad (3.4.5)
\end{aligned}
$$

The map δ is an $\mathcal{H}^\flat$–homomorphism.

The inner product corresponding to the anti-involution $\star$ of $\mathcal{H}^\flat$ is given by the discretization of the function $\mu_\bullet(X) = \mu(X)/\mu(q^\xi)$:

$$\mu_\bullet(\widehat{w}) = \prod_{\alpha,j}\Big(\frac{t_\alpha^{-1/2} - q_\alpha^j t_\alpha^{1/2} X_\alpha(q^\xi)}{t_\alpha^{1/2} - q_\alpha^j t_\alpha^{-1/2} X_\alpha(q^\xi)}\Big), \ [\alpha, \nu_\alpha j] \in \lambda(\widehat{w}). \qquad (3.4.6)$$

Actually, we will mainly use $\delta(\mu_\bullet)$, not $\mu_\bullet$ itself. The δ will be dropped in the formulas with $\delta(\mu_\bullet)$.

We put formally $\xi^* = \xi$. To be more exact, $(q^{\xi_i})^* \stackrel{\mathbf{def}}{=} q^{-\xi_i}$ for $1 \le i \le n$. Then $\mu_\bullet(\widehat{w})^* = \mu_\bullet(\widehat{w})$. The counterpart of $\langle\,\rangle$ is the sum

$$\langle f\rangle_\xi \stackrel{\mathbf{def}}{=} \sum_{\widehat{w}\in\widehat{W}^\flat} f(\widehat{w}).$$

The corresponding pairing is as follows:

$$\langle f,g\rangle_\bullet \;=\; \langle fg^*\mu_\bullet\rangle_\xi \;=\sum_{\widehat{w}\in\widehat{W}^\flat} \mu_\bullet(\widehat{w})f(\widehat{w})\; g(\widehat{w})^* \;=\; \langle g,f\rangle_\bullet^*. \tag{3.4.7}$$

Here f,g are from the $\mathcal{H}^\flat$–submodule of finitely supported functions $\mathbf{F}[\xi] \subset \mathfrak{F}[\xi]$.

Given $H \in \mathcal{H}^\flat$,

$$\langle H(f),g\rangle_\bullet \;=\; \langle f, H^\star(g)\rangle_\bullet,\;\; \langle H(f)\,g\mu_\bullet\rangle_\xi = \;\langle f\,\eta(H)^\star(g)\mu_\bullet\rangle_\xi. \tag{3.4.8}$$

To justify (3.4.8), one can follow the considerations in the case of $\langle\,\rangle$ or simply use the homomorphism δ. Also, (3.4.8) can be formally deduced from Proposition 3.3.1.

Note that the operators T_i, X_b are unitary for $\langle f,g\rangle_\bullet$ and self-adjoint with respect to the pairing $\langle fg\mu_\bullet\rangle_\xi$.

The **characteristic functions** $\chi_{\widehat{w}}$ and **delta functions** $\delta_{\widehat{w}}$ in $\mathbf{F}[\xi](\widehat{w} \in \widehat{W}^\flat)$ are defined from the relations

$$\chi_{\widehat{w}}(\hat{u}) \;=\; \delta_{\widehat{w},\hat{u}},\; \delta_{\widehat{w}}(\hat{u}) \;=\; \mu_\bullet(\widehat{w})^{-1}\chi_{\widehat{w}}$$

for the Kronecker delta. We have the formulas:

$$X_b(\chi_{\widehat{w}}) \stackrel{\mathbf{def}}{=} \delta(X_b)(\chi_{\widehat{w}}) \;=\; X_b(\widehat{w})\chi_{\widehat{w}},\;\; b\in B, \widehat{w}\in\widehat{W}^\flat.$$

When considering concrete (non-generic) $\xi \in \mathbb{C}^n$, one needs to check that all the above formulas are well defined (the denominators in (3.4.5) and (3.4.6) are nonzero).

Note that $\mu_\bullet(\widehat{w}) = 1$ if $q^{b+w(\xi)} = q^\xi$, i.e., $q^{(b+w(\xi),a)} = q^{(\xi,a)}$ for all $a \in B$, provided that the function $\mu_\bullet$ is well defined at $q^{b+w(\xi)}$. However, it is not true, generally speaking, if (3.4.6) is used as a definition of $\mu_\bullet(\widehat{w})$, without any reference to the function $\mu_\bullet(X)$.

The main example discussed in this chapter is the following specialization: $\xi = -\rho_k$. More precisely, $q^{(\xi,b)} \mapsto q^{-(\rho_k,b)}$ for $b \in B$. The $\mathcal{H}^\flat$–module $\mathbf{F}(-\rho_k)$ becomes reducible. Namely,

$$\mathbf{I}_\sharp \stackrel{\mathbf{def}}{=} \oplus_{\widehat{w}\notin\pi_B}\mathbb{Q}_{q,t}\chi_{\widehat{w}} \subset \mathbf{F}(-\rho_k) \;\text{ and }\; \mathbf{F}_\sharp \stackrel{\mathbf{def}}{=} \mathbf{F}(-\rho_k)/\mathbf{I}_\sharp \tag{3.4.9}$$

are $\mathcal{H}^\flat$–modules. We denote $\langle\ ,\ \rangle_\bullet$ by $\langle\ ,\ \rangle_\sharp$ in this case. Respectively, $\langle\ \rangle_\sharp \stackrel{\mathbf{def}}{=\!=} \langle\ \rangle_{-\rho_k}$. Since $\mathbf{I}_\sharp$ is the radical of this pairing, it is an $\mathcal{H}^\flat$–submodule, and so is $\mathbf{F}_\sharp$.

Here $\mathbf{F}_\sharp$ can be identified with the space of finitely supported $\mathbb{Q}_{q,t}$–valued functions on $\pi_B \stackrel{\mathbf{def}}{=\!=} \{\pi_b, b \in B\}$. It is irreducible for generic q,t. The $\mathcal{H}^\flat$–module $\mathfrak{F}_\sharp = \mathrm{Funct}(\pi_B, \mathbb{Q}_{q,t})$ is its natural completion.

Note that $X_a(\pi_b) = X_a(q^{b_\sharp})$. Therefore π_B can be naturally identified with the set $B_\sharp = \{b_\sharp = b - u_b^{-1}(\rho_k)\}$ for generic k.

The $\mathcal{H}^\flat$–invariance of $\mathbf{I}_\sharp$ (see [C23] and [C20]) can be seen directly from (3.1.31) and the following general (any ξ) relations:

$$
\begin{aligned}
T_i(\chi_{\widehat{w}}) &= \frac{t_i^{1/2} X_{\alpha_i}^{-1}(q^{w(\xi)})q^{-(\alpha_i,b)} - t^{-1/2}}{X_{\alpha_i}^{-1}(q^{w(\xi)})q^{-(\alpha_i,b)} - 1}\, \chi_{s_i\widehat{w}} \\
&\quad - \frac{t_i^{1/2} - t_i^{-1/2}}{X_{\alpha_i}(q^{w(\xi)})q^{(\alpha_i,b)} - 1}\, \chi_{\widehat{w}} \ \text{ for } \ 0 \le i \le n, \\
\pi_r(\chi_{\widehat{w}}) &= \chi_{\pi_r\widehat{w}}, \quad \text{for} \ \ \pi_r \in \Pi^\flat, \ \widehat{w} = bw \in \widehat{W}^\flat. \qquad (3.4.10)
\end{aligned}
$$

See Lemma 3.1.6 and formula (3.1.32) there to check that the expressions in the right-hand side are well defined.

For further references, the delta-counterparts of these formulas are

$$
\begin{aligned}
T_i(\delta_{\widehat{w}}) &= \frac{t_i^{1/2} X_{\alpha_i}(q^{w(\xi)})q^{(\alpha_i,b)} - t^{-1/2}}{X_{\alpha_i}(q^{w(\xi)})q^{(\alpha_i,b)} - 1}\, \delta_{s_i\widehat{w}} \\
&\quad - \frac{t_i^{1/2} - t_i^{-1/2}}{X_{\alpha_i}(q^{w(\xi)})q^{(\alpha_i,b)} - 1}\, \delta_{\widehat{w}} \ \text{ for } \ 0 \le i \le n, \qquad (3.4.11)
\end{aligned}
$$

and $\pi_r(\delta_{\widehat{w}}) = \delta_{\pi_r\widehat{w}}$. The action of π_r at the delta functions remains the same as for $\{\chi\}$ because $\mu_\bullet(\pi_r\widehat{w}) = \mu_\bullet(\widehat{w})$.

3.4.3 Basic transforms

We introduce the **Fourier transform** $\varphi_\circ$ and the **skew Fourier transform** $\psi_\circ$ as follows:

$$
\begin{aligned}
\varphi_\circ(f)(\pi_b) &= \langle f(X)\mathcal{E}_b(X)\mu_\circ\rangle, \ b \in B, \\
\psi_\circ(f)(\pi_b) &= \langle f(X)^*\mathcal{E}_b(X)\mu_\circ\rangle = \langle \mathcal{E}_b, f\rangle_\circ. \qquad (3.4.12)
\end{aligned}
$$

They are isomorphisms from the space of polynomials $\mathbb{Q}_{q,t}[X_b] \ni f$ onto $\mathbf{F}_\sharp$. The $\varphi_\circ$ is $\mathbb{Q}_{q,t}$–linear; the $\psi_\circ$ conjugates q,t (sends $q \mapsto q^{-1}$, $t \mapsto t^{-1}$).

We will denote the characteristic and delta functions $\chi_{\widehat{w}}, \delta_{\widehat{w}} \in \mathbf{F}_\sharp$ for $\widehat{w} = \pi_b$ by χ_b^π and δ_b^π, respectively.

Theorem 3.4.2. *The Fourier transform $\varphi_\circ$ is unitary, i.e., it sends $\langle\ ,\ \rangle_\circ$ to $\langle\ ,\ \rangle_\sharp$, and induces the automorphism σ on $\mathcal{H\!H}^\flat$: $\varphi_\circ(H(p)) = \delta(\sigma(H))(\varphi_\circ(p))$ for $p(X) \in \mathbb{Q}_{q,t}[X_b]$, $H \in \mathcal{H\!H}^\flat$. Respectively, $\psi_\circ$ is unitary too and induces the involution ε. Explicitly,*

$$\psi_\circ : \sum g_b \mathcal{E}_b(X) \mapsto \sum g_b^* \delta_b^\pi \in \mathbf{F}_\sharp, \quad \text{for } g_b \in \mathbb{Q}_{q,t}, \tag{3.4.13}$$

which is equivalent to the relations

$$\mathcal{E}_b(X) = \varepsilon(G_{\pi_b})(1) \ \text{ for } b \in B, \tag{3.4.14}$$

and results in the formula

$$\varphi_\circ : \sum g_b \mathcal{E}_b(X) \mapsto \sum g_b \prod_\nu t_\nu^{l_\nu(w_0)/2} T_{w_0}^{-1}(\delta^\pi_{\varsigma(b)})$$
$$\text{for } g_b \in \mathbb{Q}_{q,t},\ \varsigma(b) = -w_0(b). \tag{3.4.15}$$

Proof. Let us begin with ψ (the subscript "$\circ$" will be dropped until the "noncompact" Fourier transforms appear). We use the definition of $\mathcal{E}_b$, the duality formula (3.3.19), and the fact that the polynomial representation is $\star$–unitary (see (3.2.18),(3.3.5)). The duality relations are needed in the form of the Pieri rules from [C20], Theorem 5.4. Namely, we use that, given $a \in B$, $X_a^{-1}\mathcal{E}_b(X)$ is a transformation of $Y_a(\delta_b^\pi)$ when the delta functions δ_c^π are replaced by $\mathcal{E}_c$ (see (3.4.11)). This readily gives that ψ induces ε on the double Hecke algebra. Since ε is unitary and both the polynomial representation and $\mathbf{F}_\sharp$ are irreducible (for generic q,t), we obtain that ψ is unitary up to a constant. The constant is 1 because the image of $\mathcal{E}_1 = 1$ is $\delta_0^\pi = \delta_{\mathrm{id}} = \chi_0^\pi$.

Thus (3.4.13) is only a reformulation of (3.4.1). There is another way of checking it that is based on the technique of intertwiners. See [C23] (Corollary 5.2), Proposition 3.3.5 above, and formula (3.3.44).

Turning to φ, we will use the involution $\eta = \varepsilon\sigma$ sending $T_i \mapsto T_i^{-1}$ $(0 \le i \le n)$, $X_b \mapsto X_b^{-1}$, and $\pi_r \mapsto \pi_r$. It is an automorphism of $\mathcal{H\!H}^\flat$, conjugating q,t.

Recall that η extends the conjugation $*$ in the polynomial representation. Therefore the transformation of $\mathcal{H\!H}^\flat$ corresponding to φ is the composition $\varepsilon\eta = \sigma$. This gives that φ is unitary.

Concerning (3.4.14), we apply ψ to the relations

$$\delta_b^\pi = G_{\pi_b}(\delta_0^\pi) \ \text{ for } b \in B. \tag{3.4.16}$$

To obtain (3.4.15), we can combine (3.4.13) with (3.3.26), which states that

$$\mathcal{E}_b^* = \prod_\nu t_\nu^{-l_\nu(w_0)/2} T_{w_0}(\mathcal{E}_{\varsigma(b)}). \tag{3.4.17}$$

Let us give another proof of (3.4.14) which does not require knowing the exact value of the coefficient of proportionality in $\mathcal{E}_b^* = C_b T_{w_0}(\mathcal{E}_{\varsigma(b)})$ and, moreover, automatically gives its value.

What we have proved so far can be represented as follows:

$$\begin{aligned}\langle \mathcal{E}_b\, \mathcal{E}_c \mu_\circ \rangle &= C_b^{-1} \langle (T_{w_0}(\mathcal{E}_b))^* \mathcal{E}_c \mu_\circ \rangle \\ &= C_b^{-1} T_{w_0}^{-1}(\delta^\pi_{\varsigma(b)})(\pi_c). \end{aligned} \tag{3.4.18}$$

Here the left-hand side is $b \leftrightarrow c$–symmetric.

On the other hand,

$$T_{w_0}^{-1}(\delta^\pi_b)(\pi_c) = \langle T_{w_0}^{-1}(\delta^\pi_b)\delta^\pi_c \mu_\bullet \rangle_\sharp, \tag{3.4.19}$$

where the last expression is $b \leftrightarrow c$–symmetric too, thanks to formula (3.4.8).

Therefore the multiplier C_b does not depend on b and must be exactly as in (3.4.17). ❑

Corollary 3.4.3. *Given $b, c \in P$, the* **second norm formula**, *without the conjugation, reads as*

$$\langle \mathcal{E}_b\, \mathcal{E}_c \mu_\circ \rangle = \prod_\nu t_\nu^{l_\nu(w_0)/2}\, T_{w_0}^{-1}(\delta^\pi_{\varsigma(b)})(\pi_c). \tag{3.4.20}$$

The **second inner product** $\langle\langle f\,,\, g\rangle\rangle^\top \overset{\text{def}}{=\!=} \prod_\nu t_\nu^{-l_\nu(w_0)/2} \langle f\,,\, T_{w_0}(g^\varsigma)\mu_\circ\rangle$ *is symmetric for $f, g \in \mathbb{Q}_{q,t}[X]$. Here $g^\varsigma(X) = g(w_0(X)^{-1})$. The corresponding anti-involution $\heartsuit$ is the composition*

$$\heartsuit = \eta \cdot \star \cdot T_{w_0} \cdot \varsigma = \sigma^2 \cdot \eta \cdot \star = \sigma \cdot \varepsilon \cdot \star = \sigma \cdot \phi \tag{3.4.21}$$

in the notation from (2.12.8), (2.12.10), (3.3.20), and (3.3.21), where by T_{w_0} we mean the conjugation $T_{w_0}(\cdot)T_{w_0}^{-1}$. It sends

$$T_i \mapsto T_i \ (i \ge 0),\ \pi_r \mapsto \pi_r,\ Y_b \mapsto Y_b,\ X_b \mapsto T_{w_0}^{-1} X_{\varsigma(b)} T_{w_0}. \tag{3.4.22}$$

One then has:

$$\langle\langle \mathcal{E}_b\,,\, \mathcal{E}_c\rangle\rangle^\top = \delta_{bc}\mu^{-1}(q^{b_\sharp})\mu(q^{-\rho_k}).$$

Assuming that $0 < q < 1$ and imposing the inequalities $1+(\rho_k, \alpha^\vee)+k_\alpha > 0$ for all $\alpha \in R_+$, this paring is positive and therefore the polynomial representation is irreducible. If $k_{sht} = k = k_{lng}$, then the latter inequalities mean that $k > -1/h$ for the Coxeter number $h = (\rho, \vartheta) + 1$. ❑

We will introduce the **conjugated Fourier transforms** by "conjugating" (3.4.12):

$$\begin{aligned}\bar\varphi_\circ(f)(\pi_b) &\overset{\text{def}}{=\!=} \langle f(X)\mathcal{E}_b(X)^*\mu_\circ\rangle = \langle f\,,\, \mathcal{E}_b\rangle_\circ, \\ \bar\psi_\circ(f)(\pi_b) &\overset{\text{def}}{=\!=} \langle f(X)^*\mathcal{E}_b(X)^*\mu_\circ\rangle,\ b \in B. \end{aligned} \tag{3.4.23}$$

Let us "conjugate" Theorem 3.4.2.

Theorem 3.4.4. *The Fourier transforms $\bar{\varphi}$, $\bar{\psi}_\circ$ are unitary, i.e., transform $\langle\ ,\ \rangle_\circ$ to $\langle\ ,\ \rangle_\sharp$. They induce the automorphisms σ^{-1} and $\eta\sigma$ on $\mathcal{H}\!\!\mathcal{H}^\flat$ respectively, and send*

$$\bar{\varphi}_\circ : \sum g_b \mathcal{E}_b(X) \mapsto \sum g_b \delta_b^\pi \ \text{ for } b \in B, g_b \in \mathbb{Q}_{q,t},$$
$$\bar{\psi}_\circ : \sum g_b \mathcal{E}_b(X) \mapsto \sum g_b^* \prod_\nu t_\nu^{-l_\nu(w_0)/2} T_{w_0}(\delta_b^\pi). \tag{3.4.24}$$

□

3.4.4 Gauss integrals

We introduce the **Gaussians** $\gamma^{\pm 1}$ as W–invariant solutions of the following system of difference equations:

$$\omega_j(\gamma) = q^{(\omega_i,\omega_i)/2} X_i^{-1}\gamma, \ \omega_i(\gamma^{-1}) = q^{-(\omega_i,\omega_i)/2} X_i \gamma^{-1} \tag{3.4.25}$$

for $1 \le i \le n$. In the current setting, we need the Gaussian as a Laurent series:

$$\tilde{\gamma}^{-1} \stackrel{\mathbf{def}}{=\!=} \sum_{b\in B} q^{(b,b)/2} X_b. \tag{3.4.26}$$

Analytically, the values of this series at any q^z ($z \in \mathbb{C}$) are well defined if $|q| < 1$. However, we will not use this fact and instead treat $\tilde{\gamma}^{-1}$ as a formal series in this section.

The multiplication by $\tilde{\gamma}^{-1}$ preserves the space of Laurent series with coefficients from $\mathbb{Q}[[q^{1/(2\tilde{m})}]]$. Recall that $\tilde{m} \in \mathbb{N}$ is the smallest positive integer such that $\tilde{m}(B, B) \subset \mathbb{Z}$. The $\tilde{\gamma}^{-1}$ will be later multiplied by the q–expansions of $\mathcal{E}_b(X)$ and that of the $\mu_\circ$. The latter is a Laurent series with coefficients in $\mathbb{Q}[t][[q]]$, which makes the coefficients of $\tilde{\gamma}^{-1}\mu_\circ$ well defined. The q–expansions of the coefficients of $\mathcal{E}_b$ are rational functions in terms of q, t and their natural expansions belong to $\mathbb{Q}[t^{\pm 1}][[q]]$.

Theorem 3.4.5. *Given $b, c \in B$ and the corresponding spherical polynomials $\mathcal{E}_b, \mathcal{E}_c$,*

$$\langle \mathcal{E}_b \mathcal{E}_c \tilde{\gamma}^{-1}\mu_\circ\rangle = q^{(b_\sharp,b_\sharp)/2+(c_\sharp,c_\sharp)/2-(\rho_k,\rho_k)} \mathcal{E}_c(q^{b_\sharp}) \langle \tilde{\gamma}^{-1}\mu_\circ\rangle, \tag{3.4.27}$$
$$\langle \mathcal{E}_b \mathcal{E}_c^* \tilde{\gamma}^{-1}\mu_\circ\rangle = q^{(b_\sharp,b_\sharp)/2+(c_\sharp,c_\sharp)/2-(\rho_k,\rho_k)} \mathcal{E}_c^*(q^{b_\sharp}) \langle \tilde{\gamma}^{-1}\mu_\circ\rangle, \tag{3.4.28}$$
$$\langle \mathcal{E}_b^* \mathcal{E}_c^* \tilde{\gamma}^{-1}\mu_\circ\rangle = q^{(b_\sharp,b_\sharp)/2+(c_\sharp,c_\sharp)/2-(\rho_k,\rho_k)} \times \prod_\nu t_\nu^{-l_\nu(w_0)/2} T_{w_0}(\mathcal{E}_c^*)(q^{b_\sharp}) \langle \tilde{\gamma}^{-1}\mu_\circ\rangle, \tag{3.4.29}$$

where the coefficients of $\mu_\circ$, $\mathcal{E}_b$, $\mathcal{E}_c$, and $\mathcal{E}_c^*$ are expanded in terms of positive powers of q. Here the coefficient of proportionality, a q–generalization of the **Mehta–Macdonald integral**, *can be calculated explicitly:*

$$\langle\tilde{\gamma}^{-1}\mu_\circ\rangle = \prod_{\alpha\in R_+}\prod_{j=1}^{\infty}\Big(\frac{1-t_\alpha^{-1}q_\alpha^{(\rho_k,\alpha^\vee)+j}}{1-q_\alpha^{(\rho_k,\alpha^\vee)+j}}\Big). \tag{3.4.30}$$

The proof is based on the following fact: $\hat{H}\tilde{\gamma}^{-1} = \tilde{\gamma}^{-1}\tau_+(\hat{H})$ in the polynomial representation extended by the Gaussian, where $H \in \mathcal{H}$ and the mapping $H \mapsto \hat{H}$ is from Theorem 3.2.1. Indeed, the conjugation by $\tilde{\gamma}$ corresponds to τ_+ on the standard generators $X_b(b \in B), T_i(0 \le i \le n), \pi_r(r \in O')$. To be more exact, $\hat{H}\tilde{\gamma}^{-1}$ coincides with $\tilde{\gamma}^{-1}(\tau_+(H))\hat{}$ for all $H \in \mathcal{H}^\flat$. It can be readily deduced from the W–invariance of $\tilde{\gamma}$ and (3.4.25).

The same holds in the functional representations $\mathfrak{F}[\xi]$ if we take

$$\gamma^{\pm 1}(bw) \stackrel{\mathbf{def}}{=\!=} q^{\pm(b+w(\xi),b+w(\xi))/2},\ b \in B,\ w \in W. \tag{3.4.31}$$

Here we need to extend the field of constants $\mathbb{Q}_\xi$ by $q^{\pm(\xi,\xi)/2}$. We can avoid such an extension by removing $\pm(\xi,\xi)/2$ from the exponent of (3.4.31) because we need the Gaussian only up to proportionality. However, we prefer to stick to the standard formula $\gamma(q^z) = q^{(z,z)/2}$ in this section.

Involving Theorem 3.4.2, we conclude that the map

$$\varphi_\gamma : f(X) \mapsto f'(\pi_b) = \gamma^{-1}(b_\sharp)\langle f\mathcal{E}_b\tilde{\gamma}^{-1}\mu_\circ\rangle$$

induces the involution $\tau_+^{-1}\sigma\tau_+^{-1} = \tau_-^{-1}$ on $\mathcal{H}^\flat$. Cf. Proposition 3.3.4.

Note that $\gamma(b_\sharp)$ is nothing but $\gamma(\pi_b)$ evaluated at $\xi = -\rho_k$. The map φ_γ acts from $\mathbb{Q}_{q,t}[X_b]$ to the $\mathcal{H}^\flat$–module $\mathfrak{F}_\sharp$, where the field of constants is extended by $q^{-(\xi,\xi)/2}$.

The automorphism τ_- fixes the Y–operators. Hence the image of $f = \mathcal{E}_c$ is an eigenfunction of the discretizations $\delta(Y_a)$ of the $Y_a(a \in B)$ corresponding to the same set of eigenvalues as for $\mathcal{E}_c$. Let us prove that $\varphi_\gamma(\mathcal{E}_c)$ has to be proportional to the discretization $\delta(\mathcal{E}_c)$ of $\mathcal{E}_c$.

In the first place, it suffices to consider $c = 0$ because we may employ the Y–intertwiners (see (3.3.42)). Second, the images of the Y–intertwiners with respect to τ_-^{-1} can be calculated exactly. We do not need explicit formulas here. It is sufficient to know that the intertwiners are invertible operators acting in $\mathfrak{F}_\sharp$. Third, the function $g = \varphi_\gamma(1)$ has additional symmetries: $t_i^{-1/2}\delta(T_i)(g) = g = \delta(\pi_r)(g)$. We have used that τ_- fixes T_i and π_r for all i, r (see (3.2.16)). This readily leads to the $\widehat{W}^\flat$–invariance of g, which means that it has to be constant.

Now, setting $\varphi_\gamma(\mathcal{E}_c) = h_c\gamma(\pi_c)q^{-(\rho_k,\rho_k)}\delta(\mathcal{E}_c)$ for $h_c \in \mathbb{Q}_{q,t}$, we need to check that $h_c = 1$. It obviously holds for h_0. Since the left-hand and the right-hand sides of (3.4.27) are $b \leftrightarrow c$ symmetric, $h_c = h_b$ for all $c, b \in B$ and $h_c = 1$. Here we used the duality formula (3.3.19).

The second formula can be easily deduced from the first thanks to (3.4.17):

$$\mathcal{E}_c^* = \prod_\nu t_\nu^{-l_\nu(w_0)/2} T_{w_0}(\mathcal{E}_{\varsigma(c)}),\ \varsigma(c) = -w_0(c).$$

Indeed, T_{w_0} commutes with γ and its Fourier transform $\varphi_\gamma(T_{w_0})$ is the discretization of $\sigma(T_{w_0}) = T_{w_0}$, which leads to (3.4.29).

The formula (3.4.17) and its conjugation will be important when we construct the inverse Fourier transform in the next section. A direct proof of (3.4.28), without the conjugation formula, is not difficult either. Here it is.

First, $\mathcal{E}_c^*$ are eigenfunctions of the operators $\eta(Y_a)$, where $\eta = \varepsilon\sigma$ (see above). The images of these operators under φ_γ are $\tau_-^{-1}(\eta(Y_a))$. One has:

$$\tau_-^{-1}\eta = \tau_-^{-1}\varepsilon\sigma = \varepsilon\tau_+^{-1}\sigma = \varepsilon\tau_-^{-1}\tau_+ = \varepsilon\sigma\tau_- = \eta\tau_-.$$

Therefore φ_γ fixes $\eta(Y_a)$ and $\varphi_\gamma(\mathcal{E}_c^*)$ is proportional to $\delta(\mathcal{E}_c^*)$ for any $a, c \in B$.

Second, we introduce $\hat{h}$ from $\varphi_\gamma(\mathcal{E}_c^*) = \hat{h}_c\gamma(\pi_c)q^{-(\rho_k,\rho_k)}\delta(\mathcal{E}_c^*)$. However, now it is not obvious that the left-hand side of (3.4.28) is symmetric ($a \leftrightarrow b$), as it was for (3.4.27).

We need to involve

$$\psi_\gamma : f(X) \mapsto \gamma^{-1}(b)\langle f^*\mathcal{E}_b\tilde{\gamma}^{-1}\mu_\circ\rangle.$$

It induces the automorphism

$$\tilde{\varepsilon} = \tau_+^{-1}\varepsilon\tau_+.$$

Indeed, $\langle f^*\mathcal{E}_b\tilde{\gamma}^{-1}\mu_\circ\rangle = \langle (f\tilde{\gamma})^*\mathcal{E}_b\mu_\circ\rangle$. Since

$$\eta(Y_a) = \varepsilon\tau_-^{-1}\tau_+\tau_-^{-1}(Y_a) = \tau_+^{-1}\varepsilon\tau_+(Y_a) = \tilde{\varepsilon}(Y_a),$$

$\psi_\gamma(\mathcal{E}_c)$ is proportional to $\delta(\mathcal{E}_c^*)$. Therefore, letting

$$\psi_\gamma(\mathcal{E}_c^*) = \tilde{h}_c\gamma(\pi_c)q^{-(\rho_k,\rho_k)}\delta(\mathcal{E}_c^*),$$

we conclude that $\hat{h}_c = \tilde{h}_b$ for all $b, c \in B$. Hence both functions $\hat{h}, \tilde{h}$ are constants and, moreover, equal 1, thanks to the normalization of $\mathcal{E}_b$.

The explicit formula for $\langle\tilde{\gamma}^{-1}\mu_\circ\rangle$ was calculated in [C21] using the shift operators and the analytic continuation. ❑

Comment. (i) The product $\mu\tilde{\gamma}^{-1}$ generalizes the (radial) Gaussian measure in the theory of Lie groups and symmetric spaces. To be more exact, the "noncompact" case, which will be considered next, is such a generalization. The above setup can be called "compact," because taking the constant term corresponds to integration with respect to the imaginary period.

(ii) We would like to mention that the appearance of the nonsymmetric polynomials has no known counterparts in the classical representation theory even in the so-called group case when $k = 1$, i.e., $t = q$. In this case, the symmetric Macdonald polynomials become the characters of the compact Lie groups, and one can expect the nonsymmetric polynomials to be somehow connected with these representations. However, it does not happen. There are certain relations of the (degenerate) nonsymmetric polynomials with the Demazure character formulas (the Kac–Moody case, essentially basic representations). However, it looks more accidental than conceptual. It merely reflects the fact that both constructions are based on the Demazure operations.

(iii) In the group case, the symmetrizations of (3.4.27) and (3.4.28) can be readily deduced from the Weyl character formula. It seems that these formulas were not used in harmonic analysis, possibly, because they cannot be extended to the Harish-Chandra zonal spherical functions on general symmetric spaces.

3.5 Jackson integrals

The formulas from the previous sections can be generalized for Jackson integrals taken instead of the constant term functional (corresponding to the imaginary integration over the period). It is a variant of the classical noncompact case. Another variant is a straightforward analytic integration in the real direction, which will not be discussed. We keep the same notation, expend all functions in terms of non-negative powers of q, and consider $q^{(\xi,b_i)}$ as independent parameters. One may also treat ξ as a complex vector in a general position, assuming that $|q| < 1$.

The Jackson integral or **Jackson sum** of $f(bw) \in \mathfrak{F}[\xi]$ is

$$\langle f\rangle_\xi \stackrel{\mathbf{def}}{=\!=} \sum f(bw), \quad \text{where } b \in B,\ w \in W.$$

Recall that $X_a(q^z) = q^{(a,z)}$, $X_a(bw) = q^{(a,w(\xi)+b)}$, $\gamma(q^z) = q^{(z,z)/2}$, and $\gamma(bw) = \gamma(q^{w(\xi)+b})$. Thus

$$|W|^{-1}\langle\gamma\rangle_\xi = \sum_{a\in B} q^{(\xi+a,\xi+a)/2} = \tilde{\gamma}^{-1}(q^\xi)q^{(\xi,\xi)/2}$$

for $\tilde{\gamma}$ from (3.4.26).

The Jackson inner product already appeared in (3.4.7). It is $\langle f,g\rangle_\bullet = \langle fg^*\mu_\bullet\rangle_\xi$ for $\mu_\bullet$ from (3.4.6). To be more precise, here we need the discretization $\delta(\mu_\bullet)$.

3.5.1 Jackson transforms

The **Fourier–Jackson** transform $\varphi_\bullet$, the **skew** Fourier–Jackson transform $\psi_\bullet$, and their bar-counterparts are as follows:

$$\begin{aligned}
\varphi_\bullet(f)(\pi_b) &= \langle f(\widehat{w})\mathcal{E}_b(\widehat{w})\mu_\bullet\rangle_\xi,\ b\in B,\\
\psi_\bullet(f)(\pi_b) &= \langle f(\widehat{w})^*\mathcal{E}_b(\widehat{w})\mu_\bullet\rangle_\xi = \langle \mathcal{E}_b, f\rangle_\bullet, \qquad (3.5.1)\\
\bar{\varphi}_\bullet(f)(\pi_b) &= \langle f(\widehat{w})\mathcal{E}_b^*(\widehat{w})\mu_\bullet(\widehat{w})\rangle_\xi,\ b\in B,\\
\bar{\psi}_\bullet(f)(\pi_b) &= \langle f(\widehat{w})^*\mathcal{E}_b^*(\widehat{w})\mu_\bullet(\widehat{w})\rangle_\xi = \langle f, \mathcal{E}_b\rangle_\bullet. \qquad (3.5.2)
\end{aligned}$$

These transforms act from subspaces of $\mathfrak{F}[\xi]$ to proper completions of the space $\mathbf{F}_\sharp$, provided that we have the convergence. The involution $*$ is the conjugation of the values of functions $f\in\mathfrak{F}[\xi]$. It is well defined because the values are rational functions in terms of q,t (and their certain fractional powers). One has: $(f^*)(\widehat{w}) = (\delta(f^*))(\widehat{w}) = (\delta(f)(\widehat{w}))^* = (f(\widehat{w}))^*$. For instance, $\chi^*_{\widehat{w}} = \chi_{\widehat{w}}$, and $\delta^*_{\widehat{w}} = \delta_{\widehat{w}}$ since $\mu_\bullet(\widehat{w})^* = \mu_\bullet(\widehat{w})$. Thus in (3.4.12) and (3.5.1), we can replace $\mathcal{E}_b^*(\widehat{w})$ by $\mathcal{E}_b(\widehat{w})^*$.

It is straightforward to check that the corresponding automorphisms of $\mathcal{H}\!\!\mathcal{H}^\flat$ are the same as in Theorems 3.4.2 and 3.4.4. We simply replace $\langle\ \rangle$ by $\langle\ \rangle_\xi$:

$$\varphi \leftrightarrow \sigma,\ \bar{\varphi}\leftrightarrow\sigma^{-1}, \psi\leftrightarrow\varepsilon,\ \bar{\psi}\leftrightarrow\eta\sigma. \qquad (3.5.3)$$

These automorphisms commute with the anti-involution $\star$, so all transforms are $\star$–unitary up to a coefficient of proportionality provided we have the convergence and the irreducibility of the corresponding $\mathcal{H}\!\!\mathcal{H}^\flat$–modules. Here ξ is generic. We can be more precise for the spherical specialization.

Theorem 3.5.1. *(i) For $\xi = -\rho_k$, the Fourier transforms from (3.4.12) and (3.5.1) act from the space of the delta functions $\mathbf{F}_\sharp$ to the discretization of the space $\mathbb{Q}_{q,t}[X_b]$ upon restriction to the set π_B. They send the form $\langle\, ,\,\rangle_\sharp$ to $\langle\, ,\,\rangle_\circ$ and satisfy the relations*

$$\begin{aligned}
\bar{\varphi}_\circ\cdot\varphi_\bullet &= id = \varphi_\circ\cdot\bar{\varphi}_\bullet,\\
\bar{\varphi}_\bullet\cdot\varphi_\circ &= id = \varphi_\bullet\cdot\bar{\varphi}_\circ,\\
\psi_\circ\cdot\psi_\bullet &= id = \bar{\psi}_\circ\cdot\bar{\psi}_\bullet,\\
\psi_\bullet\cdot\psi_\circ &= id = \bar{\psi}_\bullet\cdot\bar{\psi}_\circ. \qquad (3.5.4)
\end{aligned}$$

(ii) Setting $\widehat{f} = \varphi_\circ(f)\in\mathbf{F}_\sharp$ for $f(X)\in\mathbb{Q}_{q,t}[X_b]$, the inverse transform reads as

$$f(X) = \prod_\nu t_\nu^{-l_\nu(w_0)/2}\langle\,\widehat{f}\ T^\bullet_{w_0}(\mathcal{E}_{\varsigma(b)})\mu_\bullet\rangle_\sharp, \qquad (3.5.5)$$

where $T^\bullet$ *acts on* $\mathcal{E}_b$ *via formula (3.4.11) for* $\delta^\pi_b = \delta_{\pi_b}$*:*

$$T^\bullet_i(\mathcal{E}_b) = \frac{t_i^{1/2}q^{(\alpha_i,b_\sharp)} - t^{-1/2}}{q^{(\alpha_i,b_\sharp)}-1}\mathcal{E}_{s_i(b)} - \frac{t_i^{1/2}-t_i^{-1/2}}{q^{(\alpha_i,b_\sharp)}-1}\mathcal{E}_b, \tag{3.5.6}$$

extended to W *naturally. Similarly, we have the* **Plancherel formula**

$$\begin{aligned}&\langle f\, g\mu_\circ\rangle = t_\nu^{-l_\nu(w_0)/2}\langle \widehat{f}\, T_{w_0}(\widehat{g}^\varsigma)\mu_\bullet\rangle_\sharp\\ &\text{for } f,g\in \mathbb{Q}_{q,t}[X_b],\ h(\widehat{w})^\varsigma = h(\varsigma(\widehat{w})).\end{aligned} \tag{3.5.7}$$

Proof of (i) is based on (3.5.3) and the irreducibility of the polynomial representation and its delta-counterpart $\mathbf{F}_\sharp$. The same argument gives the unitarity of the Fourier transforms under consideration up to proportionality. Thanks to the normalization of $\mu_\circ$ and $\mu_\bullet$,

$$\langle \delta_0,\delta_0\rangle_\sharp \;=\; 1 \;=\; \langle 1,1\rangle_\circ.$$

So the coefficient of proportionality is 1.

The explicit inversion formula from (ii) is straightforward:

$$\begin{aligned}&\varphi_\circ\Big(\prod_\nu t_\nu^{-l_\nu(w_0)/2}\langle \widehat{f}\, T^\bullet_{w_0}(\mathcal{E}_{\varsigma(b)})\mu_\bullet\rangle_\sharp\Big)\\ &= \prod_\nu t_\nu^{-l_\nu(w_0)/2}\langle \widehat{f}\, T^\bullet_{w_0}\varphi_\circ(\mathcal{E}_{\varsigma(b)})\mu_\bullet\rangle_\sharp\\ &= \langle \widehat{f}\, T^\bullet_{w_0}\Big(\prod_\nu t_\nu^{-l_\nu(w_0)/2}\varphi_\circ(\mathcal{E}_{\varsigma(b)})\Big)\mu_\bullet\rangle_\sharp\\ &= \langle \widehat{f}\, T_{w_0}T_{w_0}^{-1}(\delta_b)\mu_\bullet\rangle_\sharp = \widehat{f}(\pi_b).\end{aligned} \tag{3.5.8}$$

Here we use (3.4.15). Let us check the Plancherel formula. The anti-involution corresponding to the symmetric form $\langle f\, g\mu_\circ\rangle$ is

$$\diamond \overset{\mathbf{def}}{=\!=} \eta\cdot\star = \star\cdot\eta.$$

The one for the form $\langle \widehat{f}\, T_{w_0}(\widehat{g}^\varsigma)\mu_\bullet\rangle_\sharp$ can be readily calculated as follows:

$$\begin{aligned}&\langle\, \widehat{f}\, T_{w_0}(\varphi_\circ(H(g))^\varsigma)\mu_\bullet\rangle_\sharp = \langle\, \widehat{f}\, T_{w_0}\sigma(H^\varsigma)(\widehat{g}^\varsigma)\mu_\bullet\rangle_\sharp\\ =&\langle\, (T_{w_0}\sigma\varsigma(H)T_{w_0}^{-1})^\diamond(\widehat{f})\ T_{w_0}(\widehat{g}^\varsigma)\mu_\bullet\rangle_\sharp\\ =&\langle\, (\sigma^{-1}(H))^\diamond(\widehat{f})\ T_{w_0}(\widehat{g}^\varsigma)\mu_\bullet\rangle_\sharp\\ =&\langle\, \sigma(H^\diamond)(\widehat{f})\ T_{w_0}(\widehat{g}^\varsigma)\mu_\bullet\rangle_\sharp = \langle\, (H^\diamond(f))^\wedge\, T_{w_0}(\widehat{g}^\varsigma)\mu_\bullet\rangle_\sharp.\end{aligned}$$

Here we use that the Fourier transform φ induces σ on $\mathcal{H}$, the relations (3.2.22):

$$T_{w_0}\sigma\varsigma(H)T_{w_0}^{-1} = \sigma^{-1}(H),$$

and (see the last line) the formula

$$\sigma^\diamond = \diamond \cdot \sigma \cdot \diamond = \eta\sigma\eta = \varepsilon\sigma\varepsilon = \sigma^{-1}.$$

❑

Part (ii) of the theorem generalizes the main theorem about the p–adic spherical transform due to Macdonald (symmetric case) and Matsumoto (see [Ma] and the recent [O4, O5]). The classical p–adic spherical transform acts from $\mathbf{F}_\sharp$ to the polynomials. So we need to reverse (ii). The $\mathcal{E}_b^*$ generalize the p–adic (nonsymmetric) spherical functions due to Matsumoto. The limit $q \to \infty$ of the following corollary is exactly the theory of the spherical Fourier transform. The case of generic ξ (in place of $-\rho_k$) is presumably connected with the general (non-spherical) Fourier transform on affine Hecke algebras [KL1] and is expected to be directly related to [HO2].

Note that the latter paper is devoted to affine Hecke algebras with arbitrary labels. Only equal labels are considered in [KL1]. There are Lusztig's papers towards nonequal labels. However, the general labels are difficult to interpret geometrically.

Corollary 3.5.2. *Setting $\widehat{f} = \bar{\varphi}_\bullet(f) \in \mathbb{Q}_{q,t}[X_b]$ for $f \in \mathbf{F}_\sharp$, its* **inversion** *and the* **Plancherel formula** *are as follows:*

$$f = \prod_\nu t_\nu^{l_\nu(w_0)/2} \langle \widehat{f}\, T_{w_0}^{-1}(\mathcal{E}^*_{\varsigma(b)})\mu_\circ \rangle, \tag{3.5.9}$$

$$\langle f\, g\mu_\bullet \rangle_\sharp = t_\nu^{l_\nu(w_0)/2} \langle \widehat{f}\, T_{w_0}^{-1}(\widehat{g}^\varsigma)\mu_\circ \rangle \quad \text{for } f, g \in \mathbb{Q}_{q,t}[X_b],\ h(X_b)^\varsigma = h(X^{-1}_{w_0(b)}). \tag{3.5.10}$$

3.5.2 Gauss–Jackson integrals

In a sense, these integrals are the missing part of the Harish-Chandra theory of spherical functions and the theory of the p–adic spherical transform. The exact formulas for the integrals of the Gaussians with respect to the Harish-Chandra (zonal) transform exist in the group case only ($k = 1$). The Gaussians cannot be added to the p–adic theory too.

Let us transfer Theorem 3.4.5 from the compact case, with $\langle\rangle$ as the integration, to the Jackson case. The proof remains essentially unchanged. In fact, the formulas below are $*$–conjugations of (3.4.27) with a minor reservation about (3.4.30). See also Theorem 7.1 from [C21].

Theorem 3.5.1. *Given $b, c \in P$ and the corresponding spherical polynomials $\mathcal{E}_b, \mathcal{E}_c$,*

$$\langle \mathcal{E}_b\, \mathcal{E}_c^*\, \gamma\mu_\bullet \rangle_\xi = q^{-(b_\sharp,b_\sharp)/2-(c_\sharp,c_\sharp)/2+(\rho_k,\rho_k)} \mathcal{E}_c(q^{b_\sharp}) \langle \gamma\mu_\bullet \rangle_\xi, \tag{3.5.11}$$

$$\langle \mathcal{E}_b^*\, \mathcal{E}_c^*\, \gamma\mu_\bullet \rangle_\xi = q^{-(b_\sharp,b_\sharp)/2-(c_\sharp,c_\sharp)/2+(\rho_k,\rho_k)} \mathcal{E}_c^*(q^{b_\sharp}) \langle \gamma\mu_\bullet \rangle_\xi, \tag{3.5.12}$$

$$\langle \mathcal{E}_b\,\mathcal{E}_c\gamma\mu_\bullet\rangle_\xi \;=\; q^{-(b_\sharp,b_\sharp)/2-(c_\sharp,c_\sharp)/2+(\rho_k,\rho_k)} \times \prod_\nu t_\nu^{l_\nu(w_0)/2} T_{w_0}^{-1}(\mathcal{E}_c)(q^{b_\sharp})\langle\gamma\mu_\bullet\rangle_\xi, \tag{3.5.13}$$

$$\langle\gamma\mu_\bullet\rangle_\xi = \mu(q^\xi, t^{-1})\,|W|^{-1}\langle\gamma\rangle_\xi \prod_{\alpha\in R_+}\prod_{j=0}^{\infty}\Big(\frac{1-t_\alpha^{-1}q_\alpha^{-(\rho_k,\alpha^\vee)+j}}{1-q_\alpha^{-(\rho_k,\alpha^\vee)+j}}\Big). \tag{3.5.14}$$

Here by $\mu(q^\xi, t^{-1})$ we mean the right-hand side of (3.3.1) evaluated at $X = q^\xi$ where all t_α are replaced by t_α^{-1}. ❑

In these formulas, t is generic or complex, provided that the discretization $\delta(\mu_\bullet) = \mu_\bullet(\widehat{w})$ is well defined and the polynomials $\mathcal{E}_b, \mathcal{E}_c$ exist. Note that the right-hand side of (3.5.14) has to be replaced by the limit when $k_\nu \in \mathbb{Z}_+$.

Let us state the inversion theorem and the corresponding **Plancherel formulas**. The pairings $\langle\, f\, g\mu_\circ\rangle$ and $\langle\, f\, g\mu_\bullet\rangle_\xi$ can be naturally extended to the $\mathcal{H}^\flat$-modules $\mathbb{Q}_{q,t}[X_b]\tilde{\gamma}^{-1}$ and the discretization $\delta(\mathbb{Q}_{q,t}[X_b]\gamma) \subset \mathfrak{F}[\xi]$ of $\mathbb{Q}_{q,t}[X_b]\gamma$. We will use these pairings in the Plancherel formulas.

We set $c_\xi = \langle\gamma\mu_\bullet\rangle_\xi$. The Fourier transforms

$$\varphi^c \;=\; c_\xi^{-1}\varphi_\bullet \;\text{ and }\; \bar{\varphi}^c_\bullet \;=\; c_\xi^{-1}\bar{\varphi}_\bullet$$

transfer

$$\delta(Q_{q,t}[X_b]\gamma) \;\to\; \mathbb{Q}_{q,t}[X_b]\tilde{\gamma}^{-1}.$$

Respectively, $\psi^c_\bullet, \bar{\psi}^c_\bullet$ preserve the first module.

In the compact case, setting $c = \langle\gamma\mu_\circ\rangle$,

$$\varphi^c_\circ = c^{-1}\varphi_\circ,\;\; \bar{\varphi}^c_\circ = c^{-1}\bar{\varphi}_\circ,\;\; \psi^c_\circ = c^{-1}\psi_\circ,\;\; \bar{\psi}^c_\circ = c^{-1}\bar{\psi}_\circ,$$

where the first two transforms act from the second module $\mathbb{Q}_{q,t}[X_b]\tilde{\gamma}^{-1}$ to the first and the last two preserve $\mathbb{Q}_{q,t}[X_b]\tilde{\gamma}^{-1}$.

They satisfy the counterparts of the **inversion formulas** (3.5.4):

$$\begin{aligned}
\bar{\varphi}^c_\circ\cdot\varphi^c_\bullet &= \mathrm{id} = \varphi^c_\circ\cdot\bar{\varphi}^c_\bullet,\\
\bar{\varphi}^c_\bullet\cdot\varphi^c_\circ &= \mathrm{id} = \varphi^c_\bullet\cdot\bar{\varphi}^c_\circ,\\
\psi^c_\circ\cdot\psi^c_\circ &= \mathrm{id} = \bar{\psi}^c_\circ\cdot\bar{\psi}^c_\circ,\\
\psi^c_\bullet\cdot\psi^c_\bullet &= \mathrm{id} = \bar{\psi}^c_\bullet\cdot\bar{\psi}^c_\bullet.
\end{aligned} \tag{3.5.15}$$

It is straightforward to reformulate part (ii) of Theorem 3.5.1 for φ^c and the pairings $\langle\, f\, g\mu_\circ\rangle$ on $\mathbb{Q}_{q,t}[X_b]\tilde{\gamma}^{-1}$ and $\langle\, f\, g\mu_\bullet\rangle_\xi$ on $\delta(\mathbb{Q}_{q,t}[X_b]\gamma)$.

Theorem 3.5.1 and the above facts remain valid when $\xi = -\rho_k$. In this case,

$$\mu \;=\mu^{(k)} = \prod_{\alpha\in R_+}\prod_{i=0}^{\infty}\frac{(1-X_\alpha q_\alpha^i)(1-X_\alpha^{-1}q_\alpha^{i+1})}{(1-X_\alpha t_\alpha q_\alpha^i)(1-X_\alpha^{-1}t_\alpha q_\alpha^{i+1})},$$

$$\mu(q^{-\rho_k}, t^{-1}) = \prod_{\alpha \in R_+} \prod_{i=0}^{\infty} \frac{(1 - q_\alpha^{-(\rho_k, \alpha^\vee)+i})(1 - q_\alpha^{(\rho_k, \alpha^\vee)+i+1})}{(1 - t_\alpha^{-1} q_\alpha^{-(\rho_k, \alpha^\vee)+i})(1 - t_\alpha^{-1} q_\alpha^{(\rho_k, \alpha^\vee)+i+1})}, \tag{3.5.16}$$

$$\mu_\bullet(bw) = \mu_\bullet(q^{-w(\rho_k)+b}) = \prod_{[\alpha, j] \in \lambda'(bw)} \Big(\frac{t_\alpha^{-1/2} - t_\alpha^{1/2} q_\alpha^{(\alpha^\vee, \rho_k)+j}}{t_\alpha^{1/2} - t_\alpha^{-1/2} q_\alpha^{(\alpha^\vee, \rho_k)+j}} \Big), \tag{3.5.17}$$

for $\lambda'(bw) = \{[-\alpha, j] \mid [\alpha, \nu_\alpha j] \in \lambda(bw)\}$, $\lambda(bw) = \tilde{R}_+ \cap (bw)^{-1}(\tilde{R}_+)$. The function $\mu_\bullet(bw)$ is well defined. It is nonzero only as $\pi_b = bu_b^{-1}$ (for generic q, t). See (3.3.18) and (3.4.6). For $\widehat{w} = \pi_b$, all α in the product (3.5.17) are positive.

Formula (3.5.14) now reads as follows:

$$\langle \gamma \mu_\bullet \rangle_\sharp \; = \; |W|^{-1} \langle \gamma \rangle_\sharp \prod_{\alpha \in R_+} \prod_{j=1}^{\infty} \Big(\frac{1 - q_\alpha^{(\rho_k, \alpha^\vee)+j}}{1 - t_\alpha^{-1} q_\alpha^{(\rho_k, \alpha^\vee)+j}} \Big). \tag{3.5.18}$$

Recall that $\langle\ \rangle_\xi$ for $\xi = -\rho_k$ is denoted by $\langle\ \rangle_\sharp$. The **inversion formulas** for φ are

$$\begin{aligned} \bar{\varphi}_\circ \cdot \varphi_\bullet \; &= \; |W|^{-1} \langle \gamma \rangle_\sharp \, \mathrm{id} \; = \; \varphi_\circ \cdot \bar{\varphi}_\bullet, \\ \bar{\varphi}_\bullet \cdot \varphi_\circ \; &= \; |W|^{-1} \langle \gamma \rangle_\sharp \, \mathrm{id} \; = \; \varphi_\bullet \cdot \bar{\varphi}_\circ. \end{aligned} \tag{3.5.19}$$

3.5.3 Macdonald's eta-identities

As a by-product, we can represent $\langle \gamma \rangle_\sharp$ as a theta-like product times a certain finite sum, provided that the left-hand side of (3.5.18) contains finitely many nonzero terms. It happens when a certain $\mathbb{Z}_+$–linear combination of $k_{\text{sht}}, k_{\text{lng}}$ is from $-\mathbb{N}$. The main example (which will be used later) is as follows. Let q be generic such that $|q| < 1$.

We call a root $\alpha \in R_+$ **extreme** if $h_\alpha(k) \overset{\text{def}}{=\!=} (\rho_k, \alpha^\vee) + k_\alpha$ does not coincide with any $(\rho_k, \beta^\vee)$ for positive roots β, and **strongly extreme** if also $(\alpha, \omega_i) > 0$ for all $1 \le i \le n$. Note that $h_\vartheta(k)$ is kh for the **Coxeter number** $h = (\rho, \vartheta) + 1$ if $k_{\text{sht}} = k_{\text{lng}}$.

Here the $k_{\text{sht}}, k_{\text{lng}}$ are treated as independent parameters. We omit the subscript of k in the simply-laced case: $\rho_k = k\rho$, $h_\alpha(k) = ((\rho, \alpha) + 1)k$. In this case, there is only one extreme root, namely, ϑ. Let us list the extreme roots for the other root systems (the notation is from [Bo]). All of them are short:

B_n) all short roots ϵ_i $(1 \le i \le n)$, where $h_{\epsilon_i}(k) = 2k_{\text{sht}} + 2(n - i)k_{\text{lng}}$;

C_n) $\vartheta = \epsilon_1 + \epsilon_2$ with $h_\vartheta(k) = 2(n-1)k_{\text{sht}} + 2k_{\text{lng}}$ and $\alpha = \epsilon_1 - \epsilon_n$ with $h_\alpha(k) = n\, k_{\text{sht}}$;

F_4) $0011, 0121, 1121, \vartheta = 1232$ with h equal to $3k_{\text{sht}}$, $4k_{\text{sht}} + 2k_{\text{lng}}$ $4k_{\text{sht}} + 4k_{\text{lng}}$, $6k_{\text{sht}} + 6k_{\text{lng}}$ respectively;

G_2) α_1 $(h_{\alpha_1}(k) = 2k_{\text{sht}})$, $\vartheta = 2\alpha_1 + \alpha_2$ $(h_\vartheta(k) = 3(k_{\text{sht}} + k_{\text{lng}}))$.

Thus ϑ is a unique strongly extreme root for all root systems except for F_4. The root 1121 for F_4 is strongly extreme too.

There are three special cases: (a) $k_{\rm lng} = k_{\rm sht}$, (b) $k_{\rm sht} = 0$, (c) $k_{\rm lng} = 0$. Under either constraint, ϑ becomes a unique strongly extreme root. The next theorem can be extended to these cases.

Theorem 3.5.3. *(i) Let us assume that q is generic, $|q| < 1$, and*

$$(\rho_k, \alpha^\vee) \not\in \mathbb{Z} \setminus \{0\} \text{ for all } \alpha \in R_+. \tag{3.5.20}$$

These conditions result in

$$(\rho_k\,,\,\alpha^\vee) - k_\alpha \not\in \mathbb{Z} \setminus \{0\} \;\; \text{for} \;\; \alpha \in R_+,$$

and therefore are sufficient for the right-hand side of (3.5.18) and the sum in the definition of $\langle \gamma\mu_\bullet \rangle_\sharp$ to exist and be nonzero. The sum $\langle \gamma\mu_\bullet \rangle_\sharp$ is finite if and only if $h_\alpha(k) \in -\mathbb{N}$ for a strongly extreme root α.

(ii) If $h_\vartheta(k) = -1$ and (3.5.20) is satisfied (which automatically holds as $k_{sht} = k_{lng}$), the measure $\mu_\bullet(bw)$ takes exactly $|\Pi^\flat|$ nonzero values. All of them equal 1 and

$$\langle \gamma \rangle_\sharp \;=\; \sum_{b\in B} q^{(b+\rho_k,b+\rho_k)/2} \;=\; \Big(\sum_{\pi_r\in\Pi^\flat} q^{(\omega_r+\rho_k,\omega_r+\rho_k)/2}\Big)\times$$

$$\prod_{\alpha\in R_+}\prod_{j=1}^{\infty}\Big(\frac{1-t_\alpha^{-1}q_\alpha^{(\rho_k,\alpha^\vee)+j}}{1-q_\alpha^{(\rho_k,\alpha^\vee)+j}}\Big), \;\; \text{where} \;\; \omega_0 = 0. \tag{3.5.21}$$

In the exceptional case $\{R = F_4,\ \alpha = 1121,\ h_\alpha(k) = -1\}$, the number of nonzero values of $\mu_\bullet(bw)$ is greater than $|\Pi| = 1$.

Proof. Concerning the first claim, $(\rho_k\,,\,\alpha^\vee) - k_\alpha$ is zero for simple $\alpha = \alpha_i$. If α is not simple, then there exists a simple root α_i of the same length as $\alpha > 0$ such that $\alpha^\vee - \alpha_i^\vee$ is a positive coroot $\beta^\vee$. Then $(\rho_k, \beta^\vee) = (\rho_k\,,\,\alpha^\vee) - k_\alpha$ cannot be from $\mathbb{Z}\setminus\{0\}$ due to (3.5.20).

Generally, there must be at least one root α satisfying $h_\alpha(k) \in -\mathbb{N}$ to make the Jackson sum $\langle \gamma\mu_\bullet \rangle_\sharp \;=\; \sum_{b\in B} q^{(b_\sharp,b_\sharp)/2}\mu_\bullet(\pi_b)$ finite. Assumption (3.5.20) gives that at least one such α has to be strongly extreme.

Note that if there are two such roots $\alpha \neq \beta$ (this may happen for F_4 only), then $h_\beta(k) - h_\alpha(k) \in \mathbb{Z}$, which contradicts assumption (3.5.20).

We obtain that the above sum is finite. Indeed, recall that $\mu_\bullet(\mathrm{id}) \;=\; 1$ and

$$\mu_\bullet(\pi_b) \;=\; \prod\Big(\frac{t_\alpha^{-1/2} - t_\alpha^{1/2} q_\alpha^{(\alpha^\vee,\rho_k)+j}}{t_\alpha^{1/2} - t_\alpha^{-1/2} q_\alpha^{(\alpha^\vee,\rho_k)+j}}\Big) \;\text{ for }\; \alpha \in R_+, \tag{3.5.22}$$

$$-(b_-,\alpha^\vee) > j > 0 \;\text{ if }\; u_b^{-1}(\alpha) \in R_-,\;\; -(b_-,\alpha^\vee) \geq j > 0 \text{ otherwise.}$$

See (3.5.17) and (3.3.18).

Let us assume now that $\alpha = \vartheta$, $h_\vartheta(k) = -1$, and $b \neq 0$. The case of $F_4, \alpha = 1121$ is left to the reader.

Then $\mu_\bullet(\pi_b)$ can be nonzero only for b from the W–orbits of the minuscule weights $b_- = -\omega_r$. Indeed, otherwise $-(b_-, \vartheta) \geq 2$, $[\vartheta, 1] \in \lambda'(\pi_b)$, and the product (3.5.22) is zero. Moreover, u_b^{-1} sends the roots $\{\alpha_i^\vee, i \neq r, i > 0\}$ to $R_+^\vee$, since it is minimal such that $u_b^{-1}(\omega_r) = b_-$. Also, $u_b^{-1}(-\vartheta) \in R_+^\vee$ to make the product (3.5.22) nonzero.

However, $\{-\vartheta, \alpha_i^\vee, i \neq r, i > 0\}$ form a basis of $R^\vee$. Therefore these conditions determine u_b^{-1} uniquely, and it has to coincide with $u_{b_+}^{-1}$. Thus $b = b_+ = w_0(-b_-) = \omega_{r^*}$, the λ'–set of $\pi_{b_+} = \pi_{r^*}$ is empty, and $\mu_\bullet(\pi_{r^*}) = 1$. See (3.2.5).

To conclude, we need to evaluate the Gaussian, which is performed as follows:

$$((\omega_r)_\sharp, (\omega_r)_\sharp) = (\omega_r - u_r^{-1}(\rho_k)\,,\, \omega_r - u_r^{-1}(\rho_k)) = (\omega_{r^*} + \rho_k\,,\, \omega_{r^*} + \rho_k).$$

❑

Comment. (i) There is another method of proving (3.5.21). It is based on the symmetrization of (3.5.18) due to [C21], formula (1.11). Namely, we use the formulas:

$$\sum_{b_- \in B_-} q^{(b_- - \rho_k\,,\, b_- - \rho_k)/2} \Delta_\bullet(b_-) \;=\; \Big(\sum_{b \in B} q^{(b+\rho_k\,,\, b+\rho_k)/2}\Big) \tag{3.5.23}$$

$$\times \prod_{\alpha \in R_+} \prod_{j=1}^{\infty} \Big(\frac{1 - q_\alpha^{(\rho_k, \alpha^\vee)+j}}{1 - t_\alpha^{-1} q_\alpha^{(\rho_k, \alpha^\vee)+j}}\Big), \quad \Delta_\bullet(b_-) \;=$$

$$= \prod \Big(\frac{(t_\alpha^{-1/2} - t_\alpha^{1/2} q_\alpha^{(\alpha^\vee, \rho_k)+j-1})(1 - q_\alpha^{(\alpha^\vee, \rho_k - b_-)})}{(t_\alpha^{1/2} - t_\alpha^{-1/2} q_\alpha^{(\alpha^\vee, \rho_k)+j})(1 - q_\alpha^{(\alpha^\vee, \rho_k)})}\Big). \tag{3.5.24}$$

The latter product is over the set from (3.5.22) for $b = b_-$, i.e., the relations $-(b_-, \alpha^\vee) \geq j > 0$ for all $\alpha \in R_+$ must hold. In (3.5.24), only b_- that equals $-\omega_r$ have nonzero $\Delta_\bullet(b_-)$. Thus, it is sufficient to check that the latter expression is 1 for such a b_- (which is true). This approach seems convenient for computing explicit formulas when more general $h_\vartheta(k) = -m\,(m > 1)$ are considered. Such formulas are generalizations of (3.5.21).

(ii) We mention that when $h_\alpha(k) = -m$ for an extreme but not strongly extreme root α, then the sum for $\langle \gamma \mu_\bullet \rangle_\sharp$ and, equivalently, the sum in (3.5.23) are not finite. However, the corresponding variant of the theorem allows us to reduce the summation dramatically.

Also note that we have one *free* parameter k_{sht} in the case of B, C, G, F, provided that it is in a general position. For instance, $\vartheta = \epsilon_1$ for B_n and there is only one constraint $h_\vartheta(k) = 2k_{\text{sht}} + 2(n-1)k_{\text{lng}} = -1$, so k_{sht} can be arbitrary.

(iii) Formula (3.5.21) is one of the modifications of the Macdonald identities [M1] closely related to the Kac–Moody algebras (cf. [Ka], Ch.12). In Section 3.10 of this chapter, we will connect it with one-dimensional representations of $\mathcal{HH}^\flat$.

3.6 Semisimple representations

In this section we begin with representations $\mathcal{HH}^\flat$ with cyclic vectors and find out when they are semisimple and pseudo-unitary. The latter means the existence of a nondegenerate hermitian $\star$–invariant form. We add here "pseudo" because the form is not supposed to be positive, and, moreover, the involution $*$ acts on the constants via the true complex conjugation only as $|q| = 1$ for real k. Otherwise this conjugation is understood formally.

By definite, we mean a form with nonzero squares of the eigenfunctions with respect to either $X-$ or Y–operators. So it is a certain substitute for the positivity. We give necessary and sufficient conditions for induced modules to possess nonzero semisimple and pseudo-unitary quotients.

We will begin this section with several general definitions, mainly concerning the semisimplicity. To put things in perspective, let us now comment on the relations to the theory of affine Hecke algebras.

Comment. It is worth mentioning that the main theorem generalizes the classification of semisimple representations of affine Hecke algebra of type A for generic t. See, e.g., [C5](Section 3), references therein, and [Na1, Na2]. In the A–case, the classification of the semisimple representations is clarified in full, especially for the *degenerate* affine Hecke algebra. The theory is directly related to the classical Young's bases.

There are also partial results for other classical root systems in the author's works and in [Ram] (A. Ram also considered some special systems, for instance, G_2, F_4). Still there is no complete description of the semisimple representation apart from the A–case. Generally, one can use the classification of *all* irreducible representations from [KL1]. However, checking the semisimplicity is far from immediate if the "geometric" approach is used.

Actually, the importance of semisimple representations is somewhat doubtful from the p–adic viewpoint, in spite of the interesting combinatorial theory. The main reason is that the Bernstein–Zelevinsky Y–operators are not normal with respect to the natural unitary structure that comes from the p–adic theory.

The theory of double affine Hecke algebras, especially at $|q| = 1$, does require semisimple representations. In many interesting representations, either X–operators or Y–operators are normal, although there are important representations that are neither. The case of GL_n will be considered below. ❑

The notation is from the previous sections. All $\mathcal{H}^b$–modules (the lattice B is fixed) will be defined over the field $\mathbb{Q}_{q,t}$ or over its extension $\mathbb{Q}_\xi = \mathbb{Q}_{q,t}(q^{\xi_1}, \ldots, q^{\xi_n})$, where $q^{\xi_i} = q^{(b_i,\xi)}$ for a basis $\{b_i\}$ of B. Here q, t, ξ can be generic as above or arbitrary complex numbers. In the latter case, $\mathbb{Q}_\xi$ becomes a proper subfield of $\mathbb{C}$. We must provided the existence of the involution $*$ in the algebras $\mathbb{Q}_{q,t} \subset \mathbb{Q}_\xi$ taking

$$q^{1/2\tilde{m}} \mapsto q^{-1/2\tilde{m}},\ t_\nu^{1/2} \mapsto t_\nu^{-1/2},\ q^{\xi_i} \mapsto q^{-\xi_i}.$$

This involution is a restriction of the anti-involution (3.2.18) of $\mathcal{H}^\flat$.

3.6.1 Eigenvectors and semisimplicity

A vector v satisfying

$$X_a(v) \;=\; q^{(a,\xi)}\, v \;\text{ for }\; \text{all } a \in B \tag{3.6.1}$$

is called an X–**eigenvector** of weight ξ. Respectively, a vector v satisfying

$$Y_a(v) \;=\; q^{(a,\xi)}\, v \;\text{ for }\; \text{all } a \in B \tag{3.6.2}$$

is called a Y–**eigenvector** of weight ξ.

We set

$$\begin{aligned} V_X^s(\xi) &\stackrel{\mathbf{def}}{=\!=} \{v \in V \mid (X_a - q^{(a,\xi)})^s(v) = 0 \;\text{ for }\; a \in B\}, \\ V_X(\xi) &= V_X^1(\xi),\ V_X^\infty(\xi) = \cup_{s>0} V_X^s(\xi). \end{aligned} \tag{3.6.3}$$

They are called the spaces of generalized eigenvectors. Taking Y_a instead of X_a and (3.6.2) in place of (3.6.1), we introduce the spaces $V_Y^s(\xi)$.

The action of the group $\widehat{W}$ on the weights is affine: $\widehat{w}(\xi) = \widehat{w}((\xi))$. See (3.1.19). The $\widehat{W}^\flat$–**stabilizer** of ξ is

$$\widehat{W}^\flat[\xi] \stackrel{\mathbf{def}}{=\!=} \{\widehat{w} \in \widehat{W}^\flat,\ q^{(\widehat{w}((\xi))-\xi\,,\,a)} = 1 \;\text{ for }\; \text{all } a \in B\}. \tag{3.6.4}$$

From now on weights ξ and ξ' will be identified if $q^{(a,\xi'-\xi)} = 1$ for all $a \in B$. We put $q^\xi = q^{\xi'}$ if it is true.

The **category** $\mathcal{O}_X$ is formed by the modules V such that $V = \oplus_\xi V_X^\infty(\xi)$ and the latter spaces are finite dimensional, where the summation is over different q^ξ. This is supposed to hold for a proper extension of the field of constants. In the definition of $\mathcal{O}_Y$, we substitute Y for X.

Using the decomposition (3.2.25) from the Poincare–Birkhoff–Witt (PBW) property from Theorem 3.2.1, one can check that $\mathcal{H}^\flat$–modules V from $\mathcal{O}_X \cap \mathcal{O}_Y$ with finitely many generators are finite dimensional. Indeed, the space $\tilde{U} \stackrel{\mathbf{def}}{=\!=} \mathbb{Q}_{q,t}[Y_b]U$ is finite dimensional for any finite dimensional subspace U of such a module. If U is preserved by the affine Hecke subalgebra $\mathcal{H}_X^\flat$ generated by $\mathbb{Q}_{q,t}[X_b]$ and $T_i\,(1 \le i \le n)$ then $\tilde{U}$ is a finite sum of the spaces $\mathcal{HH}^\flat u$ for proper $u \in U$. Therefore it is an $\mathcal{HH}^\flat$–submodule of V. Since V is finitely generated, we obtain the required fact.

As a matter of fact, it is not necessary to assume in this reasoning that the spaces $V_X^\infty(\xi)$ and $V_Y^\infty(\xi)$ are finite dimensional. If we know that V is finitely generated, then it is sufficient to use that

$$\oplus_\xi V_X^\infty(\xi) \;=\; V \;=\; \oplus_\xi V_Y^\infty(\xi).$$

Intertwiners. If the X–weight of v is ξ, then $\Phi_{\widehat{w}}v$ is an X–eigenvector of weight $\widehat{w}((\xi))$ for the X–intertwiners Φ from (3.3.27); we also can take $S_{\widehat{w}}$ or $G_{\widehat{w}}$ instead of $\Phi_{\widehat{w}}$ because both are proportional to $\Phi_{\widehat{w}}$. Here the denominators of Φ, S, or G have to be nonzero upon evaluation at q^ξ if the latter is not supposed generic.

Given a reduced decomposition $\widehat{w} = \pi_r s_{i_l} \ldots s_{i_1}$,

$$\begin{aligned}
&\Phi_{\widehat{w}}v = \pi_r \Phi_{i_l}(q^{\xi\{l-1\}}) \cdots \Phi_{i_1}(q^{\xi\{0\}})(v)\\
&\text{for } \xi\{0\} = \xi,\ \xi\{p\} = s_{i_p}((\xi\{p-1\})),\ 1 \le p \le l,\\
&\Phi_i(q^\xi) \;=\; T_i + (t_i^{1/2} - t_i^{-1/2})(q^{(\xi+d,\alpha_i)} - 1)^{-1},\ i \ge 0.
\end{aligned} \tag{3.6.5}$$

See (3.3.37) and (3.3.42).

In the Y–case, the Φ_i is replaced by $\varepsilon(\Phi_i)$, respectively, the "eigenvalue" q^ξ is replaced by its ε–image $q^{-\xi}$. For instance, the $\varepsilon(\Phi_i(q^\xi))$ for $i > 0$ is given by the right-hand side of the last formula from (3.6.5) after the substitution $q^\xi \mapsto q^{-\xi}$, without touching T_i, t_i. This requires a simple algebraic manipulation. Only the $\varepsilon(\Phi_0(q^\xi))$ becomes really different. The process of producing $\xi\{p\}$ in terms of the initial Y–weight ξ remains unchanged.

There is another variant of the system of Y–intertwiners, technically more convenient. It has been applied in Proposition 3.3.5. We will use such intertwiners in the main theorem below. They are $\tau_+\big(\Phi_{\widehat{w}}(q^{-\xi})\big)$ instead of $\varepsilon(\Phi_{\widehat{w}})(q^\xi)$. These operators intertwine the Y–eigenvectors as follows.

Given an eigenvector v of Y–weight ξ, the vector $v' = \tau_+\big(\Phi_{\widehat{w}}(q^{-\xi})\big)v$ is a Y–eigenvector of weight $\widehat{w}((\xi))$. Notice the opposite sign of ξ here. Recall that τ_+ can be interpreted as a formal conjugation by the Gaussian, fixing X, $T_i(1 \le i \le n)$, and preserving the anti-involution $\star$.

Explicitly, the formula for v' reads

$$v' \;=\; \tau\big(\pi_r \Phi_{i_l}(q^{-\xi\{l-1\}}) \cdots \Phi_{i_1}(q^{-\xi\{0\}})\big)(v), \tag{3.6.6}$$

where the notation from (3.6.5) is used.

Induced and semisimple modules. An $\mathcal{HH}^\flat$–module V over $\mathbb{Q}_\xi$ is called X–**cyclic** of weight ξ if it is generated by an element $v \in V$ satisfying (3.6.1).

The universal cyclic module of weight ξ is called **induced** module and is denoted by $\mathcal{I}_X[\xi]$. It is generated by a vector v satisfying (3.6.1), that is regarded as the defining relation of $\mathcal{I}_X[\xi]$.

The weights of $\mathcal{I}_X[\xi]$ constitute the orbit $\widehat{W}^\flat((\xi))$ with respect to the affine action of $\widehat{W}$. Respectively, the weights of any cyclic modules of weight ξ belong to this orbit. The map $H \mapsto Hv$ naturally identifies $\mathcal{I}_X[\xi]$ with the affine Hecke algebra $\mathcal{H}_Y^\flat \subset \mathcal{HH}^\flat$. It sends 1 to the cyclic vector v.

Similarly, the Y–**cyclic** modules of weight ξ are introduced for Y_b instead of X_b, using (3.6.2) in place of (3.6.1).

Respectively, the Y–induced modules satisfy

$$\mathcal{I}_Y[\xi] \cong \mathcal{H}_X^\flat = \langle X_b, T_i \mid b \in B, i > 0 \rangle.$$

By X–**semisimple**, we mean an $\mathcal{HH}^\flat$–module V coinciding with $\oplus_\xi V_X(\xi)$ over an algebraic closure of the field of constants. Here ξ constitute the set of different X–weights of V, denoted by $\mathrm{Spec}_X(V)$ and called the X–**spectrum**. If V is a cyclic module of weight ξ, then the spectrum is defined over the field $\mathbb{Q}_\xi$. Indeed the spectrum of $\mathcal{I}_X[\xi]$ is the orbit $W^\flat((\xi))$. The definition of a Y–**semisimple** module is for Y taken instead of X.

The functional representation $\mathbf{F}[\xi]$ introduced in Section 3.4 is X–cyclic for generic ξ. It is isomorphic to $\mathcal{I}_X[\xi_o]$ for the weight $\xi_o \in \widehat{W}((\xi))$, satisfying the inequalities

$$t_\alpha^{-1} X_{\tilde{\alpha}}(q^{\xi_o}) \neq 1 \ \text{ for } \ \tilde{\alpha} \in \tilde{R}_+, \tag{3.6.7}$$

provided that $\widehat{W}^\flat[\xi] = \{1\}$. It is proved in [C23] (after Corollary 6.5).

The polynomial representation, which will be denoted by $\mathcal{V}$, is Y–cyclic for generic q, t. Any E_b can be taken as its cyclic vector. One has $\mathrm{Spec}_Y(\mathcal{V}) = \{b_\sharp \mid b \in B\}$.

Pseudo-unitary structure. A quadratic form (,) is called, respectively, $*$–bilinear and **pseudo-hermitian** if

$$(ru, v) \ = \ r(u, v) \ = \ (u, r^* v) \ \text{ for } \ r \in \mathbb{Q}_{q,t} \ \text{ and } \ (u, v) \ = \ (v, u)^*.$$

We will always consider nondegenerate forms unless stated otherwise.

An $\mathcal{HH}^\flat$–module V is **pseudo-unitary** if it is equipped with a $\star$–**invariant** pseudo-hermitian form:

$$(H(u), v) \ = \ (u, H^\star(v)) \ \text{ for } \ \text{all } u, v \in V, \ H \in \mathcal{HH}^\flat.$$

We call it **definite** if $(v, v) \neq 0$ for all v, and X–**definite** if $(v, v) \neq 0$ for all X–eigenvectors $v \in V$, assuming that the field of constants contains all the eigenvalues. By X–**unitary**, we mean X–semisimple V with an X–definite form. Strictly speaking, they should be called X–pseudo-unitary, but we drop the "pseudo" in the presence of X.

The definitions of Y–**definite** and Y–**unitary** are parallel to the above ones.

We do not suppose X–definite or Y–definite forms to be positive (or negative) hermitian forms. Note that X–unitary V from the category $\mathcal{O}_X$ is X–semisimple. Conversely, pseudo-unitary X–semisimple V with the simple X–spectrum is X–unitary.

When $\mathbb{Q}_\xi \subset \mathbb{C}$, the involution $*$ is the restriction of the complex conjugation, and $(u,u) > 0$ for all $0 \neq u \in V$, then we call the form positive unitary without adding "pseudo," X, or Y.

Any irreducible $\mathcal{HH}^\flat$–quotients $\mathcal{I}_X[\xi] \to V$ can be defined over $\mathbb{C}$ assuming that $|q| = 1, k_\alpha \in \mathbb{R}, \xi \in \mathbb{R}^n$ in the following sense. We claim that there is a continuous deformation of any triple q, k, ξ to such a triple, which preserves the structure of $\mathcal{I}_X[\xi]$, including quotients, submodules, semisimplicity, and the pseudo-hermitian invariant form. This simply means that the variety of all triples $\{q, k, \xi\}$ has points satisfying the above reality conditions. Thus $*$ can be assumed to be a restriction of the complex conjugation to the field of constants upon such deformation.

Semisimple range. Given a weight ξ, its **plus–range** $\Upsilon_+[\xi]$ is the set of all elements $\widehat{w} \in \widehat{W}^\flat$ satisfying the following condition:

$$\tilde{\alpha} = [\alpha, \nu_\alpha j] \in \lambda(\widehat{w}) \;\Rightarrow\; t_\alpha X_{\tilde{\alpha}}(q^\xi) = q_\alpha^{(\alpha^\vee,\xi)+k_\alpha+j} \neq 1. \tag{3.6.8}$$

Equivalently, given a reduced decomposition $\widehat{w} = \pi_r s_{i_l} \cdots s_{i_1}$ of an arbitrary $\widehat{w} \in \widehat{W}$, we need to check the conditions

$$\begin{aligned} &q_{i_p}^{(\alpha_{i_p}^\vee,\xi\{p-1\}+d)+k_{i_p}} \neq 1 \text{ for } p = 1, \ldots, l, \\ &\text{where} \quad \xi\{0\} = \xi,\; \xi\{p\} = s_{i_p}((\xi\{p-1\})). \end{aligned} \tag{3.6.9}$$

Obviously the definition (in the second form) does not depend on π_r, so $\pi_r \Upsilon_+[\xi] = \Upsilon_+[\xi]$. Also,

$$\Upsilon_+[\pi_r((\xi))] = \pi_r \Upsilon_+[\xi] \pi_r^{-1} = \Upsilon_+[\xi] \pi_r^{-1}.$$

Reading $\widehat{w}$ in the opposite order, the following equivalent set of relations must be checked:

$$q_{i_p}^{(\alpha_{i_p}^\vee,\xi\{p\}+d)-k_{i_p}} \neq 1 \;\text{ for }\; p = l, \ldots, 1. \tag{3.6.10}$$

Notice the opposite sign of k_{i_p}.

Let us introduce the **arrows in** $\widehat{W}$ by setting $\hat{u} \mapsto \widehat{w}$ if (a) either $\widehat{w} = s_i \hat{u}$ for $0 \leq i \leq n$ provided that $l(\widehat{w}) = l(\hat{u}) + 1$ or (b) $\widehat{w} = \pi_r \hat{u}$ for $\pi_r \in \Pi^\flat$. The **boundary** $\ddot{\Upsilon}_+[\xi]$ of $\Upsilon_+[\xi]$ is defined as the set of all $\widehat{w} \notin \Upsilon_+[\xi]$ such that

$\hat{u} \mapsto \widehat{w}$ for an element $\hat{u} \in \Upsilon_+[\xi]$. It may happen in case (a) only. Actually, if $\widehat{w} = s_i\hat{u}$ and $\widehat{w} \notin \Upsilon_+[\xi] \ni \hat{u}$, then automatically $l(\widehat{w}) = l(\hat{u}) + 1$, because otherwise $\lambda(\widehat{w}) \subset \lambda(\hat{u})$, which is impossible.

The **minus–range** $\Upsilon_-(\xi)$ is defined for t_α^{-1} in place of t_α in (3.6.8) and with k replaced by $-k$ in (3.6.9) and (3.6.10).

The **semisimple range** $\Upsilon_*[\xi]$ is the intersection $\Upsilon_+[\xi] \cap \Upsilon_-[\xi]$. The **zero–range** $\Upsilon_0[\xi]$ is introduced without t_α in (3.6.8) and without k in the equivalent conditions (3.6.9)–(3.6.10). The minus-boundary and zero-boundary are defined respectively.

Notice that

$$\Upsilon_-[\xi] = \Upsilon_+[-\xi],\ \Upsilon_0[\xi] = \Upsilon_0[-\xi],$$
$$\ddot{\Upsilon}_*[\xi] \subset \ddot{\Upsilon}_+[\xi] \cup \ddot{\Upsilon}_-[\xi]. \tag{3.6.11}$$

Also, we will use the compatibility of the semisimple and zero ranges with the right multiplications:

$$\Upsilon_*[\widehat{w}((\xi))] \;=\; \Upsilon_*[\xi]\,\widehat{w}^{-1} \text{ for every } \widehat{w} \in \Upsilon_*[\xi]. \tag{3.6.12}$$

It readily results from (3.1.11). Let us prove it.

It is suffices to consider $\widetilde{w} = s_{i_l} \cdots s_{i_1}$ assuming that the decomposition is reduced. We can start with $\widetilde{w}((\xi))$, read $\widetilde{w}$ in the opposit order following (3.6.10), and then continue following (3.6.9) for a reduced decomposition of $\hat{u} \in \Upsilon_*[\xi]$. Since we are doing the semisimple range, changing the sign of k in (3.6.10) does not matter, and all weights from the (opposit) ξ–sequence for $\widetilde{w}$ and the (direct) ξ–sequence for $\hat{u}$ will satisfy the required inequalities. So $\tilde{\lambda}(\hat{u}\widetilde{w}^{-1})$ satisfies the relation (3.6.8) with $t^{\pm 1}$. Here we use $\tilde{\lambda}$ instead of λ because the resulting decomposition of $\hat{u}\widetilde{w}^{-1}$ may be nonreduced. However, the reduced set $\lambda(\hat{u}\widetilde{w}^{-1})$, which is needed in the definition, belongs to $\tilde{\lambda}(\hat{u}\widetilde{w}^{-1})$ due to (3.1.11).

We arrive at $\Upsilon_*[\xi]\,\widehat{w}^{-1} \subset \Upsilon_*[\widehat{w}((\xi))]$. The inclusion obviously can be reversed:

$$\Upsilon_*[\widehat{w}((\xi))]\,\widehat{w} \subset \Upsilon_*[\xi].$$

This means the coincidence and proves (3.6.12).

Note that this argument cannot be used separately for Υ_+, Υ_- because of the change of sign of k in (3.6.10). However, it can be applied to Υ_0:

$$\Upsilon_0[\widehat{w}((\xi))] \;=\; \Upsilon_0[\xi]\,\widehat{w}^{-1} \text{ for every } \widehat{w} \in \Upsilon_0[\xi]. \tag{3.6.13}$$

Semisimple stabilizer. The following **semisimple stabilizer** of ξ

$$\widehat{W}^\flat_*[\xi] \overset{\mathbf{def}}{=\!=} \{\widehat{w} \in \widehat{W}^\flat[\xi] \mid \widehat{w} \in \Upsilon_*[\xi]\} \;=\; \widehat{W}^\flat_*[-\xi] \tag{3.6.14}$$

will play an important role in the main theorem of this section. It is a *subgroup* of $\widehat{W}^\flat[\xi]$. Indeed, if $\hat{u}, \widehat{w} \in \widehat{W}^\flat_*[\xi]$, then the set

$$\tilde{\lambda}(\widehat{w}\hat{u}) = \lambda(\hat{u}) \cup \hat{u}^{-1}(\lambda(\widehat{w}))$$

satisfies (3.6.8) and its counterpart for $\Upsilon_-[\xi]$ because $\hat{u}$ does not change q^ξ. However, $\lambda(\widehat{w}\hat{u}) \subset \tilde{\lambda}(\widehat{w}\hat{u})$ thanks to (3.1.11).

The same argument proves that

$$\Upsilon_*[\xi]\widehat{w} = \Upsilon_*[\xi] \text{ for every } \widehat{w} \in \widehat{W}^\flat_*[\xi]. \tag{3.6.15}$$

In contrast to (3.6.12), we do not need to use (3.6.10) here. Combining (3.6.15) with (3.6.12), we arrive at

$$\widehat{W}^\flat_*[\widehat{w}((\xi))] = \widehat{w}\,\widehat{W}^\flat_*[\xi]\,\widehat{w}^{-1} \text{ for all } \widehat{w} \in \Upsilon_*[\xi]. \tag{3.6.16}$$

Similarly, one can introduce the *groups* $\widehat{W}^\flat_\pm[\xi]$ and $\widehat{W}^\flat_0[\xi]$. They preserve the corresponding $\Upsilon[\xi]$ acting via the right multiplication, i.e., (3.6.15) holds for all three of them. In fact, this property can be used as another (equivalent) definition of the corresponding stabilizer. Note that the relation (3.6.16) holds, generally speaking, for $\Upsilon[\xi]_*$ and $\Upsilon[\xi]_0$ only.

Sharp case. When considering spherical and self-dual representations, the **sharp range** $\Upsilon_+[-\rho_k]$ will be used, which is simply $\Upsilon_+[\xi]$ as $\xi = -\rho_k$. The other Υ are defined correspondingly.

It follows from Lemma 3.1.6 that $\Upsilon_+[-\rho_k] = \pi_B$ for generic q and t_ν. Indeed, formula (3.1.32) there gives that

$$k_\nu + (\alpha_i^\vee, \pi_c((-\rho_k)) + d) \neq 0 \text{ for } \nu = \nu_i = \nu_{\alpha_i},\ i \geq 0,$$

if and only if $s_i\pi_c \in \pi_B$, assuming that $l(s_i\pi_c) = 1 + l(\pi_c)$. It is equivalent to

$$q_i^{k_i + (\alpha_i^\vee, \pi_c((-\rho_k)) + d)} \neq 1,$$

which we are supposed to check to conclude that $\widehat{w} = s_i\pi_c$ (under the same assumption) belongs to $\Upsilon_+[-\rho_k]$ provided that π_c is already in this set. The equivalence is of course for generic q, t only.

Let us connect the arrows in $\widehat{W}$ with those used in B. Recall that by $c \twoheadrightarrow b$, we mean that $b \in B$ can be obtained from $c \in B$ by a chain of transformations (i)–(ii) (the simple arrows) from Proposition 3.1.7. The $c \leftrightarrow b$ indicates a simple transformation of type (iii), corresponding to the application of $\pi_r \in \Pi^\flat$. It is always invertible. If such a transformation is involved, then we set $c \twoheadleftarrow\!\!\twoheadrightarrow b$.

The meaning of the simple arrows *coincides* with that of the $\hat{u} \mapsto \widehat{w}$ introduced above upon the restriction to $\Upsilon_+(-\rho_k)$. Indeed, the latter is the set of all $\widehat{w} = \pi_b$ satisfying (3.6.8). If $\pi_c \in \Upsilon_+(-\rho_k)$ and $c \to b$ for $\pi_b = s_i\pi_c$, then $\pi_b \in \Upsilon_+(-\rho_k)$ if and only if

$$t_\alpha q^{(\alpha, c_- - \rho_k)} \neq 1 \text{ for } i > 0, \ \alpha = u_c(\alpha_i),$$
$$t_\alpha q^{1+(\alpha, c_- - \rho_k)} \neq 1 \text{ for } i = 0, \ \alpha = u_c(-\vartheta). \qquad (3.6.17)$$

Recall that $c_\sharp = c - u_c^{-1}(\rho_k) = u_c^{-1}(-\rho_k + c_-)$. See (3.1.31). In the remaining case $c \leftrightarrow b$ for $\pi_b = \pi_r\pi_c$, $\pi_b \in \Upsilon_+(-\rho_k)$ if and only if $\pi_c \in \Upsilon_+(-\rho_k)$.

Generally, in terms of $\lambda(\pi_c) = \{[\alpha, \nu_\alpha j]\}$, the inclusion $\pi_c \in \Upsilon_+(-\rho_k)$ is equivalent to the conditions

$$q_\alpha^{j+k_\alpha-(\alpha^\vee,\rho_k+d)} \neq 1 \text{ for such } \tilde{\alpha}.$$

Concerning applications to the E–polynomials, these conditions ensure that there will be no intertwiners proportional to $(\tau_+(T_i) - t_i^{1/2})$ in the product formula for the polynomial $\widehat{E}_c$ from (3.3.42). The simple intertwiners proportional to $(\tau_+(T_i) + t_i^{-1/2})$ cannot appear for generic q, t in the corresponding chain.

3.6.2 Main theorem

We turn to a general description of semisimple and pseudo-unitary representations. It will be reduced to a certain combinatorial problem that can be managed in several cases. We permanently identify the cyclic generator $v \in \mathcal{I}_X[\xi]$ with $1 \in \mathcal{H}_Y^\flat$, which can be uniquely extended to an $\mathcal{H}_Y^\flat$–isomorphism $\mathcal{I}_X[\xi] \cong \mathcal{H}_Y^\flat$.

Theorem 3.6.1. *(i) Assuming that*

$$\Upsilon_*[\xi] \subset \Upsilon_0[\xi] \ \text{ and } \ \mathcal{J}_\xi \stackrel{\text{def}}{=\!=} \sum \mathcal{H}_Y \Phi_{\widehat{w}}(q^\xi) \neq \mathcal{H}_Y, \qquad (3.6.18)$$

where the summation is over $\widehat{w} \in \ddot{\Upsilon}_*[\xi]$, the quotient $U = U_X^\xi \stackrel{\text{def}}{=\!=} \mathcal{H}_Y^\flat/\mathcal{J}_\xi$ is an X–semisimple $\mathcal{HH}^\flat$–module with a basis of X–eigenvectors $\{\Phi_{\widehat{w}}(q^\xi)\}$ for $\widehat{w} \in \Upsilon_*[\xi]$. Its X–spectrum is $\{\widehat{w}(\xi)\}$ for such $\widehat{w}$.

If V is an irreducible X–semisimple $\mathcal{HH}^\flat$–module with a cyclic vector of weight ξ, then (3.6.18) holds and V is a quotient of U, provided that $t_{\text{sht}} \neq 1 \neq t_{\text{lng}}$.

(ii) The elements $\Phi_{\widehat{w}}v'$, $S_{\widehat{w}}v'$, and $G_{\widehat{w}}v'$ are well defined for all v' from the ξ–eigenspace $U(\xi)$ of U and $\widehat{w} \in \Upsilon_*[\xi]$. They induce isomorphisms $U(\xi) \cong U(\widehat{w}(\xi))$.

The group $\widehat{W}_*^\flat[\xi]$ acts in $U(\xi)$ by automorphisms via $\widehat{w} \mapsto S_{\widehat{w}}$. A quotient V of U is an irreducible $\mathcal{HH}^\flat$–module if and only if $V(\xi)$ is an irreducible $\widehat{W}_*^\flat[\xi]$–module. Moreover, provided we have (3.6.18), an arbitrary irreducible *representation V' of this group can be uniquely extended to an irreducible X–semisimple $\mathcal{H\!H}^\flat$–quotient V of $\mathcal{I}_X[\xi]$ such that $V' = V(\xi)$. Here G can be used instead of S.*

The above statements hold for $\mathcal{I}_Y[\xi]$ and its semisimple quotients when the $\Phi_{\widehat{w}}(q^\xi)$ are replaced by $\tau_+\big(\Phi_{\widehat{w}}(q^{-\xi})\big)$ everywhere.

(iii) Under the same constraint (3.6.18), the function

$$\mu_\bullet(\widehat{w}) \stackrel{\mathbf{def}}{=\!=} \prod_{[\alpha,\nu_\alpha j]\in\lambda(\widehat{w})} \Big(\frac{t_\alpha^{-1/2} - q_\alpha^j t_\alpha^{1/2} X_\alpha(q^\xi)}{t_\alpha^{1/2} - q_\alpha^j t_\alpha^{-1/2} X_\alpha(q^\xi)}\Big) \tag{3.6.19}$$

has neither poles nor zeros at $\widehat{w} \in \Upsilon_*[\xi]$. Moreover,

$$\mu_\bullet(\hat{u})\mu_\bullet(\widehat{w}) \;=\; \mu_\bullet(\hat{u}\widehat{w}) \text{ whenever } \widehat{w} \in \widehat{W}^\flat_*[\xi] \tag{3.6.20}$$

for $\hat{u}, \widehat{w} \in \Upsilon_*[\xi]$,

An $\mathcal{H\!H}^\flat$–quotient V of U is pseudo-unitary if and only if the eigenspace $V(\xi)$ is a pseudo-unitary $\widehat{W}^\flat_*[\xi]$–module. The latter means that the form $(\,,\,)$ is nondegenerate on $V(\xi)$ and

$$(S^\xi_{\widehat{w}})^\star\, S^\xi_{\widehat{w}} = \mu_\bullet(\widehat{w}) \;\; for \;\; \widehat{w} \in \widehat{W}^\flat_*[\xi],\; S^\xi_{\widehat{w}} \stackrel{\mathbf{def}}{=\!=} S_{\widehat{w}}(q^\xi). \tag{3.6.21}$$

(iv) A pseudo-hermitian form on $e', e'' \in V(\xi)$, can be uniquely extended to a pseudo-hermitian invariant form on V:

$$(S_{\hat{u}}e'\,,\, S_{\widehat{w}}e'') \;=\; \delta'_{\hat{u},\widehat{w}}\, \mu_\bullet(\hat{u})\, (e'\,,\, S_{\hat{u}^{-1}\widehat{w}}e''), \quad where \tag{3.6.22}$$
$$\hat{u}, \hat{v} \in \Upsilon_*[\xi],\; \delta'_{\hat{u},\widehat{w}} = 1 \;\; for \;\; \hat{u}^{-1}\widehat{w} \in \widehat{W}^\flat_*[\xi] \;\; and \;\; = 0 \text{ otherwise.}$$

If $(e', e') \neq 0$ as $V(\xi) \ni e' \neq 0$, then V is X–unitary.

An arbitrary irreducible pseudo-unitary (X–unitary) $\mathcal{H\!H}^\flat$–module with a cyclic vector of weight ξ can be obtained by this construction as $t_{sht} \neq 1 \neq t_{lng}$.

In the Y–case,

$$\tilde{G}^\xi_{\widehat{w}} \stackrel{\mathbf{def}}{=\!=} \tau_+\big(G_{\widehat{w}}(q^{-\xi})\big)$$

is taken instead of $S_{\widehat{w}}(q^\xi)$; these operators send $V(\xi)$ to $V(\widehat{w}((\xi)))$. The extension from a pseudo-hermitian form on $V(\xi)$ is

$$(\tilde{G}^\xi_{\hat{u}}e'\,,\, \tilde{G}^\xi_{\widehat{w}}e'') \;=\; \delta'_{\hat{u},\widehat{w}}\, \mu_\bullet(\hat{u})^{-1}\, (e'\,,\, \tilde{G}^\xi_{\hat{u}^{-1}\widehat{w}}e''). \tag{3.6.23}$$

Proof. We set $e_{\widehat{w}} = \Phi_{\widehat{w}}(1)$ for $\widehat{w} \in \widehat{W}^\flat$ provided that $\Phi_{\widehat{w}}(1)$ is well defined. It is an X–eigenvector of weight $\widehat{w}(\xi)$. Recall that we permanently identify v and 1. Then $\mathcal{H}_Y e_{\widehat{w}}$ is an $\mathcal{H\!H}^\flat$–submodule because $\mathcal{H\!H}^\flat = \mathcal{H}_Y \cdot \mathbb{Q}_{q,t}[X_b]$.

Assuming (3.6.18), $\mathcal{J}_\xi$ and U are $\mathcal{H\!H}^\flat$–modules. Moreover, all $e_{\widehat{w}}$ are invertible in $\mathcal{H}^\flat_Y$ and their images linearly generate U as $\widehat{w} \in \Upsilon_*[\xi]$. Indeed, if $\Phi_i e_{\widehat{w}}$ loses the invertibility, then $s_i\widehat{w}$ leaves $\Upsilon_*[\xi]$, $l(s_i\widehat{w}) = l(\widehat{w})+1$, and $s_i\widehat{w} \in \ddot{\Upsilon}_*[\xi]$. By (3.6.18), $\ddot{\Upsilon}_*[\xi] \subset \Upsilon_0[\xi]$. We conclude that the element $T_i(e_{\widehat{w}})$ is proportional to $e_{\widehat{w}}$ in this case.

This means that each $T_i(e_{\widehat{w}})$ is a linear combination of $e_{\hat{u}}$ for any $n \geq i \geq 0$, where $\widehat{w}, \hat{u} \in \Upsilon_*[\xi]$. Of course it holds for $\pi_r \in \Pi^\flat$ too. Thus $\{e_{\widehat{w}'}\}$ linearly generate U. Moreover, given $\widehat{w} \in \Upsilon_*[\xi]$, the map $v' \mapsto \Phi_{\widehat{w}} v'$ induces an isomorphism $U(\xi) \cong U(\widehat{w}(\xi))$.

Summarizing, (3.6.18) implies that $\{e_{\widehat{w}\hat{u}}\}$ form a basis of U when
(a) $\widehat{w}$ are representatives of all classes $\Upsilon_*[\xi]/\widehat{W}^\flat[\xi]$,
(b) $\hat{u} \in \widehat{W}^\flat[\xi]$ and $\{e_{\hat{u}}\}$ form a basis of $U(\xi)$.

Let us discuss the usage of S and G. The elements $S_{\widehat{w}}(1) \in \mathcal{H}_Y^\flat$ are well defined for $\widehat{w} \in \Upsilon_+[\xi]$. The sum $\mathcal{J}_\xi^+ \stackrel{\mathbf{def}}{=} \sum \mathcal{H}_Y S_{\widehat{w}}$, $\widehat{w} \in \ddot{\Upsilon}_+[\xi]$, is an $\mathcal{H}\!\!\mathcal{H}^\flat$–submodule of $\mathcal{I}_X[\xi]$. Finally, $S_{\widehat{w}}(1)$ are invertible for $\widehat{w} \in \Upsilon_*[\xi]$, which results directly from the definition of the latter set. The same holds for $\Upsilon_-[\xi]$ with S being replaced by G. Turning to $\mathcal{I}_Y[\xi]$, G and S serve, respectively, $\Upsilon_+[\xi]$ and $\Upsilon_-[\xi]$.

Now let us check that the existence of irreducible X–semisimple quotients $V \neq \{0\}$ of $\mathcal{I}_X[\xi]$ implies (3.6.18), and all such V are quotients of U. We suppose that $t_{\mathrm{sht}} \neq 1 \neq t_{\mathrm{lng}}$. Actually, we will impose $t_{\mathrm{sht}} \neq \pm 1 \neq t_{\mathrm{lng}}$ in the argument below, leaving the consideration of $t=-1$ to the reader. This allows us to separate $T_i - t_i^{1/2}$ and $T_i + t_i^{-1}$. When $t_i = -1$ the T_i is not semisimple, but the theorem still holds in this case.

The images $e_{\widehat{w}}$ of the vectors $\Phi_{\widehat{w}}(1)$ in V are nonzero as $\widehat{w} \in \Upsilon_*[\xi]$. If $\Upsilon_0[\xi] \not\supset \Upsilon_*[\xi]$, we can pick $e' = e_{\widehat{w}}$ such that $S_i(e') = e'$ and $X_{\alpha_i}(e') = e'$. This gives that X_{α_i} is not semisimple in the two-dimensional module generated by e' and $e'' = T_i(e')$. The dimension is two because the characters of $\mathcal{H}_X^i \stackrel{\mathbf{def}}{=} \langle T_i, X_{\alpha_i}\rangle$ send $X_{\alpha_i} \mapsto t_i^{\pm} \neq 1$. So it cannot be one.

Now let $\hat{u}$ belong to $\Upsilon_*[\xi]$ and $\widehat{w} = s_i\hat{u} \in \ddot{\Upsilon}_+[\xi]$. The case $\widehat{w} = s_i\hat{u} \in \ddot{\Upsilon}_-[\xi]$ is completely analogous, so we will skip it. Then $l(\widehat{w}) = l(\hat{u})+1$. Both $e' = \Phi_{\hat{u}}(e)$ and $e'' = \Phi_i(e') = (T_i - t_i^{1/2})(e')$ are X–eigenvectors. Here by e, we mean the image of 1 in V. Let us prove that $e''=0$.

Supposing that $e'' \neq 0$, it cannot be proportional to e'. Indeed, this may only happen if $(T_i + t_i^{-1/2})(e') = 0$, which is impossible, since $\hat{u} \in \Upsilon_*[\xi] \subset \Upsilon_-[\xi]$.

Note that if V is assumed to be X–unitary, then restricting its X–definite form to the linear span of e', e'' readily leads to a contradiction with the relation $T_i^\star = T_i^{-1}$. This argument does not require the irreducibility of V but involves the X–unitary structure.

If $e'' \neq 0$, then it generates V as an $\mathcal{H}\!\!\mathcal{H}^\flat$–module and as an $\mathcal{H}_Y^\flat$–module, thanks to the irreducibility. This means that

$$\mathcal{H}_Y^\flat(T_i - t_i^{1/2})\Phi_{\widehat{w}}(1) + J_V = \mathcal{H}_Y^\flat, \text{ where } V = \mathcal{H}_Y^\flat/J_V,$$

for a proper ideal $J_V \subset \mathcal{H}_Y^\flat$ that is an $\mathcal{HH}^\flat$–submodule of $\mathcal{I}_X[\xi]$. Hence, J_V contains $\mathcal{H}_Y^\flat(T_i + t_i^{-1/2})\Phi_{\widehat{w}}(1)$ and e'' has to be proportional to e'. This is impossible because $X_{\alpha_i}(e'') = t_i^{-1}$ whereas $X_{\alpha_i}(e') \neq t_i^{\pm 1}$.

Summarizing, we have almost completed parts (i)–(ii) from the theorem. The remaining claims there are
(a) the equivalence of the irreducibility of the $\mathcal{HH}^\flat$–module V and the $\widehat{W}_*^\flat[\xi]$–module $V(\xi)$,
(b) the possibility of taking an arbitrary irreducible representation of $\widehat{W}_*^\flat[\xi]$ as $V(\xi)$.

Applying the intertwining operators, both claims readily follow from the statements that have been checked.

Let us turn to (iii) and (iv). We take $e', e'' \in V(\xi)$ and denote $e'_{\widehat{w}} = S_{\widehat{w}}e'$ and $e''_{\widehat{w}} = S_{\widehat{w}}e''$, where $\widehat{w} \in \Upsilon_*[\xi]$, for a certain pseudo-unitary irreducible quotient V of $\mathcal{I}_X^\flat$. They are nonzero X–eigenvectors. If the eigenvalues of $e'_{\hat{u}}$ and $e''_{\widehat{w}}$ are different, i.e., $\hat{u}^{-1}\widehat{w} \notin \widehat{W}_*^\flat[\xi]$, then these vectors are orthogonal with respect to the invariant form of V.

Applying (3.3.31),

$$(e'_{\widehat{w}}, e''_{\widehat{w}}) = (S_{\widehat{w}}e', S_{\widehat{w}}e'') = (S^\star_{\widehat{w}} S_{\widehat{w}}e', e'') = \qquad (3.6.24)$$
$$(e', e'')\prod_{\alpha,j}\Big(\frac{t_\alpha^{-1/2} - q_\alpha^j t_\alpha^{1/2}X_\alpha(q^\xi)}{t_\alpha^{1/2} - q_\alpha^j t_\alpha^{-1/2}X_\alpha(q^\xi)}\Big),\ [\alpha, \nu_\alpha j] \in \lambda(\widehat{w}).$$

The latter product is just $\mu_\bullet(\widehat{w})$ from (3.6.19). Thus the extension of (,) from $V(\xi)$ to V is unique if it exists. Explicitly:

$$(S_{\hat{u}}e', S_{\widehat{w}}e'') = \delta'_{\hat{u},\widehat{w}}\,\mu_\bullet(\hat{u})\,(e', S_{\hat{u}^{-1}\widehat{w}}e''), \text{ where } e', e'' \in V(\xi), \qquad (3.6.25)$$
$$\hat{u}, \hat{v} \in \Upsilon_*[\xi],\ \ \delta'_{\hat{u},\widehat{w}} = 1 \text{ for } \hat{u}^{-1}\widehat{w} \in \widehat{W}_*^\flat[\xi] \text{ and } 0 \text{ otherwise.}$$

To check that this formula correctly extends an arbitrarily given definite form on $V(\xi)$, it suffices to compare

$$(S_{\widehat{w}}e'', S_{\hat{u}}e')^* \text{ and } (S_{\hat{u}\hat{v}}(S_{\hat{v}}^{-1}e'), S_{\widehat{w}}e'')$$

and to show that they coincide whenever $\hat{v} \in \widehat{W}_*^\flat[\xi]$. Using (3.6.25), we need the relations

$$\mu_\bullet(\widehat{w}^{-1}\hat{u})\mu_\bullet(\widehat{w}) = \mu_\bullet(\hat{u}),\ \mu_\bullet(\hat{u}\hat{v})\mu_\bullet(\hat{v}^{-1}) = \mu_\bullet(\hat{u}),$$

which are nothing but (3.6.20). Claim iv) is verified.

Relation (3.6.20) obviously holds when $l(\hat{u}\widehat{w}) = l(\hat{u}) + l(\widehat{w})$, because the $\mu_\bullet$-product for $\hat{u}\widehat{w}$ is taken over $\lambda(\hat{u}\widehat{w})$, which is a union of $\lambda(\widehat{w})$ and the set $\widehat{w}^{-1}(\lambda(\hat{u}))$. The $\mu_\bullet$-product over the latter set coincides with the product over $\lambda(\hat{u})$).

Thanks to (3.1.11), we can omit the constraint $l(\hat{u}\widehat{w}) = l(\hat{u}) + l(\widehat{w})$ in this argument. Indeed, the extra positive affine roots $\tilde{\alpha} = [\alpha, \nu_\alpha j] \in \tilde{\lambda}^+(\hat{u}\widehat{w})$ appear in $\tilde{\lambda}(\hat{u}\widehat{w})$ together with $-\tilde{\lambda} \in -\tilde{\lambda}^+(\hat{u}\widehat{w})$. Here we take the product of the reduced decompositions of $\hat{u}$ and $\widehat{w}$, which may be nonreduced, to construct $\tilde{\lambda}(\hat{u}\widehat{w})$. However, the product of the $\mu_\bullet$-factors corresponding to $\tilde{\alpha} = [\alpha, \nu_\alpha j]$ and $-\tilde{\alpha}$ is

$$\Bigl(\frac{t_\alpha^{-1/2} - q_\alpha^j t_\alpha^{1/2} X_\alpha(q^\xi)}{t_\alpha^{1/2} - q_\alpha^j t_\alpha^{-1/2} X_\alpha(q^\xi)}\Bigr)\Bigl(\frac{t_\alpha^{-1/2} - q_\alpha^{-j} t_\alpha^{1/2} X_\alpha^{-1}(q^\xi)}{t_\alpha^{1/2} - q_\alpha^{-j} t_\alpha^{-1/2} X_\alpha^{-1}(q^\xi)}\Bigr) = 1.$$

This concludes (iii) and the proof of the theorem. ❑

Thanks to Proposition 3.1.3, there is a reasonably simple and explicit combinatorial reformulation of the first part of condition (3.6.18). However, it does not guarantee the second part.

Proposition 3.6.2. *The relation $\Upsilon_*[\xi] \subset \Upsilon_0[\xi]$ is equivalent to the following property of ξ. Let $\ddot{\Upsilon} = \{\tilde{\beta} = [\beta, \nu_\beta j]\}$ be the set of all affine positive roots such that $t_\beta^{\pm 1} X_{\tilde{\beta}}(q^\xi) = 1$. Then an arbitrary positive affine $\tilde{\gamma} = [\gamma, \nu_\gamma i]$ such that $X_{\tilde{\gamma}}(q^\xi) = 1$ is a linear combination of some $\tilde{\beta} \in \ddot{\Upsilon}$ with positive rational (not always integral) coefficients. Using the closure $\bar{\mathfrak{C}}^a$ of the affine Weyl chamber $\mathfrak{C}^a$ from Section 3.1 of this chapter,*

$$\cup_{\widehat{w}}\, \widehat{w}((-\bar{\mathfrak{C}}^a)) = \{z \in \mathbb{R}^n \mid (z, \beta) + j < 0, \tilde{\beta} \in \ddot{\Upsilon}\} \quad \textit{for} \quad \widehat{w} \in \Upsilon_*[\xi]. \tag{3.6.26}$$

❑

3.6.3 Finite dimensional modules

We will start the discussion of the theorem with the irreducible $\mathcal{I}_X[\xi]$. Following [C23], Theorem 6.1, let us assume that $\widehat{W}^\flat_*[\xi]$ is conjugated in $\widehat{W}$ to the span of a proper subset of $\{s_0, s_1, \ldots, s_n\}$, possibly $\emptyset$, but smaller than the whole set of affine simple reflections. It is possible only if q is not a root of unity, because otherwise $\widehat{W}_*[\xi]$ is infinite. By the way, this constraint results in the implication

$$\Upsilon_*[\xi] \subset \Upsilon_0[\xi] \Rightarrow \mathcal{J}_\xi = \sum \mathcal{H}_Y \Phi_{\widehat{w}} \neq \mathcal{H}_Y, \quad \text{where} \quad \widehat{w} \in \ddot{\Upsilon}_*[\xi].$$

We do not prove/use this claim.

Apart from roots of unity, $\mathcal{I}_X[\xi]$ is irreducible if and only if $\Upsilon_+[\xi] = \widehat{W}^\flat = \Upsilon_-[\xi]$. Thus its semisimplicity is equivalent to the triviality of the stabilizer $\widehat{W}^\flat_*[\xi]$.

The theorem can be applied to a general description of the finite-dimensignal semisimple representations. We continue to suppose that q is generic. If q, t are treated as complex numbers, then we assume that $(q^{1/(2m)})^i(t^{1/2})^j$ do not represent nontrivial roots of unity for $i, j \in \mathbb{Z}$.

Proposition 3.6.3. *i) Under condition (3.6.18), the induced module* $\mathcal{I}_X[\xi]$ *has a nonzero finite dimensional* X*-semisimple quotient if and only if there exists a set of roots* $T \subset R$ *such that for any* $b \in B$ *there exist* $\beta \in T$ *with* $(b, \beta) > 0$, *i.e., the* β *do not belong to any half-plane in* $\mathbb{R}^n$, *and for every* $\beta \in T$,

$$\begin{aligned}(\xi, \beta^\vee) + k_\beta \in -1 - \mathbb{Z}_+ \ \text{or}\ (\xi, \beta^\vee) - k_\beta \in -1 - \mathbb{Z}_+ \ \text{for}\ \beta < 0,\\ (\xi, \beta^\vee) + k_\beta \in -\mathbb{Z}_+ \ \text{or}\ (\xi, \beta^\vee) - k_\beta \in -\mathbb{Z}_+ \ \text{for}\ \beta > 0.\end{aligned} \quad (3.6.27)$$

Proof. Let us assume that for all $\alpha \in R$ such that $(b, \alpha) > 0$, either $(\xi, \alpha^\vee) \pm k_\alpha \notin -1 - \mathbb{Z}_+$ for both signs of k_α as $\alpha < 0$ or the same inequalities hold with $-\mathbb{Z}_+$ as $\alpha > 0$. Then $\lambda(lb)$ for $l \in \mathbb{Z}_+$ consists of $\tilde{\alpha} = [\alpha, \nu_\beta j]$ for such α with integers j satisfying inequalities $0 < j \le l(b, \alpha^\vee)$ as $\alpha < 0$ and with $0 \le j$ otherwise. We take b from $\Upsilon_*[\xi]$, so all these $\tilde{\alpha}$ satisfy (3.6.8) and its counterpart with t_α^{-1} instead of t_α. Since l is arbitrary positive, we obtain that $\Upsilon_*[\xi]$ is infinite if the relations (3.6.27) don't hold.

Let us check that this condition guarantees that $\Upsilon_*[\xi]$ is finite.

First of all, any given infinite sequence of pairwise distinct $b^i \in B$ is a (finite) union of subsequences $\{\tilde{b}^i\}$ such that there exist $\beta_\pm \in T$ satisfying $\pm(\tilde{b}^i, \beta_\pm) > 0$. Taking one such infinite subsequence, we may assume that there exist $\beta_\pm \in T$ such that $\pm(b^i, \beta_\pm) > 0$. If the latter scalar products are bounded, we switch to a new sequence b^i formed by proper differences of the old $\{b^i\}$, ensuring that all $(b^i, \beta_\pm) = 0$. Then we find the next pair $\beta_\pm{}'$ such that $\pm(b^i, \beta_\pm{}') > 0$, form the next sequence of differences, and so on, until all the T are exhausted.

We conclude that any sequence $\{b^i\}$ is a union of subsequences $\{\tilde{b}^i\}$ such that $|(\tilde{b}^i, \beta)| \to \infty$ as $i \to \infty$ for certain $\beta \in T$. Following the proof of Proposition 3.1.4 (see (3.1.25), (3.1.26)), given $w \in W$, the sets $\lambda(w\tilde{b}^i)$ contain the roots $[\beta, j\nu_\beta]$, where $\max\{j\} \to \infty$ as $i \to \infty$.

Note that the proof gets more transparent via the geometric interpretation from Proposition 3.1.3. ❑

The generalized Macdonald identities correspond to the following pair. The set $T = \{\alpha_1, \ldots, \alpha_n, -\vartheta\}$ must be taken together with $\xi = -\rho_k$. The sign of k_β in the relations (3.6.27) from the proposition is plus for ϑ and minus otherwise. See the end of Section 3.10 of this chapter.

The constraint (3.6.18) will be fulfilled, for instance, if $(\xi, \alpha^\vee) \notin \mathbb{Z}$ for all $\alpha \in R$. The simplest example is as follows. Letting $k_{\text{lng}} = k = k_{\text{sht}}$, $\xi = -k\rho$, we take $k = -m/h$ for the Coxeter number h provided that $(m, h) = 1$. This representation is spherical. In the A_n–case, one can follow [C5] to describe all irreducible representations under consideration (for generic q). See the next section.

3.6.4 Roots of unity

We go back to arbitrary root systems. Let q be a primitive root of unity of order N. Then $\widehat{W}^\flat_*[\xi]$ always contains

$$A(N) \stackrel{\mathbf{def}}{=\!=} (N\cdot A)\cap B, \ \text{ where } \ A \stackrel{\mathbf{def}}{=\!=} \{a\in P^\vee \mid (a,B)\subset \mathbb{Z}\}, \qquad (3.6.28)$$
$$\text{so } A = Q^\vee \ \text{ for } \ B = P, \text{ and always } A\supset Q^\vee.$$

Recall that $Q\subset Q^\vee$ and $P\subset P^\vee$; it may occur that $A\not\subset B$ and $A(N)\neq N\cdot A$.

From here on, $q' = q^{1/(2\tilde m)}$ is assumed to be a primitive root of order $2\tilde m N$ unless otherwise stated. Recall that $\tilde m\in\mathbb{N}$ is the smallest positive integer such that $\tilde m(B,B)\subset\mathbb{Z}$. If q' is a primitive root of order $m'N$ for $m' \mid 2\tilde m$, then we shall replace A by $m'(2\tilde m)^{-1}A$ in the definition of $A(N)$.

Let us describe semisimple irreducible finite dimensional quotients V of $\mathcal{I}_X[\xi]$ assuming that $\widehat{W}^\flat_*[\xi] = A(N)$, i.e., the stabilizer of ξ is the smallest possible, and $\Upsilon_+[\xi] = \widehat{W}^\flat = \Upsilon_-[\xi]$. The main constraint (3.6.18) becomes $\Upsilon_0[\xi] = \widehat{W}^\flat$ in this case.

Theorem 3.6.1 states that the spectrum of $\{X_b\}$ in any V has to be simple, the dimension is always $|W|\cdot|B/A(N)|$, and the $\{V\}$ are in one-to-one correspondence with the one-dimensional characters $\varrho(a)$ of the group $A(N)$, which determine the action via S of $A(N)$ on $V(\xi)$. To be more exact, the V–quotients up to isomorphisms are in one-to-one correspondence with the pairs $\{\,\varrho,\text{ orbit } \ q^{\widehat{W}((\xi))}\,\}$. We denote them

$$V \stackrel{\mathbf{def}}{=\!=} V[\xi,\varrho],\ V[\xi,\varrho]\simeq V[\xi',\varrho] \ \text{ if } \ q^{\xi'} = q^{\widehat{w}((\xi))} \ \text{ for } \ \widehat{w}\in\widehat{W}. \qquad (3.6.29)$$

The isomorphisms here are established using the intertwiners $\Phi_{\widehat{w}}$.

Concerning the unitary structure, let us calculate $\mu_\bullet$ on $A(N)$. Setting $N_\alpha = N/(N,\nu_\alpha)$, the scalar products $(a,\alpha^\vee) = (a,\alpha)/\nu_\alpha$ are divisible by N_α as $a\in A(N)$, $\alpha\in R$. On the other hand, $q_\alpha = q^{\nu_\alpha}$ is a primitive root of order N_α. Using formula (3.1.14),

$$\mu_\bullet(a) \stackrel{\mathbf{def}}{=\!=} \prod_{\alpha\in R|(a,\alpha)>0}\Big(\frac{t_\alpha^{-N_\alpha/2} - t_\alpha^{N_\alpha/2}q^{N_\alpha(\alpha,\xi)}}{t_\alpha^{N_\alpha/2} - t_\alpha^{-N_\alpha/2}q^{N_\alpha(\alpha,\xi)}}\Big)^{((a,\alpha^\vee)/N_\alpha)} =$$
$$\prod_{\alpha\in R_+}\Big(\frac{t_\alpha^{-N_\alpha/2} - t_\alpha^{N_\alpha/2}q^{N_\alpha(\alpha,\xi)}}{t_\alpha^{N_\alpha/2} - t_\alpha^{-N_\alpha/2}q^{N_\alpha(\alpha,\xi)}}\Big)^{((a,\alpha^\vee)/N_\alpha)} \qquad (3.6.30)$$

for $a \in A(N)$. By the way, it is clear from this formula that $\mu_\bullet$ is a homomorphism from $A(N)$ to the multiplicative group of the field $\mathbb{Q}(t_\alpha^N, q^{N(\xi,\alpha)})$. It was already checked in the theorem. We come to the following proposition.

Proposition 3.6.4. *Assuming that q is a primitive root of unity of degree N and*

$$\widehat{W}^\flat_*[\xi] = A(N),\ \Upsilon_+[\xi] = \widehat{W}^\flat = \Upsilon_-[\xi],$$

X-semisimple irreducible quotients of $\mathcal{I}_X[\xi]$ exist if and only if

$$\Upsilon_0[\xi] = \widehat{W}^\flat \ \text{i.e., } X_{\tilde{\alpha}}(q^\xi) \neq 1 \text{ for all } \tilde{\alpha} \in \tilde{R}. \tag{3.6.31}$$

Such quotients $V[\xi, \varrho]$ are described by the characters ϱ of $A(N)$. The pseudo-unitary structure exists if and only if

$$\varrho(a)^*\varrho(a) = \mu_\bullet(a) \text{ for the generators of } A(N).$$

3.6.5 Comment on finite stabilizers

When q is a root of unity, the case of finite stabilizators is, in a sense, opposite to the case of representations $V[\xi, \varrho]$. Indeed, the group $\widehat{W}^\flat_*[\xi]$ is finite if and only if the set $\Upsilon_*[\xi]$ is finite. The check is simple. If the latter set is infinite, then it contains infinitely many elements in the form $a\widehat{w}$ for a fixed $\widehat{w}$ and $a \in A(N)$, because the quotient $\widehat{W}/A(N)$ is finite. Note that $A(N)$ is normal in $\widehat{W}$, so this quotient is a group. We conclude that $\widehat{W}^\flat_*[\widehat{w}((\xi))]$ is infinite. However, it is conjugated to $\widehat{W}^\flat_*[\xi]$ due to (3.6.16), so the latter has to be infinite as well.

Recall that the existence of the pseudo-unitary structure on V reads as

$$S^\star_{\widehat{w}} S_{\widehat{w}} = \mu_\bullet(\widehat{w}) \text{ in } V(\xi) \ \text{ for } \ \widehat{w} \in \widehat{W}^\flat_*[\xi] \text{ of finite order.} \tag{3.6.32}$$

We are going to check that here $\mu_\bullet(\widehat{w})$ is ± 1 and, moreover, automatically 1 in many case.

Proposition 3.6.5. *For an element $\widehat{w}$ in $\widehat{W}^\flat_*[\xi]$ of finite order, $\mu_\bullet(\widehat{w}) = \pm 1$ without any assumptions on q, t, ξ. Moreover, $\mu_\bullet(\widehat{w}) = 1$ if the order of $\widehat{w}$ is $2M+1$, i.e., odd, or $2(2M+1)$.*

Proof. First we consider the case $\widehat{w}^2 = 1$. Then $\lambda(\widehat{w}) = -\widehat{w}(\lambda(\widehat{w}))$ and each $X_{\tilde{\alpha}}(q^\xi)$ appears in the product (3.6.19) together with $X^{-1}_{\tilde{\alpha}}(q^\xi)$, thanks to $\widehat{w}(q^\xi) = q^\xi$. If $X_{\tilde{\alpha}}(q^\xi) \neq X^{-1}_{\tilde{\alpha}}(q^\xi)$ then the corresponding binomials will annihilate each other, i.e., their contribution to the product will be 1. However, they can coincide. There are two possibilities: $X_{\tilde{\alpha}}(q^\xi) = \pm 1$. If $X_{\tilde{\alpha}}(q^\xi) = -1$, then

$$\Big(\frac{t_\alpha^{-1/2} - q_\alpha^j t_\alpha^{1/2} X_\alpha(q^\xi)}{t_\alpha^{1/2} - q_\alpha^j t_\alpha^{-1/2} X_\alpha(q^\xi)}\Big) = 1.$$

The equality $X_{\tilde{\alpha}}(q^{\xi}) = 1$ is impossible, thanks to the constraint (3.6.18). So we obtain the complete annihilation of all factors in $\mu_\bullet(\widehat{w})$.

Now we assume that $\widehat{w}$ is of finite order. Then $\mu_\bullet(\widehat{w})$ has to be a root of unity. Under the reality conditions $|q| = 1$, $k_\alpha \in \mathbb{R}$, $\xi \in \mathbb{R}^n$, $\mu_\bullet(\widehat{w})$ is real and can be $\pm$ only. An arbitrary triple $\{q, k, \xi\}$ can be continuously deformed to the triple above, without changing the group $\widehat{W}^\flat_*[\xi]$ and the value of $\mu_\bullet(\widehat{w})$. Indeed, the structure of $\widehat{W}^\flat_*[\xi]$ depends on certain multiplicative relations among $q_{(\alpha,\xi)}$ and $q_\alpha^{k_\alpha}$. Therefore we need to deform $\{q, k, \xi\}$ within some subvariety, which is possible. We obtain the claim about ± 1.

It gives that the orders of

$$\mu_\bullet(\widehat{w}) \ \text{ and } \ \mu_\bullet(\widehat{w}^{2M+1}) = \mu_\bullet(\widehat{w})^{2M+1}$$

always coincide. If the order of $\widehat{w}$ is odd or $2(2M+1)$, then $\mu_\bullet(\widehat{w})$ has to be 1, thanks to the claim about the elements of the second order. ❑

3.7 The *GL*–case

This section is somewhat exceptional. It does not conatin proofs, although the main result is not too difficult to justify using the previous theory. We give a description of the X–semisimple representations for the **double Hecke algebra associated to GL_n** for *fractional* k in terms of the periodic skew Young diagrams.

The DAHA of type GL_n is defined as follows:

$$\mathcal{H}_n \overset{\text{def}}{=\!=} \langle X_1^{\pm 1}, \ldots, X_n^{\pm 1}, \pi, T_1, \ldots, T_{n-1}\rangle, \ \text{ where} \tag{3.7.1}$$

(i) $\pi X_i = X_{i+1}\pi \ (i = 1, \ldots n-1)$ and $\pi^n X_i = q^{-1} X_i \pi^n \ (i = 1, \ldots, n)$,
(ii) $\pi T_i = T_{i+1}\pi \ (i = 1, \ldots n-2)$ and $\pi^n T_i = T_i \pi^n \ (i = 1, \ldots, n-1)$,
(iii) $X_i X_j = X_j X_i \ (1 \le i, j \le n), \ T_i X_i T_i = X_{i+1}, \ i < n$,
(iv) $T_i T_{i+1} T_i = T_{i+1} T_i T_{i+1}, \ T_i T_j = T_j T_i$ as $i < n-1, \ j > i+1$,
(v) $(T_i - t^{1/2})(T_i + t^{-1/2}) = 0$ for $T_1, \ldots, T_{n-1}, \ t = q^k$.

See Chapter 1, Section 1.4.3. Note that the formulas for the generators Y in terms of $\pi, T_1, \ldots, T_n$ in this chapter are dual to those considered in Chapter 1. For instance, now $Y_1 = \pi T_{n-1} \cdots T_1$.

Actually, we we do not need the formulas for Y in this section. The classification will be produced in terms of the X only. It triggers a very interesting question about the description of the Y–action in the representations considered below.

For the sake of completeness, let us give the formulas for the automorphisms τ_+ and σ from (3.2.10)–(3.2.16) in the case of GL_n. Recall that they preserve $T_1, \ldots, T_{n-1}$ and do not change q, t. Also, $\tau_+(X_i) = X_i$, $\sigma(X_i) = Y_i^{-1}$ for $1 \le i \le n$. The other formulas are as follows:

$$\tau_+(Y_1 \cdots Y_i) = q^{-i/2}(X_1 \cdots X_i)(Y_1 \cdots Y_i), \ \sigma(Y_i) = T_{w_0}^{-1} X_{n-i+1} T_{w_0}. \quad (3.7.2)$$

We use that in a proper completion of $\mathcal{H}_n$,

$$\tau_+(\pi) = q^{\sum x_i^2/2} \circ \pi \circ q^{-\sum x_i^2/2} = q^{-1/2} X_1 \pi,$$

where $X_i = q^{x_i}$ and $q^{\sum x_i^2/2}$ is the Gaussian for GL_n.

Switching here from $\mathbb{C}[X_i^{\pm 1}]$ to the subalgebra $\mathbb{C}[X_i X_j^{-1}]$ of Laurent polynomials of degree zero and setting $\pi^n = 1$, we come to the definition of the double Hecke algebra $\mathcal{H}_n^o$ of type A_{n-1}, i.e., the one for SL_n. We use that π^n commutes with the X–monomials of degree zero and with T_i, thanks to (i–ii). To be exact, this definition corresponds to the root lattice Q taken as $B \ni b$ for the X_b in the definition of $\mathcal{H}^b$. In this section, we will deal with either $\mathcal{H}_n$ or $\mathcal{H}_n^o$.

When the subalgebra $\mathbb{C}[X_i X_j^{-1}] \subset \mathcal{H}_n^o$ is extended to $\mathbb{C}[(X_i \widetilde{X}^{-1/n})^{\pm 1}]$ for $\widetilde{X} = X_1 \cdots X_n$, we arrive at the case $B = P$. The $\widetilde{X}$ appears here because of the formulas for the fundamental weights in the case of A_{n-1}.

Later on, we will always consider the representations where π is invertible and *identify the representations under the automorphisms* $\pi \mapsto c\pi$ *of* $\mathcal{H}_n$ *for* $c \in \mathbb{C}^*$. Obviously all relations are preserved under this multiplication.

The following construction is a "cylindrical" counterpart of that from [C5](Section 3) and [Na1]. It generalizes the corresponding affine classification due to Bernstein and Zelevinsky. It is a direct application of the Main theorem. Fortunately, the most technical part of the construction can be borrowed from the affine Hecke theory. We will publish the details elsewhere. Examples concerning the semisimple representations of $\mathcal{H}_n$ can be found in [SuV]. See also [Suz].

3.7.1 Generic *k*

Let $\Delta = \{\delta^1, \ldots, \delta^m\}$ be an unordered collection of the skew Young diagrams without empty rows of total degree n (i.e., with n boxes in their union), $C = \{c_1, \ldots, c_m\}$ a set of complex numbers such that $q^{c_a - c_b} \notin t^{\mathbb{Z}} q^{\mathbb{Z}}$ as $a \neq b$.

A diagram is a subset of $\mathbb{Z}^n$ with the coordinate i used for the rows and the coordinate j used for the columns. We identify the points (i, j) with the centers of the corresponding unit boxes, so (i_l, j_l) are the coordinates of the center of box l. We set $p_l = p$ if this box is in δ^p.

The **skew diagram** δ is defined as follows. If box l belongs to the i–th row of δ, then $m_i < j_l \le n_i$ for the "end points" $m_i \le n_i$ of the rows (the row is empty if they coincide). The end points satisfy the inequalities

$$m_i \ge m_{i+1}, \ n_i \ge n_{i+1}, \ \ i \in (\text{set of the row-numbers in } \delta). \quad (3.7.3)$$

We always number the boxes of the collection of diagrams in the **inverse order**, i.e., from the last box of the last row of δ^m through the first box of the first row of δ^1, and set

$$\xi_l \ = \ c_{p_l} + k(i_l - j_l), \ l = 1, \ldots, n, \text{ where box } l \text{ belongs to } \delta^{p_l}. \qquad (3.7.4)$$

Concerning the notation p_l, note that $p_1 = m, \ldots, p_n = 1$.

We say that the **weight** ξ is associated with the **pair** $\{\Delta, C\}$. The weight is the main numerical invariant of the pair.

We will identify the diagrams that can be obtained from each other by translation by the diagonal vectors (c, c) (in the $\{i, j\}$–plane). This identification does not change the weight.

More generally, *translations of the diagrams with the corresponding adjustments of the c–numbers are considered identical transformations if they do not change the weight.*

Theorem 3.7.1. *Let $q^i t^j \neq 1$ for any $(i, j) \in \mathbb{Z}^2 \setminus (0, 0)$. Then condition (3.6.18) is satisfied if ξ is associated with a Δ–set of skew diagrams and the corresponding C–set of complex numbers. For such ξ, $U = U_X^\xi$ from the Main theorem is a unique irreducible X–unitary quotient of the induced module $\mathcal{I}_X[\xi]$. Its X–spectrum is simple. An arbitrary irreducible X–semisimple $\mathcal{HH}_n$–module can be constructed in this way for a proper pair $\{\Delta, C\}$. Isomorphic modules correspond to the pairs that can be obtained from each other by a permutation of the skew diagrams from the collection Δ together with permuting the corresponding numbers from C, and by adding arbitrary integers to the c–numbers.*

Proof follows [C23]. Adding intergers to the c–numbers is allowed because of the above formulas for conjugation by the element π. ❑

Given ξ, the construction of the Main theorem restricted to the nonaffine Weyl group W results in the irreducible representation V^ξ of the **affine Hecke algebra,**

$$\mathcal{H}_n \stackrel{\text{def}}{=\!=} \langle X_1^{\pm 1}, \ldots, X_n^{\pm 1}, T_1, \ldots, T_{n-1} \rangle,$$

from [C23] and [C5]. It is a unique X–semisimple nonzero irreducible quotient of the induced $\mathcal{H}_n$–module of weight ξ. Note that adding integers to the c–numbers *alters* the isomorphism class of the $\mathcal{H}_n$–representation.

If ξ corresponds to a single skew diagram, then the restriction of the corresponding $\mathcal{H}_n$–module to the **nonaffine Hecke algebra**

$$\mathbf{H}_n \stackrel{\text{def}}{=\!=} \langle T_1, \ldots, T_{n-1} \rangle$$

is isomorphic to $\mathbf{H}_n \mathbf{P}_\delta$ for the q–**Young symmetrizer** $\mathbf{P}_\delta$ associated with the skew diagram δ in a way similar to the classical definition. It is $\mathbf{H}_n$–irreducible if and only if δ is the standard Young diagram (recall that t is generic).

Given a pair $\{\Delta = \{\delta^p\}, C = \{c_p\}\}$, the corresponding irreducible $\mathcal{H}_n$–module V^ξ is parabolically induced from the modules associated to

$$\{\delta^1, c_1\}, \ldots, \{\delta^m, c_m\},$$

and so is its restriction to $\mathbf{H}_n$.

Inducing the module V^ξ up from $\mathcal{H}_n$ to $\mathcal{HH}_n$, we get exactly the U^ξ_X from the theorem.

The modules V^ξ and U^ξ_X remain irreducible upon restriction to the subalgebra $\mathcal{H}^o_n$ of $\mathcal{H}_n$ with $\mathbb{C}[X_iX_j^{-1}]$ instead of $\mathbb{C}[X_i^{\pm 1}]$ and to $\mathcal{HH}^o_n$. Concerning the last statement of the theorem, this symmetry becomes stronger. *Adding an arbitrary common complex constant to the c–numbers does not change the $\mathcal{HH}^o_n$–isomorphism classes,* as well as translations of the diagrams in Δ by a common integral vector.

3.7.2 Periodic skew diagrams

We assume that q is not a root of unity, and the products $q^a t^b$ for $a, b \in \mathbb{Z}$ do not represent nontrivial roots of unity. Let $t^r = q^s$ for integral r, s, where $r \neq 0; s > 0$, and $(r, s) = 1$. Setting $t = q^k$, we obtain that $k = s/r$. We will begin with the definitions for $r > 0$.

An infinite skew diagram δ *without empty rows* is called r–**periodic** of degree n if
(a) it is invariant with respect to translation by a vector $w = (v, v - r)$ in the (i, j)–plane for a proper positive integer v, and
(b) the **subdiagram** of δ with the rows in the range $1 \leq i \leq v$ contains exactly n boxes.

The latter subdiagram is called the **basic subdiagram**. It is skew. The subdiagrams with the rows subject to $1 + c \leq i \leq v + c$ are skew too. Note that $v \leq r$ for periodic skew diagrams.

To visualize the diagrams, here and later, we identify the centers of boxes (i, j) with the points $(x = j, y = -i)$ in the standard $\{x, y\}$–plane, i.e., the rows go down when the row-numbers increase and the column-numbers are the x-coordinates.

Figure 3.1 demonstrates how we place a Young diagram in the (x, y)–plane and number the boxes in the basic and other skew subdiagrams of δ.

This skew diagram is the basic subdiagram of the (portion) of the periodic skew diagram with $n = 11$, $r = 7$, $v = 3$, $w = (3, 3 - 7) = (3, -4)$ shown at Figure 3.2.

Any skew diagram, finite or infinite, is naturally a portion of the $\{x, y\}$–plane between its **upper-left boundary** and its **lower-right boundary**.

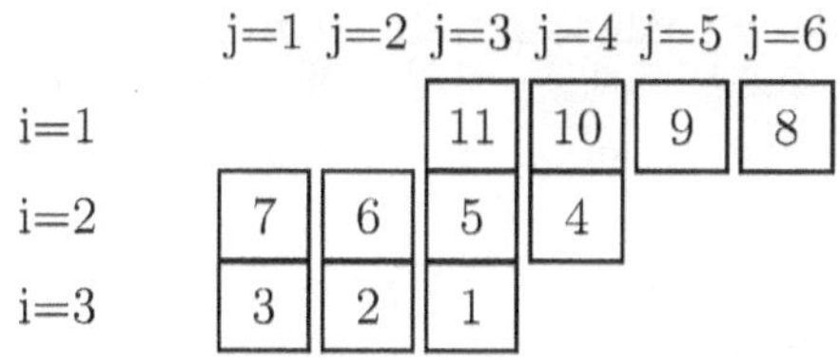

Figure 3.1: Skew Diagram

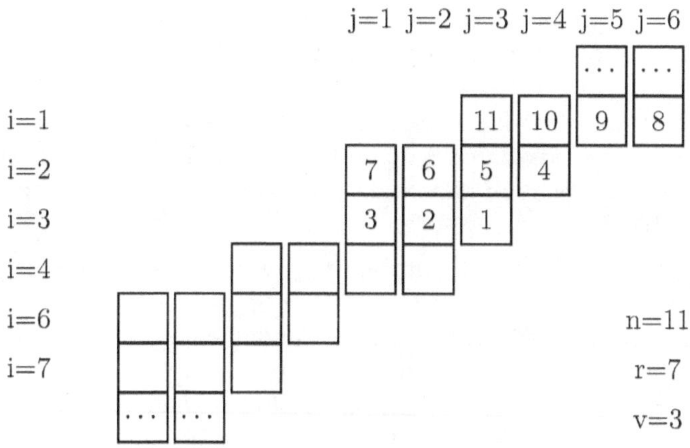

Figure 3.2: Periodic Skew Diagram

In terms of the boxes, the upper-left boundary of a single unit box is a union of its top and left sides. The bottom and right sides represent the lower-right boundary. The general definition is analogous.

The boundaries will be needed for periodic diagrams only. The intersections of any diagrams with the horizontal lines $i = const$ are segments between the left and the right end points. For skew diagrams, their intersections with vertical lines are segments too, between the lower and the upper end points.

The boundaries are infinite continuous piecewise linear paths in the coordinate net, i.e., they are "made" from the segments where either $i \in \mathbb{Z}$ or $j \in \mathbb{Z}$. They satisfy the skew condition:

$$\text{if } i' > i, \text{ then } j' \leq j, \quad \text{if } j' > j, \text{ then } i' \leq i,$$

for any pair $(i,j), (i',j')$ of points. We will call such paths the **skew paths**.

A subdiagram of δ between a given skew path Ω and its translation $w+\Omega$ is called a **fundamental subdiagram**, provided that
(a) Ω coincides with the left bondary for sufficiently big i,
(b) coincides with the right boundary for sufficiently small i,
(c) the sets of boxes above Ω and below $w + \Omega$ do not intersect.

Such Ω is called a **skew path associate with** δ. Note that the i–coordinate of the Ω associated to an r–periodic δ is infinite, i.e., changes from $-\infty$ to $+\infty$. The j–coordinate can be bounded, which happens exactly for the infinite vertical rectangles.

The fundamental subdiagram is a skew diagram that is a fundamental domian for the action of $\mathbb{Z}w$ in δ (so it contains n boxes). *It may have empty rows.*

The basic subdiagram or, more generally, a portion of δ between the horizonta lines, i.e., with the boxes satisfying the inequalities $c+1 \le i \le c+v$ for a constant $c \in \mathbb{Z}$, are examples of fundamental subdiagrams that do not have empty rows.

Note that there are infinitely many different *shapes*, diagram types up to translations, of the fundamental subdiagrams unless δ is an infinite one-column or one-row, since fundamental subdiagrams can be collections of distant disconnected pieces.

Similarly, a skew diagram δ without empty columns is called r–**periodic** for *negative* r if it is invariant with respect to $w = (v+r, v), v \in \mathbb{N}$, and the basic fundamental diagram and any other fundamental diagrams, e.g., those formed by v consecutive columns, contain n boxes.

The skew paths Ω in the definition of the fundamental diagram must coincide for $r < 0$ with the upper boundary for sufficiently big j and with the lower boundary for sufficiently small j. Respectively, the sets of boxes on the left of Ω and on the right of $w + \Omega$ do not intersect. The j coordinate changes from $-\infty$ to $+\infty$ for such Ω in the case of negative r.

In short, skew paths associated to an r–periodic δ "cross" it from left to right as $r > 0$ and from bottom to top, as $r < 0$. Here we use the (x, y)–plane under the "visualization" $(i, j) \mapsto (x = j, y = -i)$.

3.7.3 Partitions

An s–**increasing partition** of a skew diagram δ of degree n is defined as follows.

(a) It is a representation of δ as a disjoint union of $u \le s$ skew *nonempty* subdiagrams $\{\delta_0, \dots, \delta_u\}$ that are the portions of δ between a system of *consecutive skew paths without crossings.*

(b) Different paths from such a system may have coinciding subpaths, but do not cross each other. The paths are numbered upwards from the lower boundary to the upper boundary.

(c) Every δ_a is decorated with an integer $0 \le o_a < s$, called the o–number. They must form a strictly increasing sequence: $o_0 < o_1 < \cdots < o_u$.

Note that the subdiagrams in the partition cannot be empty, but some may have empty rows. Let me comment on (b).

The subdiagrams in a given partition are **partially ordered** naturally. Namely, $\delta' > \delta''$ if at least one box of δ' is in the same column and higher than a box from δ'' or is in the same row and to the left of a box from δ''.

Using this ordering, property (b) means that, first, it is not allowed that $\delta_a > \delta_b$ and at the same time $\delta_a < \delta_b$ for different a, b, and, second, $a > b$ if $\delta_a > \delta_b$.

For instance, the representation of a skew diagram δ of degree n as a union of *all* its boxes counted from the last box in the last row through the first box in the first row (i.e., with respect to the inverse order) can be taken. These boxes can be ordered in other ways unless δ is a one-column or a one-row. Indeed, the boxes in the diagonals $i + j = const$ are not ranked and can be placed in any order.

Given such a "total" decomposition of δ, the corresponding o–numbers must increase for increasing numbers of δ_a. It is possible when $s \geq n$. If $s = n$, then we must take the complete o–sequence $\{0, 1, 2, \ldots, n-1\}$.

An opposit example is when we take the whole diagram and make $u = 0$, i.e., set $\delta_0 = \delta$, and then assign an arbitrary number $0 \leq o \leq s-1$ to it. Such partitions exist for arbitrary $s > 0$.

The increasing partitions of the fundamental subdiagrams will play the key role when decomposing the semisimple representations of $\mathcal{HH}_n$ under the action of $\mathcal{H}_n$.

New (periodic) pairs. Extending the definitions above to a collection Δ, we call it r–periodic of degree n if all δ^p are r–periodic and the total degree of the fundamental domains of $\{\delta^p\}$ is n. The weights ξ associated with a **new pair** $\{\Delta, C\}$ are those for *all* choices of the fundamental subdiagrams $\{\tilde{\delta}^p \subset \delta^p\}$ calculated using the formulas above.

Note that the fundamental diagrams are defined as subsets of $\{\delta^p\}$, i.e., as subsets of the corresponding copy of $\mathbb{Z}^2$ associated with p. The embeddings $\tilde{\delta}^p \subset \mathbb{Z}^2$ are needed when calculating the corresponding weights. The c_p–numbers are taken from the corresponding δ^p.

We call the set of fundamental subdiagrams $\{\tilde{\delta}^1, \ldots, \tilde{\delta}^n\}$ with the same set of c–numbers a **fundamental subpair**.

The constraint for the c–numbers of the new (periodic) pairs is the same as for the old pairs. However, thanks to the condition $t^r = q^s$, it now reads as

$$c_a - c_b \not\in (1/r)\mathbb{Z} \text{ for all } 1 \leq a, b \leq m.$$

Note that only the "diagonal" distances matter when calculating the weights of a pair.

An s–**partition** of $\{\Delta, C\}$ is a collection of fundamental diagrams $\{\tilde{\delta}^p\}$ supplied with their (skew, increasing) s–partitions:

$$\left\{\tilde{\delta}^p_a,\ a = 0, 1, \ldots, u_p \leq s-1,\ 1 \leq p \leq m\right\},\ \tilde{\delta}^p = \cup_a \tilde{\delta}^p_a.$$

Given p, we will stick to the notation $\tilde{\delta}^p$ for a fundamental subdiagram of the component δ^p of Δ, and will use a to number the components of a partition.

Recall that the number of subdiagrams u_p can be smaller than s. The u–numbers may be different for different $\tilde{\delta}^p$.

Partition weights. The main change in the definition of the weight associated to a partition will be that *the integer* o_a *must be added to the* c_p*–number for every* $\tilde{\delta}^p_a$. The exact definition is as follows.

Given a fundamental subpair $\{\tilde{\delta}^p\}$, we number its boxes in the *totally inverse* ordering that is a union of consecutive the inverse orderings

(1) for the p–indices, (2) for the a–indices, (3) inside every $\tilde{\delta}^p_a$,

i.e., first, we take into consideration the p–indices, second, the a–indices, and, third, the (standard) numeration of the boxes inside $\tilde{\delta}^p_a$.

The **partition weight** is given by the formula

$$\xi_l = c_{p_l} + o_{a_l} + k(i_l - j_l) \ \text{ for } \ \tilde{\delta}^p = \cup_a \tilde{\delta}^p_a, \ \ 0 \le a \le u_p \le s-1, \tag{3.7.5}$$

where $1 \le l \le n$, $p_l = p, a_l = a$ if box l belongs to $\tilde{\delta}^p_a$.

Let us consider the example of the "total partition" for a single periodic δ, i.e., when $m = 1$, $\{\Delta, C\} = \{\delta, c\}$. We chose its skew fundamental subdiagram $\tilde{\delta}$ and represented it as a union of all its n boxes counted from the last box through the first one. That is, $\tilde{\delta} = \cup_{0 \le a < n} \tilde{\delta}_a$ in our notation. Let $s = n$. Then the weight of the a–th box (to be more exact, the component of the weight corresponding to the box) is $c + (i_a - j_a)k + a$, where $0 \le a \le n-1$. Here i_a, j_a are the *global* row and column numbers (inside $\delta \in \mathbb{Z}^2$).

Adding the o–numbers to the weights has the following transparent reason. It reflects the q–contribution of the intertwining operators for $\widehat{w}$ taken from the lattice. The k–part comes from the intertwiners for $w \in W$. Since $q^s = t^r$, the o–numbers can be reduced modulo s. It is why the range $0 \le o \le s-1$ is imposed. If they are not in the range, then we can recaluclate them using k.

Applying the interwiners to the weights constructed from the partitions matches exactly the combinatorial transformations of the partitions (including changing the o–numbers). All partitions can be obtained from each other using the intertwiners.

3.7.4 Equivalence

New pairs $\{\Delta, C\}$ and $\{\Delta', C'\}$ are called **equivalent pairs** if they can be obtained from each other by a permutation of the Δ–components together with the C–components and/or by adding *arbitrary* integers to the c–numbers. Only $c_i \bmod \mathbb{Z}$ matter in the (new) equivalence classes. We may also translate a diagram δ in the (i, j)–plane by an integral vector (z_1, z_2) replacing at the same time its c by $c - k(z_1 - z_2)$. Recall that we can always permute the components $\delta^1, \ldots, \delta^m$ of Δ and the corresponding c–numbers (together with the corresponding partitions).

Given a new pair, the fundamental subpairs and their s–partitions are called equivalent if they coincide pointwise inside Δ and their weights coincide too. The latter means that the c–numbers and the o–numbers must be the same.

Conversely, given an s–partition of a fundamental subpair, one checks that the corresponding weight uniquely determines the position of an individual skew diagram $\tilde{\delta}^p_a$ as a subdiagram of $\{\delta^p\}$ and the corresponding o–number.

Recall that the indices are in the range

$$0 \leq a \leq u_p, \quad \text{where} \quad u_p < s, \quad \text{and} \quad 1 \leq p \leq m.$$

Given a partition of a fundamental subpair, treated as a set of skew diagrams with the c–numbers, o–numbers, and the weights constructed in (3.7.5), one can naturally define the corresponding *irreducible* representations of $\mathcal{H}_n$. It is induced from the irreducible modules corresponding to the individual skew diagrams $\tilde{\delta}^p_a$ in the partition.

Isomorphic $\mathcal{H}_n$–modules correspond to equivalent increasing partitions. It holds for $\mathcal{H}^o_n$ generated by $T_1, \ldots, T_{n-1}$ and $X_i X_j^{-1}$ too, provided that one uses the SL–equivalence.

SL–**Equivalence.** The above equivalence, the GL–equivalence, means essentialy the coincidence. The fundamental subpairs will be called SL–**equivalent** if the corresponding weights can be obtained from each other by adding a common constant. Respectively, the increasing partitions are SL–equivalent if they can be obtained by adding a common constant c to the weight, possibly, *together with the proper cyclic permutations of* $\{\tilde{\delta}^p_a\}$ *with respect to the index* a for some of the p.

Such permutation may appear (only) for integral c because we add o_a to the weight of $\tilde{\delta}^p_a$ always assuming that $0 \leq a \leq s-1$. Therefore the reduction modulo s may be needed followed by the corresponding cyclic permutation. That is, given p, the o–numbers for the segment $\{\tilde{\delta}^p_{a'} \mid o_{a'} + c \geq s\}$ at the end of the partition must be changed to $\{o_{a'} + c - s\}$, and then this segment must be cut and placed in front of the partition.

The $\mathcal{H}_n$–module associated with the partition is irreducible when restricted to $\mathcal{H}^o_n$ and remains in the same isomorphism class under the SL–equivalence of the partitions.

3.7.5 The classification

We are ready to formulate the main result of this section, a combinatorial description of the X–semisimple irreducible modules in the GL–case and in the SL–case in terms of the sets of periodic skew diagrams and the corresponding complex numbers. The theorem also solves the problem of restriction of such modules to the affine Hecke subalgebra.

Theorem 3.7.2. *(i) Given r, s, and n, irreducible X–semisimple representations of $\mathcal{HH}_n$ up to multiplication of π by a nonzero constant are in one-to-one correspondence with the equivalence classes of new pairs $\{\Delta, C\}$ consisting of the r–periodic sets Δ of degree n and the c–numbers. Relations (3.6.18) hold for the weight ξ associated with $\{\Delta, C\}$, and the corresponding representation is $U = U_X^\xi$ from the Main theorem. It is X–unitary and its X–spectrum is simple.*

(ii) Let $s = 1$. Upon the restriction to the affine Hecke algebra $\mathcal{H}_n$, the above module $U = U_X^\xi$ associated with $\{\Delta, C\}$ is a direct sum of the pairwise non-isomorphic irreducible $\mathcal{H}_n$–modules associated to the fundamental subpairs, which are, we recall, the fundamental skew subdiagrams of $\{\delta^p, 1 \le p \le m\}$ supplied with the corresponding c–numbers (the o–numbers are zero). Each of these representations appears exactly once in the decomposition.

(iii) For arbitrary $s > 0$, the module U is isomorphic to a direct sum of the pairwise non-isomorphic irreducible $\mathcal{H}_n$–modules associated with all increasing s–partitions of the fundamental subdiagrams $\{\tilde{\delta}^p\}$, where, recall, the integer o_a is added to the weight of the partition component $\tilde{\delta}^p_a$. We treat such a partition as a system of $u_1+u_2+\cdots+u_p$ skew diagrams of total degree n equipped with the c–numbers and o–numbers and define the $\mathcal{H}_n$–modules correspondingly.

(iv) The previous claim holds for the algebra $\mathcal{HH}_n^\circ$ and the irreducible module $U = U_X^\xi$ defined for this algebra in terms of the restriction of the above weight ξ to $\mathbb{C}[X_iX_j^{-1}]$ (this U is not a restriction of the above U to $\mathcal{HH}_n^\circ$). The module U under the action of $\mathcal{H}_n^\circ$ is a direct sum of pairwise non-isomorphic irreducible modules over all classes of SL–equivalent s–partitions of fundamental subpairs.

(v) The $\mathcal{HH}_n^\circ$–module U_X^ξ is finite dimensional if and only if $\Delta = \{\delta\}$ and δ is either the infinite column as $r > 0$ or the infinite row as $r < 0$. In either case, $r = \pm n$ and s is an arbitrary natural number relatively prime to n. When $s = 1$, the dimension of this representation is 1. Generally, it equals s^{n-1}. ❑

An important part of the proof is that transposing the two consecutive segments of the weight associated with a pair $\{\Delta, C\}$ coming, respectively, from $\tilde{\delta}^p_a$ and $\tilde{\delta}^p_{a+1}$ gives another X–weight of U_X^ξ. The corresponding X–eigenvector generates this module and can be obtained from the initial one by a chain of *invertible* intertwiners acting in U_X^ξ.

Part (v) readily follows from (ii) because the number of nonequivalent fundamental subdiagrams of a periodic diagram is infinite unless it is a column (infinite) or a row. Let us describe the s–partitions in this case and calculate the dimension formula.

3.7.6 The column-row modules

An s–partition of the n–column or the n–row is the set $\underline{n} = \{n_0, \ldots, n_u\}$ such that

$$n = n_0 + \cdots + n_u \text{ for } u < s, \ n_a \in \mathbb{Z}_+. \tag{3.7.6}$$

The numbers n_a are identified with the consecutive segments of the column (or the row), counted from the last box through the first one. Recall that the δ_a^p–component of the weight is increased by o_a (depending on p) by definition. The dimension of the corresponding irreducible $\mathcal{H}_n^o$–module is $n!/(n_0! \cdots n_u!)$. Upon restriction to the $\mathbf{H}_n$, it is the standard induced module corresponding to the decomposition $n = n_0 + \cdots + n_u$.

Given a decomposition (3.7.6), we need to assign an increasing sequence of the o–integers $\underline{o} = \{0 \le o_1 < o_2 < \cdots < o_u \le s\}$ to the n–numbers. Therefore the total contribution of the decomposition $\underline{n}$ to the dimension of U for all o–sequences will be the number of such o–sequences times $n!/(n_0! \cdots n_u!)$.

Let us ignore the SL–equivalence for a while and calculate the desired dimension Dim. It is the sum of such products over all n–decompositions. There is a simple interpretation of this sum.

One can naturally identify a pair

$$n = n_0 + \cdots + n_u \text{ for } u < s, \ 0 \le o_1 < o_2 < \cdots < o_u \le s-1,$$

with the tensor product

$$E(\underline{o}, \underline{n}) = e_{o_0}^{\otimes n_0} \otimes e_{o_1}^{\otimes n_1} \otimes \cdots \otimes e_{o_u}^{\otimes n_u} \tag{3.7.7}$$

from $(\mathbb{C}^s)^{\otimes n}$ of the basic vectors $e_0, e_1, \ldots, e_{s-1} \in \mathbb{C}^s$.

Multiplication by $n!/(n_0! \cdots n_u!)$ simply means that the o–sequence is not assumed to be increasing, and, moreover, the o–numbers may coincide for different indices. Upon this relaxation, *all* tensor products of the basic vectors can be represented in the form (3.7.7). We obtain that the above Dim equals s^n.

The map to the linear space of the initial "o–increasing" tensor products is just the symmetrization homomorphism $(\mathbb{C}^s)^{\otimes n} \to S^n(\mathbb{C}^s)$ to the n–th symmetric power of $\mathbb{C}^s$. The linear span of $E(\underline{o}, \underline{n})$ for increasing $\underline{o} = \{o_0, \ldots, o_{s-1}\}$ becomes isomorphic to $S^n(\mathbb{C}^s)$ under this mapping.

The SL–equivalence means that we need to identify the vectors (3.7.7) by the folowing action of $\mathbb{Z}_s = \mathbb{Z}/(s) \ni c$:

$$e_j \mapsto e_{j'} \ \text{ for } \ j' = (j + c) \bmod s.$$

This readily gives that $\dim U_X^\xi = s^{n-1}$ in the "column-row" case.

Actually, we have obtained more than just the dimension formula. This construction gives that the U–module from (v), when restricted to the **non-affine Hecke subalgebra** $\mathbf{H}_n = \langle T_i, \ 0 < i < n \rangle$, becomes isomorphic to the

t–counterpart of the $\mathbb{S}_n$–module $\mathbb{C}[Q/sQ]$ =Funct(Q/sQ) for the root lattice Q of type A_{n-1} with the action of $\mathbb{S}_n$:

$$Q \;=\; \oplus_{i=1}^{n-1}\mathbb{Z}(e_i - e_{i+1}) \;\subset\; \mathbb{Z}^n = \oplus_i \mathbb{Z} e_i.$$

Recall that $(n, s) = 1$, so here one can take the weight lattice P instead of Q.

There is another, much simpler, proof of this fact together with the dimension formula without using the $\mathcal{H}_n^o$–decomposition, which can be readily extended to perfect finite dimensional representations of the double affine Hecke algebras for *arbitrary* reduced root systems. See Theorem 3.5.3, formula (3.6.26), and Theorem 3.10.6 below.

Comment. (i) Berest, Etingof, and Ginzburg verified that the relation $k = \pm s/n$, provided that $s \in \mathbb{Z}_+$ is not divisible by n, is necessary for the existence of finite dimensional irreducible representations of $\mathcal{H}\!\!\mathcal{H}_n^o$ in the *rational limit.* They also found that if such a representation exists, then its dimension is divisible by s^{n-1} and its $\mathbb{S}_n$–character is divisible by the one for $\mathbb{C}[Q/sQ]$ for the root lattice Q. Later they justified the coincidence. See [BEG]. The weaker statement above, with an unknown multiplier, requires relatively elementary algebraic considerations. In particular, it gives that the dimension of this representation must be greater than or equal s^{n-1}.

Our q, t–representation U is in the space $\mathbb{C}[Q/sQ]$ almost by construction. It is not difficult to see that it has a rational limit. Thus the standard deformation argument combined with the inequality from [BEG] readily shows that our representation remains irreducible upon the rational degeneration and that the "unknown multiplier" in the rational theory is actually 1.

(ii) Generally, to establish the connection between the DAHA representations in the q, t–case and the "rational" ones, there is a general method based on the existence of an isomorphism from $\mathcal{H}\!\!\mathcal{H}$ to its rational degeneration $\mathcal{H}\!\!\mathcal{H}''$ upon proper completion (this was conjectured by Etingof). We set $q = e^h, t = q^k$, and $k \neq 0$, and operate over the formal series in terms of $\sqrt{h}$. The completion is not needed in the *finite dimensional case.*

This connection was established in [C29] by combining the Lusztig-type isomorphism discussed in [C23](Appendix) and the straightforward formula connecting the trigonometric and rational *differential* Dunkl operators. The first map connects $\mathcal{H}\!\!\mathcal{H}$ and its trigonometric degeneration $\mathcal{H}\!\!\mathcal{H}'$; the second one leads to an identification of $\mathcal{H}\!\!\mathcal{H}'$ and $\mathcal{H}\!\!\mathcal{H}''$ (upon proper completion). It is not just a formal isomorphism. It establishes an isomorphism between finite dimensional representations of $\mathcal{H}\!\!\mathcal{H}$ for *generic* q and those for $\mathcal{H}\!\!\mathcal{H}''$. By generic, we mean that $q^a t^b$ do not represent nontrivial ($\neq 1$) roots of unity for integral a, b. See Section 3.11 below.

(iii) We also mention an important recent work [Vas] devoted to the K–theoretic classification of irreducible representations of $\mathcal{H}\!\!\mathcal{H}$ for arbitrary roots systems (with coinciding k), including those subject to the constraint $t^r =$

q^s from the theorem. In a sense, it is a continuation of [KL1]. However, it involves more sophisticated geometric methods and related combinatorial problems.

In the case of GL_n, Vasserot proves that irreducible $\mathcal{HH}_n$–modules are described in terms of *cylic quivers*. Combinatorially, his answer is equivalent to that from Theorem 3.7.3. There is an important continuation [VV2]. Under minor technical constraints, DAHA of type A is Morita-equivalent to the quantum affine Schur algebras at roots of unity. The latter are known to be also related to representation of the cyclic quivers.

Vasserot also gives a geometric description of the perfect representations, including the "homological" deduction of the dimension formula for arbitrary root systems; see Theorem 3.10.6 below. ❑

The representations from (v) are unique finite dimensional representations among all irreducible representations, not only among the X–semisimple ones. To see this we will generalize the Bernstein-Zelevinsky classification. Here we need the considerations that are beyond the Main theorem (the semisimple case), although the technique essentially remains the same, the intertwiners.

3.7.7 General representations

The classification will be given in terms of periodic diagrams, but with the following relaxation of the conditions from (3.7.3):

$$n_i \geq n_{i+1} \text{ or } \{n_i = n_{i+1} - 1 \text{ and } m_i \geq m_{i+1} - 1\}. \qquad (3.7.8)$$

We call such diagrams **generalized**. Respectively, we define the **generalized pairs** and then construct the corresponding weights using the same formulas as above. For generic q, t, an arbitrary irreducible representation of $\mathcal{HH}_n$ is induced from an irreducible representation of $\mathcal{H}_n$. This can be proved following [C23]. Therefore all of them are infinite dimensional for such q, t.

This subsection does not contain justifications. It is somewhat beyond the framework of this chapter. We hope that the details will be published elsewhere.

Theorem 3.7.3. *Imposing the relation $t^r = q^s$ for $r > 0, s > 0, (r, s) = 1$, all X–cyclic irreducible $\mathcal{HH}_n$–modules up to isomorphism and multiplication π by a constant are in one-to-one correspondence with the generalized new pairs $\{\Delta, C\}$ up to the equivalence, defined in the same way as above. We can skip the case of $r < 0$ because it follows formally from the positive case thanks to the automorphism of $\mathcal{HH}_n$ sending*

$$X_i \mapsto X_i,\ \pi \mapsto \pi,\ q \mapsto q,\ T_i \mapsto -T_i,\ t \mapsto t^{-1}.$$

❑

These modules are quotients of the induced modules $I = I_{\Delta,C}$ corresponding to the weights associated with $\{\Delta, C\}$. As above, the construction of such weights involves the basic fundamental subdiagrams formed by consecutive rows of $\{\tilde{\delta}^p\}$. The latter depends on the position of the zero row, but different choices can be compensated for proper changes of the c–numbers.

The induced modules I are isomorphic to the affine Hecke Y–algebra $\mathcal{H}_n^y \stackrel{\mathbf{def}}{=\!=} \langle \pi,\ T_i,\ 1 \le i \le n-1 \rangle$. Let $I' = I/(\sum \mathcal{H}_n^y(1-s_i))$ summed over all simple (nonaffine) reflections s_i corresponding to adjacent boxes in the rows of the basic fundamental subdiagrams of Δ.

The I' is an $\mathcal{HH}_n$–module, and the module from the theorem is its unique irreducible nonzero quotient; cf. [C4] and the references therein.

The w–periodicity and other definitions remain unchanged. The following reservation about the fundamental subdiagrams and their s–partitions is needed.

We assume that the fundamental subdiagrams are intersections of a given periodic diagram with sufficiently big *skew* diagrams, i.e., they are the portions of a given δ between two skew paths interesecting δ from left to right. They give some (but not all!) submodules of the $\mathcal{HH}_n$–module under consideration, provided that they are generalized.

Similarly, the partitions are portions of fundamental subdiagrams between consecutive skew paths intersecting δ "from left to right" ($r > 0$.) The s–partitions of fundamental subpairs of $\{\Delta, C\}$ describe (some but not all) submodules of the corresponding irreducible representations upon the restriction to $\mathcal{H}_n$. Here we consider the increasing partitions only. The partial ordering of the diagrams is as follows. We set $\delta > \delta'$ if and only if it holds for two *disjoint* skew diagrams containing δ and δ', respectively.

These partial results on the restriction to the affine Hecke algebra $\mathcal{H}_n$ are obviously weaker, than what can be achieved in the semisimple case, however, they are sufficient to prove that finite dimensional irreducible representations are X–semisimple. Because of the duality $X \leftrightarrow Y$, they are also Y–semisimple.

Indeed, it is clear that the irreducible X–cyclic representations of $\mathcal{HH}_n^o$ have infinitely many non-isomorphic $\mathcal{H}_n^o$–submodules, except in the case of the infinite column (recall that $r > 0$).

Comment. (i) More explicitly, it is not difficult to construct an infinite chain of X–eigenvectors with pairwise distinct eigenvalues if Δ is not a column. This argument can be applied to arbitrary root systems; however, the finite dimensional representations are not described yet in such generality.

(ii) There is a general "geometric" approach, the Vasserot theory mentioned above. It is based on the advanced K–theory of affine flag varieties. There are technical difficulties when dealing with the corresponding Springer fibers and the corresponding geometric classification is far from simple. However, we think that the K–theoretic methods are more promising for general representations, even in the GL_n–case (where the intertwiners can be used).

(iii) Our tools based on the technique of intertwiners are good enough for semisimple and spherical representations. However, they are not sufficient to describe more general representaions. In a sense, the technique of intertwiners is similar to the classical approach based on the highest vectors and the creation and annihilation e, f–operators (raising and lowering operators) in semisimple Lie algebras. The disadvantages are similar, indeed. ❑

3.8 Spherical representations

It is a continuation of Section 3.6 of this chapter. The parameters q, t, ξ are still arbitrary (i.e., nongeneric). We assume that $t_{\text{sht}} \neq \pm 1 \neq t_{\text{lng}}$ unless otherwise stated. Let us continue the list of basic definitions.

3.8.1 Spherical and cospherical modules

An $\mathcal{H}^\flat$–module V will be called X–**spherical** if it is generated by an element v, a **spherical vector**, such that

$$\begin{aligned} T_i(v) &= X_{\alpha_i}(q^{\rho_k})v = t_i v \text{ for } 1 \leq i \leq n, \\ X_a(v) &= X_a^{-1}(q^{\rho_k})v = q^{-(a,\rho_k)}v \text{ for } a \in B. \end{aligned} \tag{3.8.1}$$

It is an X–**cospherical** module if the space of linear form ϖ satisfying

$$\begin{aligned} (T_i(u)) &= t_i^{1/2}\varpi(u) \text{ for } 1 \leq i \leq n, \\ \varpi(X_a(u)) &= q^{-(a,\rho_k)}\varpi(u) \text{ for } a \in B,\ u \in V, \end{aligned} \tag{3.8.2}$$

is nonzero and at least one of the ϖ, called a **cospherical form**, does not vanish identically on each $\mathcal{H}^\flat$–submodule of V.

We will mainly call the vectors v satisfying (3.8.1) X–**invariants** and the cospherical forms X–**coinvariants**. It will be assumed that the spaces of invariants and coinvariants are finite dimensional. It is true for all modules in the category $\mathcal{O}_X$.

Note that $\mathcal{H}^\flat$–quotient of a spherical representation is spherical, and nonzero submodules of a cospherical module are cospherical.

An X–spherical module is X–cospherical and vice-versa for pseudo-unitary modules V in the category $\mathcal{O}_X$. The bi-form identifies the spaces of invariants v and coinvariants ϖ. Recall that it is nondegenerate $\star$–invariant and $*$–hermitian. One can drop the last condition in this definition.

The cospherical form is constructed as follows. If v is a given spherical vector, then the linear form (u, v) is a coinvariant and has no $\mathcal{H}^\flat$–submodules in its kernel because the latter have to be orthogonal to $\mathcal{H}\!\!\mathcal{H}^{\flat}v = V$. Therefore (u, v) is a cospherical form.

To construct an X–spherical vector from an X–cospherical form, we use that the coinvariants vanish on $V_X^{\infty}(\xi)$ (these spaces are finite dimensional) unless the X–character ξ is from the second line of (3.8.1). Indeed, they obviously vanish on $V_X(\xi)$ for such ξ, so we can go to $V_X^{\infty}(\xi)/V_X(\xi)$ and continue by induction. Therefore an arbitrary coinvariant $\varpi(u)$ can be represented as (u, v) for an invariant v. The latter has to be a spherical vector for a cospherical form ϖ. Otherwise the orthogonal complement of $\mathcal{H}\!\!\mathcal{H}^{\flat}v \neq V$ would belong to the kernel of ϖ, which is impossible. This complement is nonzero since $V \in \mathcal{O}_X$.

Concerning the coincidence of the space of invariants and coinvariants for $V \in \mathcal{O}_X$, it is not necessary to assume that the form is $\mathcal{H}\!\!\mathcal{H}^{\flat}$–invariant. It is sufficient to have a nondegenerate $*$–bilinear and $\mathcal{H}_X^{\flat}$–invariant form.

Similarly, a Y–**spherical vector**, a **cospherical form** ϖ, Y–**invariants**, and Y–**coinvariants** are defined by the relations:

$$
\begin{aligned}
&T_i(v) = t_i\, v, \ \pi_r(v) = v, \quad \text{for } \ i > 0, \ \pi_r \in \Pi^{\flat},\\
&Y_a(v) \ = \ q^{(a\,,\,\rho_k)}v \ \text{ for } \ a \in B, \\
&\varpi(T_i(u)) = t_i^{1/2}\,\varpi(u), \ \varpi(Y_a(u)) = q^{(a\,,\,\rho_k)}\varpi(u).
\end{aligned}
\tag{3.8.3}
$$

Note the different signs of $(a\,,\,\rho_k)$ for X and Y.

A module V with a one-dimensional space of invariants v has a unique nonzero spherical submodule. It is the $\mathcal{H}\!\!\mathcal{H}^{\flat}$–span of v. Dualizing, a module V with a one-dimensional space of coinvariants ϖ has a unique nonzero cospherical quotient. It is V divided by the sum of all $\mathcal{H}\!\!\mathcal{H}^{\flat}$–submodules of V inside the kernel of ϖ. If V is already (co)spherical, then it has no proper (co)spherical submodules or, respectively, quotients.

The main example is $\mathcal{I}_X[\xi]$, which has a unique Y–cospherical quotient because the space of Y–coinvariants ϖ from (3.8.3) for the module $\mathcal{I}_X[\xi]$ is one-dimensional. Cf. [C23](Lemma 6.2).

The **polynomial representation** $\mathcal{V} \stackrel{\mathbf{def}}{=\!=} \mathbb{Q}_{q,t}[X_b] = \mathbb{Q}_{q,t}[X_b, b \in B]$ is Y–spherical and maps onto an arbitrary given Y–spherical module. The constant term is its Y–coinvariant. If q, t are generic, then $\mathcal{V}$ is irreducible and automatically Y–cospherical. When q is a root of unity there are infinitely many irreducible spherical modules.

Comment. (i) We note that in the definition of spherical and cospherical modules, one may take any one-dimensional character of the affine Hecke subalgebra with respect to X or Y in place of (3.8.1), (3.8.2), and (3.8.3). The spherical representation for generic q, t remains isomorphic to the space of polynomials, however, with different formulas for the $\hat{T}$–operators depending on the choice of the character. The functional representations, which are defined via the discretization of these operators, also change. We will stick to the "plus-character" in this section, leaving a straightforward generalization to the reader.

(ii) It is worth mentioning that relations between the spherical and co-spherical representations constructed for different one-dimensional characters play an important role in the theory of $\mathcal{H}$ as well as for the classical representations of affine Hecke algebras. An example is a general theory of the shift operators (Opdam, Felder, Veselov, Macdonald, and the author). We will not consider these questions.

3.8.2 Primitive modules

The definition is of a general nature, but these modules are mainly needed for finite dimensional representations. Recall that finite dimensional $\mathcal{H}^\flat$–modules exist only when either q is a root of unity or q is generic and k is *singular*, a special rational number. By generic, we mean that the products $(q^{1/(2m)})^i(t^{1/2})^j$ do not represent nontrivial roots of unity for $i, j \in \mathbb{Z}$.

We define Y–**principal** spherical $\mathcal{H}^\flat$–modules as Y–spherical $\mathcal{H}^\flat$–modules equipped with nondegenerate $\star$–invariant $*$–bilinear forms. So these modules are also Y–cospherical. Conversely, the modules that are spherical and at the same time cospherical have to be principal, thanks to Proposition 3.3.2. In this definition, we do not assume the pairing to be $*$–hermitian. We only need it to be nondegenerate.

A Y–principal module is called Y–**primitive** if the dimension of the space of its Y–coinvariants is one. Using the nondegenerate form, we conclude that the dimension of the space of Y–invariants has to be one as well. Note that these dimensions always coincide for Y–principal spherical modules in the category $\mathcal{O}_Y$, as was discussed above (for X instead of Y).

A primitive module has no nontrivial spherical submodules and cospherical quotients. The key property of such modules, a variant of the Schur lemma, is that a nonzero $\mathcal{H}^\flat$–homomorphism between two primitive modules $V_1 \to V_2$ is an isomorphism. Indeed, the image of V_1 is spherical and therefore must coincide with V_2. Hence V_2 is a cospherical quotient of V_1, which is impossible unless the kernel is zero. We also have the following proposition.

Proposition 3.8.1. *Let V be a Y–principal module from $\mathcal{O}_Y$ embedded into a finite direct sum $V_{sum} = \oplus V_i$ of Y–primitive modules such that its intersections with all V_i are nonzero. Then V coincides with V_{sum}.*

Proof. The Y–invariants of $V_{\rm sum}$ are linear combinations $\sum c_i v_i$ of the spherical vectors $v_i \in V_i$. The space of Y–invariants of V is smaller than the space of invariants of $V_{\rm sum}$ if $V \neq V_{\rm sum}$.

Similarly, Y–coinvariants of $V_{\rm sum}$ are linear combinations $\varpi = \sum c_i \varpi_i$ of the cospherical forms ϖ_i of V_i naturally extended to the sum. Since V has nonzero intersections with V_i, all such $\varpi \neq 0$ remain nonzero when restricted to V. Indeed, $c_i \neq 0$ for certain i and if $\varpi(V) = 0$, then the cospherical form ϖ_i of V_i vanishes on $V \cap V_i$, which is impossible.

Using the coincidence of the dimensions of the spaces of invariants and coinvariants for Y–principal modules in $\mathcal{O}_Y$, we conclude that $V = V_{\text{sum}}$. ❑

Later we will see that a Y–primitive X–semisimple module is irreducible. Also, a Y–principal X–cyclic module, i.e., generated by an X–eigenvector, is Y–primitive and a Y–spherical X–cyclic irreducible module which has at least one coinvariant is primitive (the next proposition).

We note that a Y–spherical module such that $V = \oplus_\xi V_X^\infty(\xi)$ has to be finite dimensional (in particular, one can take $V \in \mathcal{O}_X$). Indeed, $V = \oplus_\xi V_Y^\infty(\xi)$ because it is a quotient of the Y–induced module with the "plus-character" as the weight. Finitely generated modules that have such decompositions for X and Y together are finite dimensional. See Section 3.6.1.

Assuming that a *finite dimensional* Y–spherical module V possesses at least one Y–coinvariant $\varpi \neq 0$, let $\tilde{V}$ be the quotient of V with respect to the intersection of all radicals of all $*$–bilinear $\star$–invariant forms. Since V is finite dimensional there exists a Y–coinvariant with the kernel precisely coinciding with the intersection of the kernels of all coinvariants. This coinvariant makes $\tilde{V}$ cospherical and therefore Y–principal. Obviously it is the universal Y–principal quotient of V.

Comment. It is worth mentioning that spherical (finite dimensional) representations in the classical theory of the affine Hecke algebra are defined with respect to its nonaffine Hecke subalgebra and always have isomorphic spaces of invariants and coinvariants provided that the latter algebra is semisimple. Therefore irreducible spherical modules are primitive in the above sense, which readily gives that (affine) primitive modules are simply irreducible quotients of the polynomial representation of the affine Hecke algebra. ❑

In the next proposition, the field of constants is assumed algebraically closed. Note that the definition of principal and primitive modules does not depend on the particular choice of the field of constants. They remain principal or primitive over any extension of $\mathbb{Q}_{q,t}$, and vice versa. If they are defined over $\mathbb{Q}_{q,t}$ and become principal/primitive over its extension, then they are already principal/primitive over $\mathbb{Q}_{q,t}$. Recall that the $*$–bilinear forms above and below are not supposed to be $*$–hermitian. By the radicals, we mean the left radicals of such forms.

Proposition 3.8.2. *A Y–spherical module $V \in \mathcal{O}_X$ is principal if and only if the dimensions of its X–eigenspaces $V(\xi)$ are no greater than one and there exists a Y–coinvariant of V, that is nonzero at all nonzero X–eigenvectors. The sum $V_{sum} = \oplus V_i$ of Y–principal modules from $\mathcal{O}_X$ is Y–principal if and* *only if*

$$Spec_X(V_i) \cap Spec_X(V_j) \;=\; \emptyset \;\; for \;\; i \neq j. \tag{3.8.4}$$

If Y–principal $V \in \mathcal{O}_X$ is generated by its X–eigenvectors, then it is a direct sum of primitive modules.

Proof. In the first place, given a Y–spherical vector $v \in V$ (by definition, it generates V), the coinvariants ϖ are in one-to-one correspondence with $\star$–invariant $*$–bilinear forms on V. Namely, any such form can be represented as $\varpi'(f_1 f_2^*)$ for $f_1, f_2 \in \mathcal{V}$, where ϖ' is the pullback of ϖ with respect to the $\mathcal{H}\!\!\!\mathcal{H}^\flat$–homomorphism $\mathcal{V} \to V$, sending 1 to v. That is, $(v_1, v_2)_\varpi = \varpi'(f_1 f_2^*)$, where

$$v_1 = f_1(v) = f_1(X)v, \;\; v_2 = f_2(v) = f_2(X)v.$$

It follows from Proposition 3.3.2.

The form may be degenerate and non-hermitian. Recall that hermitian forms (satisfying $(u, v) = (v, u)^*$) correspond to coinvariants of $\mathcal{V}$ such that $\varpi'(f^*) = \varpi'(f)^*$. We do not use hermitian forms in the proposition.

The radical of $(\;,\;)_\varpi$ is the greatest $\mathcal{H}\!\!\!\mathcal{H}^\flat$–submodule of V inside the kernel of ϖ. Hence this form is nondegenerate if and only if the coinvariant ϖ is a cospherical vector.

Let us check that in the category $\mathcal{O}_X$, a Y–coinvariant ϖ is a cospherical form if and only if

$$\varpi(e) \neq 0 \;\; \text{for all } X - \text{eigenvectors } e \in V. \tag{3.8.5}$$

These relations are obviously sufficient, since any submodule of V has at least one nonzero X–eigenvector. Let us verify that (3.8.5) is necessary. If $\varpi(e) = 0$ for such e, then $\varpi(\mathcal{H}_Y^\flat \mathcal{X}(e)) \;=\; 0 \;=\; \varpi(\mathcal{H}\!\!\!\mathcal{H}^\flat(e))$, where by $\mathcal{X}$ we mean the algebra generated by X_b for $b \in B$.

Now, if there exists a weight ξ such that $\dim V(\xi) > 1$, then V cannot be cospherical because any given coinvariant vanishes on a proper linear combination of e_1 and e_2. Thus the first claim holds as well as relations (3.8.4).

Comment. Note that the same argument proves that the dimension of the space of Y–coinvariants of an X–cyclic module is no greater than one. In particular, Y–principal X–cyclic modules are Y–primitive. Indeed, otherwise $\mathcal{H}\!\!\!\mathcal{H}^\flat v \neq V$ for an arbitrary X–eigenvector v of V because $\mathcal{H}\!\!\!\mathcal{H}^\flat v$ belongs to the kernel of a proper coinvarinat of V vanishing at v, that always exists if there are at least two non-proportional coinvariants. ❑

Coming to the last claim, for each X–eigenvector $e \neq 0$ in V, let $\mathcal{W}_e$ be the hyperplane of all coinvariants of V vanishing at e in the space $\mathcal{W}$ of all coinvariants. The codimensions of $\mathcal{W}_e$ are one because V is principal. The intersection of all $\mathcal{W}_e$ consists of the coinvariants vanishing at all X–eigenvectors of V, so it is zero because V is generated by its X–eigenvectors.

Since we have at least two different $\mathcal{W}_e$, we may pick finitely many coinvariants ϖ_i (two are sufficient for the induction step) that do not belong (all of them) to any particular $\mathcal{W}_e$ and consider the direct sum of the quotients V_i of V by the corresponding radicals. The resulting map $V \to \oplus V_i$ is injective. Indeed, for any X–eigenvector $e^o \neq 0$ in its kernel, the hyperplane $\mathcal{W}_{e^o}$ would contain all ϖ_i, which is impossible.

Applying this procedure repeatedly to the resulting factors V_i (the quotients of V are generated by X–eigenvectors as well as V) we can eventually make all V_i in the sum primitive. We may also assume that neither V_i can be removed from the sum $\oplus V_i$, i.e., that the intersections $V \cap V_i$ are all nonzero, and use the previous proposition. ❑

3.8.3 Semisimple spherical modules

Generally speaking, primitive modules can be reducible. However, in this chapter, we will mainly stick to the semisimple modules. Let us check that Y–primitive X–semisimple modules are irreducible.

If such V has an $\mathcal{H}\!\!\!\mathcal{H}^\flat$–submodule U, then its orthogonal complement $U^\perp$ with respect to the (unique) $\star$–invariant $*$–bilinear form has zero intersection with U. Here we use that the X–spectrum of Y–principal module is simple. Hence $U \oplus U^\perp = V$. However, this is impossible because the space of Y–invariants of V is one-dimensional and generates V as an $\mathcal{H}\!\!\!\mathcal{H}^\flat$–module.

Let V be an irreducible X–semisimple $\mathcal{H}\!\!\!\mathcal{H}^\flat$–module that is also Y–cospherical. Then $V = \mathcal{I}_X[\xi]/\mathcal{J}$ for proper ξ and proper $\mathcal{H}\!\!\!\mathcal{H}^\flat$–submodule $\mathcal{J}$ such that $\mathcal{J}_\xi \subset \mathcal{J} \subset \mathcal{I}_X[\xi]$ for $\mathcal{J}_\xi = \sum \mathcal{H}_Y \Phi_{\widehat{w}}$, where the summation is over $\widehat{w} \in \ddot{\Upsilon}_*[\xi]$.

It is Y–cospherical if and only if $\varpi_+(\mathcal{J}) = 0$ for the plus-character of $\mathcal{I}_X[\xi] \simeq \mathcal{H}_Y^\flat$:

$$\varpi_+(T_i H) = t_i^{1/2}\varpi_+(H),\ \varpi_+(\pi_r H) = \varpi_+(H),\ \varpi_+(1) = 1. \tag{3.8.6}$$

The condition $\varpi_+(\mathcal{J}) = 0$ readily results in $\Upsilon_*[\xi] = \Upsilon_+[\xi]$ since $\varpi_+(\Phi_{\widehat{w}}) \neq 0$ for $\widehat{w} \in \ddot{\Upsilon}_-[\xi]$. We see that the condition $\Upsilon_+[\xi] \subset \Upsilon_0[\xi]$ is fulfilled for ξ, and for any other X–weight ξ', under the assumption that V is irreducible, X–semisimple, and Y–cospherical.

In such V, the eigenspace $V(\xi)$ is one-dimensional and $S_{\widehat{w}} = 1$ in $V(\xi)$ for $\widehat{w} \in \widehat{W}^\flat[\xi]$. Indeed, $\varpi_+(S_{\widehat{w}}(1_\xi)) = 1$ in $\mathcal{H}_Y^\flat$ due to the normalization of S from (3.3.27). Here 1_ξ is 1 from $\mathcal{H}_Y^\flat$ with the action of $\{X\}$ via ξ. Therefore the same holds in any quotients of $\mathcal{H}_Y^\flat$ by submodules belonging to $\operatorname{Ker}\varpi_+$. Since $V(\xi)$ is an irreducible representation of $\widehat{W}^\flat[\xi]$, we obtain that $S_{\widehat{w}} \mapsto 1$.

Let us demonstrate that the conditions (a) $\Upsilon_+[\xi] \subset \Upsilon_0[\xi]$, (b) $\dim_{\mathbb{C}} V(\xi) = 1$, and (c) $S_{\widehat{w}} = 1$ in $V(\xi)$ whenever $\widehat{w} \in \widehat{W}^\flat[\xi]$, are sufficient to make V irreducible and Y–cospherical.

Given a weight ξ, first we need to check that

$$\mathcal{J}_\xi = \sum \mathcal{H}_Y \Phi_{\widehat{w}} \neq \mathcal{H}_Y, \text{ summed over } \ddot{\Upsilon}_+[\xi],$$

assuming that $\Phi_{\widehat{w}}$ are well defined. It is obvious since $\varpi_+(\mathcal{J}_\xi) = 0$. Moreover, the latter relation gives that $U = \mathcal{I}_X[\xi]/\mathcal{J}_\xi$ has a unique cospherical quotient. It is the quotient by the span of all $\mathcal{H}\!\!\mathcal{H}^\flat$–submodules in the kernel of ϖ_+ on U. Let us denote it by V. The restriction of ϖ_+ to the eigenspace $V(\xi)$ is nonzero; otherwise it is identically zero on the whole V. The quotient of V by the $\mathcal{H}\!\!\mathcal{H}^\flat$–span of the kernel of ϖ_+ on $V(\xi)$ is irreducible and cospherical. Hence it coincides with V due to the uniqueness of the cospherical quotients. Moreover, $\dim V(\xi) = 1$, and the action of $\widehat{W}^\flat[\xi]$ is trivial. We come to the following theorem.

Theorem 3.8.3. *i) The induced representation $\mathcal{I}_X[\xi]$ possesses a nonzero irreducible Y–cospherical X–semisimple quotient V if and only if*

$$\Upsilon_0[\xi] \supset \Upsilon_+[\xi] \subset \Upsilon_-[\xi]. \tag{3.8.7}$$

Such a quotient is unique and has one-dimensional $V(\xi)$ with the trivial ($S_{\widehat{w}}$ act as 1) action of the group $\widehat{W}^\flat[\xi]$. Its X–spectrum is simple.

This V is X–unitary if and only if $\mu_\bullet(\widehat{w}) = 1$ for all $\widehat{w} \in \widehat{W}^\flat_*[\xi]$. If such a V is finite dimensional, then it is spherical and, moreover, primitive, i.e., it has a unique nonzero Y–invariant and a unique Y–coinvariant up to proportionality. Setting $V = V_{sph}[\xi]$, the modules $V_{sph}[\xi']$ and $V_{sph}[\xi]$ are isomorphic if and only if $q^{\xi'} = q^{\widehat{w}((\xi))}$ for $\widehat{w} \in \Upsilon_+[\xi]$.

ii) Finite dimensional $V_{sph}[\xi]$ can be identified with

$$\mathfrak{F}'[\xi] \stackrel{\mathbf{def}}{=\!=} Funct(\Upsilon'_+[\xi], \mathbb{Q}_\xi) \quad where \quad \Upsilon'_+[\xi] \stackrel{\mathbf{def}}{=\!=} \Upsilon_+[\xi]/\widehat{W}^\flat_+[\xi]. \tag{3.8.8}$$

Here the action of $\mathcal{H}\!\!\mathcal{H}^\flat$ is introduced via formulas (3.4.10) and is well defined on $\mathfrak{F}'[\xi]$. The characteristic functions $\{\chi_{\widehat{w}}, \widehat{w} \in \Upsilon'_+[\xi]\}$ form a basis of $\mathfrak{F}'[\xi]$ and are permuted by the intertwiners $S_{\widehat{w}}$, which become $\widehat{w}$ in this module.

The pseudo-hermitian form of V is proportional to

$$\langle f, g\rangle' \stackrel{\mathbf{def}}{=\!=} \sum_{\widehat{w}\in\Upsilon'_+[\xi]} \mu_\bullet(\widehat{w}) f(\widehat{w})\, g(\widehat{w})^*. \tag{3.8.9}$$

Proof. The "cospherical" part of (i) has been already checked. The spectrum of V is simple, thanks to Theorem 3.6.1. The condition (3.6.21), which is necessary and sufficient for the existence of an invariant X–hermitian form on V, means that $\mu_\bullet(\widehat{w}) = 1$ for all $\widehat{w} \in \widehat{W}^\flat_*[\xi]$. See also (3.6.32).

Since the space of Y–coinvariants of $V_{sph}[\xi]$ is one-dimensional, then any $\star$–invariant $*$–bilinear form on V is hermitian (skew-symmetric) up to proportionality. This is because V has the "real" structure by the construction; we assume that $*$ fixes q^{ξ}.

Note that any pseudo-hermitian form on V is X–definite. The scalar squares of X–eigenvectors are nonzero because the X–eigenspaces are one-dimensional. Indeed, if one of these squares vanishes, then the radical of the form is a nonzero $\mathcal{H}^{\flat}$–submodule of V, which contradicts its irreducibility.

If such a V is also finite dimensional, then it is spherical and principal. The space of spherical vectors of V is one-dimensional because so is the space of ϖ. See Proposition 3.8.2.

By the way, we obtain a surjective $\mathcal{H}^{\flat}$–homomorphism $\mathcal{V} \to V$ sending 1 to the spherical generator 1_{sph} of V. Its kernel has to be an intersection of maximal ideals of $\mathcal{V}$. This argument gives another demonstration of the simplicity of the X–spectrum of V.

Part (ii) is nothing but a reformulation of (i). Really, we may introduce (the action of) s_i and $\widehat{w}$ in $V_{sph}[\xi]$ as S_i and $S_{\widehat{w}}$. Then the formulas from (ii) will follow from (i). ❑

As an immediate application of the proposition, we obtain that the irreducible representations $V[\xi, \varrho]$ from Proposition 3.6.4 cannot be, generally speaking, principal (spherical and cospherical). They are cospherical if and only if $\varrho = 1$. For such ϱ, $\mu_{\bullet}(a)$ have to be 1 on $A(N)$ to provide the existence of the X–unitary structure. Formulas (3.6.30) combined with the constraint (3.6.31) indicate that it may happen only for very special values of the parameters.

3.8.4 Spherical modules at roots of unity

Let us give a general description of spherical modules as q is a primitive root of unity of order $N \geq 0$.

We will use the lattices $A(N) = (N \cdot A) \cap B$:

$$A = \frac{m'}{2\tilde{m}} \{a \in P^{\vee} \mid (a, B) \subset \mathbb{Z}\},\ m' \text{ divides } 2\tilde{m}. \tag{3.8.10}$$

Here $q' = q^{1/(2\tilde{m})}$ is a primitive root of order $m'N$. See formula (3.6.28) and the remark after it concerning m' and q'.

If $(\tilde{m}, N) = 1$, then $A(N) = N(A \cap B)$. Recall that $Q \subset A = Q^{\vee} \subset P^{\vee}$ for $B = P$. Also, $A = P^{\vee}$ for $B = Q$.

To start with, let us assume that t_{ν} are generic and establish that the quotient $\widehat{\mathcal{V}}$ of $\mathcal{V}$ by the radical of the standard form of the polynomial representation is well defined, Y–spherical, Y–cospherical, Y–semisimple, X–semisimple, and irreducible.

In the first place, all polynomials $\mathcal{E}_b$ are well defined, thanks to the intertwining formulas. See (3.3.45). Then $e_b = E_b(q^{-\rho_k})\mathcal{E}_b$ are well defined because $E_b(q^{-\rho_k}) \neq 0$ due to (3.3.16). The polynomials E_b $(b \in B)$ form a basis of $\mathcal{V}$; so do $\mathcal{E}_b$, and the formulas (3.4.1) for $\langle \mathcal{E}_c, \mathcal{E}_c \rangle_\circ$, which were obtained for generic q, hold in the considered case as well, i.e., define the form $\langle\, , \rangle_\circ$ on $\mathcal{V}$. Its radical $\mathcal{V}_o = \mathrm{Rad}\langle\, , \rangle_\circ$ can be readily calculated.

Proposition 3.8.4. *(i) For generic t_ν, the $\mathcal{H}^\flat$–module $\widehat{\mathcal{V}} = \mathcal{V}/\mathcal{V}_0$ is well defined and*

$$\widehat{\mathcal{V}} \simeq \oplus_b \mathbb{Q}_{q,t}\mathcal{E}_b \ \ as \ -(\alpha_i^\vee, b_-) \le N_i \stackrel{\mathbf{def}}{=} N/(N, \nu_i),\ 1 \le i \le n, \qquad (3.8.11)$$
$$\{-(\alpha_i^\vee, b_-) = N_i\} \Rightarrow \{u_b^{-1}(\alpha_i) \in R_- \ \ i.e.,\ \alpha_i \in \lambda(u_b^{-1})\}.$$

The other $\mathcal{E}_b$ lineraly generate Rad.

(ii) Given b_- from (3.8.11), let W' and $W^\bullet$ be subgroups of W generated, respectively, by s_i such that $(\alpha_i^\vee, b_-) = 0$ and $-(\alpha_i^\vee, b_-) = N_i$. Their longest elements will be denoted by w_0' and $w_0^\bullet$. Then all possible u_b^{-1} from (3.8.11) are represented in the form

$$w\,w_0^\bullet \ \ for \ \ w \in W \ \ such \ \ that\ l(w\,w_0^\bullet\,w_0') - l(w) + l(w_0^\bullet) + l(w_0').$$

The Y–spectrum of $\widehat{\mathcal{V}}$ is simple, and this module is irreducible.

Proof. The conditions (3.8.11) are necessary and sufficient to ensure that $\langle \mathcal{E}_b, \mathcal{E}_b \rangle_\circ \neq 0$.

Given such $b_- \in B_-$, formula (3.1.13) provides the following general description of all possible u_b:

$$\alpha \in \lambda(u^{-1}) \Rightarrow (\alpha, b_-) \neq 0. \qquad (3.8.12)$$

Equivalently, $l(u^{-1}w_0') = l(u^{-1}) + l(w_0')$ for the longest element w_0' in the centralizer W_0' of b_- in W.

If $-(\alpha_i, b_-) = N_i$, then $\alpha_i \in \lambda(u_b^{-1})$ for such i, i.e., u_b^{-1} is divisible by $w_0^\bullet$ in the following exact sense: $l(u_b^{-1}) = l(u_b^{-1}\, w_0^\bullet) + l(w_0^\bullet)$. Here $w_0^\bullet \in W_0^\bullet$ is the longest element. This gives the description of u_b from the proposition.

Since $\{k_\nu\}$ is generic, $\mathcal{E}_b$ and $\mathcal{E}_c$ have coinciding $q^{b_\sharp}$ and $q^{c_\sharp}$ if and only if $b_- = c_-$ mod $A(N)$ and $u_b = u_c$. Let us check this.

Recall that $A \subset B \subset P$. So $b_- - c_- \in N \cdot P$ in this case. Provided (3.8.11) for b and c, we obtain that $b_- = c_- + \sum \pm N_j\,\omega_j$ for some indices j and proper signs. Here $(b_-, \alpha_j^\vee) = N_j$ and $(c_-, \alpha_j^\vee) = 0$ for j such that the term $+N_j\omega_j$ appears in the sum, and vice versa for $-N_j\omega_j$.

Let us assume that $u_b = u = u_c$. Transposing b_- and c_-, we can find j with the plus–sign of $N_j\omega_j$ in this sum. Then $u^{-1} = u_b$ is divisible by s_j. However, $l(u^{-1}s_j) = l(u^{-1}) + 1$ because $u = u_c$. This contradiction gives that u_b and u_c cannot coincide for different b, c. Hence, the spectrum $\mathrm{Spec}_Y(\widehat{\mathcal{V}})$ is simple.

Since $\widehat{\mathcal{V}}$ is spherical, i.e., generated by $\mathcal{E}_0 = 1$, and also Y–pseudo-unitary, it must be irreducible. ❑

The next theorem gives a different (and more general) construction of spherical representations involving the central characters.

Theorem 3.8.5. *(i) For generic k_ν, the center $\mathcal{Z}$ of $\mathcal{H}\!\!\!\mathcal{H}^\flat$ contains the finite sums*

$$H = \sum_{a\in A(N)} F_a\, S_a \text{ such that } F_a \in \mathbb{Q}_{q,t}[X_b], \tag{3.8.13}$$

$$b(F_a) = F_a \text{ for } b \in A(N),\ F_{w(a)} = F_a \text{ for } w \in W, \text{ and}$$

$$F_a \cdot \prod_{[\alpha,j]} (t_\alpha^{1/2} q_\alpha^j X_\alpha - t_\alpha^{-1/2})^{-1} \in \mathbb{Q}_{q,t}[X_b],\ [\alpha, \nu_\alpha j] \in \lambda(a).$$

(ii) The algebra $\mathcal{Z}_X \in \mathbb{Q}_{q,t}[X_b]$ consisting of $W\ltimes A(N)$–invariant polynomials is central as well as the algebra $\mathcal{Z}_Y$ of $W\ltimes A(N)$–invariant polynomials in terms of Y. The algebra $\mathcal{H}\!\!\!\mathcal{H}^\flat$ is a free module of dimension $|W|\,|\,W^\flat/A(N)\,|^2$ over $\mathcal{Z}_X \cdot \mathcal{Z}_Y$.

The elements (3.8.13) for rational $F_a \in \mathbb{Q}_{q,t}(X_b)$ constitute the center $\mathcal{Z}_{loc}$ of the localization $\mathcal{H}\!\!\!\mathcal{H}^\flat_{loc}$ of $\mathcal{H}\!\!\!\mathcal{H}^\flat$ by nonzero elements of $\mathcal{Z}_X$. The algebra $\mathcal{H}\!\!\!\mathcal{H}^\flat_{loc}$ is of dimension $|\,W^\flat/A(N)\,|^2$ over $\mathcal{Z}_{loc}$.

(iii) Given a central X–character, a homomorphism of algebras $\varsigma\colon \mathcal{Z}_X \to \mathbb{Q}_{q,t}$, let us denote its kernel by $Ker\varsigma$. Then

$$\mathcal{V}_\varsigma \overset{\text{def}}{=\!=} \mathcal{V}\,/\,(\mathcal{V}\cdot Ker\,\varsigma) \tag{3.8.14}$$

is a Y–spherical module of dimension $|\,W^\flat/A(N)\,|$. It is X–semisimple if and only if $\mathcal{V}\cdot Ker\varsigma$ is an intersection of maximal ideals of $\mathcal{V}$.

(iv) The module $\mathcal{V}_\varsigma$ is irreducible for generic ς. If it is irreducible, then the space $\mathcal{V}_\varsigma^\circ = Hom(\mathcal{V}_\varsigma, \mathbb{Q}_\xi)$ is isomorphic to $V[\xi, 1]$ from Proposition 3.6.4, where $Ker\varsigma \in Ker\xi$. The action of $\mathcal{H}\!\!\!\mathcal{H}^\flat$ on $\mathcal{V}_\varsigma^\circ$ is via the $\mathbb{Q}_{q,t}$–linear X–trivial anti-involution $H \mapsto H^\circ$ fixing $T_i\ (0 \le i \le n)$, fixing $X_b\ (b \in B)$, and sending $\pi_r \mapsto \pi_r^{-1}$.

Proof. Upon X–localization, the defining property of the intertwiners $S_{\widehat{w}}$ readily gives that the elements (3.8.13) for rational $F_a \in \mathbb{Q}_{q,t}(X_b)$ form the center of the localization $\mathcal{H}\!\!\!\mathcal{H}^\flat_{\text{loc}}$. Any element of the latter can be represented in this form with the summation over all $\widehat{w} \in \widehat{W}^\flat$ and arbitrary rational coefficients. Conjugating such sums by X_b and $S_{\widehat{w}}$, we get (3.8.13). The integrality condition is sufficient (but not necessary) to go back to $\mathcal{H}\!\!\!\mathcal{H}^\flat$.

We readily obtain that $\mathcal{Z}_X$ is central. The automorphisms of $PGL_2^c(\mathbb{Z})$ preserve the center of $\mathcal{H}\!\!\!\mathcal{H}^\flat$. In particular, $\mathcal{Z}_Y$ is central. The calculation of the rank of this algebra over $\mathcal{Z}_X \cdot \mathcal{Z}_Y$ is a direct corollary of the PBW theorem (see (3.2.25)).

We leave the calculation of the rank of $\mathcal{H}\!\!\mathcal{H}^\flat_{\mathrm{loc}}$ over $\mathcal{Z}_{\mathrm{loc}}$ to the reader; it will not be used in the book. Note that the calculation of the rank of $\mathcal{H}\!\!\mathcal{H}^\flat$ over the center without the X–localization is a difficult problem even for $q = 1 = t_{\mathrm{sht}} = t_{\mathrm{lng}}$.

Switching to $\mathcal{V}_\zeta$, its dimension is $|\,W^\flat/A(N)\,|$ because it is a quotient of $\mathcal{H}\!\!\mathcal{H}^\flat$ by the left ideal generated by $\mathrm{Ker}\,\zeta$ and $\mathrm{Ker}\,\varpi_+ \subset \mathcal{H}^\flat_Y$ for ϖ_+ from (3.8.6).

The X–semisimplicity of $\mathcal{V}_\zeta$ readily implies that the X–spectrum of this module is simple and ζ is associated with ξ. The module $\mathcal{V}^\circ_\zeta = \mathrm{Hom}(\mathcal{V}_\zeta, \mathbb{Q}_\xi)$ is Y–cospherical and X–semisimple with the same spectrum. The action of $\mathcal{H}\!\!\mathcal{H}^\flat$ via $H \mapsto H^\circ$ is well defined because this anti-involution preserves ζ. This module is irreducible if and only if $\mathcal{V}_\zeta$ is irreducible. We can apply Theorem 3.8.3 in this case. It results in $\mathcal{V}^\circ_\zeta = V[\xi, 1]$. ❑

An immediate application of this theorem is that the central elements are given by formulas that are uniform when represented in terms of the N–th powers of X and S. Let us formulate it letting

$$(N, \tilde{m}) = 1 = (N, \nu_\alpha),\ \ \alpha \in R_+,$$

for the sake of simplicity. Then $A(N) = N(A \cap B)$ and $N_\alpha = N/(N, \nu_\alpha) = N$.

Note that

$$X_b^N S_a = S_a X_b^N,\ \ X_b S_a^N = S_a^N X_b$$

for $a \in A \cap B$, $b \in B$.

We follow the calculation performed when checking formula (3.6.30).

Corollary 3.8.6. *(i) For $a \in A \cap B_+$ and $b \in B$, the elements*

$$Z_{a,b} = \sum_{w \in W} S^N_{w(a)} X^N_{w(b)} = \sum_{w \in W} X^N_{w(b)} S^N_{w(a)}$$

linearly generate $\mathcal{Z}_{loc}$ over the field of fractions of $\mathcal{Z}_X$.

(ii) For the same a and b, the following elements belong to the center $\mathcal{Z} \subset \mathcal{H}\!\!\mathcal{H}^\flat$:

$$\sum_{w \in W} S^N_{w(a)}\, w(X_b^N \prod_{\alpha \in R_+} (t_\alpha^{N/2} X_\alpha^N - t_\alpha^{-N/2})^{(a,\alpha^\vee)}).$$

(iii) The action of $Z_{a,b}$ in the polynomial representation $\mathcal{V}$ is well defined and is multiplication by the symmetric monomial function $\sum_{w \in W} X^N_{w(b)}$. The symmetric monomial functions in terms of Y^N act in $\mathcal{V}$ as constants. ❑

Comment. (i) The centers of the counterpart of $\mathcal{H}\!\!\mathcal{H}$ for $C^\vee C_1$ at roots of unity were recently calculated by Oblomkov. The answer is N–uniform in the above sense and is given in terms of cubic surfaces.

(ii) Similar to the theorem, one can describe the algebra of the elements of $\mathcal{HH}^\flat$ commuting with the *nonaffine* Hecke subalgebra upon a proper localization. Without the localization, it is a difficult problem even apart of roots of unity. This is connected with the results on the structure of the algebra of W–invariant polynomial differential operators due to Wallach, Levasseur, Stafford, and Joseph.

3.9 Induced and cospherical

In this section, we will figure out when the induced representations for generic q are irreducible and/or cospherical following the paper[C23]. These two problems are natural starting points in the classical theory of affine Hecke algebras. We assume that the products $(q^{1/(2m)})^i(t^{1/2})^j$ for $i,j \in \mathbb{Z}$ do not represent nontrivial roots of unity.

3.9.1 Notation

Let q and $\{t_\nu\}$ be nonzero complex numbers and $\xi \in \mathbb{C}^n$. We continue using notation from (3.4.4), that is,

$$X_{\tilde{a}}(bw) \;=\; q^{\zeta}X_a(q^{b+w(\xi)}) = q^{\zeta+(a,b)}X_{w^{-1}(a)}(q^{\xi}),$$
$$\text{where } \tilde{a}=[a,\zeta],\ a\in B. \tag{3.9.1}$$

The corresponding weight, a $\mathbb{C}$–valued character of $\mathbb{C}[X]$, will be denoted by ξ_{bw}. It can be naturally identified with the characteristic function χ_{bw} introduced in Section 3.4 of this chapter. We set $\hat{u}(\xi_{\widehat{w}}) \stackrel{\text{def}}{=\!=} \xi_{\hat{u}\widehat{w}}$ for $\hat{u},\widehat{w} \in \widehat{W}^\flat$. So such weights constitute the $\widehat{W}^\flat$–orbit of $\xi_1 = \xi$. This action is just the *affine* action of $\widehat{W}$ on the weights from the previous sections. Note that the weights $\xi_{\hat{u}}$, $\xi_{\widehat{w}}$ with different indices $\hat{u},\widehat{w}$ can coincide as characters of $\mathbb{C}[X]$ for certain q,ξ; *we will identify them in this case.*

From now on, q is not a root of unity. This hypothesis is necessary and sufficient to make the stabilizers from (3.6.4)

$$\widehat{W}^\flat_\xi(\hat{u}) \stackrel{\text{def}}{=\!=} \widehat{W}^\flat[\xi_{\hat{u}}] = \{\widehat{w}\in\widehat{W}^\flat,\ \xi_{\widehat{w}\hat{u}} = \xi_{\hat{u}}\}$$

finite for all $\hat{u}\in\widehat{W}^\flat$.

We also assume in this section that there exists a **primitive** weight $\xi_o = \xi_{\hat{u}_o}$ such that $W^o \stackrel{\text{def}}{=\!=} \widehat{W}^\flat_\xi(\hat{u}_o)$ is generated by the elements from $S^o \stackrel{\text{def}}{=\!=} W^o \cap \{s_0,\ldots,s_n\}$. This means that W^o is the Weyl group of the nonaffine Dynkin diagram $\Gamma^o \subset \tilde{\Gamma}$ (not necessarily connected) with S^o as the set of vertices. The existence of the primitive weight in the $\widehat{W}^\flat$–orbit of ξ always holds true for sufficiently general q,t, say, for the degenerate double affine Hecke algebras from Chapter 1. We denote the **nonaffine Hecke algebra** corresponding to Γ^o by $\mathbf{H}^o$. Adding $\{X_{\alpha_i}$ as $s_i \in S^o\}$ to $\mathbf{H}^o$, one obtains the corresponding affine Hecke algebra $\mathcal{H}_X^o$, which will be considered as a subalgebra of $\mathcal{HH}^\flat$.

Recall that the space of finitely supported functions on $\widehat{W}^\flat$ has a natural structure of an $\mathcal{HH}^\flat$–module defined by formulas (3.4.10) in the basis of χ–functions. We call it the functional representation and denote it by the $\mathbf{F}[\xi]$. It is well defined if the following holds:

$$X_\alpha(q^\xi)q_\alpha^j \neq 1 \quad \text{for} \quad \text{all } \alpha \in R, j \in \mathbb{Z}. \tag{3.9.2}$$

These conditions automatically imply that all $\widehat{W}_\xi^\flat(\hat{u})$ are trivial and all $\xi_{\widehat{w}}$ are pairwise different for different indices and, moreover, linearly independent.

Note that $\mathbf{F}[\xi]$ is isomorphic to $\mathbf{F}[\xi']$ for any weight $\xi' = \xi_{\widehat{w}}$. The corresponding isomorphism is the right multiplication by $\widehat{w}$ $(\xi_{\hat{u}} \to \xi_{\hat{u}\widehat{w}})$.

Similarly, one can introduce the **delta–representation** $\Delta[\xi]$ linearly generated by $\delta_{\widehat{w}}$ with the action of $\mathcal{HH}^\flat$ via (3.4.11). We need to impose the same condition (3.9.2). The delta-representations for $\widehat{W}^\flat$–conjugated weights are isomorphic too.

There is a natural nondegenerate pairing between $\mathbf{F}[\xi]$ and $\Delta[\xi]$:

$$\begin{aligned} \{g, \delta_{\widehat{w}}\} &\stackrel{\mathbf{def}}{=} g(\widehat{w}),\ \{H(g), \delta_{\widehat{w}}\} = \{g, H^\diamond(\delta_{\widehat{w}})\}, \\ &\text{for } \ H \in \mathcal{HH}^\flat,\ g \in \mathbf{F}[\xi],\ \widehat{w} \in \widehat{W}^\flat \end{aligned} \tag{3.9.3}$$

for the following X–**trivial anti-involution** of $\mathcal{HH}^\flat$:

$$T_j^\diamond = T_j (0 \le j \le n),\ \pi_r^\diamond = \pi_r^{-1} (r \in O),\ X_i^\diamond = X_i (0 \le i \le n), \tag{3.9.4}$$

sending q, t to q, t (and AB to $B^\diamond A^\diamond$). It also gives a nondegenerate pairing between $\Delta[\xi]$ and the polynomial representation $\mathbb{C}[X] \ni g$. Recall that the functional representation is the discretization of the polynomial representation.

Generally speaking, $\mathbf{F}[\xi]$ and $\Delta[\xi]$ are not isomorphic. For instance, let us consider an important specialization $\xi = -\rho_k$ assuming that t is generic. Then the module Δ_ξ has a unique nonzero irreducible submodule $\Delta_\sharp$ that is isomorphic to the unique irreducible quotient $\mathbf{F}_\sharp$ of $F[\xi]$ from (3.4.9).

Recall that an $\mathcal{HH}^\flat$–module $\mathcal{I}$ is called Y–cospherical if there exists at least one Y–cospherical form (the coinvariant) that does not vanish identically on every $\mathcal{HH}^\flat$–submodule $V \neq \{0\}$ of $\mathcal{I}$. We will consider only Y–cospherical modules in this section, so Y will be dropped. The module Δ_ξ is cospherical. Use the pairing $\{g, \delta_{\widehat{w}}\} = g(\widehat{w})$ with the polynomial representation, which is spherical.

By Y–cospherical forms, we mean the linear functions ϖ such that

$$\varpi(T_j(v)) = t_j^{1/2} v,\ \varpi(\pi_r(v)) = v \ \text{ for } \ 0 \le j \le n,\ r \in O,\ v \in \mathcal{I}. \tag{3.9.5}$$

We will consider only the modules where the space of such forms $\{\varpi\}$ is of dimension no greater than one, adding this to the definition of cospherical modules in this section.

3.9.2 When are induced cospherical?

Let us figure out when the induced representations (X–induced, to be more precise) are irreducible and/or cospherical. We use the simplified notation $\mathcal{I}_\xi$ instead of $\mathcal{I}_X[\xi]$ for the induced representation of weight ξ, which is the universal $\mathcal{HH}^\flat$–module generated by the element v_ξ such that $X_a(v_\xi) = q^{\xi_a} v_\xi$ for all $a \in B$. As an $\mathcal{H}_Y^\flat$–module, it is isomorphic to $\mathcal{H}_Y^\flat$ with the left regular action (v_ξ is identified with $1 \in \mathcal{H}_Y$). Hence, $\mathcal{I}_\xi$ posses only one $\mathcal{H}_Y^\flat$–invariant ϖ up to proportionality. It sends $T_j \to t_j^{1/2}, \pi_r \to 1$ upon identification $v_\xi = 1$.

Theorem 3.9.1. *(a) The module $\mathcal{I}_\xi$ is irreducible if and only if*

$$X_{\tilde{a}}(q^\xi) \neq t_\alpha^{\pm 1} \quad \textit{for} \quad \textit{all } \tilde{\alpha} = [\alpha, \nu_\alpha j] \in \tilde{R}_+. \tag{3.9.6}$$

Irreducible $\mathcal{I}_\xi$ are isomorphic for the weights ξ from the same $\widehat{W}^\flat$–orbit (and only for such weights).

(b) The module $\mathcal{I}_\xi$ is cospherical if and only if

$$X_{\tilde{a}}(q^\xi) \neq t_\alpha^{-1} \quad \textit{for} \quad \textit{all } \tilde{\alpha} \in \tilde{R}_+. \tag{3.9.7}$$

If (3.9.7) holds, then the induced modules associated with the weights in the form of $\xi_{\hat{w}}$ and satisfying the same inequalities are isomorphic to $\mathcal{I}_\xi$.

(c) Under the same conditions, $\mathcal{I}_\xi$ contains a unique nonzero irreducible submodule U_ξ. It is cospherical, i.e., it has a nonzero coinvariant ϖ. Any other irreducible constituents of $\mathcal{I}_\xi$ are not cospherical. An arbitrary irreducible cospherical module possessing an eigenvector of weight $\xi_{\hat{u}}$ from the orbit $\widehat{W}^\flat(\xi)$ is isomorphic to U_ξ.

Proof. The statements are parallel to the corresponding affine ones. The first one is essentially due to Bernstein–Zelevinsky (the GL–case), the second is from [Kat]. There were previous papers on spherical Hecke modules and related p–adic theory by Muller, Casselman, Steinberg, and others. In the affine theory, both statements result from the irreducibility of the induced modules with the simplest weight trivial on the root lattice. There is a simple proof of this fact in [C10](Lemma 2.12), which uses the reduction to the degenerate (graded) affine Hecke algebra. Note that in the affine case, the definitions of spherical and cospherical modules coincide.

Let us first renormalize the intertwining operators Φ from (3.3.27) to avoid the denominators:

$$\begin{aligned} \tilde{\Phi}_j &= T_j(X_{\alpha_j} - 1) + (t_j^{1/2} - t_j^{-1/2}),\ 0 \leq j \leq n, \\ \tilde{\Phi}_j^{\hat{u}} = \tilde{\Phi}_j(\hat{u}) &= T_j(X_{\alpha_j}(\hat{u}) - 1) + (t_j^{1/2} - t_j^{-1/2}),\ \hat{u} \in \widehat{W}^\flat. \end{aligned} \tag{3.9.8}$$

The corresponding $\tilde{\Phi}_{\widehat{w}}$ are well defined and enjoy the main property of the intertwiners (3.3.29). Recall that there are also the elements π_r for $r \in O$ in the product formulas for $\tilde{\Phi}$.

The multiplication on the right by the element $\tilde{\Phi}_{\hat{u}}(\widehat{w})$ (which belongs to $\mathcal{H}_Y^\flat$) is an $\mathcal{H}\!\!\mathcal{H}^\flat$–homomorphism from the module $\mathcal{I}_{\widehat{w}} \simeq \mathcal{H}_Y^\flat$ into the module $\mathcal{I}_{\widehat{w}\hat{u}}$. Here $\mathcal{I}_{\widehat{w}}$ is the induced representation corresponding to the weight (character) $X_a \to X_a(\widehat{w}^{-1}) = X_a(q^{\xi_{\widehat{w}^{-1}}})$. Similarly, if v is an X–eigenvector corresponding to $\xi_{\hat{u}}$, then $\tilde{\Phi}^{\hat{u}}_{\widehat{w}}(v)$ is that associated with $\xi_{\widehat{w}\hat{u}}$.

We will use the spaces of generalized eigenvectors (3.6.3) in $\mathcal{I}_\xi$:

$$\begin{aligned} \mathcal{I}_\xi^s(\xi_{\hat{u}}) &= \{v \in \mathcal{I}_\xi,\ (X_i - X_i(\hat{u}))^s(v) = 0 \ \text{ for } \ s \in \mathbb{N}\}, \\ \mathcal{I}_\xi^\infty(\xi_{\hat{u}}) &= \mathcal{I}_\xi^s(\hat{u}) \text{ as } s \to \infty,\ \mathcal{I}_\xi^s(\hat{u}) \stackrel{\mathbf{def}}{=\!=} \mathcal{I}_\xi^s(\xi_{\hat{u}}). \end{aligned} \tag{3.9.9}$$

The spaces $\{\mathcal{I}_\xi^\infty(\hat{u})\}$ are finite dimensional and $\mathcal{I}_\xi = \oplus \mathcal{I}_\xi^\infty(\hat{u})$ for pairwise different weights $\xi_{\hat{u}}$. Recall that they are identified if they induce the same characters.

Generally, $\mathcal{H}\!\!\mathcal{H}^\flat$–modules $\mathcal{I}$ such that $\mathcal{I}^\infty(\mu)$ are finite dimensional and $\mathcal{I} = \oplus \mathcal{I}^\infty(\mu)$ for all characters μ of $\mathbb{C}[x]$ constitute the category $\mathcal{O}_X$. All irreducible modules possessing an X–eigenvector belong to this category. If $\mathcal{I}^\infty(\mu) \neq \{0\}$, then $\mathcal{I}^1(\mu) \neq \{0\}$.

The immediate application of the generalized eigenvectors is that the set of all weights associated with X–eigenvectors of $\mathcal{I}_\xi$ for any ξ is exactly the orbit $\{\xi_{\widehat{w}}\}$. Indeed, if sufficiently general ξ' tends to ξ, then $\mathcal{I}_\xi^\infty(\xi_{\hat{u}})$ is the image of the direct sum of $\mathcal{I}_{\xi'}^\infty(\xi_{\hat{u}'})$ over all $\xi_{\hat{u}'} \to \xi_{\hat{u}}$.

If $\hat{u}_o$ is primitive, the weight $\xi_{\hat{u}_o}$, to be more precise, then

$$\mathcal{I}_\xi^\infty(\hat{u}_o) = \mathbf{H}^o \subset \mathcal{H}_Y^\flat. \tag{3.9.10}$$

This space is the induced $\mathcal{H}_X^o$–module corresponding to the trivial character $\{X_{\alpha_i} \to 1,\ s_i \in S^o\}$. Moreover, it is irreducible and contains a unique X–eigenvector that is, $1 \in \mathbf{H}^o$ (Lemma 2.12 from [C10]). Using this, we see that the dimension of the eigenspace $\mathcal{I}_\xi^1(\hat{u}_o)$ for primitive $\hat{u}_o$ is no greater than one, as it is for any submodules and subquotients (called **constituents**) of $\mathcal{I}_\xi$.

Let us check (a). If $X_{\tilde{\alpha}}(q^\xi) = t_\alpha^{\pm 1}$ for some $\tilde{\alpha}$, then there exists $\widehat{w} \in \widehat{W}^\flat$ such that the operator of the right multiplication by $\Phi_{\widehat{w}}(1)$ has a nontrivial kernel. Hence $\mathcal{I}_1 = \mathcal{I}_\xi$ is reducible. Let us assume that the inequalities (3.9.6) hold true.

All the elements $\tilde{\Phi}_j^{\hat{u}} \in \mathcal{H}_Y^\flat$ are invertible. Applying them to v_ξ, we can obtain an eigenvector v_o corresponding to $\xi_{\hat{u}_o}$. The latter generates the whole $\mathcal{I}_\xi$, since we can go back to v_ξ using $(\tilde{\Phi}_j^{\hat{u}})^{-1}$ in the opposite order. Hence we can assume that $\hat{u}_o = 1$.

If $\mathcal{I}_\xi$ contains an $\mathcal{H}\!\!\mathcal{H}^\flat$–submodule $V \neq \{0\}$, then there exists at least one X–eigenvector $v \in V$ (any vectors from $\mathcal{I}_\xi$ belong to finite dimensional X–invariant subspaces). A proper chain of the intertwiners applied to v produces a nonzero eigenvector v'_ξ corresponding to $\xi_1 = \xi$. It is proportional to v_ξ, since it has to belong to $\mathbf{H}^o$ (see above). Hence v generates $\mathcal{I}_\xi$ and $V = \mathcal{I}_\xi$.

Now, the "only if" statement from (b). If $X_{\tilde{a}}(q^\xi) = t_\alpha^{-1}$ for a certain $\tilde{\alpha} \in \tilde{R}_+$, then, using a chain of invertible intertwiners, we can replace v_ξ by an eigenvector $v \in \mathcal{I}_\xi$ associated to $\xi_{\widehat{w}}$ such that $X_j(\widehat{w}) = t_j^{-1}$ for some index $0 \le j \le n$. Therefore we can assume that $\xi_j = t_j^{-1}$, which implies that $\varpi = 0$ on the nonzero $\mathcal{H}^\flat$–submodule

$$\mathcal{I}_\xi \tilde{\Phi}_j^1 \;=\; (t_j^{1/2} - t_j^{-1/2})\mathcal{H}_Y^\flat(1 - t_j^{-1/2}T_j),$$

so it is not cospherical.

Let us suppose that ξ will satisfy (3.9.7). Applying the chains of intertwiners corresponding to the reduced decompositions of elements $\widehat{w}$ to $1 = v_\xi$, we can obtain nonzero eigenvectors corresponding to all $\xi_{\widehat{w}}$. Moreover, ϖ is nonzero at all of them, that can be obtained by this construction. Let us check this.

One has that $X_{\alpha_j}(\hat{u}) = X_{\hat{u}^{-1}(\alpha_j)}(q^\xi) \neq t_j^{-1}$ in all the factors

$$\tilde{\Phi}_j^{\hat{u}} \;=\; T_j(X_{\alpha_j}(\hat{u}) - 1) + (t_j^{1/2} - t_j^{-1/2})$$

that may appear in the chains, since the roots $\hat{u}^{-1}(\alpha_j)$ are positive there. However, ϖ vanishes after the application of such $\tilde{\Phi}_j^{\hat{u}}$ only for such values. It gives the desired fact.

Now, we can always pick a primitive weight $\xi_o = \xi_{\hat{u}_o}$ satisfying the inequalities (3.9.7). It will be called **plus-primitive,** as are the corresponding eigenvectors and $\hat{u}_o$.

The chain of intertwiners used above gives more if we start with $\xi = \xi_o$. Since the eigenvector $1 \in \mathcal{I}_{\xi_o}$ is of multiplicity one in $\mathbf{H}^o \;=\; \mathcal{I}_{\xi_o}^\infty(\xi_1)$, then

$$\tilde{\Phi}_{\widehat{w}}(\mathcal{I}_{\xi_o}^\infty(\xi_1)) \;=\; \mathcal{I}_{\xi_o}^\infty(\xi_{\widehat{w}}).$$

Indeed, the dimensions are the same and the image of 1 is nonzero, since it belongs to any X–submodules of $\mathcal{I}_{\xi_o}^\infty(\xi_1)$. Actually, this argument may be applied to any ξ (not only plus-primitive) if we know that the corresponding eigenvector is simple.

Thus, all eigenvectors of $\mathcal{I}_{\xi_o}$ are exactly the $\tilde{\Phi}$–images of 1, in particular, they are simple and ϖ is nonzero at them. If $\mathcal{I}_{\xi_o}$ contains a submodule where ϖ vanishes, then the latter possesses at least one eigenvector. However, it is impossible and $\mathcal{I}_{\xi_o}$ is cospherical.

To go from $\mathcal{I}_{\xi_o}$ to $\mathcal{I}_\xi$, we need the following general lemma, where all the modules are from the category $\mathcal{O}$ (we will apply it to subquotients of induced representations). Actually, it was already used before.

Lemma 3.9.2. *Any submodule V of cospherical module $\mathcal{I}$ is cospherical. If here $V \neq \{0\}$, then $\mathcal{I}/V$ is not cospherical. There exists a unique irreducible* *nonzero submodule $U \subset \mathcal{I}$. A module with at least one Y–cospherical form $\varpi \neq 0$ possesses a unique nonzero cospherical quotient.*

Proof. The first and second claims readily follow from the definition. If there are two irreducible submodules $U, U' \subset \mathcal{I}$, then cospherical U is contained in the cospherical $\hat{U} = U \oplus U' \subset \mathcal{I}$. Hence $\hat{U}/U \simeq U'$ could not be cospherical. The last statement is obvious as well. The kernel of the homomorphism to any cospherical module contains (the sum of) all submodules belonging to $\mathrm{Ker}(\varpi)$. It cannot be greater because of the second assertion. ❑

We are ready to check that $\mathcal{I}_\xi$ satisfying the inequalities under consideration is cospherical. There exists a map $\mathcal{I}_\xi \longrightarrow \mathcal{I}_{\xi_o}$ sending $v_\xi = 1$ to an eigenvector of $\mathcal{I}_{\xi_o}$ with ξ as the weight. From now on, all *maps* will be $\mathcal{H}^\flat$–homomorphisms. Since $\mathcal{I}_{\xi_o}$ is cospherical, the kernel of this map belongs to $\mathrm{Ker}(\varpi)$. For every $\xi_{\widehat{w}}$, there exist at least one eigenvector in $\mathcal{I}_\xi$ apart from $\mathrm{Ker}(\varpi)$ (we have already established this). Hence the image of $\mathcal{I}_\xi$ contains all eigenvectors of $\mathcal{I}_{\xi_o}$ and this map is surjective. Since the spaces $\mathcal{I}^\infty(\mu)$ for fixed μ have the same dimensions in all induced modules with the weights from the same orbit, this map has to be an isomorphism. This also proves that all cospherical modules (with the weights from the same orbit) are isomorphic.

Coming to (c), the lemma gives the uniqueness of the irreducible submodule U_ξ and that there are no submodules in $\mathcal{I}_\xi$ greater than U_ξ that have cospherical quotients. Any cospherical irreducible module U containing an eigenvector of weight $\xi_{\hat{u}}$ is the image of a surjective homomorphism $\mathcal{I}_{\hat{u}} \longrightarrow U$. Here we use the universality of the induced representations. On the other hand, we can map $\mathcal{I}_{\hat{u}}$ to $\mathcal{I}_\xi$, since the latter contains an eigenvector corresponding to $\xi_{\hat{u}}$. The image will be cospherical. Therefore this map goes through U (the last claim of the lemma) and U is isomorphic to U_ξ. ❑

Corollary 3.9.3. *Imposing conditions (3.9.2), the module $\Delta[\xi]$ is isomorphic to the module $\mathcal{I}_{\xi_o}$ for plus-primitive ξ_o. In particular, it is generated by any eigenvector with the weight satisfying (3.9.7). For instance, setting $\xi = -\rho$, the delta-representation $\Delta[-\rho]$ is generated by the eigenvector of weight $\xi_{w_0} = t^\rho$ for the longest element $w_0 \in W$, and $\Delta_\sharp$ (isomorphic to $\mathbf{F}_\sharp$ from (3.4.9)) is a unique nonzero irreducible submodule of $\Delta[-\rho]$, provided that t is generic.*

Proof. There exists a nonzero homomorphism $\mathcal{I}_{\xi_o} \longrightarrow \Delta[\xi]$. Since both are cospherical, it has to be an isomorphism. ❑

3.9.3 Irreducible cospherical modules

There are examples when cospherical irreducible modules exist, but there are no cospherical induced representations at all. Therefore it is not always reasonable to represent irreducible cospherical modules as *submodules* of the induced ones even if the orbit contains a primitive weight, which is always imposed in this section. Somitemes it is better to describe them as *quotients* of induced modules. The following theorem makes the picture more complete and it will be used to endow irreducible cospherical representations with $\star$–invariant $*$–hermitian forms.

Theorem 3.9.1. *(a) Let us assume that there exists a weight $\xi' \in \widehat{W}^\flat(\xi)$ satisfying the condition dual to (3.9.7), that is,*

$$X_{\tilde{a}}(q^{\xi'}) \neq t_\alpha \quad \text{for} \quad \text{all } \tilde{\alpha} \in \tilde{R}_+. \tag{3.9.11}$$

Then $\mathcal{I}_{\xi'}$ has a unique nonzero irreducible quotient $U_{\xi'}$. It is automatically cospherical. Other irreducible constituents (i.e., subquotients) of $\mathcal{I}_{\xi'}$ are not cospherical and do not contain eigenvectors associated with the weight ξ'.

(b) All $\mathcal{I}_{\tilde{\xi}'}$ are isomorphic to $\mathcal{I}_{\xi'}$ for the characters $\tilde{\xi}'$ satisfying (3.9.11) and from the ξ'-orbit, i.e., when $\tilde{\xi}'$ is $\widehat{W}^\flat$-conjugated to ξ'. If ξ from this orbit does not satisfy these inequalities, then $\mathcal{I}_{\xi}$ has an irreducible quotient that is not cospherical. Imposing (3.9.11), the eigenvectors of $\mathcal{I}_{\xi'}$ are simple. An eigenvector generates $\mathcal{I}_{\xi'}$ (and $U_{\xi'}$) if and only if its weight satisfies these inequalities. A cosperical irreducible representation U with a weight from the orbit of ξ' is unique up to isomorphisms and is isomorphic to $U_{\xi'}$.

Proof. Primitive weights satisfying (3.9.11) will be referred to as **minus-primitive**. They always exist in the orbit of the weight ξ' if we impose (3.9.11) and assume that ξ' is conjugated to a primitive one. Both conditions are permanently imposed in this section. The same terminology will be used for the corresponding eigenvectors.

Let v be an X–eigenvector of an irreducible cospherical U with the weight $\xi_{\hat{u}}$. Starting with v, we can construct the eigenvector $v_1 = \tilde{\Phi}^{\hat{u}}_{j_1}(v)$ with weight $\xi_{\hat{u}_1}$ for $\hat{u}_1 = s_{j_1}\hat{u}$, then $v_2 = \tilde{\Phi}^{\hat{u}_1}_{j_2}(v_1)$, and so on, until we obtain a minus-primitive eigenvector $v' = v_l$ associated with some ξ'_o. As in Lemma 2.13 from [C10], the values

$$X_{j_1}(\hat{u}),\ X_{j_2}(\hat{u}_1),\ldots,\ X_{j_l}(\hat{u}_{l-1})$$

can be chosen avoiding $t_{j_r}^{-1}$ $(1 \le r \le l)$.

Let us check that $v' \neq 0$. Otherwise $v_r \neq 0$ in this chain for some $r < l$ and

$$0 = v_{r+1} \;=\; (t_{j_r} - 1)(T_{j_r} + t_{j_r}^{-1/2})(v_r), \quad \text{where } t_{j_r} \neq -1.$$

However, $U = \mathcal{H}^\flat_Y(v_r)$, since v_r is an eigenvector and U is irreducible. Hence, $\varpi(v_r) = 0$ and $\varpi(U) = 0$, which is impossible.

This gives a surjective map $\mathcal{I}_{\xi'_o} \to U$. It establishes an isomorphism between $\mathbf{H}^o = \mathcal{I}^\infty_{\xi'_o}(\xi_1)$ and $U^\infty(\xi'_o)$ because the first is an irreducible $\mathcal{H}^o_X$–module. The surjectivity also holds because the eigenvector 1 is simple in

$\mathbf{H}^o$. We see that the submodule that is the kernel of this map (and any of its subquotients) cannot contain ξ'_o–eigenvectors. Hence U is a unique irreducible quotient of $\mathcal{I}_{\xi'_o}$.

By the way, here the combinatorial part can be simplified somewhat. We can finish this process with v' associated with an arbitrary primitive weight. Then (cf. the proof of Theorem 3.9.1(b)) it has to satisfy (3.9.11) automatically because otherwise we can construct a quotient of $\mathcal{I}_{\xi'_o}$ that contains no $\varpi \neq 0$ (use a proper intertwiner).

Coming to the inverse statement, if ξ'_o is minus-primitive, then $\mathcal{I}_{\xi'_o}$ has a unique nonzero irreducible quotient $U_{\xi'_o}$ (it holds true for all primitive characters). On the other hand, it has a unique nonzero cospherical quotient V, which contains a cospherical irreducible submodule U. It is a quotient of a proper $\mathcal{I}_{\tilde{\xi}'_o}$. Due to [C10], Lemma 2.8, the latter is isomorphic to $\mathcal{I}_{\xi'_o}$ because all primitive eigenvectors are connected by invertible intertwiners. Hence we obtain a nonzero homomorphism from $\mathcal{I}_{\xi'_o}$ onto U, which results in the coincidence $V = U$ and the isomorphism $U \simeq U_{\xi'_o}$.

To establish the required isomorphism without any reference to [C10] and also check the remaining statements of the theorem, it is convenient to apply Theorem 3.9.1 for the character of $\mathcal{H}^\flat_Y$

$$T_i \to -t_i^{-1/2},\ \pi_r \to 1,\ 0 \le i \le n,\ r \in O,$$

replacing the character used in the definition of coinvariants ϖ. The modules under consideration become cospherical for such a character.

It gives, first, that all eigenvectors of $\mathcal{I}_{\xi'_o}$ are simple and can be obtained from 1 by the intertwiners. Here, if the intertwiner is not invertible, then its image belongs to $\mathrm{Ker}(\varpi)$. Second, all $\mathcal{I}_{\xi'}$ satisfying (3.9.11) are isomorphic to $\mathcal{I}_{\xi'_o}$ (and to each other). Moreover, all eigenvectors in $\mathcal{I}_{\xi'_o}$ corresponding to such weights ξ' are connected with 1 by invertible intertwiners (and only such eigenvectors). Third, if an irreducible constituent of $\mathcal{I}_{\xi'_o}$ contains an eigenvector associated with ξ', then $\mathcal{I}_{\xi'_o}$ maps onto it through $\mathcal{I}_{\xi'}$. It can happen for $U_{\xi'_o}$ only. ❑

Plus-minus duality. Let us now assume that both plus-primitive and minus-primitive characters ξ_o, ξ'_o can be found in the orbit of ξ.

In this case, $U_\xi \simeq U_{\xi'}$, since they are unique cospherical irreducible constituents, and, moreover, the first one contains the whole space $\mathcal{I}^\infty_\xi(\xi')$. The modules $\mathcal{I}_{\xi_0}, \mathcal{I}_{\xi'_o}$ are dual to each other in the following sense.

Given a module $\mathcal{I} = \oplus\mathcal{I}^\infty(\mu) \in \mathcal{O}$, we combine the anti-involution $\diamond$ from (3.9.4) with the natural anti-action of $\mathcal{H}^\flat$ on the

$$\mathcal{I}^\diamond \overset{\mathbf{def}}{=\!=} \{f \in \mathrm{Hom}(\mathcal{I}, \mathbb{Q}_{q,t})\ \text{ s.t. } f(\mathcal{I}^\infty(\mu)) = 0 \text{ for almost all } \mu\}. \qquad (3.9.12)$$

It makes $\mathcal{I}^\diamond$ an $\mathcal{H}^\flat$–module. We claim that $\mathcal{I}^\diamond_{\xi_o} \simeq \mathcal{I}_{\xi'_o}$.

First, $U^\circ_{\xi_o} \simeq U_{\xi'_o}$. Indeed, they have the same set of characters and coinciding dimensions of the spaces of generalized vectors. Use the duality. Hence $U^\circ_{\xi_o}$ can be covered by $\mathcal{I}_{\xi'_o}$. However, the latter has a unique irreducible quotient, which is $U_{\xi'_o}$. Second, the map $\nu : \mathcal{I}_{\xi'_o} \to \mathcal{I}^\circ_{\xi_o}$ sending 1 to an eigenvector corresponding to ξ'_o composed with the map $\mathcal{I}^\circ_{\xi_o} \to U^\circ_{\xi_o}$ is obviously nonzero. This map is a dualization of the natural embedding $U_{\xi_o} \subset \mathcal{I}_{\xi_o}$ (Theorem 3.9.1). Third, the module $\mathcal{I}^\circ_{\xi_o}$ has a unique nonzero irreducible quotient. We conclude that ν is surjective and has to be an isomorphism because the dimensions of the spaces $\mathcal{I}^\infty(\mu)$ are the same for $\mathcal{I}_{\xi'_o}$ and $\mathcal{I}_{\xi_o}$ (their characters are from the same orbit).

Invariant form. The following corollary shows that, generally speaking, induced modules have invariant $*$–hermitian forms.

Corollary 3.9.4. *In the setup of Theorem 3.9.1, the module $\mathcal{I}_{\xi'}$ possesses a unique up to proportionality $\star$–invariant $*$–hermitian form in the sense of Section 3.6 in this chapter. Its radical $\mathcal{R}$ coincides with the kernel $\mathcal{K}$ of the map $\mathcal{I}_{\xi'} \to U_{\xi'}$ from Theorem 3.9.1. It induces a unique nondegenerate $\star$–invariant form on $U_{\xi'}$.*

Proof. Any $\mathcal{I}_{\xi'}$ can be considered as a limit of a one-parametric family of proper $F[\tilde{\xi}']$ satisfying the inequalities of the theorem for $\tilde{\xi}'$. Therefore $\mathcal{I}_{\xi'}$ has a nonzero $*$–invariant form. If its radical $\mathcal{R}$ is smaller than $\mathcal{K}$, then $\mathcal{I}_{\xi'}/\mathcal{R}$ contains an irreducible submodule $V \not\simeq U_{\xi'}$. All generalized X–eigenvectors with the characters not from V belong to its orthogonal compliment V'. Since V has no eigenvectors associated with ξ', the image of 1 belongs to V'. A contradiction. The uniqueness of the $*$–invariant form on $U_{\xi'}$ follows from the irreducibility. ❑

Provided we have (3.9.2), we claim that $\mathcal{I}_{\xi'_o} \simeq \mathbf{F}[\xi]$ for the minus-primitive ξ'_o weight from the orbit of ξ (cf. Corollary 3.9.3). Indeed, one can map $\mathcal{I}_{\xi'_o}$ into $\mathbf{F}[\xi]$. We can use that $\Delta[\xi]$ is isomorphic to $\mathcal{I}_{\xi_o}$ for the plus-primitive ξ_o. It results in the pairing $\{\ ,\ \}$ in $\mathcal{I}_{\xi'_o} \times \mathcal{I}_{\xi_o}$. If this pairing is degenerate, then its right radical ($\subset \mathcal{I}_{\xi_o}$) has to contain U_{ξ_o} because the latter is the smallest irreducible submodule. However, this is impossible because the image of $\mathcal{I}_{\xi'_o}$ in $F[\xi]$ contains $U_{\xi'_o} \simeq U_{\xi_o}$ as a constituent.

3.9.4 Irreducibility of induced modules

Let us generalize the first part of Theorem 3.9.1 and check that, generally speaking, cyclic irreducible modules of $\mathcal{HH}^\flat$ are induced from the irreducibles of $\mathcal{H}_X$.

Proposition 3.9.5. *(a) Given a finite dimensional irreducible $\mathcal{H}_X$–module U let as assume that*

$$1 \neq X_{\tilde{\alpha}}(q^\xi) \neq t_\alpha^{\pm 1} \quad \textit{for} \quad \textit{all } \tilde{\alpha} \in \tilde{R} \setminus R, \tag{3.9.13}$$

where ξ is any weight of U (it does not matter which because they are W–conjugated). Then the induced $\mathcal{H}\!\!\mathcal{H}^\flat$–module $M_U = Ind_{\mathcal{H}_X}^{\mathcal{H}\!\!\mathcal{H}^\flat}(U)$ is irreducible.

(b) Any irreducible $\mathcal{H}\!\!\mathcal{H}^\flat$–module M posessing an X–eigenvector with the weight $\widehat{W}^\flat$–conjugated to ξ from (3.9.13) is induced from its irreducible $\mathcal{H}_X$–submodule U generated by an eigenvector with weight from $W(\xi)$. Given the orbit $W(\xi)$, such a submodule U is unique up to isomorphisms: $M_U \simeq M_{U'}$ $\Rightarrow U \simeq U'$.

Proof. Let us check that the left-hand side inequality gives that $M^\infty(\xi) = U^\infty(\xi)$ for $M = M_U$, where (see (3.9.9)) by $M^\infty(\xi)$, we mean the space of all generalized eigenvectors in M associated with a weight ξ. It holds for $\mathcal{I}_\xi$ and for the induced $\mathcal{H}_X$–module $\mathcal{I}_\xi^X = \mathrm{Ind}_{\mathbb{C}[X]}{}^{\mathcal{H}_X}(q^\xi)$ instead of M and U. However, $\mathcal{I}_\xi$ and $\mathcal{I}_\xi^X$ cover M, U and, correspondingly, the spaces of generalized eigenvectors.

Once the coincidence is checked at the ∞–level, it is valid for all levels of the spaces of generalized eigenvectors. In particular, $M^1(w(\xi)) = U^1(w(\xi))$ for any $w \in W$. We obtain that

$$U \;=\; \oplus_{\zeta\in W(\xi)} M_U^\infty(\zeta). \tag{3.9.14}$$

Given $b \in B$ and $w \in W$,

$$(\tilde{\Phi}_{\pi_b}^\zeta)^{-1}(M^1(\pi_b(\zeta)) \;=\; M^1(\zeta),$$
$$\text{where } \tilde{\Phi}_{\pi_b}^\zeta = \tilde{\Phi}_{\pi_b}(\zeta),\ \zeta \in W(\xi). \tag{3.9.15}$$

We follow the notation from (3.9.8) and use the invertibility of $\tilde{\Phi}_{\pi_b}^{w(\xi)}$ in the space $M^1(w(\xi))$ (and in $M^\infty(w(\xi))$), thanks to the right-hand side inequality.

The elements $\{\pi_b(\zeta)\}$ for the above b and ζ constitute the whole orbit $\widehat{W}^\flat(\xi)$. Any nonzero irreducible submodule of M has at least one X–eigenvector for a weight from the orbit $\widehat{W}^\flat(\xi)$. Due to (3.9.15), it generates the whole M.

An arbitrary irreducible M possessing a weight $\widehat{W}^\flat$–conjugated to ξ from (3.9.13) can be represented as M_U, where U is any irreducible $\mathcal{H}_X$–submodule of M with an eigenvalue from $W(\xi)$. The existence of U results from the same formula (3.9.15). Moreover, U can be reconstructed uniquely as a submodule of M by means of (3.9.14). Therefore $M \simeq M' \Rightarrow U \simeq U'$. ❑

Restriction to $\mathcal{H}_X$. Let us describe the structure of M_U upon the restriction to $\mathcal{H}_X$, provided we have (3.9.13). In the first place, $M_U = \oplus J_{c,\zeta}$, where $c \in B_-$, ζ runs over a fixed set of representatives of $W(\xi) \bmod W_c$ for the centralizer W_c of c in W,

$$J_{c,\zeta} \;=\; \oplus_{\tilde{\xi}} M_U^\infty(\tilde{\xi}),\ \tilde{\xi} \in W(c(\zeta)). \tag{3.9.16}$$

All $\{J\}$ are $\mathcal{H}_X$–submodules. Their structure can be described as follows:

$$J_{c,\varsigma} = \oplus_\omega \tilde{\Phi}_\omega(\tilde{U}_c),\ \tilde{U}_c = \tilde{\Phi}_c(U_c),$$
$$U_c \stackrel{\text{def}}{=} \oplus_{\xi'} U^\infty(\xi')),\ \text{ for } \xi' \in W_c(\varsigma). \tag{3.9.17}$$

Here

$$\{\omega = u_b^{-1},\ b \in W(c)\} = \{\omega \in W,\ \lambda(\omega) \subset \{\alpha \in R_+ \text{ s.t. } (\alpha, c) < 0\}\}.$$

The intertwiners $\tilde{\Phi}_\omega, \tilde{\Phi}_c$ are invertible because $c = c_- = \pi_c$ and $(\alpha, c) \neq 0$ for all $\alpha \in \lambda(\varpi)$. The space U_c is a module over the subalgebra $\mathcal{P}_c$ of $\mathcal{H}_X$ generated by $\{T_i, s_i(c) = c\}$ and the whole $\mathbb{Q}_{q,t}[X]$. So is $\tilde{U}_c$. Indeed, applying Φ_c means replacing the action of X_j by that of $q^{(c,b_j)}X_j$ for all $1 \le j \le n,\ s_j(c) \neq c$ without touching $\{T_i\}$.

We conclude that $J_{c,\varsigma}$ is isomorphic to the representation of $\mathcal{H}_X$ induced from the $\mathcal{P}_c$–module $\tilde{U}_c$. Its irreducibllity is equivalent to the irreducibility of the $\mathcal{P}_c$–module U_c.

The simplest application of this construction is decomposing the polynomial representation considered as an $\mathcal{H}_Y^\flat$–module. See (3.2.33), (3.2.34), (3.2.35), and also [C20](formulas (3.15)–(3.17)).

Summarizing, in the case of generic q, the classification of irreducible representations of $\mathcal{HH}^\flat$ is not too far from that in the affine Hecke case. If all t_ν coincide, then we can directly use the main theorem from [KL1]. Of course, there are representations that do not satisfy (3.9.13) and can be expected to be associated with more sophisticated "Springer fibers."

3.10 Gaussian and self-duality

We follow the notation from Section 3.8 of this chapter. An $\mathcal{HH}^\flat$–module V is **self-dual** if the involution ε from (3.2.10) comes from a certain $*$–linear involution of V upon the lift to End(V). A module is called $PGL^c(2,\mathbb{Z})$–**invariant** if $\varepsilon, \tau_\pm$ act there projectively, i.e., if there are automorphisms of V satisfying

$$\tau_+\tau_-^{-1}\tau_+ \stackrel{\text{def}}{=} \sigma = \tau_-^{-1}\tau_+\tau_-^{-1},\ \tau_\pm = \varepsilon\tau_\mp\varepsilon^{-1}, \tag{3.10.1}$$

and inducing the $\varepsilon, \tau_\pm$ from Section 3.2 on End(V). By $PGL^c(2,\mathbb{Z})$, we mean the **central extension of** $PGL(2,\mathbb{Z})$ due to Steinberg, introduced by the relations (3.10.1).

Neither functional nor polynomial representations are self-dual. Either τ_- or τ_+ act there respectively, but not all together. Actually, self-dual semisimple irreducible modules (from $\mathcal{O}_X$) have to be finite dimensional. Indeed, they belong to both $\mathcal{O}_X$ and $\mathcal{O}_Y$, and one can readily check that they are finite dimensional.

The non-cyclic module $\mathcal{V}_1 \stackrel{\mathbf{def}}{=\!=} \tilde{\gamma}^{-1}\mathcal{V}$ is self-dual. Namely, the following $*$–linear involution

$$\psi(\mathcal{E}_b\tilde{\gamma}^{-1}) \;=\; q^{-(b_\sharp\,,\,b_\sharp)/2+(\rho_k\,,\,\rho_k)/2}\mathcal{E}_b\tilde{\gamma}^{-1} \text{ for } b\in B \tag{3.10.2}$$

corresponds to ε from the double Hecke algebra. It is formula (5.8) from [C21], which is equivalent to (3.4.28) above. Note that the module $\mathcal{V}_1$ is not $PGL^c(2,\mathbb{Z})$–invariant because any extension of the automorphism τ_+ has to be proportional to multiplication by $\tilde{\gamma}(X)$, that does not preserve $\mathcal{V}_1$.

3.10.1 Gaussians

We turn to the Gaussians in $V_{\mathrm{sph}}[\xi]$, which will be called **restricted Gaussians.** Generally speaking, the problem is to make τ_+ an inner automorphism. The analysis is not difficult, thanks to the Main theorem of Section 3.6. The Gaussian commutes with the X–operators, so it preserves the X–eigenspaces. This theorem provides their description. The Gaussian, if it exists in a spherical irreducible representation, has to be its element, not just an operator; the X–action makes such representations commutative algebras. Thus, the spherical case is especially interesting.

According to the Main theorem, τ_+ is inner in $V=V_{\mathrm{sph}}[\xi]$ if and only if the map

$$S^{\xi}_{\widehat{w}} \mapsto \tau_+(S^{\xi}_{\widehat{w}}), \text{ where } \widehat{w}\in \widehat{W}^{\flat}_*[\xi] \text{ and } S^{\xi}_{\widehat{w}} = S_{\widehat{w}}(q^{\xi}), \tag{3.10.3}$$

is inner (i.e., multiplication by a matrix) in the eigenspace $V(\xi)$. In the spherical case, the $S_{\widehat{w}}$ become simply $\widehat{w}$ in the realization $\mathfrak{F}'[\xi]$, and the automorphisms $\tau_+(\widehat{w})$ can be readily calculated. We come to the following proposition.

Proposition 3.10.1. *The restricted Gaussian defined by the relations*

$$\gamma H\gamma^{-1} \;=\; \tau_+(H) \text{ in } V \tag{3.10.4}$$

exists in $V=\mathfrak{F}'[\xi]$ *if and only if*

$$q^{(b,b+2\xi)/2} \;=\; 1 \text{ whenever } bw\in \widehat{W}^{\flat}_*[\xi] \text{ for } b\in B,\ w\in W. \tag{3.10.5}$$

Then it is proportional to $\gamma_*(bw) \stackrel{\mathbf{def}}{=\!=} q^{(b+2w(\xi),b)/2}$.

Proof. Let v be the cyclic generator of V (of weight ξ). We identify γ with the operator of multiplication by γ. If it exists, then

$$\gamma(S_{\widehat{w}}(v)) \;=\; S_{\widehat{w}}(S_{\widehat{w}}^{-1}\gamma S_{\widehat{w}})(v) \text{ for all } \widehat{w}\in \Upsilon_+[\xi]. \tag{3.10.6}$$

Here $\gamma'_{\widehat{w}} = S_{\widehat{w}}^{-1}\gamma S_{\widehat{w}}\gamma^{-1}$ is a function that can be readily calculated due to the definition of τ_+. However, it is more convenient to involve $\mathfrak{F}'[\xi]$, where $S_{\widehat{w}}$

is nothing but $\widehat{w}$. Recall that τ_+ is interpreted as the formal conjugation by $\gamma(q^z) = q^{(z,z)/2}$ in any functional space. See (3.4.25) and (3.4.31). Therefore

$$\gamma'_{\widehat{w}}(\hat{u}) \;=\; q^{(\xi'',\xi'')/2-(\xi',\xi')/2} \;\text{ for }\; \xi' = \hat{u}((\xi)),\; \xi'' = \widehat{w}((\xi)). \tag{3.10.7}$$

Its action on v is $\gamma'_{\widehat{w}}(\mathrm{id}) = \gamma_*(\widehat{w})$.

Thanks to the argument that has been used several times, formula (3.10.6) defines γ correctly if and only if $\gamma'_{\widehat{w}} = 1$ for $\widehat{w} \in \widehat{W}^\flat_*[\xi]$. This is exactly (3.10.5). ❑

3.10.2 Perfect representations

Following the same lines, let us find out when $V = V_{\mathrm{sph}}[\xi]$ is self-dual i.e., ε can be made an inner involution in V. For the sake of simplicity, we use the functional realization of $\mathfrak{F}'[\xi]$ of V. So 1 is the spherical vector, $\delta_0 = \delta_{\mathrm{id}}$ is the X–cyclic generator, and $\delta_0 = \delta_{\mathrm{id}}$.

Recall that

$$\chi_{\widehat{w}}(\hat{u}) \;=\; \delta_{\widehat{w},\hat{u}},\;\; \delta_{\widehat{w}}(\hat{u}) \;=\; \mu_\bullet(\widehat{w})^{-1}\chi_{\widehat{w}} \tag{3.10.8}$$

for the Kronecker $\delta_{\widehat{w},\hat{u}}$. We also set $\delta^\pi_b = \delta_{\pi_b}$ in the case $\xi = -\rho_k$.

Theorem 3.10.2. *(i) Provided that we have (3.8.7), the finite dimensional representation $V_{sph}[\xi]$ is self-dual if and only if it is X–unitary and $q^{-\rho_k} = q^{\widehat{w}((\xi))}$ for a certain $\widehat{w} \in \Upsilon_+[\xi]$. The representation $V_{sph}[-\rho_k]$, is $\mathcal{H}^\flat$–isomorphic to the irreducible X–cospherical Y–semisimple nonzero quotient of $\mathcal{I}_Y[\rho_k]$, which is unique. The X–spectrum of $V_{sph}[\xi]$, which is $\Upsilon_+[-\rho_k]$, belongs to $\pi_B = \{\pi_b \mid b \in B\}$ and is the negative of its Y–spectrum. The X–unitary structure of $V_{sph}[-\rho_k]$ is also Y–unitary.*

(ii) The module $\mathfrak{F}'[-\rho_k] \simeq V_{sph}[-\rho_k]$ has a basis formed by the discretizations of the spherical polynomials $\mathcal{E}_b$:

$$\mathcal{E}'_b(\widehat{w}) \stackrel{\mathbf{def}}{=\!=} \mathcal{E}_b(q^{-\widehat{w}((\rho_k))}) \;\; \textit{for} \;\; \widehat{w} \in \Upsilon_+[-\rho_k]. \tag{3.10.9}$$

They are well defined assuming that $\pi_b \in \Upsilon_+[-\rho_k]$. Explicitly,

$$\mathfrak{F}'[-\rho_k] \;=\; \oplus_{b^\bullet}\, \mathbb{Q}_{q,t}\, \mathcal{E}'_{b^\bullet},$$

where the summation is over the set of representatives

$$\pi_{b^\bullet} \in \Upsilon_+[-\rho_k]/\widehat{W}^\flat_*[\rho_k] \stackrel{\mathbf{def}}{=\!=} \Upsilon'_+[-\rho_k].$$

Here we use that $\widehat{W}^\flat_*[-\rho_k] = \widehat{W}^\flat_*[\rho_k]$. Moreover

$$\mathcal{E}'_b = \mathcal{E}'_c \textit{ whenever } \pi_b((\rho_k)) \;= \pi_c((\rho_k)),\;\; \textit{i.e.,}\;\; \pi_b^{-1}\pi_c \in \widehat{W}^\flat_*[\rho_k]. \tag{3.10.10}$$

(iii) The map

$$\psi' : \sum g_b \mathcal{E}'_b \mapsto \sum g_b^* \delta_b^\pi \ \ for \ \ g_b \in \mathbb{Q}_{q,t}, \ \pi_b \in \Upsilon_+[-\rho_k], \tag{3.10.11}$$

is an $\mathcal{H}^\flat$-automorphism of $\mathfrak{F}'[-\rho_k]$, inducing ε on $\mathcal{H}^\flat$. Here the δ_b^π depend only on the images of π_b in $\Upsilon'_+[-\rho_k]$. Equivalently,

$$\begin{aligned}\langle 1\rangle'\,\psi'(f)\,(\pi_b) \ &= \ \langle \mathcal{E}'_b\,,\, f\rangle' \ = \ \langle\, f^*\,\mathcal{E}'_b\,\rangle', \\ where \ \langle f\rangle' \ &\stackrel{\mathbf{def}}{=\!=} \sum_{\widehat{w}\in\Upsilon'_+[-\rho_k]} \mu_\bullet(\widehat{w})f(\widehat{w}).\end{aligned} \tag{3.10.12}$$

Proof. The condition $q^{-\rho_k} = q^{\widehat{w}((\xi))}$ is necessary because ε sends X_b to Y_b, conjugating the eigenvalues, and the self-duality implies that $\mathrm{Spec}_X = -\mathrm{Spec}_Y$. So if the involution of V, let us denote it by ψ', exists, then $\psi'(1)$ has to be proportional to δ_0. Its X–weight is $-\rho_k$. Here we use the $\mathfrak{F}'$–realization of V. Recall that the Y–weight of 1 is ρ_k.

Now let $\xi = -\rho_k$ and

$$\Upsilon_0[\rho_k] \supset \Upsilon_-[\rho_k] \subset \Upsilon_+[\rho_k]. \tag{3.10.13}$$

It is a reformulation of (3.8.7) using the obvious relations $\Upsilon_\pm[\xi] = \Upsilon_\mp[-\xi]$. We know how to describe nonzero irreducible Y–semisimple quotients of $\mathcal{I}_Y[\rho_k]$. The latter module has such quotients thanks to (3.10.13) and Theorem 3.6.1. The universal semisimple quotient is $U_Y[\rho_k] \stackrel{\mathbf{def}}{=\!=} \mathcal{H}^\flat_X/\mathcal{J}$ for

$$\mathcal{J} \stackrel{\mathbf{def}}{=\!=} \sum \mathcal{H}_X \cdot \tau_+\big(\Phi_{\widehat{w}}(q^{-\rho_k})\big), \tag{3.10.14}$$

where the summation is over $\widehat{w} \in \ddot{\Upsilon}_+[-\rho_k]$ and 1 is identified with the cyclic generator of $\mathcal{I}_Y[\rho_k]$. Using Theorem 3.8.3, we obtain that $U_Y[\rho_k]$ has a unique nonzero irreducible X–cospherical quotient. This implies that the latter is a unique irreducible X–cospherical quotient of $\mathcal{I}_Y[\rho_k]$. Cf. Lemma 6.2 from [C23]. We will denote it by $V^Y_{\mathrm{sph}}[\rho_k]$.

Let us check that the module $V_{\mathrm{sph}}[-\rho_k] \simeq \mathfrak{F}'[-\rho_k]$ is not only Y–cosherical but also X–cospherical. Since it is irreducible, we need to find a vector v satisfying (3.8.1):

$$T_i(v) = t_i\, v \ \text{ for } \ i > 0, \ X_a(v) = q^{-(a\,,\,\rho_k)}v \ \text{ for } \ a \in B. \tag{3.10.15}$$

Such a v is exactly the generator δ_0. Note that the first formula in (3.10.15) holds because $\{s_1, \ldots, s_n\} \subset \ddot{\Upsilon}_+[-\rho_k]$.

Summarizing, the $\mathcal{H}^\flat$–homomorphism $\mathcal{I}_Y[\rho_k] \to \mathfrak{F}'[-\rho_k]$ sending $1 \mapsto 1$, which exists because the former module is induced, identifies the latter with the $V^Y_{\mathrm{sph}}[\rho_k]$ constructed above. There are several immediate corollaries.

First, the operators Y_a are diagonalizable in the module $\mathfrak{F}'[-\rho_k]$. Second, the Y–spectrum is simple there and coincides with the negative of the X–spectrum. Third, the invariant Y–definite $*$–bilinear form on $V^Y_{\text{sph}}[\rho_k]$ has to be proportional to the X–definite invariant form on $\mathfrak{F}'[-\rho_k]$ upon this identification. Fourth, the polynomials $\mathcal{E}_b \in \mathcal{V}$ are well defined for $\pi_b \in \Upsilon_+[-\rho_k]$. Let us check the latter claim.

Using the standard $\mathcal{H}^\flat$–homomorphism from $\mathcal{V}$ to $\mathfrak{F}'[-\rho_k]$ and the simplicity of the Y–spectrum of the target, we diagonalize the Y–operators in the subspaces

$$\Sigma(b) \;=\; \oplus_{c\succ b}\,\mathbb{Q}_{q,t}X_c \oplus \mathbb{Q}_{q,t}X_b, \quad \text{for } B \ni c \succ b, \tag{3.10.16}$$

starting with $b=0$ and "decreasing" b as far as $\pi_b \in \Upsilon_+[-\rho_k]$. See (3.3.4). The leading monomial X_b always contributes to E_b. Thus the Macdonald polynomials $\{E_b\}$ are well defined in this range and their images E'_b in $\mathfrak{F}'[-\rho_k]$ are nonzero. Let us verify that $E_b(q^{-\rho_k}) \neq 0$ for such b.

Following [C19, C20], we set

$$\begin{aligned} \{f,g\} \;&=\; \{L_{\imath(f)}(g(X))\}(q^{-\rho_k}) \;\text{ for }\; f,g\in\mathcal{V}, \\ \imath(X_b) \;&=\; X_{-b} \;=\; X_b^{-1},\; \imath(z) \;=\; z \;\text{ for }\; z\in\mathbb{Q}_{q,t}\,, \end{aligned} \tag{3.10.17}$$

where L_f is from Proposition 3.3.1. Let us use the $\mathbb{Q}_{q,t}$–linear anti-involution of $\mathcal{H}^\flat$ from (3.3.20):

$$\phi \overset{\text{def}}{=\!=} \varepsilon\star = \star\varepsilon :\; X_b \mapsto Y_b^{-1},\; T_i \mapsto T_i \;(1 \le i \le n). \tag{3.10.18}$$

The basis of all duality formulas is the following lemma.

Lemma 3.10.3. *For arbitrary nonzero* q, t_{sht}, t_{lng},

$$\{f,g\} = \{g,f\} \;\text{ and }\; \{H(f),g\} = \{f, H^{\phi}(g)\},\; H\in\mathcal{H}^\flat. \tag{3.10.19}$$

The quotient $\mathcal{V}'$ of the polynomial representation $\mathcal{V}$ by the radical Rad$\{\ ,\ \}$ of the pairing $\{\ ,\ \}$ is an $\mathcal{H}^\flat$–module such that

(a) all Y–eigenspaces $\mathcal{V}'_Y(\xi)$ of $\mathcal{V}'$ are zero or one-dimensional,

(b) $E(q^{-\rho_k}) \neq 0$ if the image $E' \neq 0$ of E in $\mathcal{V}'$ is a Y–eigenvector.

Proof. Formulas (3.10.19) are from Theorem 2.2 of [C20]. See also [C23], Corollary 5.4. Thus Rad$\{\ ,\ \}$ is a submodule and the form $\{\ ,\ \}$ is well defined and nondegenerate on $\mathcal{V}'$. For any pullback $E \in \mathcal{V}$ of $E' \in \mathcal{V}'$, $E(q^{-\rho_k}) = \{E,1\} = \{E',1'\}$. If E' is a Y–eigenvector of weight ξ and $E(q^{-\rho_k})$ vanishes, then

$$\{\mathcal{Y}^\flat(E'), \mathcal{H}^\flat_Y(1')\} = 0 = \{E', \mathcal{V}\cdot\mathcal{H}^\flat_Y(1')\} \;\text{ for }\; \mathcal{Y}^\flat = \mathbb{Q}_{q,t}[Y_b].$$

Therefore $\{E', \mathcal{V}'\} = 0$, which is impossible.

We can use it to check that $\dim \mathcal{V}'_Y[\xi] \le 1$. Indeed, there is always a linear combination of two eigenvectors in $\mathcal{V}'$ of the same weight with zero value at $q^{-\rho_k}$, which has to be zero identically. ❑

The following fact is of similar nature. It will not be used in proving the theorem, but it is actually the key in the general theory of the pairing $\{\,,\,\}$ and perfect representations. Cf. also the definition of the principal and primitive modules from Proposition 3.8.1.

Proposition 3.10.4. *If the quotient $\mathcal{V}'$ of the polynomial representation $\mathcal{V}$ by the radical Rad$\{\,,\,\}$ is finite dimensional and Rad is τ_-–invariant for the canonical action of τ_- in $\mathcal{V}$, then $\mathcal{V}/Rad$ is an irreducible $\mathcal{H}^\flat$–module.*

Proof. Using $\phi\tau_-\phi = \tau_+$, the relation

$$\{\tau_+ f, g\} \;=\; \{f, \tau_- g\} \;\text{ for }\; f, g \in \mathcal{V}'$$

defines the action of τ_+ and the action of σ in $\mathcal{V}'$ satisfying

$$\tau_+\tau_-^{-1}\tau_+ \;=\; \sigma \;=\; \tau_-^{-1}\tau_+\tau_-^{-1}.$$

The pairing $\{f, g\}_\sigma \overset{\mathbf{def}}{=\!=} \{\sigma f, g\} = \{f, \sigma^{-1} g\}$ corresponds to the *anti-involution* $\heartsuit = \sigma \cdot \phi = \phi \cdot \sigma^{-1}$ of $\mathcal{H}^\flat$ from (3.3.21). It holds in either direction, from f to g and the other way round, but the form $\{f, g\}_\sigma$, generally speaking, may be non-symmetric. Actually, it is symmetric, but we do not need it for the proof.

Using this *non-degenerate* pairing, we proceed as follows. Any proper $\mathcal{H}^\flat$–submodule $\mathcal{V}''$ of $\mathcal{V}'$ contains at least one Y–eigenvector e'', so we can assume that $\mathcal{V}'' = \mathcal{H}^\flat e''$. The corresponding eigenvalue cannot coincide with that of 1, thanks to the previous lemma. Therefore $\{1', \mathcal{V}''\}_\sigma = 0$ and the orthogonal complement of $\mathcal{V}''$ in $\mathcal{V}'$ is a *proper* $\mathcal{H}^\flat$–submodule of $\mathcal{V}'$ containing $1'$, which is impossible. ❑

Let us check that the pairing $\{f, g\}_\sigma$ is symmetric, indeed. First of all,

$$\{\tau_+(1'), 1'\} = \{1', \tau_-(1')\} = \{1', 1'\} = 1 \Rightarrow$$
$$\{1', 1'\}_\sigma = \{\sigma(1'), 1'\} = \{\tau_+(1'), \tau_-(1')\} = \{\tau_+(1'), 1'\} = 1.$$

Then, $\{1', f\}_\sigma - \{f, 1'\}_\sigma = \{(\sigma - \sigma^{-1})(1'), f\} = \{(1 - \sigma^{-2})(1'), f\}_\sigma$. Recall that $\tau_-(1) = 1$ in $\mathcal{V}$. However, σ^{-2} coincides with T_{w_o} up to proportionality in *irreducible* $\mathcal{H}$–modules where σ acts. See Proposition 3.2.2.

Thus $(1 - \sigma^{-2})(1')$ is proportional to $1'$ and must be zero in $\mathcal{V}'$ due to the calculation above. We obtain that $1'$ is in the radical of the pairing $\{f, g\}_\sigma - \{g, f\}_\sigma$, which makes this difference identically zero, since $1'$ is a generator.

Comment. The radical is always τ_-–invariant if q is not a root of unity. To see this, one can use that τ_- in $\mathcal{V}$ is proportional to the multiplication by the series $\sum_{b \in B} q^{(b,b)/2} Y_b$ provided that $|q| < 1$. See Proposition 3.3.4.

If q is a root of unity, then $\mathcal{V}'$ is not τ_-–invariant for generic t_ν, i.e., in the setting of Proposition 3.8.4.

Concerning the previous considerations, note that the representation $\widehat{\mathcal{V}}$ from Proposition 3.8.4 which is the quotient of v by the radical of the pairing $\langle\, ,\, \rangle_\circ$ is *not* isomorphic to $\mathcal{V}'$ for generic k. Cf. also Theorem 2.8.5(iii) from the previuos chapter (the module $\mathcal{V}'$ was denoted $\mathcal{P}/Rad$ there). ❑

Let us go back to the proof of the theorem. Obviously the radical of the pairing $\{\, ,\, \}$ belongs to the kernel of the discretization map $\delta' : \mathcal{V} \to \mathfrak{F}'[-\rho_k]$. Since the latter module is irreducible, the radical and the kernel coincide, and we have a nondegenerate pairing on $\mathfrak{F}'[-\rho_k]$ inducing ϕ on $\mathcal{H}^\flat$. Since the image of E_b in this module is nonzero for $\pi_b \in \Upsilon_+[-\rho_k]$, we conclude that $E_b(q^{-\rho_k}) \neq 0$ and the spherical polynomials $\mathcal{E}_b = E_b/(E_b(q^{-\rho_k}))$ are well defined for such b. Moreover, if $\mathcal{E}_b$ and $\mathcal{E}_c$ are well defined and their eigenvalues coincide, then they can be different in $\mathcal{V}$, but their images in $\mathfrak{F}'[-\rho_k]$ have to coincide, thanks to the normalization $\mathcal{E}_b(q^{-\rho_k}) = 1$.

So far we have not used the X–unitary structure at all. If it exists, then

$$\begin{aligned}
&E_b(q^{-\rho_k}) \;=\; 0 \;\Rightarrow\; (\, E'_b\, ,\, \delta_0\,) \;=\; 0 \;\Rightarrow\\
&(\, Y_a(E'_b)\, ,\, T_i X_c(\delta_0)\,) = 0 \ \text{ for }\ i > 0,\ a, c \in B \Rightarrow\\
&(\, E'_b\, ,\, Y_a^{-1} T_i X_c(\delta_0)\,) \;=\; 0 \;\Rightarrow\; (\, E'_b\, ,\, \mathcal{H}^\flat(\delta_0)\,) \;=\; 0 \;\Rightarrow\\
&(\, E'_b\, ,\, \mathfrak{F}'[-\rho_k]\,) \;=\; 0 \Rightarrow E'_b = 0.
\end{aligned} \tag{3.10.20}$$

The existence of an X–unitary structure, i.e., the relation $\mu_\bullet(\widehat{w}) = 1$ for $\widehat{w} \in \widehat{W}^\flat_*[\rho_k]$, is equivalent to the existence of ψ' from the theorem. Indeed, the form $(f, g) = \{\psi(f), \psi(g)\}$ is obviously invariant pseudo-hermitian, assuming that ψ induces ε, and vice versa. Recall that by invariant pseudo-hermitian forms, we mean $\star$–invariant $*$–skew-symmetric nondegenerate forms. As we already used, such a form is automatically X–definite because the X–eigenspaces of $\mathfrak{F}'[-\rho_k]$ are one-dimensional. It is equally applicable to Y instead of X.

Recall that we do not suppose X–definite or Y–definite forms to be positive (or negative) hermitian forms. It only means that $(v, v) \neq 0$ for all the corresponding eigenvectors.

If $\mu_\bullet = 1$ on $\widehat{W}^\flat_*[\rho_k]$, then the delta functions δ^π_b depend only on the classes of π_b modulo the latter group. Hence, the map ψ' from (iii) is a well defined restriction of $\psi_\circ$ from formula (3.4.13). It induces ε on $\mathcal{H}^\flat$ (Theorem 3.4.2) and

$$\mathcal{E}_b(X) \;=\; \varepsilon(G_{\pi_b})(1) \ \text{ for }\ b \in B. \tag{3.10.21}$$

The latter holds until the elements π_b sit in $\Upsilon_+[-\rho_k]$. ❑

Let us combine together all the structures discussed above. Our aim is to describe finite dimensional irreducible self-dual semisimple pseudo-unitary representations with the action of $PGL^c(2,\mathbb{Z})$. We call them **perfect** in the book.

General **quasi-perfect** representations are $\mathcal{H}\!\!\mathcal{H}^\flat$–modules which have a non-degenerate form $\{\ ,\ \}$ inducing the duality anti-involution ψ. The greatest quasi-perfect quotient of the polynomial representation is $\mathcal{V}/Rad$. Indeed, any such quotient V of $\mathcal{V}$ supplies it with a form $\{f,g\}_V = \{f',g'\}$ for the images f',g' of f,g in V. Then a proper linear combination $\{\,,\,\}_o$ of the canonical form $\{\,,\,\}$ from (3.10.17) and $\{\,,\,\}_V$ will satisfy $\{1,1\}_o = 0$, which immediately makes it zero identically.

Corollary 3.10.5. *(i) A self-dual spherical X,Y–semisimple pseudo-unitary irreducible representation of $\mathcal{H}\!\!\mathcal{H}^\flat$ is finite dimensional and possesses an X–eigenvector of weight $-\rho_k$ and a Y–eigenvector of weight ρ_k. It exists if and only if*

$$\Upsilon_0[-\rho_k] \supset \Upsilon_+[-\rho_k] \subset \Upsilon_-[-\rho_k],\ \ |\,\Upsilon_+[-\rho_k]\,| \le \infty,$$
$$\text{and}\ \ \mu_\bullet(\widehat{w}) = 1\ \ \text{for}\ \ \widehat{w} \in \widehat{W}^\flat_*[\rho_k]. \tag{3.10.22}$$

There is only one such representation up to isomorphisms, namely, $\mathfrak{F}'[-\rho_k]$.

(ii) It is $PGL^c(2,\mathbb{Z})$–invariant if and only it contains the restricted Gaussian

$$\gamma_*(\pi_b) \stackrel{\mathbf{def}}{=\!=} q^{(b-2u_b^{-1}(\rho_k)\,,\,b)/2}\ \ \text{for}\ \ b_\sharp = b - u_b^{-1}(\rho_k),$$

which means the relations

$$q^{(b,b-2\rho_k)/2}\ =\ 1\ \text{for all}\ \ \pi_b \in \widehat{W}^\flat_*[\rho_k]. \tag{3.10.23}$$

In this case, τ_+ corresponds to multiplication by γ_*, and ψ' from (3.10.11) is proportional to the involution of $\mathfrak{F}'[-\rho_k]$ sending

$$\mathcal{E}'_b\gamma^{-1} \mapsto \gamma_*(\pi_b)^{-1}\,\mathcal{E}'_b\gamma_*^{-1},\ \ \text{where}\ \ \pi_b \in \Upsilon_+[-\rho_k]\,. \tag{3.10.24}$$

Proof. Only the last formula requires some comment. It is a straightforward specialization of (3.10.2). ❑

3.10.3 Generic q, singular k

The above corollary gives an approach to the classification of the triples $\{q,t,B\subset P\}$ such that $\mathcal{H}\!\!\mathcal{H}^\flat$ possesses a *perfect representation*, i.e., a finite dimensional nonzero irreducible self-dual semisimple pseudo-unitary representation with the action of $PGL^c(2,\mathbb{Z})$. If it exists, then this representation is isomorphic to $\mathfrak{F}'[-\rho_k]$.

The importance of these representations is obvious. They carry all properties of the classical Fourier transform and directly generalize the truncated

Bessel functions; see [CM],[CO], and Chapter 2. The Verlinde algebras and irreducible representations of the Weyl algebras are examples of such representations at the roots of unity

Here we will discuss only two "main sectors" of perfect representations, negative $k = -e/h$ for arbitrary q as $(e, h) = 1$ and, later, the case of positive integral $hk < N$ as q is a primitive N–root of unity. Here h is the Coxeter number. Actually, the theorems below are somewhat more general because we do not assume that the k–parameters coincide.

The root system A_1 is considered in detail in Chapter 2 and also in [C27] and [CO], including the explicit formulas for the Gauss–Selberg sums.

We begin with the case of generic q: $(q^{1/(2m)})^i(t^{1/2})^j$ do not represent nontrivial roots of unity for $i, j \in \mathbb{Z}$.

Theorem 3.10.6. *(i) In notation from Theorem 3.5.3, let us impose a somewhat stronger variant of condition (3.5.20), assuming that q is generic and*

$$\begin{aligned}(\rho_k, \alpha^\vee) \ &\notin \ \mathbb{Z}\setminus\{0\} \ \textit{for all } \alpha \in R_+, \qquad (3.10.25)\\ h_\alpha(k) = (\rho_k, \alpha^\vee) + k_\alpha \ &\notin \ \mathbb{Z} \ \textit{for all extreme } \alpha \neq \vartheta, \ h_\vartheta(k) \in \ -\mathbb{N}.\end{aligned}$$

Then $\mathfrak{F}'[-\rho_k]$ is perfect. It exists and remains perfect if q is an N–th root of unity, provided that $h_\vartheta(k) \geq -N$ and all fractional powers of q that may appear in the formulas are primitive roots of unity of maximal possible order.

(ii) Setting $e \stackrel{\mathbf{def}}{=} -h_\vartheta(k)$,

$$\widehat{W}^\flat_*[\rho_k] \ = \ \{(e\omega_r)u_r^{-1} \mid e\omega_r \in B, \, r \in O\}.$$

Identifying b upon the symmetries

$$\pi_b \mapsto \pi_b \cdot u_r \cdot (e\omega_r), \ \textit{i.e.}, \ b \mapsto b + e u_b^{-1}(\omega_r) \ \ \textit{for} \ \ (e\omega_r)\cdot u_r^{-1} \in \widehat{W}^\flat_*[\rho_k],$$

$$\begin{aligned}\mathfrak{F}'[-\rho_k] \ &= \ \sum_{b\in B} \mathbb{Q}_{q,t}\delta_b^\pi, \ \textit{where either } b = 0, \qquad (3.10.26)\\ &\textit{or } (\vartheta, \rho_k) + k_\vartheta - (b_-, \vartheta) < 0, \ \textit{or}\\ &(\vartheta, \rho_k) + k_\vartheta - (b_-, \vartheta) = 0 \ \ \textit{and} \ \ u_b^{-1}(\vartheta) \in R_-.\end{aligned}$$

(iii) For instance, we may pick $k_{\text{lng}} = k_{\text{sht}} = -e/h$ to ensure (3.10.25), where $(e, h) = 1$ for $e \in \mathbb{N}$ and $h = (\rho, \vartheta) + 1$ is the Coxeter number. In this case, the index of P/Q is relatively prime to e, so

$$\widehat{W}^\flat_*[\rho_k] \ = \ \{(e\omega_r)u_r^{-1}, \, \omega_r \in B, \, r \in O\}.$$

For example, let $\omega = \omega_1$, $k = k_{\text{lng}}$, $e = -2k \in \mathbb{N}$, and $B = P$ in the case $R = A_1$. Then assumption (3.10.25) means that e is odd and

$$\widehat{W}^\flat_*[\rho_k] = \{\pi_0 = \text{id}, \pi_{e\omega}\}, \ \mathfrak{F}'[-\rho_k] = \sum_{j=0}^{e-1} \mathbb{Q}_{q,t}\,\delta^\pi_{-j\omega}.$$

The proof is close to that of Theorem 3.5.3. The explicit description of the set $\Upsilon_+[-\rho_k]$ is nothing but the definition (3.6.8).

Let $bw \in \widehat{W}^\flat_*[\rho_k]$. See (3.6.4) and (3.6.14). Then $bw((\rho_k)) = w(\rho_k)+b = \rho_k$, $w \in W$, $b \in B$. Setting $\beta = w^{-1}(\alpha_i)$ for $1 \le i \le n$,

$$(w(\rho_k) - \rho_k\,,\, \alpha_i^\vee) = (\rho_k\,,\, \beta^\vee) - k_\beta = -(b, \alpha_i^\vee) \in \mathbb{Z}.$$

If $\beta > 0$, then there exists a simple root α_j of the same length as β such that $\beta^\vee - \alpha_j^\vee$ is a positive coroot. Hence $\beta = \alpha_j$ thanks to the first condition in (3.10.25). Similarly, β can be negative only for extreme $-\beta$. However, it contradicts the second condition unless $\beta = -\vartheta$. We see that w preserves the set $\{-\vartheta, \alpha_1, \ldots, \alpha_n\}$ and is u_r^{-1} for certain $r \in O$. We obtain that

$$b = \rho_k - u_r^{-1}(\rho_k) = e\omega_r \quad \text{and} \quad \pi_{e\omega_r} = (e\omega_r)\cdot u_r^{-1}.$$

Indeed,

$$(b, \alpha_i) = 0 \Leftrightarrow u_r(\alpha_i) \neq -\vartheta \Leftrightarrow i \neq r, \ (b, \alpha_r) = -h_\vartheta(k).$$

There is an additional condition to check, namely, (3.10.26) for $b_- = -e\omega_{r^*}$ and u_r:

$$(e\omega_{r^*}, \vartheta) < e, \text{ or } \{\, (e\omega_{r^*}, \vartheta) = e \ \text{ and } \ u_r^{-1}(\vartheta) \in R_- \,\}.$$

Since $(\omega_{r^*}, \vartheta) = 1$, only the second relation may happen. It holds because $u_r^{-1}(\vartheta) = -\alpha_r$.

Now let us establish the existence of the pseudo-unitary structure, which is equivalent to the relations $\mu_\bullet(\pi_{e\omega_r}) = 1$. Thanks to Proposition 3.6.5, only A_{4l-1} and D_{2l+1} have to be examined. In the other cases, the orders of the elements of $bw \in \widehat{W}^\flat_*[\rho_k]$ are not divisible by 4. The proof below is actually uniform, but the calculation is more relaxed in the simply-laced case.

Letting $k = k_{\text{lng}}$, the integer $e = -kh$ has to be relatively prime to h and, in particular, must be odd. Recall that $r \mapsto r^*$ for $r \in O$ describes the inversion in Π and corresponds to the automorphism of the nonaffine Dynkin diagram induced by $\varsigma = -w_0$. We will use (3.5.22):

$$\mu_\bullet(\pi_b) \;=\; \prod\Big(\frac{t_\alpha^{-1/2} - t_\alpha^{1/2} q_\alpha^{(\alpha^\vee,\rho_k)+j}}{t_\alpha^{1/2} - t_\alpha^{-1/2} q_\alpha^{(\alpha^\vee,\rho_k)+j}}\Big) \ \text{ for } \ \alpha \in R_+, \qquad (3.10.27).$$

$$-(b_-, \alpha^\vee) > j > 0 \ \text{ if } \ u_b^{-1}(\alpha) \in R_-, \ -(b_-, \alpha^\vee) \ge j > 0 \text{ otherwise.}$$

In the simply-laced case, $t_\alpha = t$, $q_\alpha = q$, and $\alpha^\vee = \alpha$. Let $b = e\omega_r$.

One has $-(b_-, \alpha^\vee) = e(\omega_{r^*}, \alpha) = e$ when α contains α_{r^*}, and $= 0$ otherwise. Using that the total number of factors here is even, namely, $2 * (e-1) * (\rho, \omega_{r^*})$, we may transpose the binomials in the numerator:

$$\mu_\bullet(\pi_{e\omega_r}) = \prod_{(\omega_{r^*},\alpha)=1}^{1<j<e} q^{2(\alpha,\rho_k)+2j} \Big(\frac{t^{1/2} - t^{-1/2} q^{-(\alpha,\rho_k)-j}}{t^{1/2} - t^{-1/2} q^{(\alpha,\rho_k)+j}}\Big). \qquad (3.10.28)$$

Here $j < e$ because $j = e$ may appear only under the condition $u_r^{-1}(\alpha) > 0$, which never holds. Indeed, let $\alpha = \alpha_{r^*} + \beta$, where β does not contain α_{r^*}. Then $u_r^{-1}(\alpha_{r^*} + \beta) = -\vartheta + \beta'$ for β' without α_r. Since ϑ contains all simple roots, we conclude that $-\vartheta + \beta'$ is always negative.

As a by-product, we have established that $-u_r^{-1}$, sending

$$\begin{aligned} &\Lambda \stackrel{\text{def}}{=} \{\alpha > 0,\, (\alpha, \omega_{r^*}) = 1\} \ni \alpha \mapsto \\ &\alpha' = -u_r^{-1}(\alpha) \in \{\alpha' > 0,\, (\alpha', \omega_r) = 1\} \stackrel{\text{def}}{=} \Lambda', \end{aligned} \tag{3.10.29}$$

is an isomorphism. Its inverse takes

$$(\alpha, \rho_k) \mapsto (-u_r(\alpha), \rho_k) = -(\alpha, u_r^{-1}(\rho_k)) = (\alpha, e\omega_r - \rho_k) = e - (\alpha, \rho_k).$$

Since the sets Λ and Λ' are isomorphic under the automorphism ς, the sets $\{(\alpha, \rho_k)\}$ and $\{(\alpha', \rho_k)\}$ coincide for $\alpha \in \Lambda$ and $\alpha' \in \Lambda'$. Finally, we conclude that $(\alpha, \rho_k) \mapsto e - (\alpha, \rho_k)$ is a symmetry of $\Lambda = \Lambda'$.

It ensures a complete cancelation of the binomials in (3.10.28) and results in

$$\begin{aligned} \mu_\bullet(\pi_{e\omega_r}) &= q^{2\Sigma},\ \Sigma = \sum_{\alpha\in\Lambda, 1<j<e} ((\alpha, \rho_k) + j) \\ &= (e-1)hk(\omega_{r^*}, \rho) + (e(e-1)/2)(2(\omega_{r^*}, \rho)) \ = \ 0. \end{aligned} \tag{3.10.30}$$

We have used that $hk = e$, $|\Lambda| = 2(\omega_{r^*}, \rho)$, and the following general formula

$$\sum_{\alpha>0} (a, \alpha)(b, \alpha^\vee) \ = \ h(a, b),\ \ a, b \in \mathbb{C}^n.$$

Therefore $\mathfrak{F}'[-\rho_k]$ is pseudo-unitary.

The last check is (3.10.23), ensuring the existence of the restricted Gaussian in this module:

$$\begin{aligned} (\omega_r, e\omega_r - 2\rho_k)/2 \ &= \ (\omega_r, -u_r^{-1}(\rho_k) - \rho_k)/2 \\ = -(\omega_r + u_r(\omega_r), \rho_k)/2 \ &= \ -(\omega_{r^*} - \omega_r, \rho_k)/2 = 0, \\ \text{where } \{(e\omega_r) \cdot u_r^{-1}\} \ &= \ \widehat{W}^\flat_*[\rho_k]. \end{aligned} \tag{3.10.31}$$

Once again we have used that the automorphism $\varsigma = -w_0$ transposes ω_r and ω_{r^*} and preserves ρ_k.

The analysis of the case of A_1 is straightforward, as is the statement about the roots of unity. ❑

Recall that $\mathfrak{F}'[-\rho_k]$ is a unique nonzero irreducible quotient of the polynomial representation $\mathcal{V}$ because the latter is spherical and the multiplicity of the character ϖ of $\mathcal{H}_Y^\flat$ in $\mathcal{V}$ is one for generic q. It is instructional to calculate the dimension of $\mathfrak{F}'[-\rho_k]$.

For the sake of simplicity, let $B = Q$, $k_{\text{lng}} = -e/h = k_{\text{sht}}$. We combine (3.10.26) with formula (3.1.22) applied to $z \in (1/e)Q$, which gives the existence and uniqueness of $u \in W$, $a \in Q$ such that

$$
\begin{aligned}
&z_- \stackrel{\text{def}}{=} ua((z)) \in (1/e)P_-, \ (z_-, \vartheta) \geq -1, \\
&(z_-, \vartheta) = -1 \Rightarrow u^{-1}(\vartheta) \in R_-, \text{ and} \\
&(\alpha_i, z_-) = 0 \Rightarrow u^{-1}(\alpha_i) \in R_+, \ i > 0.
\end{aligned} \tag{3.10.32}
$$

Multiplying z and a by e, we obtain that the elements b from (3.10.26) are in one-to-one correspondence with the elements of Q/eQ. So the dimension is e^n.

There is an immediate identification of the set of all b and Q/eQ and the dimension formula based on formula (3.6.26). Indeed, it gives that the dimension is the volume of the domain $\{z \in \mathbb{R}^n\}$ such that $(z, \alpha_i) \leq 0$ for $i > 0$ and $(z, \vartheta) \geq -e$ divided by the volume of the affine Weyl chamber.

More generally, the module $\mathfrak{F}'[-\rho_k]$ for arbitrary B is naturally isomorphic to the space of functions on B/eB because $\widehat{W}^\flat_*[\rho_k] = \{(e\omega_r)u_r^{-1}, \ \omega_r \in B\}$. We see that its dimension remains e^n.

3.10.4 Roots of unity

Let q be a primitive N-th root of unity under assumption (3.8.10). See also (3.6.28). However, we not assume now that fractional powers of q that appear in the formulas are primitive roots of unity unless otherwise stated.

We are going to describe the main "positive" sector of perfect $\mathcal{H}^\flat$-modules, generalizing the Verlinde algebras.

The original **Verlinde algebras** are subalgebras of "symmetric" elements in the perfect representations as $k = 1$, i.e., in the group case. The symmetric elements in arbitrary $\mathcal{H}^\flat$-modules are defined as follows: $T_i(v) = t_i^{1/2} v$ for $i > 0$.

For the main sector, k_{lng} and k_{sht} are positive and rational. The perfect modules exist only if further constraints are imposed.

We note that there exist other sectors of perfect representations, for instance, with negative $h_\vartheta(k)$ (see the above theorem) and with more special choices of the roots of unity, not only among the primitive ones. The complete list is unknown, although Corollary 3.10.5 seems sufficient for the complete classification.

To simplify considerations, we assume that N is greater than the Coxeter number, but this is actually not necessary. The perfect representations considered below are well defined for small N. To consider small N, one can follow the proof of the previous theorem instead of using the affine Weyl chambers (see below).

We pick

$$\hat{q} = q^{1/(2\hat{m})} \text{ for } (B, B+2\rho_k) = \hat{m}^{-1}\mathbb{Z},\ \hat{m} \in \mathbb{N},$$

setting

$$q^{(b,c+2\rho_k)} = \hat{q}^{(b,c+2\rho_k)\hat{m}}$$

unless stated otherwise. Note that q is primitive, but $\hat{q}$ is not supposed to be primitive $2\hat{m}N$–th root of unity.

Theorem 3.10.7. *(i) We assume that* $k_{lng}, k_{sht} > 0$,

$$\begin{aligned} &\mathbb{N} \ni h_\vartheta(k) = (\rho_k, \vartheta) + k_{sht} < N_{sht} = N, \qquad (3.10.33)\\ &\textit{and also either (a) } (N, \nu_{lng}) = 1, \textit{ or}\\ &\textit{(b) } \mathbb{N} \ni h_\theta(k) = (\rho_k, \theta^\vee) + k_{lng} < N_{lng} = N/(N, \nu_{lng})\\ &\textit{for the longest root } \theta \in R_+ \ (\nu_{lng} = (\theta, th)/2), \textit{ or}\\ &\textit{(c) } (\rho_k, \alpha^\vee) \notin \mathbb{Z}_+ \textit{ for all long } \alpha \in R_+. \end{aligned}$$

In case (c), we suppose that $\hat{q}$ is a primitive root of unity. Then relation (3.8.7) holds, which guarantees the existence of the Y-cospherical X-semisimple finite dimensional irreducible module $\mathfrak{F}'[-\rho_k]$. It is nonzero if $h_\vartheta(k) - N \le \ Max(B_-, \vartheta)$. In case (b), we also add here the counterpart of this inequality for θ.

(ii) We also impose either the condition

$$\begin{aligned} &\{\rho \in B,\ h_\vartheta(k) + h - 1 \le N,\ h_\theta(k) + h - 1 < N_{lng}\} \textit{ or}\\ &\{\rho \notin B,\ h_\vartheta(k) + h - 1 \le N + 1,\ h_\theta(k) + h - 1 < N_{lng} + 1\}. \qquad (3.10.34) \end{aligned}$$

Only the relations with ϑ are necessary under condition (a) or (c) from (i). Then $\mathfrak{F}'[-\rho_k]$ is pseudo-unitary and for $e \stackrel{\mathbf{def}}{=\!=} N - h_\vartheta(k)$,

$$\widehat{W}^\flat_*[\rho_k] = \{(e\omega_r)u_r^{-1} \mid e\omega_r \in B,\ q^{N(\omega_r, B)} = 1\}. \qquad (3.10.35)$$

This module is positive unitary if $q^{1/2} = \exp(\pi i/N)$. The restricted Gaussian exists in $\mathfrak{F}'[-\rho_k]$ if and only

$$q^{m_r} = \hat{q}^{2m_r\hat{m}} = 1 \ \textit{ for } \ m_r = (N - h_\vartheta(k))N(\omega_r, \omega_r)/2$$

and ω_r from (3.10.35). This may restrict the choices for $\hat{q}$.

(iii) Provided that we have the conditions from (i) and (3.10.34) from (ii),

$$\mathfrak{F}'[-\rho_k] = \mathbb{Q}_{q,t} \oplus \sum_{0 \ne b \in B} \mathbb{Q}_{q,t}\delta_b^\pi, \textit{ where} \qquad (3.10.36)$$

$$\begin{aligned} &\textit{(a) } h_\vartheta(k) - (b_-, \vartheta) < N, \textit{ or}\\ &\quad \{h_\vartheta(k) - (b_-, \vartheta) = N \textit{ and } u_b^{-1}(\vartheta) \in R_-\}, \textit{ and}\\ &\textit{(b) } h_\theta(k) - (b_-, \theta^\vee) < N_{lng}, \textit{ or}\\ &\quad \{h_\theta(k) - (b_-, \theta^\vee) = N_{lng} \textit{ and } u_b^{-1}(\theta) \in R_-\}. \end{aligned}$$

The latter is necessary only under (b) from (i). Here we identify b modulo $\widehat{W}^\flat_*[\rho_k]$:

$$b \mapsto b + u_b^{-1}(c),\ b_- \mapsto u_c(b_-) + c_- \ \text{ for } \ \pi_c \in \widehat{W}^\flat_*[\rho_k].$$

The space without this identification is also an $\mathcal{HH}^\flat$-module, which always contains the restricted Gaussian.

(iv) For example, let $R = A_1$,

$$\omega = \omega_1,\ k = k_{lng},\ e = N - 2k \in \mathbb{N},\ B = P.$$

Then

$$\mathfrak{F}'[-\rho_k] = \sum_{j=-e+1}^{e} \mathbb{Q}_{q,t}\,\delta^\pi_{j\omega} \ \text{ for primitive } q^{1/2}$$

is perfect. It is positive unitary for $q^{1/2} = \pm\exp(\pi i/N)$ if $k \in \mathbb{N}$; the sign has to be plus for a half-integral k. If $k \in \mathbb{N}$ and N is odd, we may pick $q^{1/2}$ in primitive N-th roots of unity. Then the summation above is over $1 \le j \le e$. The Gaussian exists in $\mathfrak{F}'[-\rho_k]$ in this case if either $N = 4l + 3$ for integral k or k is half-integral. This module is positive unitary for $q^{1/2} = -\exp(\pi i/N)$.

Proof. The elements π_b for b satisfying (3.10.36) constitute the set $\Upsilon_+[-\rho_k]$ due to the positivity of k_α. Here we use that $\rho_k \in P_+$, ϑ is the maximal coroot, and $\theta^\vee$ is maximal among the short coroots. The positivity also results in the inequalities

$$(\rho_k, \alpha^\vee) - (b_-, \alpha) - \upsilon k_\alpha + \epsilon^b_\alpha < N_\alpha = N/(\nu_\alpha, N)$$

for the same b and all $\alpha > 0$, where $\upsilon = 0, 1$, $\epsilon^b_\alpha = 1$ if $u_b^{-1}(\alpha) > 0$ and 0 otherwise. These inequalities are necessary and sufficient for the existence of $\mathfrak{F}'[-\rho_k]$. Let us calculate the stabilizer $\widehat{W}^\flat_*[\rho_k]$.

Lemma 3.10.8. *Under the assumptions from (ii), the group of elements π_c such that $\Upsilon_+[-\rho_k]\pi_c = \Upsilon_+[-\rho_k]$ is exactly $\Pi = \{(e\omega_r)u_r^{-1}, r \in O\}$ for $e = N - h_\vartheta(k)$.*

Proof. By definition, $\pi_c = cu_c^{-1}$ preserves the set of all element b satisfying (3.10.36) under the mapping

$$b \mapsto b', \ \text{ where } \ \pi_b\pi_c = \pi_{b'},\ b' = b + u_b^{-1}(c),\ u_{b'} = u_c u_b.$$

The corresponding set of b_- will be denoted by B^N_-. In terms of b_-:

$$\pi_b\pi_c = u_b^{-1}b_-u_c^{-1}c_- = u_b^{-1}u_c^{-1}u_c(b_-)c_- = u_{b'}^{-1} \cdot (u_c(b_-) + c_-).$$

The element $b'_- = u_c(b_-) + c_-$ has to be from B^N_-. Using the affine action (3.1.19):

$$b_- \mapsto b'_- \ = \ c_-u_c((b_-)) \ = \ u_c c((b_-)). \tag{3.10.37}$$

Due to the conditions (3.10.34) from part (ii), either $-\rho \in B_-^N$ (it happens when $\rho \in B$) or $-\rho - \omega_r \in B_-^N$ for proper minuscule ω_r otherwise. The inner points of the segment connecting 0 and such b_- in $\mathbb{R}^n$ are also inner in the polyhedron $\bar{B}_-^N \in \mathbb{R}^n$, that is defined by the same inequalities but for real $\bar{b}_-$. The set of its inner points is either $-e\mathfrak{C}^a$, where $e = N - h_\vartheta(k)$, $\mathfrak{C}^a$ is the affine Weyl chamber, or its intersection with the counterpart for θ and $N_{\text{lng}} - h_\theta(k)$ in case (b) from (i). See (3.1.21) for the definition of the affine Weyl chamber and its basic properties.

Hence, there exist points that do not leave the negative Weyl chamber $-\mathfrak{C}^a$ under

$$z \mapsto e^{-1}u_c c(\!(ez)\!).$$

Thus, all the points do not leave this chamber. The stabilizer of $-\mathfrak{C}^a = w_0(\mathfrak{C}^a)$ is $\{(-\omega_{r^*})u_{r^*}^{-1}\}$. Therefore

$$\begin{aligned} c_- u_c &= (-e\omega_{r^*})u_{r^*}^{-1} \text{ for } r \in O \text{ and } c_- = -e\omega_{r^*},\ u_c = u_{r^*}^{-1}, \\ \pi_c &= u_c^{-1}c_- = u_{r^*} \cdot (-e\omega_{r^*}) = (e\omega_r)u_r^{-1},\ c = e\omega_r. \end{aligned}$$

❑

The elements $\pi_c \in \widehat{W}_*^b[\rho_k]$ satisfy the conditions of the lemma. See (3.6.4) and (3.6.14). The definition is as follows:

$$\widehat{W}^b[\rho] = \{\widehat{w} \in \Upsilon_+[-\rho_k],\ q^{(\widehat{w}(\!(\rho_k)\!)-\rho_k\,,b)} = 1 \text{ for all } b \in B\}. \tag{3.10.38}$$

Setting $\widehat{w} = (e\omega_r)u_r^{-1}$,

$$\widehat{w}(\!(\rho_k)\!) - \rho_k = u_r^{-1}(\rho_k) - \rho_k + e\omega_r = h_\vartheta(k)\omega_r + e\omega_r = N\omega_r.$$

Cf. the proof of the previous theorem. Hence $q^{N(\omega_r,b)}$ has to be 1 for all $b \in B$. We arrive at the conditions from (ii). Now we can simply follow Theorem 3.10.6, formulas (3.10.30) and (3.10.31), to check the existence of the pseudo-unitary structure and the restricted Gaussian.

The claim about the positivity of the pairing for the "smallest" root of unity $q^{1/2} = \pm\exp(\pi i/N)$ is straightforward too. We represent the binomials in formula (3.10.28) for $\mu_\bullet$ in the form $(e^{ix} - e^{-ix})$ and use inequalities (3.10.36). In the numerator, the x do not reach π and $e^{ix} - e^{-ix} = ic$ for positive numbers c. Therefore the same holds true in the denominator. ❑

As an application, we obtain a counterpart of Theorem 3.5.1 at roots of unity, namely,

$$\langle \mathcal{E}_b' \, (\mathcal{E}_c')^* \, \gamma\mu_\bullet \rangle_\sharp' = q^{-(b_\sharp,b_\sharp)/2-(c_\sharp,c_\sharp)/2+(\rho_k,\rho_k)} \mathcal{E}_c(q^{b_\sharp}) \langle \gamma\mu_\bullet \rangle_\sharp', \tag{3.10.39}$$

$$\begin{aligned} \langle \mathcal{E}_b' \, \mathcal{E}_c' \gamma\mu_\bullet \rangle_\sharp' &= q^{-(b_\sharp,b_\sharp)/2-(c_\sharp,c_\sharp)/2+(\rho_k,\rho_k)} \\ &\times \prod_\nu t_\nu^{l_\nu(w_0)/2} T_{w_0}^{-1}(\mathcal{E}_c)(q^{b_\sharp}) \langle \gamma\mu_\bullet \rangle_\sharp'. \end{aligned} \tag{3.10.40}$$

Here $\langle f\mu_\bullet\rangle'_\sharp$ is the sum $\sum f\mu_\bullet(\pi_b)$ over $\{\pi_b\}$ for b satisfying (3.10.36) upon the identification modulo $\widehat{W}^\flat_*[\rho_k]$. The functions $\mathcal{E}'_b, \mathcal{E}'_c$ are the images of the polynomials $\mathcal{E}_b, \mathcal{E}_c$ in $\mathfrak{F}'[-\rho_k]$ for b, c satisfying the same constraint. They are well defined and nonzero. See Theorem 3.10.2.

Concerning the generalized Gauss–Selberg sums $\langle\gamma\mu_\bullet\rangle'_\sharp$, the formulas can be obtained using the shift operators, which we do not discuss. See [C25] about the integral k and [C27] with the complete list of formulas for A_1. Generally speaking, the shift operator can be used as follows.

Given k, first we calculate $\langle\gamma\mu_\bullet^\varkappa\rangle'_\sharp$ for $\varkappa = \{\varkappa_{\text{sht}}, \varkappa_{\text{lng}}\}$ taken from the sets $k_{\text{sht}} + \mathbb{Z}$ and $k_{\text{lng}} + \mathbb{Z}$ with the simplest possible perfect representations. We denote the corresponding μ by $\mu^\varkappa$. If the k are fractional, we take $\varkappa$ negative and use the Macdonald identities from Section 3.5. Then we can apply the counterpart of (3.5.18):

$$\frac{\langle\gamma\mu_\bullet^k\rangle'_\sharp}{\langle\gamma\mu_\bullet^\varkappa\rangle'_\sharp} = \prod_{\alpha\in R_+}\prod_{j=1}^{\infty}\Big(\frac{1-q_\alpha^{(\rho_k,\alpha^\vee)+j}}{1-t_\alpha^{-1}q_\alpha^{(\rho_k,\alpha^\vee)+j}}\Big)\Big(\frac{1-t_\alpha^{-1}q_\alpha^{(\rho_\varkappa,\alpha^\vee)+j}}{1-q_\alpha^{(\rho_\varkappa,\alpha^\vee)+j}}\Big). \tag{3.10.41}$$

3.11 DAHA and double polynomials

In this section we will proof the q,t–variant of the Gordon theorem justifying the Haiman conjecture about the diagonal coinvariants from [Go]. See Section 2.12.5 in Chapter 2. The last two subsections are devoted to the universal Dunkl operators. We follow [C29]. In this section, we take $B = P$, i.e., $\mathcal{H}^\flat = \mathcal{H}$.

Let $k_{\text{sht}} = -(1+1/h) = k_{\text{lng}}$ for the Coxeter number $h \stackrel{\text{def}}{=} 1 + (\rho, \vartheta)$. The polynomial representation $\mathbb{Q}[x]$ of the rational DAHA $\mathcal{H}''$ is $\{D, W\}$–spherical by construction. Indeed, 1 is its generator and is a unique W–invariant polynomial killed by all Dunkl operators D_b. Then, the general theory guarantees that the polynomial representation has a unique nonzero irreducible quotient module. Let us denote it V''. The subalgebra $(\mathcal{H}'')^W$ of the W–invariant elements from $\mathcal{H}''$ obviously preserves the space

$$\mathbb{Q}\tilde{\delta} \subset V'' \ \text{ for } \delta \stackrel{\text{def}}{=} \prod_{\alpha>0} x_\alpha;\ \tilde{\delta} = \text{image of } \delta.$$

Let $I_o \subset (\mathbf{H}'')^W$ be the ideal of the elements from the rational DAHA vanishing at $\tilde{\delta}$ in V''. Gordon proves that V'' *coincides* with the quotient $\tilde{V}''$ of the module $\mathcal{H}''(\delta) = \mathbb{Q}[x]$ by the $\mathcal{H}''$–submodule $\mathcal{H}''I_o(\delta)$, where the map is naturally $\delta \mapsto \tilde{\delta}$. Here the key step is his proof of the irreducibility of $\tilde{V}''$.

We are going to generalize the equality $\tilde{V}'' = V''$ to the q,t–case. The proof is simpler than in the rational case and leads to a new proof of Gordon's theorem via the Lusztig isomorphisms from Section 2.12.5 (Chapter 2) and the classical facts about the Weyl algebras at roots of unity.

3.11.1 Good reductions

We will begin with a generalization of the deformations of the perfect representations, from the previous chapter (A_1), also called there the nonsymmetric Verlinde algebras.

Let $k_{\rm sht} = -e/h = k_{\rm lng}$, provided that $(e,h) = 1$, $e \in \mathbb{N}$. The notation will be k for both $k_{\rm sht}$ and $k_{\rm lng}$, since we assume that they coincide in this section. The $\mathcal{H}$ is for the *generic* q $(t = q^k)$. By generic, we mean that $(q^{1/(2m)})^i(t^{1/2})^j$ do not represent nontrivial roots of unity for $i, j \in \mathbb{Z}$.

A **good cyclotomic reduction** of the pair $\{q^{1/m}, k\}$ is defined as follows:

(a) q and $q^{1/m}$ are both primitive roots of unity of order $N \in \mathbb{Z}_+$;

(b) $(N,h) = 1$ and $\widetilde{k}h + e = N$ for the Coxeter number h.

The corresponding DAHA, subject to (a) and (b), will be denoted by $\widetilde{\mathcal{H}}$ in this section. Note that $\widetilde{k} = k \bmod N$ and $\widetilde{k}h < N$, so $\widetilde{\mathcal{H}}$ is really a specialization of $\mathcal{H}$.

In contrast to Theorem 3.10.7, we choose $q^{1/m}$ in the primitive N–th roots of unity. Recall that $(P,P) = (1/m)\mathbb{Z}$; hence, the condition $(N,h) = 1$ implies $(m,h) = 1$ and such a choice is possible.

As a matter of fact, the theory for such $q^{1/m}$ is completely equivalent to that of the $\mathcal{H}^\flat$ for $B = Q$ and a primitive N–th root q, i.e., is equivalent to the case of little DAHA. However, we prefer to stick to the "standard" DAHA, with $B = P$, and the above choice of $q^{1/m}$.

The perfect representations for generic q and $k = -e/h$ remain perfect for $q, \widetilde{k}$ satisfying (a) and (b). We do not claim that such N, q, k constitute the whole set of possibile "good reductions" of perfect representations, i.e., when the latter are well defined and remain irreducible at the roots of unity.

Theorem 3.11.1. *The perfect representation of $\mathcal{H}$ for generic q and $k = -e/h$ as $(e,h) = 1, e \in \mathbb{N}$ becomes irreducible $\widetilde{\mathcal{H}}$–modules when $q^{1/m}$ is taken in the primitive N–th roots of unity and*

$$\widetilde{k} \in \mathbb{Z}_+,\ \widetilde{k}h + e = N.$$

Proof follows that of Theorem 3.10.6. We also apply Theorem 3.10.7.

Concerning the use of Theorem 3.10.7 for such a choice of $q^{1/m}$, the perfect modules become smaller. They are now quotients of the polynomial representation for $B = Q$, that is, $\mathbb{C}[X_\alpha, \alpha \in R]$. In all other respects the theory remains unchanged. We may identify the corresponding perfect representation with the space $\mathrm{Funct}(Q/eQ)$, which proves the theorem. ❑

The theorem is the main engine of our approach to the diagonal coinvariants, although we will not need this statement for the proof of the Main theorem below. The latter uses the good reduction only for trivial $\widetilde{k} = 0$ when $e = h + 1$.

3.11.2 Main theorem

Now $k_{\text{sht}} = -1 - (1/h) = k_{\text{lng}}$ and q is generic. We denote k_{sht} and k_{lng} by k. It is the simplest nontrivial (non-one-dimensional) example of a perfect module for generic q.

The t–counterpart of the element δ is

$$\Delta = \prod_{\alpha \in R_+} (t_\alpha^{1/2} X_\alpha^{1/2} - t_\alpha^{-1/2} X_\alpha^{-1/2}).$$

It plays a major role in the definition of the t–shift operator (see [C16]). One has:

$$T_i(\Delta) = -t_i^{-1/2}\Delta,\ 1 \le i \le n.$$

We extend $\varpi_-(T_i) = -t_i^{-1/2}$ to a one-dimensional representation of the **nonaffine Hecke algebra H** generated by $T_i, 1 \le i \le n$.

We come to a q,t–generalization of Gordon's theorem. The field of rationals of $\mathbb{Q}_{q,t}$ will be denoted by $\widetilde{\mathbb{Q}}_{q,t}$.

Theorem 3.11.2. *(i) The polynomial representation $\mathbb{Q}_{q,t}[X]$ of $\mathcal{H}$ has a unique nonzero quotient V that is torsion-free and irreducible over $\widetilde{\mathbb{Q}}_{q,t}$. It is of dimension $(1+h)^n$. The action of X and Y is semisimple with simple spectra. The module $V \otimes \widetilde{\mathbb{Q}}_{q,t}$, considered as an* **H***–module, contains a unique submodule isomorphic to ϖ_-.*

(ii) The module V coincides with the quotient $\tilde{V}$ of $\mathbb{Q}_{q,t}[X]$ by the $\mathcal{H}$–submodule $\mathcal{H}\mathcal{I}_o(\widetilde{\mathbb{Q}}_{q,t}\Delta)$ intersected with $\mathbb{Q}_{q,t}[X]$, where $\mathcal{I}_o$ is the kernel of the algebra homomorphism $\mathcal{H}_{inv} \ni H \mapsto H(\Delta) \in V$ for the subalgebra $\mathcal{H}_{inv}$ of the elements of $\mathcal{H}$ commuting with $T_1, \ldots, T_n$.

Proof. The existence of V and its isomorphism with the linear space $\text{Funct}[P/(1+h)P]$ follow from the previous section. See Theorem 3.10.6 and formula (3.10.32). The statements there are for general $k = -e/h$ as $(e,h) = 1$.

Comment. Actually, we need here only the self-duality of V from the general theory of this chapter, that is, the action of the involution ε of $\mathcal{H}$ from (3.2.10) in V. No other facts will be used, including the isomorphism with the $\text{Funct}[P/(1+h)P]$. The self-duality can be justified directly, without the general theory. It readily follows from the realization of V as the quotient of $Q_{q,t}[X]$ by the radical of the invariant bilinear form from Lemma 3.10.3.

Moreover, the self-duality will be needed only upon the reduction • below, where it is simple to check directly. We mention that the automorphisms $\tau_\pm$

can be defined in V as well, but we do not use it in this section. Thus the proof we give is essentially self-contained. ❑

The construction of V holds over $\mathbb{Q}_{q,t}$. Actually, it suffices to have the definition of V and to prove the theorem over $\widetilde{\mathbb{Q}}_{q,t}$. Then one can use the basic facts about the modules over PID. Note that the module $\mathbb{Q}_{q,t}[X]/\mathcal{H}\mathcal{I}_o(\Delta)$, generally speaking, has torsion.

The uniqueness of ϖ_- in $V \otimes \widetilde{\mathbb{Q}}_{q,t}$ results from the following fact. The orbit $W(\rho)$ is a unique simple W–orbit in $P/(1+h)P$. It will be checked below. ❑

Comment. There is another proof of the theorem based on the shift operator, which identifies the ϖ_-–component of $V = V^k$ with the ϖ_+–component of V^{k+1} defined for $k+1 = -1/h$, where $\varpi_+ : T_i \mapsto t_i^{1/2}$. The V^{k+1} is one-dimensional and coincides with its ϖ_+–component. The shift operator here is the division by Δ. This description is helpful when calculating the ideal $\mathcal{I}_o$; its intersections with $\mathbb{Q}_{q,t}[X]_{\text{inv}}$ and $\mathbb{Q}_{q,t}[Y]_{\text{inv}}$ are not difficult to describe. This approach gives that the $\mathcal{I}_o$ is ε–invariant and results in a "direct" verification of the self-duality of the $\tilde{V}$.

Note that the self-dualty of V combined with the uniqueness of ϖ_- in V formally imply that $\tilde{V}$ is self-dual too. Indeed, the character ϖ_- is ε–invariant. So are $\mathbb{Q}_{q,t}\Delta \subset V$, $\mathcal{I}_o$, and the kernel of the map $\mathbb{Q}_{q,t}[X] \to \tilde{V}$.

3.11.3 Weyl algebra

For $N \stackrel{\text{def}}{=\!=} 1+h$, we take $q = \exp(2\pi\imath/N)$, making $k = 0, t^{\pm 1/2} = 1$. Using that ν_α and the index of P/Q are relatively prime to N, we will pick $q^{1/m}$ in the roots of unity of the same order N. The bullet $^\bullet$ will be used to denote this reduction. The algebra $\mathcal{H}^\bullet$ is simply the semidirect product of the Weyl algebra generated by pairwise commutative X_a, Y_b for $a, b \in P$ and $\mathbf{H}^\bullet = \mathbb{Q}_q W$. The relations are

$$w\,X_a\,w^{-1} = X_{w(a)},\ w\,Y_b\,w^{-1} = Y_{w(b)},\ X_a Y_b X_a^{-1} Y_b^{-1} = q^{-(a,b)}.$$

We define $V^\bullet$ as a unique nonzero irreducible quotient of $\mathbb{Q}_q[X]$. It is self-dual. It is straightforward to check it directly, as well as the semisimplicity of X and Y, in the $\bullet$–case.

Since all eigenvalues of Y_b in $V^\bullet$ (and in the whole $\mathbb{Q}_q[X]$) are N–th roots of unity, the same holds for X_b in $V^\bullet$, thanks to the self-duality. Thus $X_b^N = 1 = Y_b^N$ in $V^\bullet$ for all $b \in P$.

Theorem 3.10.6 guarantees that V remains irreducible under such specialization, so $V^\bullet$ is a point of good reduction of V. It also can be seen from the dimension formula in the following lemma.

Lemma 3.11.3. *(i) The algebra* $\mathcal{HH}^\bullet_N \stackrel{\mathrm{def}}{=} \mathcal{HH}^\bullet/(X^N = 1 = Y^N)$ *has a unique irreducible nonzero representation* $V^\bullet$ *up to isomorphisms. Its dimension is* N^n.

(ii) It is isomorphic to $Q_q[P/NP]$ *as a* W*–module. The representation* $\varpi^\bullet_- : w \mapsto sgn(w)$ *has multiplicity one in* $V^\bullet$.

(iii) The quotient $\tilde{V}^\bullet$ *of* $\mathbb{Q}_q[X]$ *by* $\mathcal{HH}^\bullet\mathcal{I}^\bullet_o(\Delta^\bullet)$ *is an* $\mathcal{HH}^\bullet_N$*–module and coincides with* $V^\bullet$.

Proof. In the first place, $\mathcal{HH}^\bullet_N$ is a group algebra of a finite group, and therefore it is semisimple. Let us use the well-known fact that the Weyl algebra generated by X_a, Y_b modulo the (central) relations $X^N = 1 = Y^N$ has a unique irreducible representation up to isomorphisms and W acts in this representation. This representation equals $\mathbb{Q}_q[X]/(X^N = 1)$ and is just $V^\bullet$.

The multiplicity one statement from (ii) follows from the uniqueness of a simple W–orbit in P/NP. Let us check it.

We can assume that the orbit is in the form $W(b)$ for $b \in P$ such that $0 < (b, \alpha^\vee) < 1 + h$ for all $\alpha \in R_+$. Therefore b can be $\rho = \sum_i^n \omega_i$ or $\rho + \omega_r$ for $r \in O'$, i.e., for a minuscule weight ω_r. Indeed, the coefficient of $\alpha_i^\vee$ in the decomposition of ϑ in terms of simple coroots is one only for $r \in O'$. For $b = \rho + \omega_r$, let $w = u_r^{-1}$ for u_r from $\omega_r = \pi_r u_r$. Then

$$(w(\rho) - b, \alpha_r) = -(\rho, \vartheta) - (\omega_r, \alpha_r^\vee) - 1 = -(1 + h) = -N \ \text{ and}$$
$$(w(\rho) - b, \alpha_i) = (\rho, \alpha_j) - (\rho, \alpha_i) = 0 \ \text{ for } \ i \neq r, \ w^{-1}(\alpha_i) = \alpha_j.$$

Thus b and ρ generate the same W–orbit modulo NP.

The module $V^\bullet$ is obviously self-dual; so is $\tilde{V}^\bullet$ because the $\mathbf{H}$–module $\mathbb{Q}_q\Delta^\bullet$ is of multiplicity one in $\tilde{V}^\bullet$ and therefore invariant with respect to ε. It gives that $\tilde{V}^\bullet$ is a finite dimensional $\mathcal{HH}^\bullet_N$–module. It has $V^\bullet$ as a quotient and contains a unique W–submodule isomorphic to $\varpi^\bullet_-$.

Supposing that the kernel K of the map $\tilde{V}^\bullet \to V^\bullet$ is nonzero, it must contain a nonzero $\mathcal{HH}^\bullet_N$–submodule. Hence K contains at least one copy of $V^\bullet$ (the uniqueness), and the multiplicity of $\varpi^\bullet_-$ in $\tilde{V}^\bullet$ cannot be one. ❑

Actually, the proof we give is similar to that of the standard fact that the discriminant is the W–anti-invariant of the lowest possible degree in the ring of polynomials from [Bo].

Comment. Note that claims (ii)–(iii) are somewhat unusual in the general theory of Weyl algebras. Given the rank, they hold only for special choices of the roots of unity. For instance, N must be 3 in the case of A_1. Otherwise the dimension of the space of "anti-invariants" (odd characteristics in the classical theory of theta functions) is greater than one.

A general problem is to find "natural" σ–eigenfunctions in the module $V^\bullet$. Its self-duality does not garrantee that the solution is simple. Actually, it is a typical problem in the theory of automorphic functions. Recall that

σ is a cyclotomic counterpart of the classical Fourier transform, where such questions are of fundamental importance. The above approach gives a simple answer, but only for very small N. ❑

Coming back to the general case, the coincidence statement of the theorem is actually over $\tilde{\mathbb{Q}}_{q,t}$, so it suffices to check it at one special point. The lemma gives it for the $\bullet$–point. To be more exact, claim (iii) of the lemma gives the irreducibility of $\tilde{V}$ and therefore the coincidence $\tilde{V} = V$ at the common point, and proves the theorem. ❑

The application to the diagonal coinvariants goes via the universal Dunkl operators, that will be introduced later. The problem is to calculate the action of Y_b in the linear space of Laurent polynomials $\mathbb{Q}_{q,t}[X,Y]$ identified with the $\mathcal{H}$–module induced from a one-dimensional character of $\mathbf{H}$. We assume here that the X–monomials are placed before the Y–monomials. Then the action of X_b in the space $\mathbb{Q}_{q,t}[X,Y]$ is the "commutative" multiplication by X_b. The action of Y_b is by the left multiplication of the monomials in the form $X^{\cdot}Y^{\cdot}$. Hence it requires reordering and leads to nontrivial formulas.

The $\mathcal{H}$–module $\mathbb{Q}_{q,t}[X,Y]$ is obviously self-dual. However, since it is necessary to order X and Y after applying ε, the formulas for its action in $\mathbb{Q}_{q,t}[X,Y]$ are involved.

Now, the module V is the quotient of the module $\mathbb{Q}^{-}_{q,t}[X,Y]$ induced for the character ϖ_- of $\mathbf{H}$ by its submodule $\mathcal{H}\mathcal{I}_o(1)$. It is a one-parametric deformation of the polynomial ring $\mathbb{Q}_{q,t}[X,Y]$ divided by the ideal $\mathbb{Q}_{q,t}[X,Y]^W_o = \{g(X,Y)(f(X,Y) - f(1,1))\}$ for W–invariant Laurent polynomials f and arbitrary g. Therefore it can be identified with a quotient of the space of **diagonal coinvariants**

$$\mathbb{Q}[X,Y]/(\mathbb{Q}[X,Y]\mathbb{Q}[X,Y]^W_o) \simeq \mathbb{Q}[x,y]/(\mathbb{Q}[x,y]\mathbb{Q}[x,y]^W_o),$$

where the ring of definition is extended to $\mathbb{Q}_{q,t}$.

Concerning Gordon's theorem, the degenerations V', V'' of V can be introduced as quotients of the polynomial representation by the radicals of the degenerations of the bilinear form from Lemma 3.10.3. They are irreducible modules for $\mathcal{H}', \mathcal{H}''$, thanks to Lusztig's isomorphisms from Section 2.12.5, and satisfy the same multiplicity one statement. The modules $\tilde{V}', \tilde{V}''$ are defined in terms of the ideal of the W–invariant elements of the double Hecke algebra vanishing at the discriminant subspace of V', V'' (corresponding to the sign-character of W). The $\tilde{V}', \tilde{V}''$ are irreducible due to Lusztig's isomorphisms, and therefore $V' = \tilde{V}'$ and $V'' = \tilde{V}''$. The latter is the (main part of the) Gordon theorem.

Comment. One can use the specialization $\mathcal{H}^\bullet$ to obtain a "Laurent" counterpart of the Gordon theorem, without any reference to the theory of DAHA. An obvious difficulty is that there is no grading with respect to the degree of polynomials in this case. We therefore need to proceed as follows.

Let assume that N is a prime number p and make the field of constants $\mathbf{F}_p$. Then we readily obtain that the space $\mathbf{F}_p[X,Y]/(\mathbf{F}_p[X,Y]\mathbf{F}_p[X,Y]_o^W)$ must have a "natural" quotient-space isomorphic to $V^\bullet$ over $\mathbf{F}_p$. We use that $1-q$ belongs to the ideal in the ring of integers of the cyclotomic field $\mathbb{Q}[e^{\frac{2\pi i}{p}}]$, which contains (p) and has the residue field $\mathbf{F}_p$.

One can check that the formulas for the Lusztig homomorphism æ contain no denominators divisible by p, and therefore it is well defined over $\mathbf{F}_p$. Applying æ, we interpret $V^\bullet \otimes \mathbf{F}_p$ as a quotient of the space of the diagonal coinvariants considered over the field $\mathbf{F}_p$. It coincides with the quotient due to Gordon taken modulo p.

3.11.4 Universal DAHA

First, we will give a $X \leftrightarrow Y$–symmetric presentation of $\mathcal{H}\!\!\!\mathcal{H}$. It goes via the **universal double affine braid group** $\widehat{\mathfrak{B}}$. This group is defined to be generated by the pairwise commutative X_b, the pairwise commutative Y_b, the elements $\widehat{T}_i$, where $b \in P$, $0 \le i \le n$, and the group $\widehat{\Pi} = \{\widehat{\pi}_r, r \in O\} \simeq \Pi$ with the following defining relations:

(a) $\widehat{T}_i\widehat{T}_j\widehat{T}_i... = \widehat{T}_j\widehat{T}_i\widehat{T}_j...$, m_{ij} factors on each side,
$\widehat{\pi}_r\widehat{T}_i\widehat{\pi}_r^{-1} = \widehat{T}_j$ if $\widehat{\pi}_r(\alpha_i) = \alpha_j$;

(b) $\widehat{T}_iX_b\widehat{T}_i = X_bX_{\alpha_i}^{-1},\ \widehat{T}_i^{-1}Y_b\widehat{T}_i^{-1} = Y_bY_{\alpha_i}^{-1}$
if $(b,\alpha_i^\vee) = 1$ for $0 \le i \le n$;

(c) $\widehat{T}_iX_b = X_b\widehat{T}_i,\ \widehat{T}_iY_b = Y_b\widehat{T}_i$
if $(b,\alpha_i^\vee) = 0$ for $0 \le i \le n$;

(d) $\widehat{\pi}_rX_b\widehat{\pi}_r^{-1} = X_{\widehat{\pi}_r(b)},\ \widehat{\pi}_rY_b\widehat{\pi}_r^{-1} = Y_{\widehat{\pi}_r(b)},\ r \in O'$.

Note that no relations between X and Y are imposed. We continue using the notation $X_{[b,j]} = X_bq^j$, $Y_{[b,j]} = Y_bq^{-j}$. The element $q^{1/m}$ is treated as a generator that is central. Later, $q^{1/(2m)}$ will be needed.

The relations (a–d) are obviously invariant with respect to the involution

$$\widehat{\varepsilon}:\ X_b \leftrightarrow Y_b,\ \widehat{T}_i \mapsto \widehat{T}_i^{-1}(0 \le i \le n),\ \widehat{\pi}_r \mapsto \widehat{\pi}_r(r \in O). \tag{3.11.1}$$

Concerning (d), recall that π_r^{-1} is π_{r^*} for $r^* \in O$. The same holds for $u_r = \pi_r^{-1}\omega_r$ and $\widehat{\pi}_r$.

Theorem 3.11.4. *The group $\mathfrak{B}$ generated by $X, T, \Pi, q^{1/(2m)}$ subject to the relations (i–v) from Definition 3.2.1 coincides with the quotient of $\widehat{\mathfrak{B}}$ by the relations:*

$$\widehat{T}_0 = q^{-1}\,X_\vartheta\widehat{T}_{s_\vartheta}Y_\vartheta^{-1},\ \widehat{\pi}_r = q^{(\omega_r,\omega_r)/2}\,Y_r\widehat{T}_{u_r}^{-1}X_{r^*}^{-1}. \tag{3.11.2}$$

The map is

$$\begin{aligned}\mathfrak{pr}:\ & X_b \mapsto X_b,\ Y_b \mapsto Y_b,\ \widehat{T}_i \mapsto T_i\ (i>0),\ \widehat{T}_0 \mapsto q^{-1}\,X_\vartheta T_{s_\vartheta}Y_\vartheta^{-1},\\ & \widehat{\pi}_r \mapsto q^{(\omega_r,\omega_r)/2}\,Y_rT_{u_r}^{-1}X_{r^*}^{-1},\ q^{1/(2m)} \mapsto q^{1/(2m)},\end{aligned} \tag{3.11.3}$$

where the elements $Y_b \in \mathcal{H}$ are given by (3.2.8). The images of $\widehat{T}_i, \widehat{\pi}_r$ in $\mathfrak{B}$ coincide with $\tau_+(T_i), \tau_+(\pi_r)$ for τ_+ from (3.2.11). The relations (3.11.2) are invariant with respect to the involution $\widehat{\varepsilon}$. The latter becomes ε from (3.2.10) in $\mathfrak{B}$.

Proof. The key fact here is that

$$\varepsilon(\tau_+(T_i)) = (\tau_+(T_i))^{-1}\ (i \geq 0),\ \varepsilon(\tau_+(\pi_r)) = \tau_+(\pi_r), \tag{3.11.4}$$
$$\text{where } \tau_+(T_i) = T_i \text{ for } i > 0,\ \tau_+(T_0) = q^{-1} X_\vartheta T_{s_\vartheta} Y_\vartheta^{-1},$$
$$\tau_+(\pi_r) = q^{\frac{(\omega_r,\omega_r)}{2}} Y_r T_{u_r}^{-1} X_{r^*}^{-1},\ r \in O.$$

See formula (3.2.19) for the action of the involution $\eta = \varepsilon\tau_-^{-1}\tau_+\tau_-^{-1}$. We recall that this formula directly results in the invariance of the Gaussians with respect to the q–Fourier transform corresponding to the involution ε. The elements $\tau_+(T_i), \tau_+(\pi_r)$ are exactly the images of the elements $\widehat{T}_i, \widehat{\pi}_r$ in $\mathfrak{B}$ under $\mathfrak{pr}$. Relations (3.11.4) readily follow from the explicit formulas. ❑

There are important quotients of the group $\widehat{\mathfrak{B}}$ and the algebra $\widehat{\mathcal{H}}$ (defined below) obtained by imposing the commutativity of X with Y. They are essential in the theory of the q–Fourier transform. The $\{X_i\}$ are treated as the coordinates, and the $\{Y_i\}$ play the role of spectral parameters. Note the immediate projection of these quotients to $\mathfrak{B}$, $\mathcal{H}$ when we make $Y_b = X_b^{-1}$.

Comment. The realization of $\mathfrak{B}$ as a quotient of $\widehat{\mathfrak{B}}$ can be used for a direct proof that ε can be extended to an involution of $\mathfrak{B}$. It can simplify the straightforward proof due to Macdonald and the author from [M8], but the difference is not very significant. Our $\widehat{\mathfrak{B}}$ has something in common with the "triple affine Artin group" introduced recently by Ion and Sahi in [IS] for the purpose of interpreting the projective action of $PSL(2,\mathbb{Z})$. Compare our relation (3.11.2) and formula (23) in [IS], which establishes the connection of their group with the double affine braid group from [C13]. ❑

Turning to the Hecke algebras, let us define the **universal affine double Hecke algebra** $\widehat{\mathcal{H}}$ as the quotient of the group algebra $\mathbb{Q}_{q,t}\widehat{\mathfrak{B}}$ by the quadratic relations (o) from Definition 3.2.1 for $\widehat{T}_i$. Here we do not assume that $t_{\rm lng} = t_{\rm sht}$. Recall that the elements X_b and Y_b are entirely independent, so the counterpart of the PBW theorem is that an arbitrary element $\widehat{H} \in \widehat{\mathcal{H}}$ can be uniquely represented as

$$\widehat{H} = \sum_{\widehat{w}\in\widehat{W}} Q_{\widehat{w}} T_{\widehat{w}}, \quad \text{where } Q_{\widehat{w}} \text{ are noncommutative polynomials in } X, Y.$$

For applications to the Dunkl operators, this definition will be needed in the following form. We claim that the algebra $\widehat{\mathcal{H}}$ is generated over $\mathbb{Q}_{q,t}$ by the affine Hecke algebra

$$\widehat{\mathcal{H}} \stackrel{\text{def}}{=} \langle \widehat{T}_i, \widehat{\pi}_r \rangle,\ i \geq 0,\ r \in O,$$

the pairwise commutative $X_b\,(b \in P)$ and the pairwise commutative $Y_b\,(b \in P)$, satisfying the relations (d) above with the $\widehat{\pi}_r$, and the Lusztig-type relations

$$\widehat{T}_i X_b - X_{s_i(b)}\widehat{T}_i = (t_i^{1/2} - t_i^{-1/2})\frac{X_{s_i(b)} - X_b}{X_{\alpha_i} - 1},\ 0 \le i \le n, \tag{3.11.5}$$

$$\widehat{T}_i Y_b - Y_{s_i(b)}\widehat{T}_i = (t_i^{1/2} - t_i^{-1/2})\frac{Y_{s_i(b)} - Y_b}{Y_{\alpha_i}^{-1} - 1},\ 0 \le i \le n. \tag{3.11.6}$$

Imposing (3.11.2), we represent $\mathcal{H}$ as a quotient of $\widehat{\mathcal{H}}$. Note that this definition is compatible with the restriction to the lattices B between Q and P taken instead of P.

We set

$$\widehat{Y}_{b_+} \stackrel{\mathbf{def}}{=\!=} \widehat{\pi}_r \widehat{T}_{i_1} \cdots \widehat{T}_{i_l} \ \text{ for } \ b_+ = \pi_r s_{i_1} \cdots s_{i_l} \text{ as } b_+ \in P_+,\ l = l(b_+).$$

More generally, $\widehat{Y}_{b_+ - c_+} = \widehat{Y}_{b_+}\widehat{Y}_{c_+}^{-1}$ for $b_+, c_+ \in P_+$.

3.11.5 Universal Dunkl operators

Given a representation $\widehat{V}$ of $\widehat{\mathcal{H}}$, the general **universal Dunkl operators** are the images of $\widehat{Y}_b\,(b \in P)$ and $\widehat{\pi}_r\,(r \in O)$ in the $\widehat{\mathcal{H}}$–module $\mathcal{I}_{\widehat{V}}$ induced from $\widehat{V}$. As a linear space, it is isomorphic to the linear space of *noncommutative* polynomials of X, Y with the (right) coefficients in $\widehat{V}$: $\mathcal{I}_{\widehat{V}} = \cup_{l \in \mathbb{N}} \mathcal{P}_l$ for

$$\mathcal{P}_l \stackrel{\mathbf{def}}{=\!=} \{\sum_{\mathbf{b},\mathbf{c} \in \mathbf{P}} X_{b_1} Y_{c_1} \cdots X_{b_l} Y_{c_l} \widehat{v}_{\mathbf{b},\mathbf{c}}\},\ \mathbf{b} \in \mathbf{P} = P^l \ni \mathbf{c},\ \widehat{v}_{\mathbf{b},\mathbf{c}} \in \widehat{V}. \tag{3.11.7}$$

Here the sums in (3.11.7) represent different vectors in $\mathcal{I}_{\widehat{V}}$ for different v–coefficients if we assume that $b_i \neq 0 \neq c_j$ for the indices $1 < i \le l,\ 1 \le j < l$ as $\widehat{v}_{\mathbf{b},\mathbf{c}} \neq 0$.

The action of X and Y is by left multiplication. The subspaces $\mathcal{P}_l$ are $\widehat{\mathcal{H}}$–submodules and also X–submodules for $l > 0$.

The action of $\widehat{T}_i\,(i \ge 0)$ and $\widehat{\pi}_r$ in $\mathcal{P}_l$ can be calculated using formulas (3.11.5) and and (3.11.6), and the relations (d) above.

In the first interesting case $l = 1$, it is as follows:

$$\begin{aligned}\widehat{T}_i(X_b Y_c \widehat{v}) &= X_{s_i(b)} Y_{s_i(c)} \widehat{T}_i(\widehat{v}) + \\ (t_i^{1/2} - t_i^{-1/2})&\Big(\frac{X_{s_i(b)} - X_b}{X_{\alpha_i} - 1} Y_c + X_{s_i(b)} \frac{Y_{s_i(c)} - Y_c}{Y_{\alpha_i}^{-1} - 1}\Big)\,\widehat{v},\ 0 \le i \le n,\end{aligned} \tag{3.11.8}$$

$$\widehat{\pi}_r(X_b Y_c \widehat{v}) = X_{\pi_r(b)} Y_{\pi_r(c)} \widehat{\pi}_r(\widehat{v}) \quad \text{for } \ b, c \in P,\ \widehat{v} \in \widehat{V}. \tag{3.11.9}$$

When the initial representation $\widehat{V}$ is the character

$$\widehat{\varpi}_+ : \widehat{T}_i \mapsto t_i^{1/2},\ i \ge 0,\ \widehat{\pi}_r \mapsto 1,$$

we arrive at a double variant of formulas (3.2.23) and the corresponding Dunkl operators:

$$\widehat{T}_i = t_i^{1/2} s_i^x s_i^y + (t_i^{1/2} - t_i^{-1/2})\Big(\frac{s_i^x - 1}{X_{\alpha_i} - 1} + s_i^x \frac{s_i^y - 1}{Y_{\alpha_i}^{-1} - 1}\Big). \qquad (3.11.10)$$

Here s_i^x, s_i^y act, respectively, on X and Y, and the differences are applied before the division in the divided differences. Similarly, $\widehat{\pi}_r = \pi_r^x \pi_r^y$.

3.11.6 Double polynomials

The above consideration leads to the formulas for the action of the hat-operators in the $\mathcal{H\!H}$-module $\mathbb{Q}_{q,t}[X,Y]$ induced from the character ϖ_+ of the nonaffine Hecke subalgebra $\mathbf{H}$.

These operators are the images of $\{\widehat{T}_i, \widehat{\pi}_r\}$ under the projection $\mathfrak{pr}$. They coincide with $\tau_+(T_i), \tau_+(\pi_r)$. See (3.11.4). We will use the same hat-notation for them, although now they are the elements of $\mathcal{H\!H}$.

The formulas are

$$\begin{aligned}
&\widehat{T}_i(X_b Y_c) = X_{s_i(b)} Y_{s_i(c)} \widehat{T}_i(1) + \\
&(t_i^{1/2} - t_i^{-1/2})\Big(\frac{X_{s_i(b)} - X_b}{X_{\alpha_i} - 1} Y_c + X_{s_i(b)} \frac{Y_{s_i(c)} - Y_c}{Y_{\alpha_i}^{-1} - 1}\Big),\\
&\widehat{T}_i(1) = t_i^{1/2}\ (i > 0),\ \widehat{T}_0(1) = q^{-1} X_\vartheta T_{s_\vartheta} Y_\vartheta^{-1}(1)\\
&= q^{-1} X_\vartheta (Y_\vartheta T_{s_\vartheta}^{-1}(1) - (t_0^{1/2} - t_0^{-1/2})),\\
&T_{s_\vartheta}^{-1}(1) = t_{\text{sht}}^{1-(\vartheta, \sum_{\text{sht}} \alpha^\vee)}\, t_{\text{lng}}^{-(\vartheta, \sum_{\text{lng}} \alpha^\vee)},\ \alpha \in R_+.
\end{aligned} \qquad (3.11.11)$$

The X–monomials act by left multiplication, the operators $\widehat{\pi}_r$ via the relations (d) and the formula for $\widehat{\pi}_r(1)$, similar to that from (3.11.11). Knowing the action of $X_b, \widehat{T}_i$, and $\widehat{\pi}_r$ is sufficient for determining the structure of the $\mathcal{H\!H}$-module.

Recall that the involution ε acts naturally in $\mathbb{Q}_{q,t}[X,Y]$, sending $\widehat{T}_i \mapsto \widehat{T}_i^{-1}$ (all i) and $\widehat{\pi}_r \mapsto \widehat{\pi}_r$.

We note that the hat-operators can be used, for instance, to introduce the "double radial parts." As usual, the simplest ones, corresponding to the minuscule symmetric monomial functions, can be calculated explicitly.

To conclude, we note that there is an interesting group of automorphisms of $\widehat{\mathfrak{B}}, \widehat{\mathcal{H\!H}}$ generated by the tau-automorphisms

$$\tau_\pm^x,\ \ \tau_\pm^y \stackrel{\mathbf{def}}{=\!=} \widehat{\varepsilon}\tau_\mp^x\widehat{\varepsilon}.$$

They act, respectively, in $\langle \widehat{T}, \widehat{\pi}, X\rangle$ fixing Y and in $\langle \widehat{T}, \widehat{\pi}, Y\rangle$ fixing X. This group is an extension of the $PSL_3(\mathbb{Z})$. Hopefully it is "naturally" connected to the action of $PSL_3(\mathbb{Z})$ on solutions of some KZB–type equations found by Felder and Varchenko. See [FV2].

The action of $PSL_3(\mathbb{Z})$ is directly related to the multiple Gamma functions introduced and studied by Barns [Ba]. This 100 year-old theory is of great potential for the DAHAs and the modern theory of special functions.

Bibliography

[An] G. E. Andrews et al., Eds., *Ramanujan Revisited*, Academic Press, San Diego (1988).

[A1] K. Aomoto, *A note on holonomic q–difference systems*, in Algebraic Analysis 1, Eds. M. Kashiwara, T. Kawai, Academic Press, San Diego (1988), 22–28.

[A2] K. Aomoto, *On elliptic product formulas for Jackson integrals associated with reduced root systems*, J. Algebraic Combin. 8:2 (1998), 115–126.

[AKM] — , and Y. Kato, and K. Mimachi, *A solution of the Yang–Baxter equation as connection coefficients of a holonomic q–difference system*, Duke Math. J. 65 (1992), 7–15.

[AI] R. Askey, and M.E.H. Ismail, *A generalization of ultraspherical polynomials*, in *Studies in Pure Mathematics*, Ed. P. Erdös, Birkhäuser, Boston (1983), 55–78.

[AW] R. Askey, and J. Wilson, *Some basic hypergeometric orthogonal polynomials that generalize Jacobi polynomials*, Memoirs AMS 319 (1985), 1–55.

[BF] H.M. Babujan, and R. Flume, *Off shell Bethe ansatz equation for Gaudin magnets and solutions of Knizhnik–Zamolodchikov equations*, Mod. Phys. Lett. A9 (1994), 2029–2040.

[BS] E.P. van den Ban, and H. Schlichtkrull, *The most continuous part of the Plancherel decomposition for a reductive symmetric space*, Annals of Math. 145 (1997), 267–364.

[Ba] E. Barns, *On the theory of multiple gamma function*, Trans. Cambridge Phil. Soc. 19 (1904), 374–425.

[BD] A.A. Belavin, and V.G. Drinfeld, *Solutions of the classical Yang–Baxter equations for simple Lie algebras*, Funct. Anal. Appl. 16 (1982), 159–180.

[BEG] B. Berest, and P. Etingof, and V. Ginzburg, *Finite-dimensional representations of rational Cherednik algebras*, IMRN 19 (2003), 1053–1088.

[Be] D. Bernard, *On the Wess–Zumino–Witten model on the torus*, Nucl. Phys. B 303 (1988), 77–93.

[Bi] J. Birman, *On Braid groups*, Commun. Pure Appl. Math. 22 (1969), 41–72.

[Bo] N. Bourbaki, *Groupes et algèbres de Lie*, Ch. 4–6, Hermann, Paris (1969).

[BZ] D. Bressoud, and D. Zeilberger, *A proof of Andrews' q-Dyson conjecture*, Discrete Math. 54 (1985), 201–224.

[Ca] F. Calogero, *Solutions of the one-dimensional n–body quadratic and/or inversely quadratic pair potential*, J. Math. Phys. 12 (1971), 419–436.

[Ch] K. Chandrasekharan, *Elliptic Functions*, A series of Comprehensive Studies in Math. 281, Springer-Verlag, Berlin, Heidelberg, New York, Tokyo (1985).

[C1] I. Cherednik, *A definition of τ–functions for generalized affine Lie algebras*, Funct. Anal. Appl. 17:3 (1983), 93–95.

[C2] — , *Factorized particles on a half-line and root systems*, Theor. Math. Phys. 61:1 (1984), 35–43.

[C3] — , *Functional realizations of basic representations of factorizable groups and Lie algebras* Funct. Anal. Appl. 19:3 (1985), 36–52.

[C4] — , *On special bases of irreducible finite-dimensional representations of the degenerate affine Hecke algebra*, Funct. Analysis Appl. 20 (1986), 87–89.

[C5] — , *A new interpretation of Gelfand–Tzetlin bases*, Duke Math. J. 54:2 (1987), 563–577.

[C6] — , *Generalized braid groups and local r–matrix systems*, Doklady Akad. Nauk SSSR 307:1 (1989), 27–34.

[C7] — , *Affine extensions of Knizhnik–Zamolodchikov equations and Lusztig's isomorphisms*, in *Special Functions*, ICM-90 Satellite Conference Proceedings, Eds. M. Kashiwara and T. Miwa, Springer-Verlag (1991), 63–77.

[C8] — , *Monodromy representations for generalized Knizhnik–Zamolodchikov equations and Hecke algebras*, Publ. RIMS 27 (1991), 711–726.

[C9] — , *Integral solutions of trigonometric Knizhnik–Zamolodchikov equations and Kac-Moody algebras*, Publ. RIMS 27 (1991), 727–744.

[C10] — , *A unification of Knizhnik–Zamolodchikov and Dunkl operators via affine Hecke algebras*, Invent. Math. 106 (1991), 411–431.

[C11] — , *Integration of quantum many-body problems by affine Knizhnik–Zamolodchikov equations*, Preprint RIMS 776 (1991), Adv. Math. 106 (1994), 65–95.

[C12] — , *Quantum Knizhnik-Zamolodchikov equations and affine root systems*, Commun. Math. Phys. 150 (1992), 109–136.

[C13] — , *Double affine Hecke algebras, Knizhnik–Zamolodchikov equations, and Macdonald's operators*, IMRN 9 (1992), 171–180.

[C14] — , *The Macdonald constant-term conjecture*, IMRN 6 (1993), 165–177.

[C15] — , *Induced representations of double affine Hecke algebras and applications*, Math. Res. Lett. 1 (1994), 319–337.

[C16] — , *Double affine Hecke algebras and Macdonald's conjectures*, Annals of Mathematics 141 (1995), 191–216.

[C17] — , *Elliptic quantum many-body problem and double affine Knizhnik–Zamolodchikov equation*, Commun. Math. Phys. 169:2 (1995), 441–461.

[C18] — , *Difference–elliptic operators and root systems*, IMRN 1 (1995), 43–59.

[C19] — , *Macdonald's evaluation conjectures and difference Fourier transform*, Inventiones Math. 122 (1995), 119–145.

[C20] — , *Nonsymmetric Macdonald polynomials*, IMRN 10 (1995), 483–515.

[C21] — , *Difference Macdonald-Mehta conjecture*, IMRN 10 (1997), 449–467.

[C22] — , *Lectures on Knizhnik–Zamolodchikov equations and Hecke algebras*, MSJ Memoirs 1 (1998).

[C23] — , *Intertwining operators of double affine Hecke algebras*, Selecta Math. New ser. 3 (1997), 459–495.

[C24] — , *Inverse Harish–Chandra transform and difference operators* IMRN 15 (1997), 733–750.

[C25] — , *From double Hecke algebra to analysis*, Doc. Math. J. DMV, ICM 1998, Extra volume II (1998), 527–531.

[C26] — , *On q-analogues of Riemann's zeta function*, Sel. Math., New ser. 7 (2001), 1–44.

[C27] — , *One-dimensional double Hecke algebras and Gaussian sums*, Duke Math. J. 108:3 (2001), 511–538.

[C28] — , *Double affine Hecke algebras and difference Fourier transforms*, Inventiones Math. 152:2 (2003), 213–303.

[C29] — , *Diagonal coinvariants and Double Affine Hecke algebras*, IMRN 16 (2004), 769–791.

[CM] — , and Y. Markov, *Hankel transform via double Hecke algebra*, Lecture Notes in Mathematics 1804, (1999), 1–25.

[CO] — , and V. Ostrik, *From Double Hecke Algebras to Fourier Transform*, Selecta Math., New Ser. 8 (2003), 1–89.

[CG] N. Chriss, and V. Ginzburg, *Representation Theory and Complex Geometry*, Birkhüser, Boston (1997).

[DJKM1] E. Date, and M. Jimbo, and M. Kashiwara, and T. Miwa, *Operator approach to the Kadomtsev-Petviashvili equation. Transformation groups for soliton equations III*, J. Phys. Soc. Japan 50 (1981), 3806–3812.

[DJKM2] — , and — , and — , and — , *Transformation groups for soliton equations*, in Proceedings of RIMS symposium, Eds. M. Jimbo and T. Miwa (1983), 39–120.

[DJMM] E. Date, and M. Jimbo, and A. Matsuo, and T. Miwa, *Hypergeometric-type integrals and the $SL_2(C)$ Knizhnik–Zamolodchikov equation*, Int. J. Mod. Phys. B4 (1990), 1049–1057.

[De] P. Delorme, *Formule de Plancherel pour les espaces symetriques reductifs*, Ann. of Math. 147 (1998), 417–452.

[Di] J.F. van Diejen, *Self-dual Koornwinder–Macdonald polynomials*, Invent. Math. 126 (1996), 319-339.

[DS] — , and J.V. Stokman, *Multivariable q–Racah polynomials.* Duke Math. J. 91 (1998), 89–136.

[DV] R. Doran, and V.S. Varadarajan, Eds., *The Mathematical Legacy of Harish-Chandra: A Celebration of Representation Theory and Harmonic Analysis*, Proceedings of Symposia in Pure Mathematics 68, AMS (2000).

[Dr1] V. Drinfeld, *Degenerate affine Hecke algebras and Yangians*, Funct. Anal. Appl. 21:1 (1986), 69–70.

[Dr2] — , *Quantum Groups*, Proc. ICM-86 (Berkeley) Amer. Math. Soc. 1 (1987), 798–820.

[Du1] C.F. Dunkl, *Differential-difference operators associated to reflection groups*, Trans. AMS. 311 (1989), 167–183.

[Du2] — , *Hankel transforms associated to finite reflection groups*, Contemp. Math. 138 (1992), 123–138.

[Ed] H.M. Edwards, *Riemann's Zeta Function*, Academic Press, New York and London (1974).

[EG] P. Etingof, and V. Ginzburg, *Symplectic reflection algebras, Calogero–Moser space, and deformed Harish-Chandra homomorphism*, Inventiones Mat. 147 (2002), 243–348.

[EK1] P. Etingof, and A. Kirillov Jr., *Representations of affine Lie algebras, parabolic equations, and Lamé functions*, Duke Math. J. 74:3 (1994), 585–614.

[EK2] — , and — , *Macdonald's polynomials and representations of quantum groups*, Math. Res. Let. 1 (1994), 279–296.

[EK3] — , and — , *Representation-theoretic proof of the inner product and symmetry identities for Macdonald's polynomials*, Compositio Mathematica (1995).

[EOR] P. Etingof, and A. Oblomkov, and E. Rains, *Generalized double affine Hecke algebras of rank 1 and quantized Del Pezzo surfaces*, Preprint math.QA/0406480 (2004).

[ES] — , and O. Schiffman, *Lectures on Quantum Groups*, International Press, Boston (1998).

[EVa] — , and A. Varchenko, *The orthogonality and qKZB–heat equation for traces of $U_q(\mathfrak{g})$–intertwiners*, Preprint math. QA/0302071 (2003).

[Ev] R.J. Evans, *The evaluation of Selberg character sums*, L'Enseignement Math. 37 (1991), 235–248.

[FT] L.D. Faddeev, and L.A. Takhtajan, *Hamiltonian Methods in the Theory of Solitons*, Springer Verlag (1987).

[FFR] B. Feigin, and E. Frenkel, and N. Reshetikhin, *Gaudin model, Bethe ansatz and correlation functions at critical level*, Comm. Math. Phys 166 (1994), 27–62.

[FKLMM] B. Feigin, and R. Kedem, and S. Loktev, and T. Miwa, and E. Mukhin, *Combinatorics of the* $\hat{sl}_2$ *spaces of coinvariants*, Transformation Groups 6:1 (2001), 25–52.

[Fe] G. Felder, *Conformal field theory and integrable systems associated to elliptic curves*, in Proceed. of the International Congress of Mathematicians, Zürich (1994), Birkhäuser, Boston (1994), 1247–1255,

[FTV] G. Felder, and V. Tarasov, and A. Varchenko, *Monodromy of solutions of the elliptic quantum Knizhnik-Zamolodchikov-Bernard difference equations*, Int. J. Math. 10 (1999), 943–975.

[FV1] G. Felder, and A. Varchenko, *Integral representation of solutions of the elliptic Knizhnik–Zamolodchikov–Bernard equation*, IMRN 5 (1995), 221–233.

[FV2] — , and — , *The elliptic gamma function and* $SL(3,\mathbb{Z})\ltimes\mathbb{Z}^3$, Adv. Math 156:1 (2000), 44–76.

[FV3] — , and — , *Hypergeometric theta functions and elliptic Macdonald polynomials*, IMRN 21 (2004), 1037–1055.

[FW1] G. Felder, and C. Wieczerkowski, *The Knizhnik–Zamolodchikov–Bernard equation on the torus*, in Proceedings of the Vancower Summer School on Mathematical Quantum Field Theory , CRM Proceedings and lecture Notes 7 (1994), 167–171.

[FW2] — , and — , *Conformal blocks on elliptic curves and the Knizhnik–Zamolodchikov–Bernard equations*, Commun. Math. Phys. 167 (1996), 133–162.

[FGT] S. Fishel, and I. Grojnowski, and C. Teleman, *The strong Macdonald conjecture*, Preprint math.RT/0107072 (2001).

[FR] I.B. Frenkel, and N.Yu. Reshetikhin, *Quantum affine algebras and holonomic difference equations*, Commun. Math. Phys. 146 (1992), 1–60.

[FHST] J. Fuchs, and S. Hwang, and A. M. Semikhatov, and I. Yu. Tipunin, *Nonsemisimple Fusion Algebras and the Verlinde Formula*, Preprint hep-th/0306274 (2003).

[GK1] D. Gaitsgory, and D. Kazhdan, *Representations of algebraic groups over a 2-dimensional local field*, Preprint RT/ 0302174 (2003).

[GK2] D. Gaitsgory, and D. Kazhdan, *Algebraic groups over a 2-dimensional local field: some further constructions*, Preprint RT/0406282 (2004).

[GV] R. Gandolli, and V.S. Varadarajan, *Harmonic Analysis of Spherical Functions on Real Reductive Groups*, Springer-Verlag (1988).

[GG] H. Garland, and I. Grojnowski, *Affine Hecke algebras associated to Kac-Moody groups*, Preprint q-alg/9508019 (1995).

[GH] A. Garsia, and M. Haiman, *A graded representation model for Macdonald's polynomials*, Proc. Nat. Acad. Sci. USA 90:3 (1993), 3607–3610.

[GR] G. Gasper, and M. Rahman, *Basic Hypergeometric Series*, Encyclopedia of Mathematics and its Applications 35 Cambridge University Press, Cambridge, MA (1990).

[GGOR] V. Ginzburg, and N. Guay, and E. Opdam, and R. Rouquier, *On the category $\mathcal{O}$ for rational Cherednik algebras*, Preprint RT/0212036, Inventiones Math. (2003).

[GKV] — , and M. Kapranov, and E. Vasserot, *Residue construction of Hecke algebras*, Preprint (1995).

[Go] I. Gordon, *On the quotient ring by diagonal harmonics*, Preprint RT/ 0208126 (2002), Inventiones Math.

[Ha1] M. Haiman, *Conjectures on the quotient ring by diagonal invariants*, J. Algebraic Combin. 3 (1994), 17–76.

[Ha2] — , *Hilbert schemes, polygraphs, and the Macdonald positivity conjecture*, Math. Soc. 14 (2001), 941–1006.

[HC] Harish-Chandra, *Discrete series for semisimple Lie groups*, Acta Math. 116 (1986), 1–111.

[H1] G.J. Heckman, *Root systems and hypergeometric functions II*, Comp. Math. 64 (1987), 353-373.

[H2] — , *An elementary approach to the hypergeometric shift operator of Opdam*. Invent. math. 103 (1991), 341–350.

[HO1] G.J. Heckman, and E.M. Opdam, *Root systems and hypergeometric functions I*, Comp. Math. 64 (1987), 329–352.

[HO2] — , and — , *Harmonic analysis for affine Hecke algebras*, in Current Developments in Mathematics, International Press, Boston (1996).

[HS] G.J. Heckman, and H. Schlichtkrull, *Harmonic Analysis and Special Functions on Symmetric Spaces*, Persp. in Math., Acad. Press (1994).

[He1] S. Helgason, *An analogue of the Paley–Wiener theorem for the Fourier transform on certain symmetric spaces*, Math. Ann. 165 (1966), 297–308.

[He2] — , *Differential Geometry, Lie Groups and Symmetric Spaces*, Acad. Press (1978).

[He3] — , *Groups and Geometric Analysis*, Academic Press, New York (1984).

[Io] B. Ion, *Involutions of double affine Hecke algebras*, Preprint QA/0111010 (2001).

[IS] B. Ion, and S. Sahi, *Triple affine Artin groups and Cherednik algebras*, Preprint QA/0304186 (2003).

[It] M. Ito, *On a theta product formula for Jackson integrals associated with root systems of rank two*, J. Math. Anal. Appl. 216 (1997), 122–163.

[Je] M.F.E. de Jeu, *The Dunkl transform*, Invent. Math. 113 (1993), 147–162.

[Ji] M. Jimbo, *A q–analogue of $U_q(gl(N+1))$, Hecke algebras, and Yang–Baxter equations*, Lett. Math. Phys. 11 (1986), 247–252.

[Ka] V.G. Kac, *Infinite Dimensional Lie Algebras*, Cambridge University Press, Cambridge (1990).

[KP] — , and D. Peterson, *Spin and wedge representations of infinite dimensional Lie algebras and groups*, Proc. Nat. Acad. Sci. USA 78 (1981), 3308–3312.

[Kad] K. Kadell, *A proof of the q–Macdonald–Morris conjecture for BC_n*, AMS (1994).

[Kap] M. Kapranov, *Double affine Hecke algebras and 2-dimensional local fields*, J. Amer. Math. Soc. 14:1 (2001), 239–262.

[KMS] M. Kashiwara, and T. Miwa, and E. Stern, *Decomposition of q-deformed Fock spaces*, Selecta Math. 1 (1995), 787–805.

[Kat] S. Kato, *Irreducibility of principal series representations for Hecke algebras of affine type*, J. Fac. Sci. Univ. Tokyo, IA, 28:3 (1983), 929–943.

[KKS] D. Kazhdan, and B. Kostant, and S. Sternberg, *Hamiltonian group actions and dynamical systems of Calogero type*, Comm. Pure Apll. Math. 31 (1978), 481–507.

[KL1] D. Kazhdan, and G. Lusztig, *Proof of the Deligne-Langlands conjecture for Hecke algebras*, Invent.Math. 87 (1987), 153–215.

[KL2] — , and — , *Tensor structures arising from affine Lie algebras. III,* J. of AMS 7 (1994), 335–381.

[Ki1] A. Kirillov, Jr., *On inner product in modular tensor categories. I*, Jour. of AMS 9 (1996), 1135–1170.

[Ki2] — , *Inner product on conformal blocks and affine inner product identities*, Adv. Theor. Math. Phys. 2 (1998), 155–180.

[KZ] V.G. Knizhnik, and A.B. Zamolodchikov, *Current algebra and Wess-Zumino models in two-dimensions*, Nuclear Physics. B 247, (1984), 83–103.

[Kn] F. Knop, *Integrality of two variable Kostka functions*, J. Reine Angew. Math. 482 (1997), 177–189.

[KnS] — , and S. Sahi, *A recursion and a combinatorial formula for Jack polynomials*, Invent. Math. 128 (1997), 9–22.

[KS1] E. Koelink, and J. Stokman, *The big q–Jacobi function transform*, Constr. Approx. 19:2 (2003), 191–235.

[KS2] — , and — , *The Askey–Wilson function transform*, Internat. Math. Res. Notices 22 (2001), 1203–1227.

[Ko] T. Kohno, *Monodromy representations of braid groups and Yang–Baxter equations*, Ann. Inst. Fourier Grenouble 37 (1987), 139–160.

[Kos] B. Kostant, *On Laguerre polynomials, Bessel functions, Hankel transform and a series in the unitary dual of the simply-connected covering group of* $SL(2,\mathbb{R})$, Represent. Theory 4 (2000), 181–224.

[KK] — , and S. Kumar, *T-Equivariant K-theory of generalized flag varieties*, J. Diff. Geometry 32 (1990), 549–603.

[KuS] P. Kulish, and E. Sklyanin, *On solutions of the Yang–Baxeter equation*, J. Sov. Math. 19:5 (1982), 1596–1620.

[Le] H. van der Lek, *Extended Artin groups*, Proc. Symp. Pure Math. 40:2 (1983), 117–122.

[LNT] B. Leclerc, and M. Nazarov, and J.-Y. Thibon, *Induced representations of affine Hecke algebras and canonical bases of quantum groups*, Studies in Memory of Issai Schur, Progress in Mathematics, Birkhauser 210 (2003), 115–153.

[Lo] E. Looijenga, *Root systems and elliptic curves*, Inventiones Math. 38 (1976), 17–32.

[Lu] J. Luke, *Integrals of Bessel Functions*, McGraw-Hill Book Company, New York-Toronto-London (1962).

[Lus1] G. Lusztig, *Affine Hecke algebras and their graded version*, J. of the AMS 2:3 (1989), 599–685.

[Lus2] — , *Introduction to Quantum Groups*, Birkhäuser, Boston (1993).

[Lus3] — , *Quantum groups at roots of* 1, Geom. Dedicata 35 (1990), 89–114.

[M1] I.G. Macdonald, *Affine root systems and Dedekind's η-function*, Inventiones Math. 15 (1972), 91–143.

[M2] — , *Some conjectures for root systems*, SIAM J. Math. Anal. 13:6 (1982), 988–1007.

[M3] — , *A new class of symmetric functions*, Publ I.R.M.A., Strasbourg, Actes 20-e Seminaire Lotharingen, (1988), 131–171 .

[M4] — , *Orthogonal polynomials associated with root systems*, Preprint (1988).

[M5] — , *Affine Hecke algebras and orthogonal polynomials*, Séminaire Bourbaki 47:797 (1995), 1–18.

[M6] — , *A formal identity for affine root systems*, Preprint (1996).

[M7] — , *Symmetric Functions and Hall Polynomials*, Second Edition, Oxford University Press (1999).

[M8] — , *Affine Hecke algebras and orthogonal polynomials*, Cambridge University Press (2003).

[Ma] H. Matsumoto, *Analyse harmonique dans les systemes de Tits bornologiques de type affine*, Lecture Notes in Math. 590 (1977).

[Mat] A. Matsuo, *Integrable connections related to zonal spherical functions*, Invent. Math. 110 (1992), 96–121.

[Meh] M.L. Mehta, *Random Matrices*, 2nd rev. enl. ed., Academic Press, New York (1991).

[Nak] H. Nakajima, *Lectures on Hilbert schemes of points on surfaces*, University Lecture Series 18, AMS, Providence, RI (1999).

[Na1] M. Nazarov, *Young's symmetrizers for projective representations of the symmetric group*, Adv. Math. 127 (1997), 190–257.

[Na2] — , *Representations of Yangians associated with skew Young diagrams*, Proceedings of the ICM, Beijing 2002, 2, 643–654.

[NT] — , and V. Tarasov, *On irreducibility of tensor products of Yangian modules associated with skew Young diagrams*, Duke Math. J. 112 (2002), 343–378.

[No1] M. Noumi, *Macdonald's symmetric polynomials as zonal spherical functions on some quantum homogeneous spaces*, Advances in Mathematics 123 (1996), 16–77.

[No2] — , *Affine Hecke algebras and Macdonald polynomials*, in Progress in Mathematics 160, Birkhauser (1998).

[OOS] H. Ochiai, and T. Oshima, and H. Sekiguchi, *Commuting families of symmetric differential operators*, Proc. of the Japan Acad., A 70:2 (1994), 62–68.

[OP] M.A. Olshanetsky, and A.M. Perelomov, *Quantum integrable systems related to Lie algebras*, Phys. Rep. 94 (1983), 313–404.

[O1] E.M. Opdam, *Some applications of hypergeometric shift operators*, Invent. Math. 98 (1989), 1–18.

[O2] — , *Dunkl operators, Bessel functions and the discriminant of a finite Coxeter group*, Composito Mathematica 85 (1993), 333–373.

[O3] — , *Harmonic analysis for certain representations of graded Hecke algebras*, Acta Math. 175 (1995), 75–121.

[O4] — , *A generating function for the trace of the Iwahori–Hecke algebra*, in Studies in memory of Issai Schur (Joseph, Melnikov and Rentschler Eds.), Progress in Math., Birkhuser, (2001).

[O5] — , *On the spectral decomposition of affine Hecke algebras*, math. RT/0101007 (2001).

[PK] D.H. Peterson, and V.G. Kac, *Infinite flag varieties and conjugacy theorems*, Proc. Nat. Acad. Sci. USA 80, 1778–1782 (1983).

[Po] M.P. Polishchuk, *M.P. Appell's function and vector bundles of rank 2 on elliptic curve*, Preprint math. AG/9810084 (1998).

[Ra] E. Rains, *BC_n–symmetric abelian functions*, Preprint CO/0402113 (2004).

[Ram] A. Ram, *Seminormal representations of Weyl groups and Iwahori–Hecke algebras*, Proceedings of the London Math. Soc. 3:75 (1997), 99–133.

[Ro] J. Rogawski, *On modules over the Hecke algebra of a p–adic group*, Invent. Math. 79 (1985), 443–465.

[Ru] S.N.M. Ruijsenaars, *Complete integrability of relativistic Calogero-Moser systems and elliptic function identities*, Commun. Math. Phys. 110 (1987), 191–213.

[Sa] S. Sahi, *Nonsymmetric Koornwinder polynomials and duality*, Annals of Mathematics. 150:1 (1999), 267–282.

[Sai] K. Saito, *Extended affine root systems I.*, Publ. RIMS, Kyoto Univ. 21:1 (1985), 75–179.

[STU] Y. Saito, and K. Takemura, and D. Uglov, *Toroidal actions on level-1 modules of $U_q(\hat{sl}_n)$*, Transformation Groups 3 (1998), 75–102.

[Sat] M. Sato, *Soliton equations as dynamical systems on infinite-dimensional Grassman manifolds*, RIMS Kokyroku 439 (1981), 30–40.

[SV] V.V. Schechtman, and A.A. Varchenko, *Arrangements of hyperplanes and Lie algebra homology*, Inventiones Math. 106 (1991), 139–194.

[Se] M. Semenov–Tjan–Shanskii, *What is a classical r–matrix?*, Funct. Anal. Appl. 17:4 (1983), 259–272.

[Sk] E.K. Sklyanin, *On complete integrability of the Landau–Lifshitz equations*, LOMI preprint E-3-1979, Leningrad (1979).

[Sc] G.P. Scott, *Braid groups and the group of homeomorphisms of a surface*, Proc. Camb. Phil. Soc. 68 (1970), 605–617.

[Sm] F. Smirnov, *General formula for soliton formfactors in Sine–Gordon*, Model. J. Phys. A : Math. Gen. 19 (1986), L575–L580.

[St] J.V. Stokman, *Difference Fourier transforms for nonreduced root systems*, Sel. Math., New ser. 9:3 (2003), 409–494.

[Su] B. Sutherland, *Exact results for a quantum many-body problem in one-dimension*, Phys. Rev. A 4 (1971), 2019–2021, Phys. Rev. A 5 (1971), 1372–1376.

[Suz] T. Suzuki, *Classification of simple modules over degenerate double affine Hecke algebras of type A*, Int. Math. Res. Notes 43 (2003), 2313–2339.

[SuV] — , and M. Vazirani, *Tableaux theory on affine diagrams and irreducible representations of double affine Hecke algebras of type A*, Preprint 2004.

[TV] V. Tarasov, and A. Varchenko, *Geometry of q-hypergeometric functions, quantum affine algebras and elliptic quantum groups*, Astrisque 246 (1997).

[TK] A. Tsuchiya, and Y. Kanie, *Vertex operators in conformal field theory on P^1 and monodromy representations of braid groups*, Adv. Stud. Pure. Math. 16 (1988), 297–372.

[Va] A. Varchenko, *Multidimensional hypergeometric functions and representation theory of quantum groups*, Adv. Ser. Math. Phys. 21, World Sci, River Edge, NJ (1995).

[VV1] M. Varagnolo, and E. Vasserot, *Double-loop algebras and the Fock space*, Invent. Math. 133:1 (1998), 133–159.

[VV2] — , and — , *From double affine Hecke algebras to quantized affine Schur algebras*, Preprint math.RT/0307047 (2003).

[Vas] E. Vasserot, *Induced and simple modules of double affine Hecke algebras*. Preprint (2002).

[Ve] J.-L. Verdier, *Les representations des algébres de Lie affines: application a quelques problémes de physique*, Seminaire Bourbaki, exp. 596, 1–13 (1982).

[Ver] E. Verlinde, *Fusion rules and modular transformations in 2D conformal field theory*, Nucl. Phys. B 300(FS22) (1987), 360–376.

Index

www.ingramcontent.com/pod-product-compliance
Lightning Source LLC
LaVergne TN
LVHW090259210926
848480LV00001B/69
* 9 7 8 0 5 2 1 6 0 9 1 8 0 *